Intermolecular Interactions: From Diatomics to Biopolymers

PERSPECTIVES IN QUANTUM CHEMISTRY AND BIOCHEMISTRY

Edited by
Bernard Pullman

Professor of Quantum Chemistry
Institut de Biologie Physico-Chimique
Fondation Edmond de Rothschild, Paris

Volume I: Quantum Mechanics of Molecular Conformations
Edited by Bernard Pullman

Volume II: Intermolecular Interactions: From Diatomics to Biopolymers
Edited by Bernard Pullman

Intermolecular Interactions: From Diatomics to Biopolymers

Edited by
Bernard Pullman
Professor of Quantum Chemistry,
Institut de Biologie Physico-Chimique,
Fondation Edmond de Rothschild, Paris

A Wiley–Interscience Publication

JOHN WILEY & SONS

Chichester · New York · Brisbane · Toronto

Library of Congress Cataloging in Publication Data:

Main entry under title:

Intermolecular interactions, from diatomics to biopolymers.

(Perspectives in quantum chemistry and biochemistry)
"A Wiley–Interscience publication."
Includes indexes.
1. Quantum chemistry—Addresses, essays,
lectures. 2. Molecules—Addresses, essays,
lectures. I. Pullman, Bernard, 1919–
QD462.5.I57 541'.28 77-24278
ISBN 0 471 99507 X

Typeset in Northern Ireland at The Universities Press, Belfast
and printed in Great Britain at The Pitman Press, Bath

Preface to the Series

The outstanding development of quantum mechanical ideas, methods, theories and computational facilities during recent years has enlarged old roads and opened new paths for their penetration into many fields of chemistry, physics, biochemistry, biophysics and pharmacology. The harvest of results is impressive and greater now than it has ever been. The great adventure which started with the application of the basic ideas of quantum mechanics to the hydrogen molecule has reached the realm of biopolymers and their constituents.

This series of books was conceived to provide an interdisciplinary survey of the principal topics of recent developments. Its originality springs from the fact that each book is devoted to a *specific subject* of major significance. The aim is to demonstrate the importance of the quantum mechanical approach in each of the selected fields for all types of molecules. Because of this selectiveness of *topics* the series is fundamentally different from the usual 'Advances in . . .', where each volume is quite composite, and from the classical types of book which usually describe the contribution of quantum methodology to a specific branch of sciences: organic chemistry, biochemistry, pharmacology etc.

We hope that these volumes will be useful both to specialists in quantum molecular sciences, who, it is to be regretted, frequently restrict their interest to one type of molecule only, and to outsiders eager to become acquainted with the possibilities and achievements of this essential technique.

BERNARD PULLMAN

Preface to the Volume

The first volume of this series was devoted to quantum mechanics of molecular conformations, to the study of intramolecular interactions. The present volume extends the subject to intermolecular interactions.

It is frequently believed that quantum mechanical theories and computations are, in this last field, applicable only to small molecular systems. This is an erroneous conception. Recent developments in fundamental methodologies enable substantial results to be obtained for large organic and biological systems.

The four review papers presented here illustrate the achievements and the possibilities of quantum mechanical techniques in this field.

BERNARD PULLMAN

Contents

CHAPTER 1

Basic theory of intermolecular forces: Applications to small molecules

A. D. Buckingham

University Chemical Laboratory, Cambridge, U.K.

I. INTRODUCTION

Molecules attract one another when they are far apart, since liquids and solids exist. They repel one another when close, since densities are finite and have the magnitude they do under normal conditions of temperature and pressure. This well-known truth is illustrated in Figure 1.1 which shows a typical interaction energy $u(R)$ of two spherical molecules as a function of their separation R.

At long range, $u(R)$ approaches zero from below while $\mathrm{d}u/\mathrm{d}R$ approaches it from above. The intermolecular force $F(R) = -\mathrm{d}u/\mathrm{d}R$ is negative, meaning that it is attractive and that the potential energy is increased on increasing R. As R decreases towards the minimum in the energy, u also decreases, and the force goes through a minimum at the separation at which $\mathrm{d}^2u/\mathrm{d}R^2$ is zero. At the minimum in energy $u<0$, $\mathrm{d}u/\mathrm{d}R = 0$ and $\mathrm{d}^2u/\mathrm{d}R^2 > 0$; there is an exact balance between the attractive and the repulsive forces, so the force is zero and a stationary system would be at equilibrium at this point.

When two polyatomic molecules interact, the potential energy depends on their orientations as well as their separation R. There is also a dependence on the internal coordinates that prescribe the positions of the nuclei in each molecule. Thus when a diatomic molecule interacts with a spherical atom, the potential energy depends on three independent variables which may be chosen to be the separation R of the centres of mass, the angle θ between the axis of the diatomic and the line of centres, and the internuclear distance r (see Figure 1.2). If the diatomic has a centre of symmetry, as in H_2, N_2, O_2, Cl_2, then the energy must be unaffected by the change $\theta \rightarrow \theta + \pi$ and is therefore an even function of $\cos\theta$. Thus if we were to write

$$u(R, \theta, r) = \sum_n u_n(R, r) P_n(\cos\theta), \qquad (n = 0, 1, 2, \ldots) \tag{1.1}$$

where P_n is a Legendre polynomial,† then n would be restricted to even integers provided the molecular centre was chosen as the origin of coordinates in the centrosymmetric molecule. In general both odd and even values of n are required.

For some purposes it may be possible to neglect the r dependence of u; thus if one is interested in bulk properties of gases, such as the second virial coefficient or the viscosity, or in the structure of liquids and solids, or in the

† The Legendre polynomials form an orthogonal set of functions and are such that $P_n(\cos\theta) = 1$ for $\theta = 0$. The first five polynomials are:

$$\begin{aligned}
P_0(\cos\theta) &= 1 \\
P_1(\cos\theta) &= \cos\theta \\
P_2(\cos\theta) &= \tfrac{1}{2}(3\cos^2\theta - 1) \\
P_3(\cos\theta) &= \tfrac{1}{2}(5\cos^3\theta - 3\cos\theta) \\
P_4(\cos\theta) &= \tfrac{1}{8}(35\cos^4\theta - 30\cos^2\theta + 3)
\end{aligned}$$

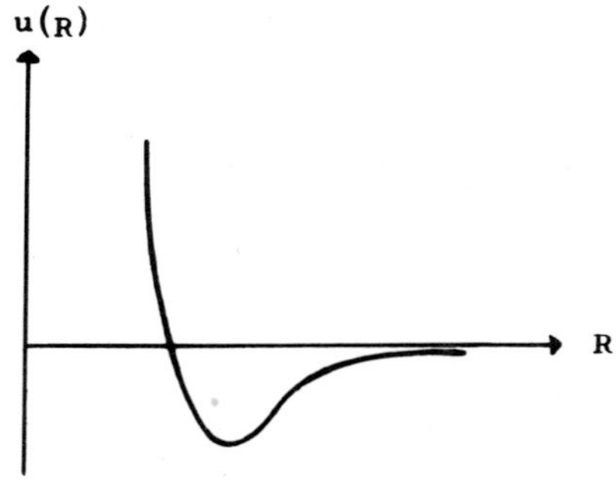

Figure 1.1. The interaction energy $u(R)$ of two spherical molecules as a function of the separation R of their centres

effect of collisions on the rotational motion of molecules, the significance of the vibrational coordinate r may be slight. However, if one is concerned with the effect of intermolecular forces on vibrational spectra, the retention of r is vital.

The number of variables upon which the intermolecular energy depends increases rapidly as the molecular size increases. Thus for two diatomic molecules, there are six variables (the intermolecular separation R, the angles θ_1 and θ_2 between the molecular axes and the line of centres, the angle ϕ between the planes containing the line of centres and the two molecules, and the two internuclear distances r_1 and r_2—see Figure 1.3. In the general case of two interacting non-linear polyatomic molecules, u depends on $3(N_1+N_2)-6$ independent variables, where N_1 and N_2 are the number of atoms in molecules 1 and 2; these could be chosen to be the separation R of the centres of mass, the angles θ_1 and θ_2 between an axis fixed in each molecule and the line of centres, the angles χ_1 and χ_2 giving the orientation of each molecule about the fixed axis, the angle ϕ between the planes containing the line of centres and the molecular axes, and the $(3N_1-6)+(3N_2-6)$ independent vibrational coordinates. However, if the

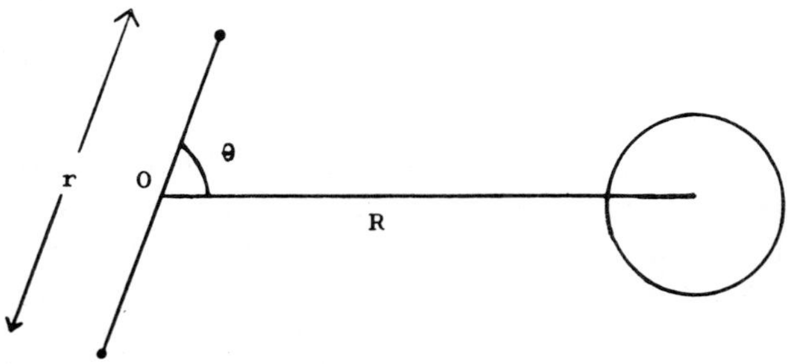

Figure 1.2. The three coordinates R, θ, r appropriate for describing the interaction of a diatomic molecule and a spherical atom

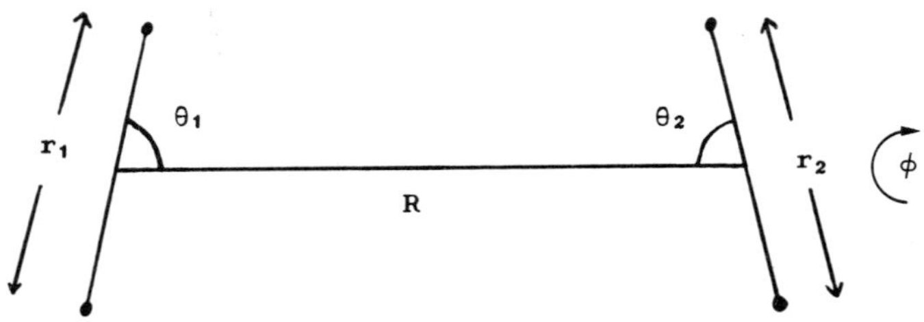

Figure 1.3. The six coordinates R, θ_1, θ_2, ϕ, r_1, r_2 appropriate for describing the interaction of two diatomic molecules

internal variables are not important, only 6 coordinates need be specified $(R, \theta_1, \theta_2, \chi_1, \chi_2, \phi)$.

In some instances it is convenient to consider a *potential of average force* A between a pair of molecules. This is the potential for fixed positions and orientations of two molecules in the presence of others, such as those of a solvent, and is averaged over all positions of these other molecules. Thus A is a temperature-dependent potential in this case, and depends on relative position and orientation as in the case of an isolated pair. It is actually a Helmholtz free energy and is comprised of an energy component and an entropic part:

$$A(R, \theta) = u(R, \theta) - TS(R, \theta) \tag{1.2}$$

Because of the averaging, both u and S are also functions of the absolute temperature T. The entropic contribution results from the change in order in the environment resulting from the interaction of the pair; thus if there is an increase in order in the solvent when molecules 1 and 2 are fixed at a particular relative configuration R, θ over the order when $R \to \infty$, the interaction entropy $S(R, \theta)$ is negative and the entropic contribution leads to a positive contribution to the effective potential $A(R, \theta)$. The well-known *hydrophobic effect* producing an effective force of attraction between hydrocarbon chains in aqueous media[1] depends on the entropic component, for the decrease in entropy on forming a cage of water molecules is apparently less in the case of a nearby pair of solutes than when they are far apart. Rubber elasticity may also be attributed to a decrease in entropy on stretching.[1a]

The concept of an intermolecular potential depends upon the Born–Oppenheimer separation[2,3] of the nuclear and electronic motion. The potential $u(R)$ in Figure 1.1 shows the interaction energy as a function of the nuclear separation R—it is therefore necessary to invoke the approximation that the electrons adjust to each nuclear position as if the momentum of the nuclei were zero. The nuclei are then allowed to move in the total potential resulting from all electrons and nuclei, giving, according to the laws of quantum mechanics, the vibrational, rotational and translational motion of the molecules. This motion is perturbed by the intermolecular force, but since the interaction energy u is normally small compared to the potential energy of the free molecules, it is possible to employ perturbation theory and to assign the molecules to the electronic and vibrational quantum states of the non-interacting molecules. The perturbation makes small changes in the energy levels and in the wavefunctions and may cause very large changes in the rotational and translational motion through the dependence of u on the angular and positional variables respectively. The basic problem is the evaluation of the energy as a function of the relative position and orientation of two or more molecules. When this has been solved, the effects of the interaction can be determined by considering the translational, rotational and vibrational motion; in some cases this is a formidable task, but often the occupied states have energy separations that are small compared to kT and a classical treatment suffices. In other instances semi-classical techniques may be useful,[4,5] but sometimes a full quantum mechanical scattering calculation may be required.[6,7]

Some of the minor deficiencies associated with the use of the Born–Oppenheimer approximation and the resultant potential energy surfaces can be overcome by employing a semi-classical theory. This involves the use of a complex dynamical variable for the interacting system; the imaginary component of the separation can be associated with the ability of real particles to tunnel into classically inaccessible regions.[5]

In this chapter the theory of intermolecular forces is reviewed. The aim is to describe the dependence of the energy on the separation and orientations of the molecules, both in the vicinity of the minimum of the potential as well as at short and long range. The topic has been covered by three well-known books.[8,9,10] The theory of the interaction potential of inert gas atoms has recently been reviewed[11,12] and the best potentials deduced from various sources, theoretical and experimental, for both the pure and mixed gases.[13]

II. CLASSIFICATION OF INTERMOLECULAR FORCES

There is now general agreement that the significant forces between molecules have an electric origin. It is true that other sources exist, such as magnetic and gravitational interactions, but these can normally be neglected. Even when the cooperative nature of ferromagnetism is evoked, the magnetic forces are weak. In considering the nature of intermolecular potentials, it is helpful to attempt to separate various contributions. The development of a general theory of interaction energies has provided a basis for this separation.

The first classification to be made is into *short-range* and *long-range* forces. The former decrease exponentially with increasing separation R and result from overlap of the electronic wavefunctions describing the separate molecules. Long-range interactions vary as R^{-n} at large R where n is a positive integer. Thus the long-range energy of interaction of two ions varies as R^{-1} and of two dipoles in fixed relative orientations as R^{-3}; the corresponding forces vary as R^{-2} and R^{-4}. At long-range, the indistinguishability of electrons plays no role in determining the interaction and it is possible to consider the electrons as if they were associated with one or other of the molecules. The n-electron wavefunction (where $n = n_1 + n_2$) need not then be anti-symmetrized with respect to exchange of electrons between the molecules 1 and 2, and a simple perturbation theory can be developed in which the unperturbed system is the non-interacting molecular pair whose wavefunction is a simple product of the n_1- and n_2-electron eigenfunctions of the Hamiltonians $\mathcal{H}_1(n_1)$ and $\mathcal{H}_2(n_2)$ of the isolated molecules. Thus the long-range forces between molecules can be related in a general way to the properties of the free molecules. However, when electron exchange is significant, the wavefunction must be fully anti-symmetrized and it is then not an eigenfunction of the unperturbed Hamiltonian $\mathcal{H}_1(n_1) + \mathcal{H}_2(n_2)$. Several attempts have been made to formulate a perturbation theory appropriate to this region, but none is entirely satisfactory. Some of the proposed schemes were applied to an interacting pair of hydrogen atoms by Certain, Hirschfelder, Kołos and Wolniewicz[14] and while some of the procedures taken to second order gave satisfying results, none appeared to offer computational advantages over the standard variational calculation.[15] It would seem that there exists no general theory relating the short-range interaction energy to the properties of the isolated molecules, and that it is necessary to treat each case separately. The energy surface is obtained by applying conventional quantum-mechanical techniques to the 'supermolecule' containing $n = n_1 + n_2$ electrons in various nuclear configurations. Although the total energy of the supermolecule, calculated using the variational principle, must be an upper bound to the true energy, the interaction energy is not bounded, since it emerges as the difference between the total energy and the sum of the calculated energies of the non-interacting molecules.

The short-range forces may be attractive or repulsive, but at small separations they are always repulsive. In the case of two H atoms interacting to form the ground ${}^1\Sigma_g^+$ state of H_2, there is a very strong short-range attractive force which produces an energy minimum of $-0{\cdot}174475$ a.u. $=$ $-38292{\cdot}8\ \text{cm}^{-1}$ at $R = 1{\cdot}401078$ a.u. $= 74{\cdot}1414$ pm; the energy is zero at $R = 0{\cdot}7758$ a.u. $= 41{\cdot}05$ pm. and rises steeply at smaller separations.[15]

The repulsive force at small separations arises from the Coulombic interaction of the electrons and nuclei and from the effects of electron exchange.[10] The Hellmann–Feynman theorem[16,17] requires that the forces on the nuclei may be precisely calculated by classical electrostatics from a knowledge of the electron distribution. The redistribution of electrons resulting from the interaction gives rise to the force. Thus at large distances the attractive dispersion force between two spherical atoms must be associated with an increase in the electron density in the region between the nuclei. Each atom acquires an electric dipole moment proportional to R^{-7};

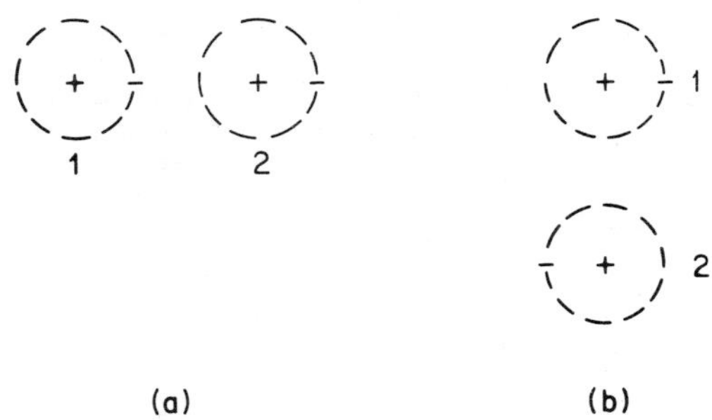

Figure 1.4. Favourable relative positions of electrons in atoms 1 and 2 separated by a large distance. (a) Correlation of the parallel components and (b) of the perpendicular components, of the fluctuating dipoles in the two atoms

in the case of a homonuclear species such as Ar_2 this atomic dipole is compensated by the opposite dipole on the other atom but the pair possesses a quadrupole moment proportional to R^{-6}. A dissimilar pair such as HeAr has a long-range dipole proportional to R^{-7}.[18] The long-range attractive force varying as R^{-7} does not result from the interaction of these two atomic dipoles (such a force would vary as R^{-18}) but rather from the force exerted on each nucleus by the distorted electron distribution in its own atom.[19,20] The long-range redistribution of charge results from an intermolecular correlation of the fluctuations in the electronic coordinates. This is illustrated in Figure 1.4 which shows the attractive effect of a correlation in the dipoles of atoms 1 and 2. The qualitative effect of the dispersion-type interaction on the atomic charge distribution is illustrated in Figure 1.5. In the short-range overlap region, it is no longer necessary to invoke a redistribution of charge to explain an interatomic force, although there must be such redistribution and, by the variation principle, it would lead to a reduction in the interaction energy of atoms in their ground electronic states. If one requires the force between two atoms or molecules it is preferable to evaluate the energy u as a function of the separation R and thereby obtain $-\partial u/\partial R$, rather than to calculate the expectation value of the force directly; thus to calculate the component of the force between two inert gas atoms, using the direct approach through the Hellmann–Feynman theorem, it is necessary to know the components of the approximate

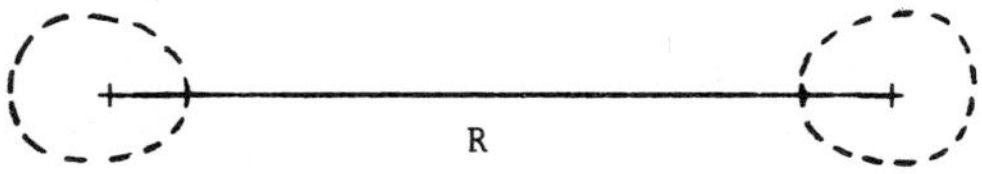

Figure 1.5. A pictorial representation of the distorted charge distribution in a pair of spherical atoms at large separations. Each atom has a dipole proportional to R^{-7}

Table 1.1. A classification of the interaction energy of a pair of molecules. The primary classification is into long and short range interaction, and the long-range energy is further divided into five distinct types. The attractive or repulsive nature of the corresponding force is shown in column 3 and the difference between u_{123} and $u_{12}+u_{23}+u_{31}$ for a cluster of three molecules is indicated in the fourth column

Range	Type	Attractive (−) or repulsive (+)	Additive or non-additive
	Electrostatic	∓	Additive
	Induction	−	Non-additive
Long	Dispersion	−	Nearly additive
	Resonance	∓	Non-additive
	Magnetic	∓	(Weak)
Short	Overlap	∓	Non-additive

wavefunction proportional to R^{-3}, R^{-4} and R^{-7}, whereas for the corresponding energy C_6R^{-6} only the component in R^{-3} is required.[20]

The long-range interaction energy between a pair of molecules can be further divided into several distinct types, as shown in Table 1.1. The perturbation theory, developed in Section III, gives a basis for this classification.

The electrostatic energy $u_{\text{electrostatic}}$ is the interaction energy of the permanent charge distributions of the molecules. It is true that the interaction modifies the charge distribution, but this is not part of $u_{\text{electrostatic}}$. Thus the electrostatic energy is determined by the charge distribution in the free molecules. It has a simple interpretation. It may be evaluated in general by performing an integration over the space of each molecule. However, if the separation between the molecules is large compared to their dimensions, the multipole expansion[8,21,22] may conveniently be employed and the energy expressed in terms of charge–charge, charge–dipole, dipole–dipole,... components. This expansion is most useful when only the first non-vanishing component need be considered, as in ionic solids when the charge–charge term in R^{-1} is dominant; in a polar fluid the dipole–dipole energy in R^{-3} may sometimes be a sufficient approximation to $u_{\text{electrostatic}}$; and for centrosymmetric molecules as in gaseous, liquid and solid hydrogen or nitrogen, the quadrupole–quadrupole interaction in R^{-5} is a good representation of $u_{\text{electrostatic}}$ (the next longest range term varies as R^{-7} in this case since the octopole moments are zero if the molecular centres of symmetry are chosen as the origins of both the multipole moments and of the expansion of $u_{\text{electrostatic}}$ in R^{-1}). The electrostatic energy is related to the potential produced at a point by the charge distribution of a single molecule. The potential has been related to the probability of electrophilic attack on various sites in an organic molecule by Bonaccorsi, Scrocco and Tomasi[23] who constructed maps of the potential for such molecules as cyclopropane

$(CH_2)_3$ and aziridine $(CH_2)_2NH$ using self-consistent-field wavefunctions. The relationship between the proton-affinity and the potential has been explored by Pullman *et al.*[23a]

The induction energy $u_{\text{induction}}$ is the interaction of the induced electric moments of each molecule with the permanent charge distribution of its neighbours. The longest range contribution is due to the induced dipoles and may be written as

$$-\tfrac{1}{2}\sum_i \boldsymbol{\alpha}^{(i)} : \mathbf{F}^{(i)2}$$

where $\boldsymbol{\alpha}^{(i)}$ is the polarizability of molecule i and $\mathbf{F}^{(i)}$ is the electric field at the origin of this molecule resulting from the permanent charge distribution of all the other molecules. Thus in the interaction of an ion and a spherical atom, the induction energy is $-\frac{1}{2}\alpha q^2R^{-4}(4\pi\epsilon_0)^{-2}$ where $qR^{-2}(4\pi\epsilon_0)^{-1}$ is the field strength produced by the ion of charge q at the atom and α is the polarizability of the atom; ϵ_0 is the permittivity of free space ($4\pi\epsilon_0 = 1{\cdot}11265\times10^{-10}\,\mathrm{C\,V^{-1}\,m^{-1}} = 1$ e.s.u.). Since the induction energy results from a distortion of a charge distribution in response to an external field it is always negative for molecules in their ground electronic states. And since the induction energy of molecule i is proportional to the square of the total electric field at the molecule, it is non-additive.

$$F^{(i)2} = \left(\sum_{j\neq i} F^{(ij)}\right)^2 = \sum_{j\neq i} F^{(ij)2} + \sum_{j\neq i}\sum_{k\neq i,j} \mathbf{F}^{(ij)}\ \mathbf{F}^{(ik)} \qquad (1.3)$$

where $\mathbf{F}^{(ij)}$ is the field at i due to the charge distribution of molecule j, and the contribution $\sum_{k\neq i}\sum_{k\neq i} \mathbf{F}^{(ij)} \cdot \mathbf{F}^{(ik)}$ leads to the non-additivity. This can be seen in Figure 1.6 which shows the induction energy of an atom in the presence of one and two ions.

(a) There is an induced dipole $\alpha qR^{-2}(4\pi\epsilon_0)^{-1}$ and a dipole induction energy $-\frac{1}{2}\alpha q^2R^{-4}(4\pi\epsilon_0)^{-2}$ where α is the polarizability of the atom. (b) In the presence of a second symmetrically placed charge the induced dipole vanishes and there is no dipole induction energy. However, there is a field-gradient $F'_{zz} = -4qR^{-3}(4\pi\epsilon_0)^{-1}$ at the atom in (b) and this induces a quadrupole moment CF'_{zz} in the atom[24] and there is an induction

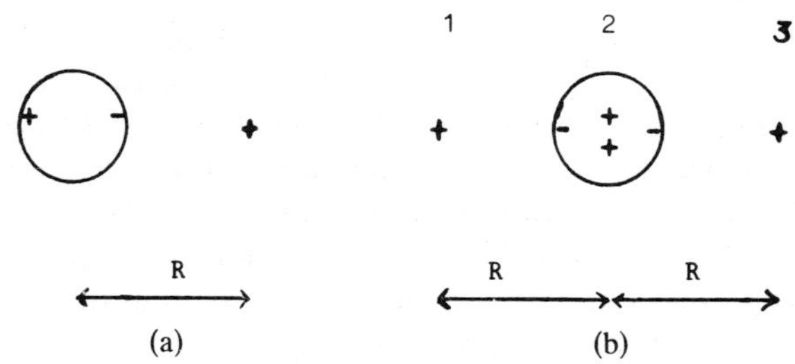

Figure 1.6. The interaction of an atom with (a) one or (b) two point charges

energy $-\frac{1}{4}CF'^2_{zz} = -4Cq^2R^{-6}(4\pi\epsilon_0)^{-2}$ for large R. Thus

$$u^{(123)}_{\text{induction}} = -4Cq^2R^{-6}(4\pi\epsilon_0)^{-2}$$
$$u^{(12)}_{\text{induction}} = u^{(23)}_{\text{induction}} = (-\tfrac{1}{2}\alpha q^2R^{-4} - Cq^2R^{-6} - \ldots)(4\pi\epsilon_0)^{-2}$$
$$u^{(13)}_{\text{induction}} = 0 \text{ (since the point charges are non-polarizable).}$$

The departure from additivity is therefore

$$u^{(123)}_{\text{induction}} - u^{(12)}_{\text{induction}} - u^{(23)}_{\text{induction}} - u^{(13)}_{\text{induction}} = (\alpha q^2R^{-4} - 2Cq^2R^{-6} - \ldots)(4\pi\epsilon_0)^{-2}$$

The dispersion energy $u_{\text{dispersion}}$ is the energy resulting from the correlation in the fluctuations in the electronic coordinates of the interacting molecules. Interatomic dipole correlation is illustrated in Figure 1.4 and the resulting distortion of the charge distribution in Figure 1.5. The dispersion energy is a consequence of the quantum-mechanical nature of the electron. If the electron were a classical particle its position could be specified and there would be an electrostatic energy for each electronic configuration. In the case of two interacting spherical atoms the classical electrostatic energy would average to zero in first order but would lead to a temperature-dependent attractive energy in second order because of the Boltzmann favouring of the configurations of lower energy; temperature-dependent forces of this nature between molecules with permanent electric moments were discussed by Keesom[25] in 1921. The origin of the substantial attractive forces between non-polar molecules was a serious problem early in the century. Much was known of these forces from the van der Waals equation of state of imperfect gases and through the study of liquids and solids. There was the rather surprising fact that the binding energy of solid argon is of the same order of magnitude as that of the highly polar species HCl. It was known that an Ar atom has no dipole moment, but Debye[26] suggested that it may have a quadrupole which would produce a temperature-independent negative induction energy proportional to R^{-8}. However, after the advent of quantum mechanics it became clear that the inert-gas atoms are spherical in their ground state and therefore have no permanent electric moments of any order.

Soon after the arrival of quantum mechanics in 1926, Wang[27] showed that there is an attractive force between two hydrogen atoms that varies as R^{-7} at large R. In 1930, London[28,29] gave a sound description of these long-range forces using second-order quantum-mechanical perturbation theory and a useful approximate formula for estimating their strength. He pointed out a link between them and optical dispersion in the gas and he called them *dispersion forces*. They arise from an intermolecular correlation in the fluctuations of the electronic coordinates, and are therefore absent from a one-electron or orbital description of the system.

A partial appreciation of the dispersion energy of a pair of atoms can be gained by relating it to the mean induction energy resulting from the fluctuating electric field and field-gradient produced by each atom at the other. The mean-square field at a point distant R from the atom is

$$\overline{F^2} = (4\pi\epsilon_0)^{-2}[2\overline{\mu^2}R^{-6} + 2\overline{\boldsymbol{\Theta}:\boldsymbol{\Theta}}R^{-8} + \ldots] \tag{1.4}$$

where

$$\boldsymbol{\mu} = \sum_i e_i \mathbf{r}_i$$

is the instantaneous dipole moment and

$$\boldsymbol{\Theta} = \tfrac{1}{2}\sum_i e_i(3\mathbf{r}_i\mathbf{r}_i - r_i^2\mathbf{1})$$

the quadrupole moment, and the summation extends over all the electrons (and over the nuclei in the case of a molecule, even though there are no fluctuations in the field of the nuclei in the clamped-nuclei approximation). If it is assumed that the interaction energy is the induction energy of each atom in the mean-square field and field-gradient of the other

$$u = -\tfrac{1}{2}\alpha^{(1)}\overline{(F^{(1)})^2} - \tfrac{1}{2}\alpha^{(2)}\overline{(F^{(2)})^2} - \tfrac{1}{6}C^{(1)}\overline{(F'^{(1)})^2} - \tfrac{1}{6}C^{(2)}\overline{(F'^{(2)})^2} - \ldots \tag{1.5}$$

where $\mathbf{F}'^{(n)}$ is the field gradient at atom n, and

$$(F')^2 = F_{\alpha\beta}F_{\alpha\beta} = (4\pi\epsilon_0)^{-2}[30\overline{\mu^2}R^{-8} + \ldots]. \tag{1.6}$$

For a pair of identical atoms

$$u = -(4\pi\epsilon_0)^{-2}[2\alpha\overline{\mu^2}R^{-6} + 2\alpha\overline{\Theta^2}R^{-8} + 10C\overline{\mu^2}R^{-8} + \ldots] \tag{1.7}$$

For a hydrogen atom $(4\pi\epsilon_0)^{-1}\alpha = \frac{9}{2}a_0^3$, $(4\pi\epsilon_0)^{-1}C = \frac{15}{2}a_0^5$, $\overline{\mu^2} = 3e^2a_0^2$ and

$$\overline{\Theta^2} = \overline{\Theta_{\alpha\beta}:\Theta_{\alpha\beta}} = \tfrac{3}{2}e^2\overline{r^4} = \tfrac{135}{4}e^2a_0^4$$

Hence with this model the dispersion energy of a pair of H atoms is

$$u = -e^2(4\pi\epsilon_0 a_0)^{-1}[27(R/a_0)^{-6} + \tfrac{2115}{4}(R/a_0)^{-8} + \ldots] \tag{1.8}$$

The accurate result for the dispersion energy has been obtained by variation-perturbation techniques and is[30–32]

$$u_{\text{dispersion}} = -e^2(4\pi\epsilon_0 a_0)^{-1}[6{\cdot}4990(R/a_0)^{-6} + 124{\cdot}399(R_0/a)^{-8} + \ldots] \tag{1.9}$$

The long-range dipole–dipole dispersion energy of two H atoms is therefore 0·241 times that calculated for the model, and the R^{-8} term is 0·235 of that for the model. The failure of the model is due to the use of the static polarizabilities α and C; the other atom cannot respond fully to the fluctuating field and field-gradient. The fact that the R^{-6} and R^{-8} terms are a similar fraction of the model values may reflect opposed effects in the R^{-8} term—the field of the quadrupole fluctuates at twice the frequency of that of

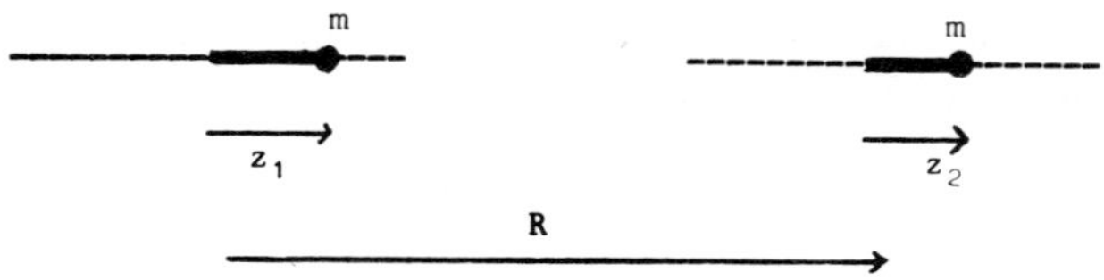

Figure 1.7. Two identical coupled linear harmonic oscillators. The Hamiltonian is $\mathcal{H}=\frac{1}{2}m(\dot{z}_1^2+\dot{z}_2^2)+\frac{1}{2}K(z_1^2+z_2^2+2az_1z_2)$, where m is the mass of each oscillator, K is the force constant and a determines the strength of the coupling. If the oscillators can be considered as dipoles of magnitude qz_1 and qz_2, then a is negative and equal to $-2(q^2/KR^3)(4\pi\varepsilon_0)^{-1}$ where R is their separation

the dipole, so the term in $\alpha^2\Theta^2R^{-8}$ is grossly exaggerated, but the quadrupole polarizability C may respond reasonably well to the field-gradient from the dipole of the other atom, since the main resonance in the frequency-dependent quadrupole polarizability is at the $1s\rightarrow3d$ transition frequency, and the main dipole resonance frequency is at the $1s\rightarrow2p$ frequency; the term in $C\mu^2$ contributes 42·6 per cent of the model energy in R^{-8}.

It is clear that an account of dispersion energy in these terms is inadequate, although the model yields the right dependence on R and an increasing energy with increasing polarizability of the interacting species.

The quantum-mechanical nature of dispersion forces can be appreciated following Longuet–Higgins,[33] by considering the motion of a pair of coupled one-dimensional harmonic oscillators, as in Figure 1.7. The Hamiltonian

$$\mathcal{H}=\tfrac{1}{2}m(\dot{z}_1^2+\dot{z}_2^2)+\tfrac{1}{2}K(z_1^2+z_2^2+2az_1z_2) \qquad (1.10)$$

contains an interaction proportional to the product of the displacements and can be rearranged to the separable form

$$\mathcal{H}=\tfrac{1}{4}m[(\dot{z}_1+\dot{z}_2)^2+(\dot{z}_1-\dot{z}_2)^2] + \tfrac{1}{4}K[(1+a)(z_1+z_2)^2+(1-a)(z_1-z_2)^2] \qquad (1.11)$$

The oscillators are therefore split into two new normal vibrations of angular frequencies ω_+ and ω_- where

$$\begin{aligned}\omega_+&=\omega_0(1+a)^{1/2}\\ \omega_-&=\omega_0(1-a)^{1/2}\end{aligned} \qquad (1.12)$$

and $\omega_0=(K/m)^{1/2}$ is the angular frequency of the isolated oscillators. In the classical limit, when the gap $\hbar\omega$ between adjacent levels of a harmonic oscillator is very small compared to the thermal energy kT, the partition function Q is equal to $kT/\hbar\omega$. Thus the classical partition function for the coupled oscillators is

$$Q=\left(\frac{kT}{\hbar\omega_+}\right)\left(\frac{kT}{\hbar\omega_-}\right)=\left(\frac{kT}{\hbar\omega_0}\right)^2(1-a^2)^{-1/2} \qquad (1.13)$$

Their Helmholtz free energy is

$$A = -kT \ln Q = A_0 + \tfrac{1}{2}kT \ln(1-a^2) \tag{1.14}$$

and the change $A - A_0$ is therefore *negative*, giving an attractive effective force $-\mathrm{d}A/\mathrm{d}R$. This force results purely from the change in entropy, for

$$\begin{aligned} S = -\left(\frac{\partial A}{\partial T}\right)_V &= k \ln Q + kT(\mathrm{d} \ln Q/\mathrm{d}T)_V \\ &= S_0 - \tfrac{1}{2}k \ln(1-a^2) \end{aligned} \tag{1.15}$$

and hence $A - A_0 = -T(S - S_0)$, so that the change in the internal energy is zero in the classical limit. In reality, the oscillators possess zero-point energies and these lead to a change in energy given by

$$u - u_0 = \tfrac{1}{2}\hbar(\omega_+ + \omega_- - 2\omega_0) = \tfrac{1}{2}\hbar\omega_0[(1+a)^{1/2} + (1-a)^{1/2} - 2] \tag{1.16}$$

This is negative when a is either positive or negative and vanishes when a is zero. For small a, $u - u_0 = -\tfrac{1}{8}a^2\hbar\omega_0$, demonstrating the second-order dependence of the energy change on the strength of the interaction. Thus the dispersion energy comes from the change in the energy of the ground state of the system resulting from correlation in the motion of the quantum particles—in the coupled oscillators these are the oscillating particles themselves, and in the atomic case they are the electrons. The electrons acquire a small asymmetry in their distribution (see Figure 1.5) but the mean displacements z_1 and z_2 of each oscillator remain zero in the coupled system; however, as in the atomic case, there is a mutual disturbance of their motion, as shown by the increase in the entropy.

If the two coupled oscillators are dissimilar, there is no degeneracy to be lifted by the interaction, but there is again a lowering of the total zero-point energy that is quadratic in the coupling parameter a. The major effect due to the similarity of the oscillators occurs in excited states, where there is a change in zero-point energy linear in a. This is an example of a *resonance energy* (see below) which results from the first-order lifting of the degeneracy by the interaction.

The quantum-mechanical nature of particles requires that there be fluctuations in their positions and these fluctuations are weakly correlated in interacting systems. Thus dispersion forces are present in all matter. In the inert gases, they are the sole cause of attraction and even in highly polar molecules such as HCl they contribute a major part of the energy of vaporization of the liquid and to the energy of sublimation of the crystal. In hydrogen-bonded systems there are fluctuations in the positions of the protons, and these are correlated because of the electrostatic interactions, leading to a type of proton dispersion energy. Such forces have been investigated by Zundel.[34,35]

The *resonance* energy $u_{\mathrm{resonance}}$ is the additional energy that results from a lifting of degeneracy by the interaction. The degeneracy may arise because one of the molecules is in a degenerate state as in the case of a hydrogen atom in an excited state with principal quantum number 2 interacting with an ion or a polar molecule. Or the degeneracy may result from the possibility of an exchange of excitation between two like molecules, as in the case of a hydrogen atom in the ground $1s$ state interacting with an excited H atom in the $2p_z$ state, when the long-range interaction energy varies as the inverse cube of the separation R. Consider two like molecules a and b described by the wavefunctions $\psi_i^{(a)}$ and $\psi_j^{(b)}$, where the subscripts i and j denote two different eigenstates of a free molecule. There is another state of the system degenerate with this described by $\psi_j^{(a)}$ and $\psi_i^{(b)}$. The degeneracy is split by the interaction of the molecules if there are non-vanishing matrix elements of the molecular electric moment operators between the states ψ_i and ψ_j, and the longest-range interaction energy $u_{\text{resonance}}$ then varies with R in the same way as it would if the molecules possessed permanent electric moments. For example, the interaction of an atom in an S state with a like atom in a P state leads to a resonance energy varying as R^{-3} even though neither free atom possesses a permanent dipole moment. This resonance energy results from the fact that the matrix elements of the dipole moment operator between S and P states are non-zero and it is proportional to the square of the transition dipole moment, that is, to the intensity of the transition $S \rightarrow P$; thus the resonance energy may be considered to arise from a photon exchange by a dipolar mechanism, the excitation energy being shared between the two atoms. Similarly the interaction of two like atoms, one in an S state and the other in a D state, leads to a resonance energy varying as R^{-5} and may be attributed to a photon exchange by a quadrupolar mechanism.

The lifting of degeneracy by the interaction produces two or more potential surfaces which may lie below or above zero. A sum over all the surfaces produces zero in the long-range limit; however, in any particular collision $u_{\text{resonance}}$ produces either an attractive or repulsive interaction according to the quantum numbers describing the state of the pair. Thus in the interaction of two like atoms in S and P states, the appropriate eigenfunctions at long range are the gerade and ungerade combinations,

$$\psi_{\substack{g\\u}} = 2^{-1/2}(S^{(a)}P^{(b)} \pm S^{(b)}P^{(a)}) \tag{1.17}$$

with energies

$$W_{\substack{g\\u}} = W_S + W_P \mp T_{\alpha\beta}\langle S\,|\mu_\alpha|\,P\rangle\langle P\,|\mu_\beta|\,S\rangle \tag{1.18}$$

where the dipole–dipole interaction tensor $T_{\alpha\beta}$ is

$$T_{\alpha\beta} = (4\pi\epsilon_0)^{-1}\nabla_\alpha\nabla_\beta(R^{-1}) = (4\pi\epsilon_0)^{-1}(3R_\alpha R_\beta - R^2\delta_{\alpha\beta})R^{-5} \tag{1.19}$$

and a repeated Greek tensor subscript denotes a summation over the three cartesian components; hence

$$T_{\alpha\beta}\mu_\alpha = T_{x\beta}\mu_x + T_{y\beta}\mu_y + T_{z\beta}\mu_z$$

and

$$T_{\alpha\beta}\mu_\alpha\mu_\beta = T_{xx}\mu_x^2 + T_{yy}\mu_y^2 + T_{zz}\mu_z^2 + 2T_{xy}\mu_x\mu_y + 2T_{xz}\mu_x\mu_z + 2T_{yz}\mu_y\mu_z$$

Resonance energy may be very important in determining the widths of rotational lines in the pure-rotation and rotation-vibration spectra of small gaseous molecules[36]; thus the width at half-height of an infrared spectral line of HF in the pure gas at a pressure of say 1 bar is larger than it is in HF in DF at the same pressure—in the pure gas there is an important contribution to the width resulting from dipolar resonance interactions in collisions involving two HF molecules that differ in their rotational quantum number J by unity. And the lifetime of a vibrationally excited C_6H_6 molecule is significantly shorter in liquid C_6H_6 than in C_6D_6.[37]

There is a resonance energy associated with an interacting pair of like oscillators when one is in an excited vibrational state. Thus for the model discussed on page 12, consider the case of one oscillator in its first excited state and the other in its ground state. When the oscillators are isolated (that is, when $a = 0$) the energy is $2\hbar\omega_0$ and when they are coupled it is either

$$\begin{aligned} \tfrac{3}{2}\hbar\omega_- + \tfrac{1}{2}\hbar\omega_+ &= \hbar\omega_0[\tfrac{3}{2}(1-a)^{1/2} + \tfrac{1}{2}(1+a)^{1/2}] \\ &= \hbar\omega_0[2 - \tfrac{1}{2}a - \tfrac{1}{4}a^2 + \ldots] \end{aligned} \tag{1.20}$$

or

$$\tfrac{1}{2}\hbar\omega_- + \tfrac{3}{2}\hbar\omega_+ = \hbar\omega_0[2 + \tfrac{1}{2}a - \tfrac{1}{4}a^2 + \ldots] \tag{1.21}$$

The interaction energy is therefore $\hbar\omega_0[\mp\tfrac{1}{2}a - \tfrac{1}{4}a^2 + \ldots]$ which includes a resonance contribution linear in a. This contribution would not be present if the interacting oscillators possessed different fundamental frequencies.

The exciton splittings that are observed in the electronic and vibrational spectra of molecular crystals are closely related to $u_{\text{resonance}}$.[38,39] So too are the 'excimers' and 'exciplexes' observed in solutions of aromatic molecules in the presence of resonance radiation.[40]

Finally, there are the magnetic interactions. These are normally neglected since they are so weak compared to the corresponding electric interactions. The magnetostatic dipole–dipole energy is typically of the order of 10^{-4} times the electrostatic dipole–dipole energy (since magnetic dipole moments are of the order of 1 Bohr magneton $= 0{\cdot}9274 \times 10^{-20}$ e.m.u. $= 0{\cdot}9274 \times 10^{-23}$ A m^2 while electric dipole moments are of the order of 1 debye $= 10^{-18}$ e.s.u. $= 3{\cdot}336 \times 10^{-30}$ C m). The interaction energies resulting from induced or fluctuating magnetic moments are relatively much smaller and typically of the order of 10^{-8} times the corresponding electrical energy. In optically active molecules there may be a coupling of the fluctuating electric and magnetic moments giving rise to a very weak dispersion energy that is dependent on the chirality of the molecules; that is, there is a component of the R^{-6} dispersion energy that changes sign on changing one of the molecules into its mirror image.[41,42] There are also other very small relativistic contributions to the interaction energy; these have been reviewed by Hirschfelder and Meath.[20]

It is possible that the cooperative nature of the magnetic moments in a ferromagnetic material could compensate, at least partially, for the relative weakness of magnetic forces. Thus one might think that a paramagnetic species, such as a lanthanide ion in aqueous solution, would experience a significant magnetic force in the vicinity of a ferromagnetic chip; such a force is appealing because it would discriminate very effectively between the chemically similar rare-earth ions. However, even here the magnetic forces are weak; it is necessary to consider a non-planar surface for the ferromagnetic chip, since the force is proportional to the *gradient* of the magnetic field, and it is difficult to achieve a sufficiently large field-gradient to cause a useful magnetic force.

III. GENERAL THEORY OF LONG-RANGE INTERMOLECULAR FORCES

In this section, we shall consider the long-range interaction of molecule a in state m_a with molecule b in state m_b. It is therefore supposed that the electrons are associated with a single molecule, or, more accurately, that the effects of electron exchange are negligible. This is justifiable in the long-range region, where one disregards effects which decrease exponentially with increasing molecular separation R (and this is true of the exchange forces). Each of the molecules a and b therefore retain their identity in the long-range region and the aim of the general theory is to relate the intermolecular potential to the properties of the isolated molecules. The subject has been treated by a number of authors.[20,24,43]

The Hamiltonian of the pair of molecules is written in the form

$$\mathcal{H} = \mathcal{H}^{(a)} + \mathcal{H}^{(b)} + \mathcal{H}' \tag{1.22}$$

where $\mathcal{H}^{(a)}$ and $\mathcal{H}^{(b)}$ are the Hamiltonians of the free molecules a and b and $\mathcal{H}'$ is the interaction. If we confine our considerations to Coulombic interactions then

$$\mathcal{H}' = (4\pi\epsilon_0)^{-1} \sum_{i,j} e_i^{(a)} e_j^{(b)} (R_{ij})^{-1} \tag{1.23}$$

where R_{ij} is the distance between the ith change $e_i^{(a)}$ in molecule a and the jth change $e_j^{(b)}$ in molecule b. Provided the distance R between the origins of each molecule is large compared to their dimensions, $\mathcal{H}'$ can be expanded as a series in the multipole moments of the molecules.[8,21,22] In this article we shall employ real rather than complex moments, and shall therefore use the usual Cartesian tensor notation.[44,45] However, it is also possible to employ spherical tensors,[22] and to enjoy their well-known transformation properties which may provide advantages in the derivation of general expressions. However, the simplicity and reality of vector transformations also have their attractions, especially for the first few multipoles, and normally the moments of a particular three-dimensional molecule are most conveniently given in an x, y, z frame fixed in the molecule.

A general theory of long-range intermolecular forces is obtained by treating $\mathcal{H}'$ in equation (1.22) as a perturbation to $\mathcal{H}^{(a)}+\mathcal{H}^{(b)}$. The unperturbed wavefunctions are the eigenfunctions of $\mathcal{H}^{(a)}+\mathcal{H}^{(b)}$ and are therefore simple products of the wavefunctions of the free molecules a and b. The perturbed wavefunction may be expressed in terms of these unperturbed states by standard quantum-mechanical perturbation theory[46–48] and is

$$|\psi_{m_a m_b}\rangle = |m_a m_b\rangle + \sum_{p_a,p_b}{}' \frac{\langle p_a p_b|\,\mathcal{H}'\,|m_a m_b\rangle}{W_{m_a}+W_{m_b}-W_{p_a}-W_{p_b}}|p_a p_b\rangle + \ldots \tag{1.24}$$

where the symbol

$$\sum_{p_a,p_b}{}'$$

denotes a sum over the complete set of unperturbed states $p_a p_b$ with the exception of the initial state $m_a m_b$. The first-order perturbed wavefunction therefore consists of the addition to the unperturbed function $|m_a m_b\rangle$ of a small amount of all those other unperturbed states of the system that are mixed with $|m_a m_b\rangle$ by the perturbation $\mathcal{H}'$. The extent of the admixture is proportional to the off-diagonal matrix element $\langle p_a p_b|\,\mathcal{H}'\,|m_a m_b\rangle$ and inversely to the difference $W_{m_a}+W_{m_b}-W_{p_a}-W_{p_b}$ in the energies of the unperturbed states. The unperturbed states will be assumed to be normalized. The energy of the pair of molecules in the perturbed state $\psi_{m_a m_b}$ is

$$\begin{aligned} W_{m_a m_b} &= \langle \psi_{m_a m_b}|\,\mathcal{H}\,|\psi_{m_a m_b}\rangle / \langle \psi_{m_a m_b} \mid \psi_{m_a m_b}\rangle \\ &= W_{m_a}+W_{m_b}+\langle m_a m_b|\,\mathcal{H}'\,|m_a m_b\rangle - \sum_{p_a,p_b}{}' \frac{|\langle p_a p_b|\,\mathcal{H}'\,|m_a m_b\rangle|^2}{W_{p_a}+W_{p_b}-W_{m_a}-W_{m_b}} + \ldots \end{aligned} \tag{1.25}$$

The first-order energy $\langle m_a m_b|\,\mathcal{H}'\,|m_a m_b\rangle$ is the unperturbed expectation value of the perturbation $\mathcal{H}'$ and is the electrostatic interaction energy $u_{\text{electrostatic}}$.

The second-order energy in equation (1.25) may be separated into two distinct contributions. The first, which is the induction energy $u_{\text{induction}}$, consists of all those terms in which either $p_a = m_a$ with $p_b \neq m_b$, or $p_b = m_b$ with $p_a \neq m_a$. The other contribution is $u_{\text{dispersion}}$ and it is comprised of the remaining terms in which $p_a \neq m_a$ and $p_b \neq m_b$. The induction energy is therefore the sum of two parts:

$$u_{\text{induction}} = u_{\text{induction}}^{(a)} + u_{\text{induction}}^{(b)} \tag{1.26}$$

where

$$u_{\text{induction}}^{(a)} = -\sum_{p_a \neq m_a} \frac{|\langle p_a m_b|\,\mathcal{H}'\,|m_a m_b\rangle|^2}{W_{p_a}-W_{m_a}} \tag{1.27}$$

$$u_{\text{induction}}^{(b)} = -\sum_{p_a \neq m_a} \frac{|\langle m_a p_b|\,\mathcal{H}'\,|m_a m_b\rangle|^2}{W_{p_b}-W_{m_b}} \tag{1.28}$$

Thus $u^{(a)}_{\text{induction}}$ arises from a distortion of the charge distribution of molecule a (that is, from a mixing of a small amount of each state p_a into the initial state m_a) through its interaction with the unperturbed molecule b. If the initial state m_a is the ground electronic state when $W_{p_a} - W_{m_a} > 0$ for all p_a and $u^{(a)}_{\text{induction}}$ is negative; this is in accord with the variation principle which ensures that if a system in its ground state is given the freedom to distort it uses that freedom to lower its energy.

Similarly

$$u_{\text{dispersion}} = -\sum_{p_a \neq m_a,\, p_b \neq m_b} \frac{|\langle p_a p_b | \mathscr{H}' | m_a m_b \rangle|^2}{W_{p_a} + W_{p_b} - W_{m_a} - W_{m_b}} \tag{1.29}$$

is necessarily negative for a system in which both molecules a and b are in their ground states. The dispersion energy results from matrix elements of $\mathscr{H}'$ that are off-diagonal in both eigenfunctions and is therefore attributable to the simultaneous fluctuations in the charge distributions in each molecule.

If the initial state $|m_a m_b\rangle$ is degenerate then it is necessary to choose as the zeroth-order states that mixture of the degenerate states which diagonalizes the Hamiltonian $\mathscr{H}$. If the degeneracy is m-fold there are m such independent states; if one of them is $|m'_a m'_b\rangle$ the corresponding energy to first-order in $\mathscr{H}'$ is

$$\langle m'_a m'_b | \mathscr{H}^{(a)} + \mathscr{H}^{(b)} + \mathscr{H}' | m'_a m'_b \rangle = W_{m_a} + W_{m_b} + u_{\text{electrostatic}} + u_{\text{resonance}} \tag{1.30}$$

The second-order energy is a sum over all the unperturbed states $|p_a p_b\rangle$ that are not degenerate with $|m_a m_b\rangle$ and is again the sum of the induction and the dispersion energy.

To proceed further and to relate $u_{\text{electrostatic}} + u_{\text{induction}} + u_{\text{dispersion}}$ to the properties of the free molecules a and b in the initial states, it is necessary to expand $\mathscr{H}'$ in a multipole series. The inverse distance $(R_{ij})^{-1}$ is expanded as a power series in R^{-1}, where R is the distance between origins O_a and O_b fixed in each molecule (see Figure 1.8). From the laws of vector addition

$$\mathbf{R}_{ij} = \mathbf{R} + \mathbf{r}_j - \mathbf{r}_i \tag{1.31}$$

The interaction Hamiltonian $\mathscr{H}'$ may be written in the form

$$\mathscr{H}' = (4\pi\epsilon_0)^{-1} \sum_{i,j} e_i^{(a)} e_j^{(b)} (R_{ij})^{-1} = \sum_i e_i^{(a)} \phi_i^{(a)} = \sum_j e_j^{(b)} \phi_j^{(b)} \tag{1.32}$$

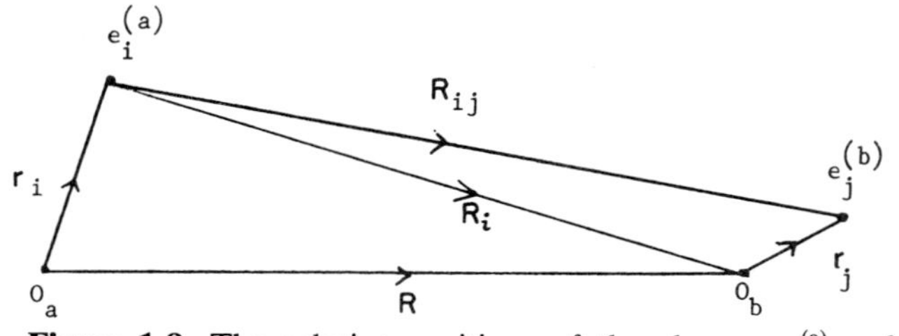

Figure 1.8. The relative positions of the charges $e_i^{(a)}$ and $e_j^{(b)}$ in molecules a and b. O_a and O_b are origins fixed in molecules a and b

where

$$\phi_j^{(b)} = (4\pi\epsilon_0)^{-1}\sum_i e_i^{(a)}(R_{ij})^{-1}$$

is the potential at the jth charge of molecule b due to all the charges in molecule a. This potential may be expanded as a Taylor series about the origin O_b,

$$\mathcal{H}' = \sum_j e_j^{(b)}\phi_j^{(b)} = \sum_j e_j^{(b)}[\phi^{(b)} + (\nabla_\alpha\phi^{(b)})_0 r_{j\alpha} + \tfrac{1}{2}(\nabla_\alpha\nabla_\beta\phi^{(b)})_0 r_{j\alpha}r_{j\beta} + \ldots]$$

$$= q^{(b)}\phi^{(b)} - \mu_\alpha^{(b)}F_\alpha^{(b)} - \tfrac{1}{3}\Theta_{\alpha\beta}^{(b)}F_{\alpha\beta}^{(b)} - \ldots - \frac{1}{1.3.5\ldots(2n-1)}$$

$$\times \xi_{\alpha\beta\gamma\ldots\nu}^{(n)(b)}F_{\alpha\beta\gamma\ldots\nu}^{(b)} - \ldots \qquad (1.33)$$

In equation (1.33), $F_\alpha^{(b)} = -(\nabla_\alpha\phi^{(b)})_0$ is the electric field at the origin of molecule b due to the instantaneous charge distribution of molecule a, $F_{\alpha\beta}^{(b)} = -(\nabla_\alpha\nabla_\beta\phi^{(b)})_0$ is the field-gradient, and the electric multipole tensors of molecule b have been used. They are defined by

$$\left.\begin{aligned}
q^{(b)} &= \xi^{(0)(b)} = \sum_j e_j^{(b)} \\
\mu_\alpha^{(b)} &= \xi_\alpha^{(1)(b)} = \sum_j e_j^{(b)} r_{j\alpha} \\
\Theta_{\alpha\beta}^{(b)} &= \xi_{\alpha\beta}^{(2)(b)} = \tfrac{1}{2}\sum_j e_j^{(b)}(3r_{j\alpha}r_{j\beta} - r_j^2\delta_{\alpha\beta}) \\
\xi_{\alpha\beta\gamma\ldots\nu}^{(n)(b)} &= (-1)^n(n!)^{-1}\sum_j e_j^{(b)} r_j^{2n+1}\frac{\partial}{\partial r_{i_\alpha}}\frac{\partial}{\partial r_{i_\beta}}\frac{\partial}{\partial r_{i_\gamma}}\cdots\frac{\partial}{\partial r_{i_\nu}}\left(\frac{1}{r_i}\right)
\end{aligned}\right\} \qquad (1.34)$$

and are symmetric with respect to interchange of any pair of suffixes (for example, $\Theta_{\alpha\beta} = \Theta_{\beta\alpha}$); they are reduced to zero on contraction (that is, $\xi_{\alpha\alpha\ldots\nu}^{(n)} = 0$; for example, $\Theta_{\alpha\alpha} = \Theta_{xx} + \Theta_{yy} + \Theta_{zz} = 0$). The zeroth moment of the charge distribution is the total charge q and this must be a positive or negative integer, or zero, times the protonic charge $e = 1{\cdot}6022 \times 10^{-19}$ C. The first moment is the electric dipole and it is independent of the choice of the origin if $q = 0$; this follows from the general equation

$$\mu'_\alpha = \sum_j e_j r'_{j\alpha} = \sum_j e_j(r_{j\alpha} - r'_\alpha) = \mu_\alpha - qr'_\alpha \qquad (1.35)$$

which describes the effect of a shift of the origin from O to O′, a point at $\mathbf{r}'$ relative to O (see Figure 1.9).

The second moment of the charge about O is the quadrupole moment $\boldsymbol{\Theta}$. It has been defined, like the other moments in equation (1.34), in such a way that it vanishes for a spherical distribution of charge if the centre of the sphere is the origin. Thus the spherical part of $3r_{j\alpha}r_{j\beta}$ is $r_j^2\,\delta_{\alpha\beta}$ and this has been subtracted from $3r_{j\alpha}r_{j\beta}$ in $\Theta_{\alpha\beta}$ in (1.34). All the multipole moments

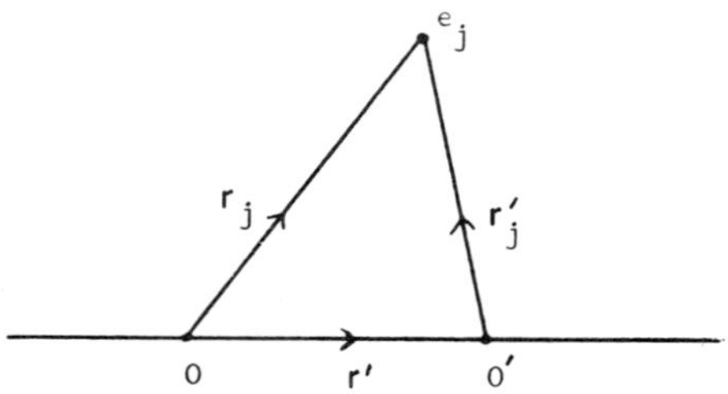

Figure 1.9. The positions of the point charge e_j relative to the origins O and O′. The origin O′ is at $\mathbf{r}'$ relative to O, so that $\mathbf{r}_j' = \mathbf{r}_j - \mathbf{r}'$

(except the zeroth) therefore represent departures from an isotropic charge distribution. Only the first non-vanishing moment is independent of the origin. The origin-dependence of the first moment is given in (1.35), and for the quadrupole

$$\Theta'_{\alpha\beta} = \tfrac{1}{2}\sum_j e_j(3r'_{j\alpha}r'_{j\beta} - r_j'^2\delta_{\alpha\beta})$$
$$= \Theta_{\alpha\beta} - \tfrac{3}{2}\mu_\alpha r'_\beta - \tfrac{3}{2}\mu_\beta r'_\alpha + \mu_\gamma r'_\gamma\delta_{\alpha\beta} + \tfrac{1}{2}q(3r'_\alpha r'_\beta - r'^2\delta_{\alpha\beta}) \qquad (1.36)$$

Every ion has a *centre-of-charge* which is the point relative to which $\boldsymbol{\mu} = \mathbf{0}$. This point must lie on the internuclear axis in a diatomic ion such as OH^-, and in the acetate ion CH_3COO^- it must lie in the plane of the carbon and oxygen nuclei; in an ion without symmetry such as $PHFClBr^+$, one could find the centre of charge only if all three independent dipoles relative to molecule-fixed axes were known—a unique translation in three dimensions could then be made to reduce all three dipoles to zero, using equation (1.35).

In general there are $2n+1$ independent multipole moments of degree n; that is $2n+1$ numbers must be given to specify $\boldsymbol{\xi}^{(n)}$ and the general tensor $\xi^{(n)}_{\alpha\beta\gamma\ldots\nu}$ may be expressed in terms of these $2n+1$ components by the usual tensor transformation law. Thus the molecule without symmetry (such as CHFClBr) has three independent dipoles, five quadrupoles, seven octopoles, and so on. One might specify the three dipoles μ_1, μ_2, μ_3 in a cartesian frame fixed in the molecule, or one could quote the total dipole $\mu = (\mu_1^2 + \mu_2^2 + \mu_3^2)^{1/2}$ and the two angles θ and ϕ which specify the orientation of the total dipole in the molecule-fixed frame 1, 2, 3. Similarly, one could quote all five independent quadrupoles

$$\boldsymbol{\Theta} = \begin{pmatrix} \Theta_{11} & \Theta_{12} & \Theta_{13} \\ \Theta_{21} = \Theta_{12} & \Theta_{22} & \Theta_{23} \\ \Theta_{31} = \Theta_{13} & \Theta_{32} = \Theta_{23} & \Theta_{33} = -(\Theta_{11} + \Theta_{22}) \end{pmatrix} \qquad (1.37)$$

or the two independent components $\Theta_{1'1'}, \Theta_{2'2'}$ in a principal-axis system $1', 2', 3'$ in which all off-diagonal elements are zero, together with the three angles needed to specify the orientation of the $1', 2', 3'$ cartesian frame relative to the $1, 2, 3$ frame; since there may be five independent components of $\mathbf{\Theta}$ and only three translational degrees of freedom, it is not possible to find a *centre of dipole* in the general case. However, in a molecule with a three-fold or higher axis of rotational symmetry, for example, CH_3Cl, there is just one independent dipole, and one quadrupole, since symmetry fixes the orientation of the dipole μ_3 and quadrupole $\Theta_{33} = -2\Theta_{11} = -2\Theta_{22}$; equation (1.36) then simplifies to

$$\Theta'_{33} = \Theta_{33} - 2\mu_3 r'_3 \tag{1.38}$$

and $\Theta'_{33} = 0$ at $r'_3 = \frac{1}{2}\Theta_{33}\mu_3^{-1}$ which is therefore the centre of dipole. In a molecule with C_{2v} symmetry, such as H_2O, there is one dipole μ_3 along the two-fold axis and two independent quadrupoles Θ_{33} and Θ_{22} where the 2 axis is a plane of symmetry. However Θ_{22} cannot be changed independently of Θ_{33}, since there is only the one non-vanishing dipole μ_3. Thus there is no centre of dipole in a molecule having C_{2v} symmetry, and $\Theta_{11} - \Theta_{22}$ is invariant to movement of the origin.

The actual number of independent multipole moments for molecules having symmetry elements can be determined by group-theoretical methods[49,50]. The key fact is that the independent components, and their linear combinations, must be invariant under all symmetry transformations; all other components and combinations must vanish. Thus each invariant linear combination transforms as the totally symmetric representation of the appropriate point group. The number of independent constants required to describe a tensor property is the number of times the totally symmetric representation occurs in the reduced form of the property appropriate to the particular symmetry group. These numbers have been evaluated for many properties[50–53,24]; Table 1.2 gives the number of independent multipole moments of order 1 to 6 for the main point-group symmetries.

We now return to equation (1.33) for the interaction Hamiltonian $\mathcal{H}'$ and express the potential $\phi^{(b)}$ and the field $\mathbf{F}^{(b)}$ and its derivatives in terms of the multipole moments of molecule a. This may be achieved by expanding R_i^{-1} in Figure 1.8 as a Taylor series,

$$\begin{aligned}
\phi^{(b)} &= (4\pi\epsilon_0)^{-1} \sum_i e_i^{(a)} R_i^{-1} \\
&= (4\pi\epsilon_0)^{-1} \sum_i e_i^{(a)} \left[\frac{1}{R} + \left(\frac{\partial}{\partial r_{i_\alpha}} \frac{1}{R_i} \right)_0 r_{i_\alpha} + \frac{1}{2} \left(\frac{\partial^2}{\partial r_{i_\alpha} \partial r_{i_\beta}} \frac{1}{R_i} \right)_0 r_{i_\alpha} r_{i_\beta} + \dots \right] \\
&= (4\pi\epsilon_0)^{-1} \sum_i e_i^{(a)} [R^{-1} - r_{i_\alpha} \nabla_\alpha R^{-1} + \tfrac{1}{2} r_{i_\alpha} r_{i_\beta} \nabla_\alpha \nabla_\beta R^{-1} \dots] \\
&= q^{(a)} T - \mu_\alpha^{(a)} T_\alpha + \tfrac{1}{3} \Theta_{\alpha\beta}^{(a)} T_{\alpha\beta} - \tfrac{1}{15} \Omega_{\alpha\beta\gamma}^{(a)} T_{\alpha\beta\gamma} + \dots + (-1)^n \\
&\quad \times \frac{1}{1.3.5\dots(2n-1)} \xi_{\alpha\beta\gamma\dots\nu}^{(n)(a)} T_{\alpha\beta\gamma\dots\nu} + \dots
\end{aligned} \tag{1.39}$$

Table 1.2. The number of independent electric multipole moments for systems of different symmetries

Group symbols		Number of constants					
International	Schoenflies	$\boldsymbol{\mu}(\xi^{(1)})$	$\boldsymbol{\Theta}(\xi^{(2)})$	$\boldsymbol{\Omega}(\xi^{(3)})$	$\boldsymbol{\Phi}(\xi^{(4)})$	$\xi^{(5)}$	$\xi^{(6)}$
1	C_1	3	5	7	9	11	13
$\bar{1}$	C_i	0	5	0	9	0	13
m	C_s	2	3	4	5	6	7
2	C_2	1	3	3	5	5	7
$2/m$	C_{2h}	0	3	0	5	0	7
$2mm$	C_{2v}	1	2	2	3	3	4
222	D_2	0	2	1	3	2	4
mmm	D_{2h}	0	2	0	3	0	4
4	C_4	1	1	1	3	3	3
$\bar{4}$	S_4	0	1	2	3	2	3
$4/m$	C_{4h}	0	1	0	3	0	3
$4mm$	C_{4v}	1	1	1	2	2	2
$\bar{4}2m$	D_{2d}	0	1	1	2	1	2
422	D_4	0	1	0	2	1	2
$4/mmm$	D_{4h}	0	1	0	2	0	2
3	C_3	1	1	3	3	3	5
$\bar{3}$	S_6	0	1	0	3	0	5
$3m$	C_{3v}	1	1	2	2	2	3
32	D_3	0	1	1	2	1	3
$\bar{3}m$	D_{3d}	0	1	0	2	0	3
$\bar{6}$	C_{3h}	0	1	2	1	2	3
6	C_6	1	1	1	1	1	3
$6/m$	C_{6h}	0	1	0	1	0	3
$\bar{6}m2$	D_{3h}	0	1	1	1	1	2
$6mm$	C_{6v}	1	1	1	1	1	2
622	D_6	0	1	0	1	0	2
$6/mmm$	D_{6h}	0	1	0	1	0	2
23	T	0	0	1	1	0	2
$m3$	T_h	0	0	0	1	0	2
$\bar{4}3m$	T_d	0	0	1	1	0	1
432	O	0	0	0	1	0	1
$m3m$	O_h	0	0	0	1	0	1
∞m	$C_{\infty v}$	1	1	1	1	1	1
∞/mm	$D_{\infty h}$	0	1	0	1	0	1
Sphere	K_h	0	0	0	0	0	0

where $T_{\alpha\beta\gamma\ldots\nu} = (4\pi\epsilon_0)^{-1}\nabla_\alpha\nabla_\beta\nabla_\gamma\ldots\nabla_\nu R^{-1}$ is a tensor which is proportional to $R^{-(n+1)}$, is symmetric with respect to interchange of any pair of suffixes and is reduced to zero on contraction (that is, $T_{\alpha\alpha\gamma\ldots\nu}=0$). The first five **T**-tensors are

$$T = (4\pi\epsilon_0)^{-1}R^{-1} \tag{1.40}$$

$$T_\alpha = (4\pi\epsilon_0)^{-1}\nabla_\alpha R^{-1} = -(4\pi\epsilon_0)^{-1}R_\alpha R^{-3} \tag{1.41}$$

$$T_{\alpha\beta} = (4\pi\epsilon_0)^{-1}(3R_\alpha R_\beta - R^2\delta_{\alpha\beta})R^{-5} \tag{1.42}$$

$$T_{\alpha\beta\gamma} = (4\pi\epsilon_0)^{-1}(-3)[5R_\alpha R_\beta R_\gamma - R^2(R_\alpha\delta_{\beta\gamma} + R_\beta\delta_{\gamma\alpha} + R_\gamma\delta_{\alpha\beta})]R^{-7} \tag{1.43}$$

$$T_{\alpha\beta\gamma\delta} = (4\pi\epsilon_0)^{-1}3[35R_\alpha R_\beta R_\gamma R_\delta - 5R^2(R_\alpha R_\beta \delta_{\gamma\delta} + R_\alpha R_\gamma \delta_{\beta\delta} + R_\alpha R_\delta \delta_{\beta\gamma} + R_\beta R_\gamma \delta_{\alpha\delta} + R_\beta R_\delta \delta_{\alpha\gamma} + R_\gamma R_\delta \delta_{\alpha\beta}) + R^4(\delta_{\alpha\beta}\delta_{\gamma\delta} + \delta_{\alpha\gamma}\delta_{\beta\delta} + \delta_{\alpha\delta}\delta_{\beta\gamma})]R^{-9} \tag{1.44}$$

Since the field $\mathbf{F}^{(b)}$ is minus the gradient of $\phi^{(b)}$ at O_b,

$$\left.\begin{aligned} F_\alpha^{(b)} &= -\nabla_\alpha \phi^{(b)} = -q^{(a)}T_\alpha + \mu_\beta^{(a)}T_{\alpha\beta} - \tfrac{1}{3}\Theta_{\beta\gamma}^{(a)}T_{\alpha\beta\gamma} + \dots \\ F_{\alpha\beta}^{(b)} &= \nabla_\alpha F_\beta^{(b)} = -\nabla_\alpha\nabla_\beta\phi^{(b)} = -q^{(a)}T_{\alpha\beta} + \mu_\gamma^{(a)}T_{\alpha\beta\gamma} - \tfrac{1}{3}\Theta_{\gamma\delta}^{(a)}T_{\alpha\beta\gamma\delta} + \dots \end{aligned}\right\} \tag{1.45}$$

and the interaction Hamiltonian $\mathscr{H}'$ in (1.33) is

$$\begin{aligned} \mathscr{H}' &= Tq^{(a)}q^{(b)} + T_\alpha(q^{(a)}\mu_\alpha^{(b)} - q^{(b)}\mu_\alpha^{(a)}) \\ &+ T_{\alpha\beta}(\tfrac{1}{3}q^{(a)}\Theta_{\alpha\beta}^{(b)} + \tfrac{1}{3}q^{(b)}\Theta_{\alpha\beta}^{(a)} - \mu_\alpha^{(a)}\mu_\beta^{(b)}) \\ &+ T_{\alpha\beta\gamma}(\tfrac{1}{15}q^{(a)}\Omega_{\alpha\beta\gamma}^{(b)} - \tfrac{1}{15}q^{(b)}\Omega_{\alpha\beta\gamma}^{(a)} - \tfrac{1}{3}\mu_\alpha^{(a)}\Theta_{\beta\gamma}^{(b)} + \tfrac{1}{3}\mu_\alpha^{(b)}\Theta_{\beta\gamma}^{(a)}) \\ &+ T_{\alpha\beta\gamma\delta}(\tfrac{1}{9}\Theta_{\alpha\beta}^{(a)}\Theta_{\gamma\delta}^{(b)} + \dots) + \dots \\ &= \sum_{n=0}^{\infty}\sum_{n'=0}^{\infty}(-1)^{n'}\frac{1}{1.3.5\dots(2n-1)}\,\frac{1}{1.3.5\dots(2n'-1)} \\ &\times T_{\alpha\beta\gamma\dots\nu\alpha'\beta'\gamma'\dots\nu'}\,\xi_{\alpha\beta\gamma\dots\nu}^{(n)(b)}\,\xi_{\alpha'\beta'\gamma'\dots\nu'}^{(n')(a)} \end{aligned} \tag{1.46}$$

A movement of either of the origins O_a and O_b changes each term in $\mathscr{H}'$ but leaves $\mathscr{H}'$ invariant. In molecules possessing symmetry elements, there may be a natural origin; thus if there is a centre of inversion, as in H_2, CO_2 and C_6H_6, this centre should be chosen as the origin, and the centre of a tetrahedral molecule such as CH_4 or of a molecule with D_{3h} symmetry such as cyclopropane, should obviously be the origin. If one is concerned with a transfer of energy between rotational and translational degrees of freedom in a collision, the centre of mass is an appropriate origin. However, in the interaction of HD with another molecule, the natural origin in HD might be either the mid-point of the line joining the nuclei (this is the electrical centre of the molecule), or the centre of mass.

The electrostatic interaction energy can easily be expressed in terms of the permanent multipole moments of molecules a and b. From equations (1.25) and (1.46)

$$\begin{aligned} u_{\text{electrostatic}} &= \langle m_a m_b|\,\mathscr{H}'\,|m_a m_b\rangle \\ &= Tq^{(a)}q^{(b)} + T_\alpha(q^{(a)}\mu_\alpha^{(m_b)} - q^{(b)}\mu_\alpha^{(m_a)}) \\ &+ T_{\alpha\beta}(\tfrac{1}{3}q^{(a)}\Theta_{\alpha\beta}^{(m_b)} + \tfrac{1}{3}q^{(b)}\Theta_{\alpha\beta}^{(m_a)} - \mu_\alpha^{(m_a)}\mu_\beta^{(m_b)}) \\ &+ T_{\alpha\beta\gamma}(\tfrac{1}{15}q^{(a)}\Omega_{\alpha\beta\gamma}^{(m_b)} - \tfrac{1}{15}q^{(b)}\Omega_{\alpha\beta\gamma}^{(m_a)} - \tfrac{1}{3}\mu_\alpha^{(m_a)}\Theta_{\beta\gamma}^{(m_b)} + \tfrac{1}{3}\mu_\alpha^{(m_b)}\Theta_{\beta\gamma}^{(m_a)}) \\ &+ T_{\alpha\beta\gamma\delta}(\tfrac{1}{9}\Theta_{\alpha\beta}^{(m_a)}\Theta_{\gamma\delta}^{(m_b)} + \dots) + \dots \\ &= \sum_{n=0}^{\infty}\sum_{n'=0}^{\infty}(-1)^{n'}\frac{2^{n+n'}(n)!(n')!}{(2n)!(2n')!}T_{\alpha\beta\gamma\dots\nu\alpha'\beta'\gamma'\dots\nu'}\,\xi_{\alpha\beta\gamma\dots\nu}^{(n)(m_b)}\,\xi_{\alpha'\beta'\gamma'\dots\nu'}^{(n')(m_a)} \end{aligned} \tag{1.47}$$

where

$$\xi^{(n)(m_a)}_{\alpha\beta\gamma\ldots\nu} = \langle m_a | \, \xi^{(n)(a)}_{\alpha\beta\gamma\ldots\nu} \, | m_a \rangle \tag{1.48}$$

is the nth permanent multipole moment of molecule a in the state m_a. For $n = 0$, 1 and 2 (that is, the charge, dipole and quadrupole) there are sound experimental techniques for determining these moments for molecules in their ground electronic states[53], and the electric dipole moments of molecules in excited electronic states can also be measured through the Stark effect in optical spectroscopy[54]. The zeroth moments or charges $q^{(a)}$ and $q^{(b)}$ do not bear the labels m_a and m_b since total charges are independent of the quantum state, that is, there are no fluctuations in q, and all off-diagonal matrix elements of the charge operator vanish.

A. Molecules in external fields—polarizability

In order to express the induction and dispersion energies in terms of the properties of the isolated molecules it is necessary to consider the interaction of a molecule with an external electric field. If this external field $\mathbf{F}$ is uniform, the Hamiltonian $\mathscr{H}(\mathbf{F}) = \mathscr{H} - \boldsymbol{\mu} \cdot \mathbf{F}$ and the energy of the molecule initially in the mth stationary state can be written as a power series in $\mathbf{F}$[24,55]:

$$W_m(\mathbf{F}) = W_m - \mu^{(m)}_\alpha F_\alpha - \tfrac{1}{2}\alpha^{(m)}_{\alpha\beta} F_\alpha F_\beta - \tfrac{1}{6}\beta^{(m)}_{\alpha\beta\gamma} F_\alpha F_\beta F_\gamma - \tfrac{1}{24}\gamma^{(m)}_{\alpha\beta\gamma\delta} F_\alpha F_\beta F_\gamma F_\delta - \ldots \tag{1.49}$$

where $\boldsymbol{\alpha}^{(m)}$ is the polarizability, $\boldsymbol{\beta}^{(m)}$ the first hyperpolarizability and $\boldsymbol{\gamma}^{(m)}$ the second hyperpolarizability of the molecule in the state m.

Perturbation theory leads to the following expressions for $\boldsymbol{\alpha}^{(m)}$, $\boldsymbol{\beta}^{(m)}$ and $\boldsymbol{\gamma}^{(m)}$,

$$\begin{aligned} \alpha^{(m)}_{\alpha\beta} &= \sum_{p \neq m} \frac{\langle m| \mu_\alpha |p\rangle\langle p| \mu_\beta |m\rangle + \langle m| \mu_\beta |p\rangle\langle p| \mu_\alpha |m\rangle}{W_p - W_m} \\ &= S(\alpha\beta) \sum_{p \neq m} \frac{\langle m| \mu_\alpha |p\rangle\langle p| \mu_\beta |m\rangle}{W_p - W_m} = \alpha^{(m)}_{\beta\alpha} \end{aligned} \tag{1.50}$$

$$\beta^{(m)}_{\alpha\beta\gamma} = S(\alpha\beta\gamma) \sum_{p \neq m} \left[\sum_{q \neq m} \frac{\langle m| \mu_\alpha |p\rangle\langle p| \mu_\beta |q\rangle\langle q| \mu_\gamma |m\rangle}{(W_p - W_m)(W_q - W_m)} - \frac{\langle m| \mu_\alpha |m\rangle\langle m| \mu_\beta |p\rangle\langle p| \mu_\delta |m\rangle}{(W_p - W_m)^2} \right. \tag{1.51}$$

$$\begin{aligned} \gamma^{(m)}_{\alpha\beta\gamma\delta} = S(\alpha\beta\gamma\delta) \sum_{p \neq m} \Bigg[\sum_{q \neq m} \Bigg\{ \sum_{r \neq m} \frac{\langle m| \mu_\alpha |p\rangle\langle p| \mu_\beta |q\rangle\langle q| \mu_\gamma |r\rangle\langle r| \mu_\delta |m\rangle}{(W_p - W_m)(W_q - W_m)(W_r - W_m)} \\ - \frac{\langle m| \mu_\alpha |p\rangle\langle p| \mu_\beta |m\rangle\langle m| \mu_\gamma |q\rangle\langle q| \mu_\delta |m\rangle + 2\langle m| \mu_\alpha |m\rangle\langle m| \mu_\beta |p\rangle\langle p| \mu_\gamma |q\rangle\langle q| \mu_\delta |m\rangle}{(W_p - W_m)^2(W_q - W_m)} \Bigg\} \\ + \frac{\langle m| \mu_\alpha |m\rangle\langle m| \mu_\beta |m\rangle\langle m| \mu_\gamma |p\rangle\langle p| \mu_\delta |m\rangle}{(W_p - W_m)^3} \Bigg] \end{aligned} \tag{1.52}$$

The symbol $S(\alpha\beta\gamma\ldots)$ implies a summation over all permutations of the tensor components; thus

$$S(\alpha\beta\gamma)X_{\alpha\beta\gamma} = X_{\alpha\beta\gamma} + X_{\alpha\gamma\beta} + X_{\beta\gamma\alpha} + X_{\beta\alpha\gamma} + X_{\gamma\alpha\beta} + X_{\gamma\beta\alpha}.$$

$S(\alpha\beta\gamma\ldots)$ is a symmetrizing operator and $\boldsymbol{\alpha}^{(m)}, \boldsymbol{\beta}^{(m)}, \boldsymbol{\gamma}^{(m)}, \ldots$ are symmetric with respect to interchange of any pair of tensor suffixes.

The polarizability and hyperpolarizabilities can be appreciated by considering the dipole moment of the molecule in the field **F**. This dipole is equal to minus the derivative of the energy with respect to **F**:

$$\frac{\partial W_m(\mathbf{F})}{\partial F_\alpha} = \frac{\partial}{\partial F_\alpha} \langle \psi_m(\mathbf{F})|\, \mathcal{H}(\mathbf{F})\, |\psi_m(\mathbf{F})\rangle$$

provided $\psi_m(\mathbf{F})$ is normalized. Thus

$$\begin{aligned}\frac{\partial W_m(\mathbf{F})}{\partial F_\alpha} &= \langle \psi_m(\mathbf{F})| \frac{\partial \mathcal{H}(\mathbf{F})}{\partial F_\alpha} |\psi_m(\mathbf{F})\rangle + \langle \frac{\partial \psi_m(\mathbf{F})}{\partial F_\alpha} |\, \mathcal{H}(\mathbf{F})\, |\psi_m(\mathbf{F})\rangle \\ &\quad + \langle \psi_m(\mathbf{F})|\, \mathcal{H}(\mathbf{F})\, | \frac{\partial \psi_m(\mathbf{F})}{\partial F_\alpha}\rangle \\ &= -\langle \psi_m(\mathbf{F})|\, \mu_\alpha\, |\psi_m(\mathbf{F})\rangle + W_m(\mathbf{F}) \Big[\langle \frac{\partial \psi_m(\mathbf{F})}{\partial F_\alpha} |\, \psi_m(\mathbf{F})\rangle \\ &\quad + \langle \psi_m(\mathbf{F})\, | \frac{\partial \psi_m(\mathbf{F})}{\partial F_\alpha}\rangle \Big] \\ &= -\mu_\alpha^{(m)}(\mathbf{F})\end{aligned} \tag{1.53}$$

since the term in square brackets multiplying $W_m(\mathbf{F})$ vanishes because of the normalization identity $\langle \psi_m(\mathbf{F}) \mid \psi_m(\mathbf{F})\rangle \equiv 1$. Hence from (1.49),

$$\mu_\alpha^{(m)}(\mathbf{F}) = \mu_\alpha^{(m)} + \alpha_{\alpha\beta}^{(m)} F_\beta + \tfrac{1}{2}\beta_{\alpha\beta\gamma}^{(m)} F_\beta F_\gamma + \tfrac{1}{6}\gamma_{\alpha\beta\gamma\delta}^{(m)} F_\beta F_\gamma F_\delta + \ldots \tag{1.54}$$

Thus $\boldsymbol{\alpha}^{(m)}$ determines the linear polarization of a molecule by an external field and $\boldsymbol{\beta}^{(m)}$ the initial departure from a linear polarization law. If the molecule is centro-symmetric, the energy is unaffected by reversing **F** so all odd powers of **F** in (1.49) would then vanish. For spherical systems, such as the inert gas atoms in their ground states, the polarizability and all even hyperpolarizabilities are isotropic and fully symmetric and are expressible in terms of a single parameter. Thus for spheres,

$$\alpha_{\alpha\beta}^{(m)} = \alpha^{(m)} \delta_{\alpha\beta} \tag{1.55}$$

$$\gamma_{\alpha\beta\gamma\delta}^{(m)} = \tfrac{1}{3}\gamma^{(m)}(\delta_{\alpha\beta}\delta_{\gamma\delta} + \delta_{\alpha\gamma}\delta_{\beta\delta} + \delta_{\alpha\delta}\delta_{\beta\gamma}) \tag{1.56}$$

where $\alpha^{(m)} = \alpha_{11}^{(m)} = \frac{1}{3}\alpha_{\beta\beta}^{(m)}$ and $\gamma^{(m)} = \gamma_{1111}^{(m)} = \frac{1}{5}\gamma_{\alpha\beta\alpha\beta}^{(m)}$.

Typical values for the various parameters in (1.54) applicable to small molecules are as follows:

$$\begin{aligned}
\mu^{(m)} &\sim 1 \text{ debye} &&= 10^{-18}\text{ e.s.u.} = 3{\cdot}33564\times 10^{-30}\text{ C m}\\
(4\pi\epsilon_0)^{-1}\alpha^{(m)} &\sim 1\text{ Å}^3 &&= 10^{-24}\text{ e.s.u.} = 10^{-30}\text{ m}^3\\
(4\pi\epsilon_0)^{-2}\beta^{(m)} &\sim &&\pm 10^{-30}\text{ e.s.u.} = \pm 0{\cdot}29979\times 10^{-30}\text{ C}^{-1}\text{ m}^5\\
(4\pi\epsilon_0)^{-3}\gamma^{(m)} &\sim && 10^{-36}\text{ e.s.u.} = 0{\cdot}089876\times 10^{-30}\text{ C}^{-2}\text{ m}^7
\end{aligned}$$

The quantity $(4\pi\epsilon_0)^{-1}\alpha^{(m)}$ may conveniently be compared to the volume of the molecule, and the various induced dipoles in (1.54) are of the same order of magnitude when the field strength F is of the order of 1 atomic unit $= 5\cdot 1423\times 10^{11}\text{ V m}^{-1} = 1\cdot 7153\times 10^{7}$ e.s.u.

When the external field is non-uniform additional distortions are induced. The Hamiltonian $\mathcal{H}(F_\alpha, F_{\alpha\beta}) = \mathcal{H} - \mu_\alpha F_\alpha - \frac{1}{3}\Theta_{\alpha\beta}F_{\alpha\beta} - \frac{1}{15}\Omega_{\alpha\beta\gamma}F_{\alpha\beta\gamma} - \ldots$ and the energy may be expressed as a power series in F_α, $F_{\alpha\beta}$, $F_{\alpha\beta\gamma}$, etc.

$$\begin{aligned}
W_m(F_\alpha, F_{\alpha\beta}, F_{\alpha\beta\gamma}\ldots) = W_m &- \mu_\alpha^{(m)}F_\alpha - \tfrac{1}{3}\Theta_{\alpha\beta}^{(m)}F_{\alpha\beta} - \tfrac{1}{15}\Omega_{\alpha\beta\gamma}^{(m)}F_{\alpha\beta\gamma} - \ldots\\
&- \tfrac{1}{2}\alpha_{\alpha\beta}^{(m)}F_\alpha F_\beta - \tfrac{1}{6}C_{\alpha\beta,\gamma\delta}^{(m)}F_{\alpha\beta}F_{\gamma\delta}\\
&- \tfrac{1}{30}R_{\alpha\beta\gamma,\delta\epsilon\phi}^{(m)}F_{\alpha\beta\gamma}F_{\delta\epsilon\phi} - \tfrac{1}{3}A_{\alpha,\beta\gamma}^{(m)}F_\alpha F_{\beta\gamma}\\
&- \tfrac{1}{15}E_{\alpha,\beta\gamma\delta}^{(m)}F_\alpha F_{\beta\gamma\delta} - \tfrac{1}{45}H_{\alpha\beta,\gamma\delta\epsilon}^{(m)}F_{\alpha\beta}F_{\gamma\delta\epsilon} - \ldots
\end{aligned} \tag{1.57}$$

In the following equations, a partial derivative with respect to $F_{\alpha\beta\gamma\ldots}$ is the mathematical operation of taking the derivative with respect to $F_{\alpha\beta\gamma\ldots}$ *while keeping all other components of the field and its derivatives constant.*

The total electric multipole moments of the molecule are

$$\begin{aligned}
\mu_\alpha^{(m)}(F_\alpha, F_{\alpha\beta}, F_{\alpha\beta\gamma}, \ldots) &= -\frac{\partial W_m(F_\alpha, F_{\alpha\beta}, \ldots)}{\partial F_\alpha}\\
&= \mu_\alpha^{(m)} + \alpha_{\alpha\beta}^{(m)}F_\beta + \tfrac{1}{3}A_{\alpha,\beta\gamma}^{(m)}F_{\beta\gamma} + \tfrac{1}{15}E_{\alpha,\beta\gamma\delta}^{(m)}F_{\beta\gamma\delta} + \ldots
\end{aligned} \tag{1.58}$$

$$\begin{aligned}
\Theta_{\alpha\beta}^{(m)}(F_\alpha, F_{\alpha\beta}, F_{\alpha\beta\gamma}, \ldots) &= -3\frac{\partial W_m(F_\alpha, F_{\alpha\beta}, \ldots)}{\partial F_{\alpha\beta}}\\
&= \Theta_{\alpha\beta}^{(m)} + A_{\gamma,\alpha\beta}^{(m)}F_\gamma + C_{\alpha\beta,\gamma\delta}^{(m)}F_{\gamma\delta} + \tfrac{1}{15}H_{\alpha\beta,\gamma\delta\epsilon}^{(m)}F_{\gamma\delta\epsilon} + \ldots
\end{aligned} \tag{1.59}$$

$$\begin{aligned}
\Omega_{\alpha\beta\gamma}^{(m)}(F_\alpha, F_{\alpha\beta}, F_{\alpha\beta\gamma}, \ldots) &= -15\frac{\partial W_m(F_\alpha, F_{\alpha\beta}, \ldots)}{\partial F_{\alpha\beta\gamma}}\\
&= \Omega_{\alpha\beta\gamma}^{(m)} + E_{\delta,\alpha\beta\gamma}^{(m)}F_\delta + \tfrac{1}{3}H_{\delta\epsilon,\alpha\beta\gamma}^{(m)}F_{\delta\epsilon} + R_{\alpha\beta\gamma,\delta\epsilon\phi}^{(m)}F_{\delta\epsilon\phi} + \ldots
\end{aligned} \tag{1.60}$$

and the quantities $\boldsymbol{\alpha}^{(m)}$, $\mathbf{C}^{(m)}$ and $\mathbf{R}^{(m)}$ are the dipole, quadrupole and octopole polarizabilities of the molecules in the state m. (Some authors, for example Dalgarno[57], define the nth order multipole polarizability as $(n!)^{-1}$ times this value. With our definition the multipole polarizability is equal to the multipole moment induced by unit magnitude of the appropriate field strength.) They have non-vanishing and positive components which always

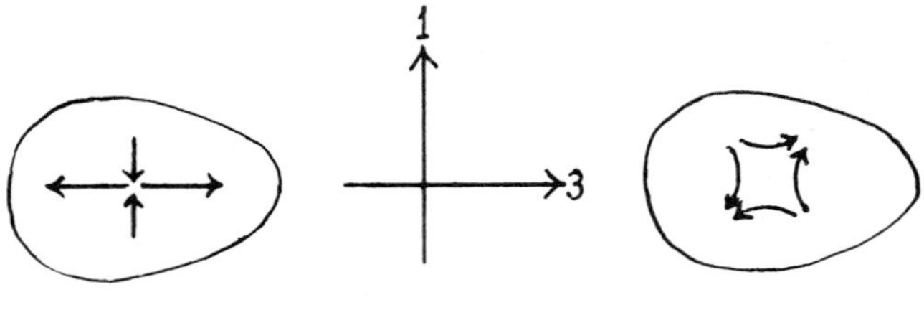

(a) (b)

Figure 1.10. The origin of the dipole induced in a linear polar molecule by an electric field-gradient. (a) An axial field-gradient $F_{33}=-2F_{11}=-2F_{22}$; $\mu_3=\frac{1}{3}A_{3,33}F_{33}+\frac{2}{3}A_{3,11}F_{11}=\frac{1}{2}A_{3,33}F_{33}$. (b) An off-diagonal field-gradient $F_{13}=F_{31}$; $\mu_1=\frac{2}{3}A_{1,13}F_{13}$

lead to a lowering of the energy of the ground state. The 'cross terms' in (1.57) in the field and its derivatives would vanish if the molecules were spherically symmetric; terms in both odd and even derivatives of the field, such as those in $A_{\alpha,\beta\gamma}$ and $H_{\alpha\beta,\gamma\delta\epsilon}$ are zero in centro-symmetric molecules if the centre is chosen as the origin. As may be seen from (1.58) and (1.59), the dipole-quadrupole polarizability $A_{\alpha,\beta\gamma}$ gives both the dipole moment induced by a field-gradient and the quadrupole moment induced by a uniform field; its origin in the asymmetry of the molecule may be appreciated from Figure 1.10.

The quantum-mechanical expressions analogous to (1.50) deduced for these higher static polarizabilities by perturbation theory are

$$A^{(m)}_{\alpha,\beta\gamma}=\sum_{p\neq m}\frac{\langle m|\,\mu_\alpha\,|p\rangle\langle p|\,\Theta_{\beta\gamma}\,|m\rangle+\langle m|\,\Theta_{\beta\gamma}\,|p\rangle\langle p|\,\mu_\alpha\,|m\rangle}{W_p-W_m} \tag{1.61}$$

$$C^{(m)}_{\alpha\beta,\gamma\delta}=\frac{1}{3}\sum_{p\neq m}\frac{\langle m|\,\Theta_{\alpha\beta}\,|p\rangle\langle p|\,\Theta_{\gamma\delta}\,|m\rangle+\langle m|\,\Theta_{\gamma\delta}\,|p\rangle\langle p|\,\Theta_{\alpha\beta}\,|m\rangle}{W_p-W_m} \tag{1.62}$$

$$E^{(m)}_{\alpha,\beta\gamma\delta}=\sum_{p\neq m}\frac{\langle m|\,\mu_\alpha\,|p\rangle\langle p|\,\Omega_{\beta\gamma\delta}\,|m\rangle+\langle m|\,\Omega_{\beta\gamma\delta}\,|p\rangle\langle p|\,\mu_\alpha\,|m\rangle}{W_p-W_m} \tag{1.63}$$

$$H^{(m)}_{\alpha\beta,\gamma\delta\epsilon}=\sum_{p\neq m}\frac{\langle m|\,\Theta_{\alpha\beta}\,|p\rangle\langle p|\,\Omega_{\gamma\delta\epsilon}\,|m\rangle+\langle m|\,\Omega_{\gamma\delta\epsilon}\,|p\rangle\langle p|\,\Theta_{\alpha\beta}\,|m\rangle}{W_p-W_m} \tag{1.64}$$

$$R^{(m)}_{\alpha\beta\gamma,\delta\epsilon\phi}=\frac{1}{15}\sum_{p\neq m}\frac{\langle m|\,\Omega_{\alpha\beta\gamma}\,|p\rangle\langle p|\,\Omega_{\delta\epsilon\phi}\,|m\rangle+\langle m|\,\Omega_{\delta\epsilon\phi}\,|p\rangle\langle p|\,\Omega_{\alpha\beta\gamma}\,|m\rangle}{W_p-W_m} \tag{1.65}$$

The dipole polarizability and hyperpolarizabilities, $\boldsymbol{\alpha}, \boldsymbol{\beta}, \boldsymbol{\gamma}, \ldots$ are independent of the origin, but all the polarizabilities involving higher multipoles are origin-dependent; they can be thought of as representing the distribution of polarizability in the molecule (see Figure 1.10). On moving the origin from O to O′ (see Figure 1.9),

$$\left.\begin{aligned} \alpha'_{\alpha\beta} &= \alpha_{\alpha\beta} \\ \beta'_{\alpha\beta\gamma} &= \beta_{\alpha\beta\gamma} \\ \gamma'_{\alpha\beta\gamma\delta} &= \gamma_{\alpha\beta\gamma\delta} \end{aligned}\right\} \tag{1.66}$$

$$\left.\begin{aligned} A'_{\alpha,\beta\gamma} &= A_{\alpha,\beta\gamma} - \tfrac{3}{2}r'_\beta\alpha_{\gamma\alpha} - \tfrac{3}{2}r'_\gamma\alpha_{\alpha\beta} + r'_\delta\alpha_{\delta\alpha}\delta_{\beta\gamma} \\ C'_{\alpha\beta,\gamma\delta} &= C_{\alpha\beta,\gamma\delta} - \tfrac{1}{2}r'_\alpha A_{\beta,\gamma\delta} - \tfrac{1}{2}r'_\beta A_{\alpha,\gamma\delta} + \tfrac{1}{3}r'_\epsilon A_{\epsilon,\gamma\delta}\delta_{\alpha\beta} \\ &\quad - \tfrac{1}{2}r'_\gamma A_{\delta,\alpha\beta} - \tfrac{1}{2}r'_\delta A_{\gamma,\alpha\beta} + \tfrac{1}{3}r'_\epsilon A_{\epsilon,\alpha\beta}\delta_{\gamma\delta} \\ &\quad + \tfrac{3}{4}r'_\alpha r'_\gamma\alpha_{\beta\delta} + \tfrac{3}{4}r'_\alpha r'_\delta\alpha_{\beta\gamma} - \tfrac{1}{2}r'_\alpha r'_\epsilon\delta_{\gamma\delta}\alpha_{\epsilon\beta} \\ &\quad + \tfrac{3}{4}r'_\beta r'_\gamma\alpha_{\delta\alpha} + \tfrac{3}{4}r'_\beta r'_\delta\alpha_{\alpha\gamma} - \tfrac{1}{2}r'_\beta r'_\epsilon\delta_{\gamma\delta}\alpha_{\epsilon\alpha} \\ &\quad - \tfrac{1}{2}r'_\epsilon r'_\gamma\delta_{\alpha\beta}\alpha_{\epsilon\delta} - \tfrac{1}{2}r'_\epsilon r'_\delta\delta_{\alpha\beta}\alpha_{\epsilon\gamma} + \tfrac{1}{3}r'_\epsilon r'_\phi\delta_{\alpha\beta}\delta_{\gamma\delta}\alpha_{\epsilon\phi} \end{aligned}\right\} \tag{1.67}$$

The number of independent components of the polarizabilities and hyperpolarizabilities **α, β, γ, A, C, E, H,** and **R** is given in Table 1.3 for the important symmetry groups. There are some connections between the numbers in Tables 1.2 and 1.3—thus the number of independent quadrupoles **Θ** is one fewer than the number of polarizabilities **α** (for both $\Theta_{\alpha\beta}$ and $\alpha_{\alpha\beta}$ are symmetric second-rank tensors, and $\Theta_{\alpha\beta}$ has the restriction that it is traceless, i.e. $\Theta_{\alpha\alpha} = 0$), the number of independent octopole moments is the number of hyperpolarizabilities **β** less the number of independent dipoles **μ**, and the number of independent hexadecapoles $\mathbf{\Phi}(\boldsymbol{\xi}^{(4)})$ is equal to the number of hyperpolarizabilities **γ** less the number of polarizabilities **α**. Also the number of independent quadrupole polarizabilities **C** is equal to the number of hyperpolarizabilities **γ**.

If the molecule is axially symmetric ($C_{\infty v}$) then all the property tensors can be expressed in terms of the orientation of the molecular axis. If **n** is the unit vector in the direction of the axis

$$\mu_\alpha = \mu n_\alpha \tag{1.68}$$

$$\Theta_{\alpha\beta} = \tfrac{1}{2}\Theta(3n_\alpha n_\beta - \delta_{\alpha\beta}) \tag{1.69}$$

$$\Omega_{\alpha\beta\gamma} = \tfrac{1}{2}\Omega(5n_\alpha n_\beta n_\gamma - n_\alpha\delta_{\beta\gamma} - n_\beta\delta_{\gamma\alpha} - n_\gamma\delta_{\alpha\beta}) \tag{1.70}$$

$$\begin{aligned} \Phi_{\alpha\beta\gamma\delta} = \tfrac{1}{8}\Phi\{&35n_\alpha n_\beta n_\gamma n_\delta - 5(n_\alpha n_\beta\delta_{\gamma\delta} + n_\alpha n_\gamma\delta_{\beta\delta} + n_\alpha n_\delta\delta_{\beta\gamma} \\ &+ n_\beta n_\gamma\delta_{\alpha\delta} + n_\beta n_\delta\delta_{\alpha\gamma} + n_\gamma n_\delta\delta_{\alpha\beta}) \\ &+ \delta_{\alpha\beta}\delta_{\gamma\delta} + \delta_{\alpha\gamma}\delta_{\beta\delta} + \delta_{\alpha\delta}\delta_{\beta\gamma}\} \end{aligned} \tag{1.71}$$

$$\begin{aligned} \alpha_{\alpha\beta} &= \alpha_\perp\delta_{\alpha\beta} + (\alpha_\parallel - \alpha_\perp)n_\alpha n_\beta \\ &= \alpha\delta_{\alpha\beta} + \tfrac{1}{3}(\alpha_\parallel - \alpha_\perp)(3n_\alpha n_\beta - \delta_{\alpha\beta}) \end{aligned} \tag{1.72}$$

$$\beta_{\alpha\beta\gamma} = \beta_\perp(n_\alpha\delta_{\beta\gamma} + n_\beta\delta_{\gamma\alpha} + n_\gamma\delta_{\alpha\beta}) + (\beta_\parallel - 3\beta_\perp)n_\alpha n_\beta n_\gamma \tag{1.73}$$

$$A_{\alpha,\beta\gamma} = \tfrac{1}{2}A_\parallel n_\alpha(3n_\beta n_\gamma - \delta_{\beta\gamma}) + A_\perp(n_\beta\delta_{\alpha\gamma} + n_\gamma\delta_{\alpha\beta} - 2n_\alpha n_\beta n_\gamma) \tag{1.74}$$

Table 1.3. The number of independent components of the polarizabilities **α**, **β**, **γ**, **A**, **C**, **E**, **H** and **R** for systems of different symmetries

Group symbols		$\alpha_{\alpha\beta}$	$\beta_{\alpha\beta\gamma}$	$\gamma_{\alpha\beta\gamma\delta}$	$A_{\alpha,\beta\gamma}$	$C_{\alpha\beta,\gamma\delta}$	$E_{\alpha,\beta\gamma\delta}$	$H_{\alpha\beta,\gamma\delta\epsilon}$	$R_{\alpha\beta\gamma,\delta\epsilon\phi}$
1	C_1	6	10	15	15	15	21	35	28
$\bar{1}$	C_i	6	0	15	0	15	21	0	28
m	C_s	4	6	9	8	9	11	18	16
2	C_2	4	4	9	7	9	11	17	16
$2/m$	C_{2h}	4	0	9	0	9	11	0	16
$2mm$	C_{2v}	3	3	6	4	6	6	9	10
222	D_2	3	1	6	3	6	6	8	10
mmm	D_{2h}	3	0	6	0	6	6	0	10
4	C_4	2	2	5	3	5	5	9	8
$\bar{4}$	S_4	2	2	5	4	5	5	8	8
$4/m$	C_{4h}	2	0	5	0	5	5	0	8
$4mm$	C_{4v}	2	2	4	2	4	3	5	6
$\bar{4}2m$	D_{2d}	2	1	4	2	4	3	4	6
422	D_4	2	0	4	1	4	3	4	6
$4/mmm$	D_{4h}	2	0	4	0	4	3	0	6
3	C_3	2	4	5	5	5	7	11	10
$\bar{3}$	S_6	2	0	5	0	5	7	0	10
$3m$	C_{3v}	2	3	4	3	4	4	6	7
32	D_3	2	1	4	2	4	4	5	7
$\bar{3}m$	D_{3d}	2	0	4	0	4	4	0	7
$\bar{6}$	C_{3h}	2	2	3	2	3	3	6	6
6	C_6	2	2	3	3	3	3	5	6
$6/m$	C_{6h}	2	0	3	0	3	3	0	6
$\bar{6}m2$	D_{3h}	2	1	3	1	3	2	3	5
$6mm$	C_{6v}	2	2	3	2	3	2	3	5
622	D_6	2	0	3	1	3	2	2	5
$6/mmm$	D_{6h}	2	0	3	0	3	2	0	5
23	T	1	1	2	1	2	2	2	4
$m3$	T_h	1	0	2	0	2	2	0	4
$\bar{4}3m$	T_d	1	1	2	1	2	1	1	3
432	O	1	0	2	0	2	1	1	3
$m3m$	O_h	1	0	2	0	2	1	0	3
∞m	$C_{\infty v}$	2	2	3	2	3	2	3	4
∞/mm	$D_{\infty h}$	2	0	3	0	3	2	0	4
Sphere	K_h	1	0	1	0	1	0	0	1

where $\mu = \mu_3$, $\Theta = \Theta_{33} = -2\Theta_{11} = -2\Theta_{22}$, $\Omega = \Omega_{333}$, $\Phi = \Phi_{3333}$, $\alpha_\parallel = \alpha_{33}$, $\alpha_\perp = \alpha_{11} = \alpha_{22}$, $\beta_\parallel = \beta_{333}$, $\beta_\perp = \beta_{311} = \beta_{322}$, $A_\parallel = A_{3,33} = -2A_{3,11} = -2A_{3,22}$, $A_\perp = A_{1,13} = A_{1,31} = A_{2,23} = A_{2,32}$.

There is just one permanent multipole moment of any order for an axially symmetric charge distribution,[58] and if there is also a centre of symmetry ($D_{\infty h}$) all the odd multipoles vanish.

For tetrahedral symmetry (T_d), there is a natural origin at the centre of the tetrahedron, e.g. at the C atom in CH_4, and it is convenient to refer the property tensors to their components in a cartesian system defined by the cube in which the corners of the tetrahedron are at the points $(1, 1, 1)$, $(1, -1, -1)$, $(-1, 1, -1)$, $(-1, -1, 1)$ (see Figure 1.11). The unit vectors

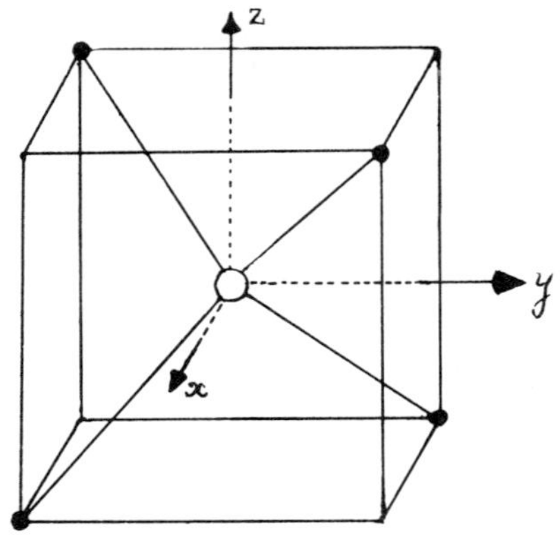

Figure 1.11. The cube defining the cartesian axes in a tetrahedron. The corners of the tetrahedron are at the points (1, 1, 1), (1, −1, −1), (−1, 1, −1) and (−1, −1, 1)

i, j, k are along the x, y, z axes of the cube. The dipole and quadrupole moments of the tetrahedron are zero, as is the fifth-order moment $\boldsymbol{\xi}^{(5)}$ (see Table 1.2). The octopole and hexadecapole moments have just one independent component and

$$\Omega_{\alpha\beta\gamma} = \Omega(i_\alpha j_\beta k_\gamma + i_\alpha j_\gamma k_\beta + i_\beta j_\gamma k_\alpha + i_\beta j_\alpha k_\gamma + i_\gamma j_\alpha k_\beta + i_\gamma j_\beta k_\alpha) \tag{1.75}$$

$$\Phi_{\alpha\beta\gamma\delta} = \Phi[\tfrac{5}{2}(i_\alpha i_\beta i_\gamma i_\delta + j_\alpha j_\beta j_\gamma j_\delta + k_\alpha k_\beta k_\gamma k_\delta) - \tfrac{1}{2}(\delta_{\alpha\beta}\delta_{\gamma\delta} + \delta_{\alpha\gamma}\delta_{\beta\delta} + \delta_{\alpha\delta}\delta_{\beta\gamma})] \tag{1.76}$$

where

$$\Omega = \Omega_{xyz} = \tfrac{5}{2}\sum_i e_i\langle x_i y_i z_i\rangle \tag{1.77}$$

$$\Phi = \Phi_{xxxx} = \tfrac{1}{8}\sum_i e_i\langle 35x_i^4 - 30x_i^2 r_i^2 + 3r_i^4\rangle \tag{1.78}$$

The polarizability $\boldsymbol{\alpha}$ is isotopic and there is just one independent component of both $\boldsymbol{\beta}$ and **A**,

$$\alpha_{\alpha\beta} = \alpha\delta_{\alpha\beta} \tag{1.79}$$

$$\beta_{\alpha\beta\gamma} = \beta(i_\alpha j_\beta k_\gamma + i_\alpha j_\gamma k_\beta + i_\beta j_\gamma k_\alpha + i_\beta j_\alpha k_\gamma + i_\gamma j_\alpha k_\beta + i_\gamma j_\beta k_\alpha) \tag{1.80}$$

$$\gamma_{\alpha\beta\gamma\delta} = \tfrac{1}{3}\gamma(\delta_{\alpha\beta}\delta_{\gamma\delta} + \delta_{\alpha\gamma}\delta_{\beta\delta} + \delta_{\alpha\delta}\delta_{\beta\gamma}) + \Delta\gamma[i_\alpha i_\beta i_\gamma i_\delta + j_\alpha j_\beta j_\gamma j_\delta + k_\alpha k_\beta k_\gamma k_\delta - \tfrac{1}{5}(\delta_{\alpha\beta}\delta_{\gamma\delta} + \delta_{\alpha\gamma}\delta_{\beta\delta} + \delta_{\alpha\delta}\delta_{\beta\gamma})] \tag{1.81}$$

$$A_{\alpha,\beta\gamma} = A(i_\alpha j_\beta k_\gamma + i_\alpha j_\gamma k_\beta + i_\beta j_\gamma k_\alpha + i_\beta j_\alpha k_\gamma + i_\gamma j_\alpha k_\beta + i_\gamma j_\beta k_\alpha) \tag{1.82}$$

$$C_{\alpha\beta,\gamma\delta} = C\{\tfrac{1}{2}(\delta_{\alpha\gamma}\delta_{\beta\delta} + \delta_{\alpha\delta}\delta_{\beta\gamma}) - \tfrac{1}{3}\delta_{\alpha\beta}\delta_{\gamma\delta}\} + \Delta C\{i_\alpha i_\beta i_\gamma i_\delta + j_\alpha j_\beta j_\gamma j_\delta + k_\alpha k_\beta k_\gamma k_\delta - \tfrac{1}{5}(\delta_{\alpha\beta}\delta_{\gamma\delta} + \delta_{\alpha\gamma}\delta_{\beta\delta} + \delta_{\alpha\delta}\delta_{\beta\gamma})\} \tag{1.83}$$

where

$$\alpha = \tfrac{1}{3}\alpha_{\alpha\beta}\delta_{\alpha\beta} = \tfrac{1}{3}\alpha_{\beta\beta} = \alpha_{xx} = \alpha_{yy} = \alpha_{zz} \tag{1.84}$$

$$\beta = \beta_{xyz} \tag{1.85}$$

$$\gamma = \tfrac{1}{5}\gamma_{\alpha\beta\alpha\beta} = \tfrac{3}{5}(\gamma_{xxxx} + 2\gamma_{xyxy}) \tag{1.86}$$

$$\Delta\gamma = \gamma_{xxxx} - 3\gamma_{xyxy} \tag{1.87}$$

$$A = A_{xyz} \tag{1.88}$$

$$C = \tfrac{1}{5}C_{\alpha\beta,\alpha\beta} = \tfrac{3}{5}(C_{xx,xx} + 2C_{xy,xy}) \tag{1.89}$$

$$\Delta C = \tfrac{3}{2}C_{xx,xx} - 2C_{xy,xy} \tag{1.90}$$

For octahedral symmetry (O_h), as in sulphur hexafluoride, the first non-vanishing multipole is the hexadecapole and it transforms in the same way as $\mathbf{\Phi}$ for a tetrahedron (see equation (1.76)). The centre of symmetry eliminates all odd permanent multipole moments as well as $\boldsymbol{\beta}$ and $\mathbf{A}$; the polarizabilities $\boldsymbol{\alpha}, \boldsymbol{\gamma}$ and $\mathbf{C}$ transform as for a tetrahedron.

B. Molecules in time-dependent fields

In the previous section, the interaction of a molecule with an electrostatic field was described in terms of permanent moments and a variety of polarizabilities. If the electric field is periodic and of angular frequency ω, the interaction is modified and the polarizabilities are functions of ω. Their evaluation may be carried out by quantum-mechanical time-dependent perturbation theory.[48,24] Some of the most accurate methods of measuring polarizabilities and hyperpolarizabilities involve optical fields, so it is important to know their frequency-dependence or dispersion. Also in the rigorous theory,[57,59] the long-range dispersion forces are related to an integral over all frequencies ω of the product of polarizabilities of each molecule at *imaginary frequencies* $i\omega$. A discussion of frequency-dependent polarizabilities is therefore appropriate in an article on intermolecular forces.

An oscillating electric field of angular frequency ω causes a periodic distortion of the charge distribution in a molecule. If ω is not close to an absorption frequency, the induced charge and current densities can conveniently be interpreted in terms of induced electric and magnetic moments. The moments may be in or out of phase with the incident oscillating electric field. When ω is near a resonant frequency, the induced moments can be very large and there may be absorption or emission of a photon, associated with a change in the quantum state of the molecule.

It is convenient to use complex polarizabilities $\tilde{\alpha}_{\alpha\beta}$, $\tilde{A}_{\alpha,\beta\gamma}$, etc. to describe the complex moments induced in the molecule by an external electric field

$$F_\beta = F_\beta^{(0)} \exp(-i\omega t) \tag{1.91}$$

and field-gradient

$$F_{\beta\gamma} = \nabla_\beta F_\gamma = F_{\beta\gamma}^{(0)} \exp(-i\omega t) \tag{1.92}$$

and by the associated magnetic field B_β which is required by Maxwell's equations[60] to be present with a time-dependent electric field. For a plane electromagnetic wave travelling with velocity c in the direction of the unit vector $\mathbf{k}$, these three fields take the form

$$F_\beta = F_\beta^{(0)} \exp\{-\mathrm{i}\omega(t-\mathbf{k}\cdot\mathbf{r}c^{-1})\} \tag{1.93}$$

$$F_{\beta\gamma} = \nabla_\beta F_\gamma = \mathrm{i}\omega c^{-1} k_\beta F_\gamma \tag{1.94}$$

$$B_\beta = c^{-1}\epsilon_{\beta\gamma\delta} k_\gamma F_\delta \tag{1.95}$$

and where $\epsilon_{\alpha\beta\gamma}$ is the anti-symmetric tensor which is equal to $+1$ or -1 if α, β, γ is an even or odd permutation of x, y, z and is zero otherwise.

The induced electric dipole and quadrupole moments, and the magnetic dipole moment $\mathbf{m}$, are the real parts of the following complex moments[61]

$$\mu_\alpha = \tilde{\alpha}_{\alpha\beta}F_\beta + \tfrac{1}{3}\tilde{A}_{\alpha,\beta\gamma}F_{\beta\gamma} + \tilde{G}_{\alpha\beta}B_\beta + \ldots \tag{1.96}$$

$$\Theta_{\alpha\beta} = \tilde{\mathscr{A}}_{\gamma,\alpha\beta}F_\gamma + \tilde{C}_{\alpha\beta,\gamma\delta}F_{\gamma\delta} + \ldots \tag{1.97}$$

$$m_\alpha = \tilde{\mathscr{G}}_{\alpha\beta}F_\beta + \tilde{\chi}_{\alpha\beta}B_\beta + \ldots \tag{1.98}$$

where, for a system in the quantum state m

$$\begin{aligned}\tilde{\alpha}_{\alpha\beta} &= 2\hbar^{-1}\sum_p Z(\omega, \omega_{pm}, \Gamma_{pm})[\omega_{pm}\,\mathrm{Re}\,\{\langle m|\,\mu_\alpha\,|p\rangle\langle p|\,\mu_\beta\,|m\rangle\} \\ &\quad + \mathrm{i}\omega\,\mathrm{Im}\,\{\langle m|\,\mu_\alpha\,|p\rangle\langle p|\,\mu_\beta\,|m\rangle\}] \\ &= \alpha_{\alpha\beta} - \mathrm{i}\alpha'_{\alpha\beta} = \alpha_{\beta\alpha} + \mathrm{i}\alpha'_{\beta\alpha}\end{aligned} \tag{1.99}$$

$$\begin{aligned}\tilde{A}_{\alpha,\beta\gamma} &= 2\hbar^{-1}\sum_p Z(\omega, \omega_{pm}, \Gamma_{pm}) \\ &\quad \times[\omega_{pm}\,\mathrm{Re}\,\{\langle m|\,\mu_\alpha\,|p\rangle\langle p|\,\Theta_{\beta\gamma}\,|m\rangle\} + \mathrm{i}\omega\,\mathrm{Im}\,\{\langle m|\,\mu_\alpha\,|p\rangle\langle p|\,\Theta_{\beta\gamma}\,|m\rangle\}] \\ &= A_{\alpha,\beta\gamma} - \mathrm{i}\omega A'_{\alpha,\beta\gamma} = A_{\alpha,\gamma\beta} - \mathrm{i}\omega A'_{\alpha,\gamma\beta}\end{aligned} \tag{1.100}$$

$$\begin{aligned}\tilde{G}_{\alpha\beta} &= 2\hbar^{-1}\sum_p Z(\omega, \omega_{pm}, \Gamma_{pm})[\omega_{pm}\,\mathrm{Re}\,\{\langle m|\,\mu_\alpha\,|p\rangle\langle p|\,m_\beta\,|m\rangle\} \\ &\quad + \mathrm{i}\omega\,\mathrm{Im}\,\{\langle m|\,\mu_\alpha\,|p\rangle\langle p|\,m_\beta\,|m\rangle\}] \\ &= G_{\alpha\beta} - \mathrm{i}G'_{\alpha\beta}\end{aligned} \tag{1.101}$$

$$\begin{aligned}\tilde{\mathscr{A}}_{\alpha,\beta\gamma} &= 2\hbar^{-1}\sum_p Z(\omega, \omega_{pm}, \Gamma_{pm}) \\ &\quad \times[\omega_{pm}\,\mathrm{Re}\,\{\langle m|\,\Theta_{\beta\gamma}\,|p\rangle\langle p|\,\mu_\alpha\,|m\rangle\} + \mathrm{i}\omega\,\mathrm{Im}\,\{\langle m|\,\Theta_{\beta\gamma}\,|p\rangle\langle p|\,\mu_\alpha\,|m\rangle\}] \\ &= A_{\alpha,\beta\gamma} + \mathrm{i}A'_{\alpha,\beta\gamma} = A_{\alpha,\gamma\beta} + \mathrm{i}A'_{\alpha,\gamma\beta}\end{aligned} \tag{1.102}$$

$$\begin{aligned}\tilde{\mathscr{G}}_{\alpha\beta} &= 2\hbar^{-1}\sum_p Z(\omega, \omega_{pm}, \Gamma_{pm})[\omega_{pm}\,\mathrm{Re}\,\{\langle m|\,m_\alpha\,|p\rangle\langle p|\,\mu_\beta\,|m\rangle\} \\ &\quad + \mathrm{i}\omega\,\mathrm{Im}\,\{\langle m|\,m_\alpha\,|p\rangle\langle p|\,\mu_\beta\,|m\rangle\}] \\ &= G_{\beta\alpha} + \mathrm{i}G'_{\beta\alpha}\end{aligned} \tag{1.103}$$

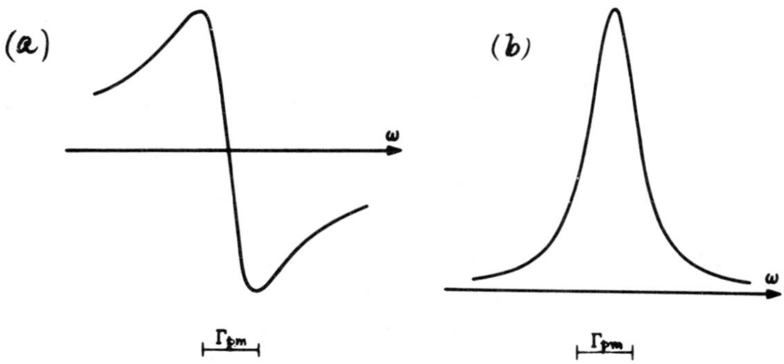

Figure 1.12. Typical dispersion (a) and absorption (b) line shape functions f and g. The resonance is at the angular frequency ω_{pm} and the width at half-height of the absorption curve g is Γ_{pm}

$$\tilde{\chi}_{\alpha\beta}=\left\langle m\left|\sum_i \frac{e_i^2}{4m_i}(r_{i_\alpha}r_{i_\beta}-r_i^2\delta_{\alpha\beta})\right|m\right\rangle+2\hbar^{-1}\sum_p Z(\omega,\omega_{pm},\Gamma_{pm})$$
$$\times[\omega_{pm}\,\mathrm{Re}\,\{\langle m|\,m_\alpha\,|p\rangle\langle p|\,m_\beta\,|m\rangle\}+\mathrm{i}\omega\,\mathrm{Im}\,\{\langle m|\,m_\alpha\,|p\rangle\langle p|\,m_\beta\,|m\rangle\}] \tag{1.104}$$

The complex function $Z(\omega, \omega_{pm}, \Gamma_{pm})$ is comprised of a dispersion function f and an absorption function g (see Figure 1.12),

$$Z(\omega,\omega_{pm},\Gamma_{pm})=f(\omega,\omega_{pm},\Gamma_{pm})+\mathrm{i}g(\omega,\omega_{pm},\Gamma_{pm}) \tag{1.105}$$

At large displacements from the resonance absorption frequency $\omega_{pm}=\hbar^{-1}(W_p-W_m)$, the absorption function g is zero and Z is real,

$$Z=f=(\omega_{pm}^2-\omega^2)^{-1}, \quad \text{for} \quad |\omega-\omega_{pm}|\gg\Gamma_{pm} \tag{1.106}$$

In the vicinity of a Lorentzian absorption line at ω_{pm},

$$Z=\frac{1}{\omega_{pm}^2-\omega^2-\mathrm{i}\omega\Gamma_{pm}}$$

and this can be rationalized to give

$$Z(\omega,\omega_{pm},\Gamma_{pm})=\frac{\omega_{pm}^2-\omega^2}{(\omega_{pm}^2-\omega^2)^2+\omega^2\Gamma_{pm}^2}+\mathrm{i}\frac{\omega\Gamma_{pm}}{(\omega_{pm}^2-\omega^2)^2+\omega^2\Gamma_{pm}^2} \tag{1.107}$$

where Γ_{pm} is the width at half the maximum height of g.

The magnetic moment is the expectation value of the operator $\mathbf{m}$ where

$$\mathbf{m}=\tfrac{1}{2}\sum_i e_i(\mathbf{r}_i\times\dot{\mathbf{r}}_i+\hbar m_i^{-1}g_i\mathbf{s}_i) \tag{1.108}$$

and $\hbar\mathbf{s}_i$ is the spin angular momentum operator of particle i; g_i is the g-factor' for particle i and is equal to 2·0023 for an electron and 5·5856 for a proton. The symmetry properties of the electric and magnetic dipole operators $\boldsymbol{\mu}$ and $\mathbf{m}$ are opposite under parity (which converts $\mathbf{r}_i$ into $-\mathbf{r}_i$) and

Table 1.4. The behaviour of various operators under parity and time-reversal

Operator	Electric dipole	Magnetic dipole	Electric quadrupole	Magnetic quadrupole	Electric octopole
Parity	Odd	Even	Even	Odd	Odd
Time-reversal	Even	Odd	Even	Odd	Even

under time-reversal. Table 1.4 shows the behaviour of electric and magnetic moment operators under parity and time-reversal. Magnetic moments are moments of the *current* density[62] and are therefore odd under time-reversal, unlike the electric moments, which are moments of the *charge* density and are even under time-reversal.

The absorption of electromagnetic radiation can be related to the polarizabilities in the following way. Consider the change $\mathrm{d}W$ in the energy of the radiation field when the electric and magnetic moments of the molecule change by $\mathrm{d}\boldsymbol{\mu}$, $\mathrm{d}\boldsymbol{\Theta}$, $\mathrm{d}\mathbf{m}$, etc.

$$\mathrm{d}W = F_\alpha^* \,\mathrm{d}\mu_\alpha + \tfrac{1}{3}F_{\alpha\beta}^* \,\mathrm{d}\Theta_{\alpha\beta} + B_\alpha^* \,\mathrm{d}m_\alpha + \ldots \tag{1.109}$$

The power absorbed from the radiation field is therefore the real part of

$$\dot{W} = \dot{\mu}_\alpha F_\alpha^* + \tfrac{1}{3}\dot{\Theta}_{\alpha\beta}F_{\alpha\beta}^* + \dot{m}_\alpha B_\alpha^* + \ldots \tag{1.110}$$

and this can be expressed in terms of the polarizabilities with the aid of equations (1.96)–(1.98). The mean energy absorbed in one second is, in the electric dipole approximation,

$$\begin{aligned}\bar{\dot{W}} &= \omega \,\mathrm{Im}\, \{\tilde{\alpha}_{\alpha\beta}F_\beta F_\alpha^*\} \\ &= \tfrac{1}{2}\omega \,\mathrm{Im}\, \{\alpha_{\alpha\beta}\}(F_\alpha F_\beta^* + F_\beta F_\alpha^*) + \tfrac{1}{2}\mathrm{i}\omega \,\mathrm{Im}\, \{\alpha'_{\alpha\beta}\}(F_\alpha F_\beta^* - F_\beta F_\alpha^*)\end{aligned} \tag{1.111}$$

The contribution in the antisymmetric polarizability $\boldsymbol{\alpha}'$ is non-vanishing only if the system is not symmetric under time-reversal, that is, if its wavefunctions are necessarily complex so that

$$\mathrm{Im}\, \{\langle m|\, \mu_\alpha \,|p\rangle\langle p|\, \mu_\beta \,|m\rangle\} \neq 0;$$

$\boldsymbol{\mu} = \sum_i e_i \mathbf{r}_i$ is a real operator. The lack of symmetry under time-reversal may arise from the presence of a magnetostatic field, as in the Faraday effect,[63] or from an intrinsic angular momentum, as in the ground state of the hydrogen or sodium atom. The atom may exist in states having components of angular momentum equal to $\frac{1}{2}$ or $-\frac{1}{2}$ in the direction of the axis of quantization and these states have equal but opposite values of $\boldsymbol{\alpha}'$; thus in an assembly of atoms, the $\boldsymbol{\alpha}'$-contributions to the refraction and absorption are non-zero only if there is an unequal population of the two angular momentum states, as, for example, in the magnetostatic field of the Faraday effect. Similarly, $\mathbf{A}'$ and $\mathbf{G}$ in (1.100) and (1.101) are non-zero only for systems that are not symmetric under time-reversal.

The use of circularly polarized radiation can lead to measurement of $\boldsymbol{\alpha}'$

and $\mathbf{G}'$. For right and left circularly polarized light travelling in the $\mathbf{k}$-direction

$$\mathbf{F}^{\pm} = 2^{-1/2}F^{(0)}(\mathbf{i} \mp \mathrm{i}\mathbf{j}) \exp\{-\mathrm{i}\omega(t - \mathbf{k} \cdot \mathbf{r}c^{-1})\} \quad (1.112)$$

where the unit vectors $\mathbf{i}$, $\mathbf{j}$, $\mathbf{k}$ form a right-handed cartesian frame, and the upper sign describes right, and the lower left, circular polarization. The absorption of right or left circularly polarized light is obtained from (1.110) and (1.112) and is the real part of

$$\begin{aligned}\bar{\dot{W}}^{\pm} &= -\tfrac{1}{2}\mathrm{i}\omega F^{(0)^2}[(i_\alpha i_\beta + j_\alpha j_\beta) \mp \mathrm{i}(i_\alpha j_\beta - i_\beta j_\alpha)][\tilde{\alpha}_{\alpha\beta} \pm \mathrm{i}c^{-1}(\tilde{G}_{\alpha\beta} - \tilde{\mathscr{G}}_{\alpha\beta}) \\ &\quad + \tfrac{1}{3}\mathrm{i}\omega c^{-1}k_\gamma(\tilde{A}_{\alpha\gamma\beta} - \tilde{\mathscr{A}}_{\beta\gamma\alpha}) + \ldots] \\ &= -\tfrac{1}{2}\mathrm{i}\omega F^{(0)^2}[\alpha_{xx} + \alpha_{yy} \mp (\alpha'_{xy} - \alpha'_{yx}) \pm 2c^{-1}(G'_{xx} + G'_{yy}) \\ &\quad + 2c^{-1}(G_{xy} - G_{yx}) \\ &\quad \pm \tfrac{2}{3}\omega c^{-1}(A_{xzy} - A_{yzx}) + \tfrac{2}{3}\omega c^{-1}(A'_{xzx} + A'_{yzy}) + \ldots]\end{aligned} \quad (1.113)$$

Hence

$$\bar{\dot{W}}^{+} + \bar{\dot{W}}^{-} = \omega F^{(0)^2}\,\mathrm{Im}\,\{\alpha_{xx} + \alpha_{yy} + \ldots\} \quad (1.114)$$

and the circular dichroism is

$$\bar{\dot{W}}^{+} - \bar{\dot{W}}^{-} = -2\omega F^{(0)^2}\,\mathrm{Im}\,\{\alpha'_{xy} - c^{-1}[G'_{xx} + G'_{yy} + \tfrac{1}{3}\omega(A_{xzy} - A_{yzx})] + \ldots\} \quad (1.115)$$

Magnetic circular dichroism (MCD) arises from the effect of a uniform magnetostatic field $B_z^{(0)}$ on α'_{xy} and its so-called 'C-term', which is proportional to $\langle m|\, m_z\, |m\rangle B_z^{(0)}/kT$ gives a measure of α'_{xy} in the unperturbed molecule.[63] Ordinary circular dichroism (CD) is due to the terms in

$$G'_{xx} + G'_{yy} + \tfrac{1}{3}\omega(A_{xzy} - A_{yzx})$$

and in an isotropic medium such as a gas or liquid, this simplifies to $\tfrac{2}{3}G'_{\alpha\alpha}$.[64]

C. The induction energy

The induction energy results from the distortion of a molecule by the electric field and field-gradients due to the undistorted charge distribution of its neighbours. It is therefore dependent upon the static molecular polarizabilities. For the pair of molecules a and b, the induction energy of molecule b in the state m_b due to molecule a in m_a can be obtained from equations (1.28), (1.33), (1.50), (1.61) and (1.62).

$$u^{(b)}_{\text{induction}} = -\tfrac{1}{2}\alpha^{(m_b)}_{\alpha\beta}F^{(b)}_\alpha F^{(b)}_\beta - \tfrac{1}{3}A^{(m_b)}_{\alpha,\beta\gamma}F_\alpha F_{\beta\gamma} - \tfrac{1}{6}C^{(m_b)}_{\alpha\beta,\gamma\delta}F_{\alpha\beta}F_{\gamma\delta} - \ldots \quad (1.116)$$

$F^{(b)}_\alpha$ and $F^{(b)}_{\alpha\beta}$ are the electric field and field-gradient at the origin in molecule b due to the permanent charge distribution of molecule a, and they may be expressed in terms of the moments of molecule a by means of equation (1.45). If more than two molecules are present $F^{(b)}_\alpha$ and $F^{(b)}_{\alpha\beta}$ are the total field and field-gradient at molecule b due to all the other molecules.

The effects of distortion of molecule a on the induction energy of molecule b could easily be incorporated into the theory by proceeding to the third order of perturbation. These higher terms were first obtained by Barker[65] and have been used in a theory of equilibrium properties of polar gases.[66]

The induction energy is easy to evaluate, since molecular static polarizabilities are accessible. As stated in Section II, it must be negative when m_b is the ground state. However, it is rarely the predominant source of attraction between molecules and can often be neglected in comparison to the electrostatic and dispersion energies.

D. The dispersion energy

Unlike the induction energy, the dispersion energy obtained from equations (1.29) and (1.46) does not easily separate into single-centre contributions. If precise formulae are required, then it is necessary to consider the frequency-dependent polarizabilities, or rather the polarizabilities at imaginary frequencies.[67] The evaluation of dispersion energies for simple systems using this approach has been well described by Dalgarno and Davison,[68] Dalgarno,[57] and Mahanty and Ninham.[59] The relationship of the dispersion energy to the molecular polarizability at imaginary frequencies $\mathrm{i}u$ emerges from the identities

$$\frac{1}{A+B}=\frac{2}{\pi}\int_0^\infty \frac{AB}{(A^2+u^2)(B^2+u^2)}\,du=\frac{2}{\pi}\int_0^\infty \frac{u^2}{(A^2+u^2)(B^2+u^2)}\,du,\quad \begin{cases}A>0\\B>0\end{cases} \tag{1.117}$$

From (1.99) and (1.106), the polarizability at the imaginary frequency $\omega=\mathrm{i}u$ of the molecule in state m is

$$\alpha_{\alpha\gamma}^{(m)}(\mathrm{i}u)=\sum_{p\neq m}\frac{\omega_{pm}[\langle m|\,\mu_\alpha\,|p\rangle\langle p|\,\mu_\gamma\,|m\rangle+\langle m|\,\mu_\gamma\,|p\rangle\langle p|\,\mu_\alpha\,|m\rangle]}{\hbar(\omega_{pm}^2+u^2)} \tag{1.118}$$

The dependence of $\alpha_{\alpha\gamma}^{(m)}(\mathrm{i}u)$ on the line width Γ has been dropped since there are no poles in $\alpha_{\alpha\gamma}^{(m)}$ at imaginary frequencies; in fact $\alpha_{\alpha\gamma}^{(m)}(\mathrm{i}u)$ is a positive, monotonically decreasing, function of u. From (1.29) and (1.46) the dipole dispersion energy is

$$u_{\text{dispersion}}(R^{-6})=-T_{\alpha\beta}T_{\gamma\delta}\sum_{\substack{p_a\neq m_a\\ p_b\neq m_b}}\frac{\langle m_am_b|\,\mu_\alpha^{(a)}\mu_\beta^{(b)}\,|p_ap_b\rangle\langle p_ap_b|\,\mu_\gamma^{(a)}\mu_\delta^{(b)}\,|m_am_b\rangle}{\hbar(\omega_{p_am_a}+\omega_{p_bm_b})} \tag{1.119}$$

The wavefunctions m_am_b and p_ap_b in (1.119) are simple products and if m_a and m_b are the ground electronic states the energy differences $W_{p_a}-W_{m_a}$ and $W_{p_b}-W_{m_b}$ are positive and equal to $\hbar\omega_{p_am_a}$ and to $\hbar\omega_{p_bm_b}$. Hence from

(1.117)–(1.119)

$$u_{\text{dispersion}}(R^{-6}) = -\frac{\hbar}{2\pi} T_{\alpha\beta}T_{\gamma\delta}\left[\int_0^\infty \alpha_{\alpha\gamma}^{(m_a)}(\text{i}u)\alpha_{\beta\delta}^{(m_b)}(\text{i}u)\,\text{d}u + \int_0^\infty \alpha'^{(m_a)}_{\alpha\gamma}(\text{i}u)\alpha'^{(m_b)}_{\beta\delta}(\text{i}u)\,\text{d}u\right]$$

$$= -\frac{\hbar}{2\pi} T_{\alpha\beta}T_{\gamma\delta}\int_0^\infty \tilde{\alpha}_{\alpha\gamma}^{(m_a)}(\text{i}u)^*\tilde{\alpha}_{\beta\delta}^{(m_b)}(\text{i}u)\,\text{d}u \qquad (1.120)$$

The term in (1.120) in the polarizability product $\alpha'^{(m_a)}_{\alpha\gamma}\alpha'^{(m_b)}_{\beta\delta}$ vanishes if either molecule is symmetric under time-reversal; it is small and negative for a pair of alkali metal atoms having the same spin, and is positive when the spins are opposed. If the molecules are isotropically polarizable, so that (1.79) applies at all frequencies,

$$u_{\text{dispersion}}(R^{-6}) = -(4\pi\epsilon_0)^{-2}\frac{3\hbar}{\pi}R^{-6}\int_0^\infty \alpha^{(m_a)}(\text{i}u)\alpha^{(m_b)}(\text{i}u)\,\text{d}u \qquad (1.121)$$

This is the expression that has been applied to the inert gases and other simple systems, using either variational calculations of the frequency-dependent polarizabilities or semiempirical methods employing observed frequencies and intensities together with sum rules.[57,69] Bounds can be imposed on the coefficients C_6 and C_8 of R^{-6} and R^{-8} giving the dipole–dipole and dipole–quadrupole dispersion energies[70,71] and reliable values are available for many simple systems (see Table 1.5). It is of interest to note that

Table 1.5. Values of the coefficients C_6, C_8 and C_{10} in the expansion of the dispersion energy, $u_{\text{dispersion}} = C_6R^{-6}[1+\Delta(\frac{3}{2}\cos^2\theta-\frac{1}{2})]+C_8R^{-8}+C_{10}R^{-10}+\ldots$ where θ is the angle between R and the axis of a homonuclear diatomic molecule. The values are for systems in their ground state and are in atomic units [1 a.u. of energy $= e^2(4\pi\epsilon_0 a_0)^{-1} = 4{\cdot}3598+10^{-18}$ J, 1 a.u. of length $= 0{\cdot}529177\times10^{-10}$ m]. The values were obtained from references 32, 69, 70 and 72

Molecules	$-C_6$	$-C_8$	$-C_{10}$	Δ
H–H	6·4990	124·399	3285·828	0
H–He	2·83	41·9	—	0
He–He	1·4618	13·9	182	0
He–Ne	3·06	—	—	0
Ne–Ne	6·6	57	700	0
Ne–Ar	20·5	—	—	0
Ar–Ar	68	1130	$2{\cdot}5\times10^4$	0
Kr–Kr	133	2500	6×10^4	0
Xe–Xe	300	—	—	0
$He–H_2$	4·00	—	—	0·105
$H_2–H_2$	12·1	—	—	0·112
$He–N_2$	10·3	—	—	0·101
$Li–H_2$	82·4	—	—	0·126
$Li–N_2$	183	—	—	0·127

the accurate values of C_6 are considerably smaller than the empirically determined coefficients $-4\epsilon\sigma^6$ of R^{-6} in the Lennard–Jones 12–6 potential[8]

$$u_{\text{L–J}} = 4\epsilon\{(\sigma/R)^{12} - (\sigma/R)^6\} \tag{1.122}$$

This is because the entire attractive potential is made to vary as R^{-6} in $u_{\text{L–J}}$ and the empirical coefficient takes some account of the shorter range attractions, including the dipole–quadrupole dispersion energy C_8R^{-8}.

The shorter-range contributions to the dispersion energy can be related to the higher-order frequency-dependent polarizabilities at imaginary frequencies. Thus for a pair of inert gas atoms

$$u_{\text{dispersion}}(R^{-8}) = -(4\pi\epsilon_0)^{-2}\frac{15\hbar}{\pi R^8}\int_0^\infty [\alpha^{(m_a)}(\text{i}u)C^{(m_b)}(\text{i}u) + C^{(m_a)}(\text{i}u)\alpha^{(m_b)}(\text{i}u)]\,\text{d}u \tag{1.123}$$

and in the general case of two anisotropic but time-symmetric molecules

$$u_{\text{dispersion}}(R^{-7}) = -\frac{\hbar}{3\pi}T_{\alpha\beta}T_{\gamma\delta\epsilon}\int_0^\infty [\alpha^{(m_a)}_{\alpha\gamma}(\text{i}u)A^{(m_b)}_{\beta,\delta\epsilon}(\text{i}u) - A^{(m_a)}_{\beta,\delta\epsilon}(\text{i}u)\alpha^{(m_b)}_{\alpha\gamma}(\text{i}u)]\,\text{d}u \tag{1.124}$$

$$\begin{aligned} u_{\text{dispersion}}(R^{-8}) = &-\frac{\hbar}{6\pi}T_{\alpha\beta\gamma}T_{\delta\epsilon\phi}\int_0^\infty [\alpha^{(m_a)}_{\alpha\delta}(\text{i}u)C^{(m_b)}_{\beta\gamma,\epsilon\phi}(\text{i}u) + C^{(m_a)}_{\beta\gamma,\epsilon\phi}(\text{i}u)\alpha^{(m_b)}_{\alpha\delta}(\text{i}u) \\ &-\tfrac{2}{3}A^{(m_a)}_{\alpha,\epsilon\phi}(\text{i}u)A^{(m_b)}_{\delta,\beta\gamma}(\text{i}u)]\,\text{d}u \\ &+\frac{\hbar}{9\pi}T_{\alpha\beta}T_{\gamma\delta\epsilon\phi}\int_0^\infty A^{(m_a)}_{\alpha,\gamma\delta}(\text{i}u)A^{(m_b)}_{\beta,\epsilon\phi}(\text{i}u)\,\text{d}u \end{aligned} \tag{1.125}$$

In the case of a tetrahedral molecule b interacting with a spherical molecule a (e.g. $CH_4 + Ar$), equations (1.79), (1.82) and (1.83) may be inserted into (1.121), (1.123) *and* (1.125) *to give*

$$\begin{aligned} u_{\text{dispersion}} = &-(4\pi\epsilon_0)^{-2}\frac{3\hbar}{\pi R^6}\int_0^\infty [\alpha^{(a)}(\text{i}u)\alpha^{(b)}(\text{i}u) + 8R^{-1}\cos\theta_x\cos\theta_y\cos\theta_z\alpha^{(a)} \\ &\times(\text{i}u)A^{(b)}(\text{i}u) \\ &+5R^{-2}\{\alpha^{(a)}(\text{i}u)C^{(b)}(\text{i}u) + C^{(a)}(\text{i}u)\alpha^{(b)}(\text{i}u) + \tfrac{1}{2}\alpha^{(a)}(\text{i}u)\Delta C^{(b)}(\text{i}u) \\ &\times(\cos^4\theta_x + \cos^4\theta_y + \cos^4\theta_z - \tfrac{3}{5})\} + \ldots]\,\text{d}u \end{aligned} \tag{1.126}$$

where θ_α is the angle between the α-axis of the cube associated with the tetrahedron (see Figure 1.11) and the vector from the centre of the tetrahedron to the sphere.

Little is known of the quadrupole intensities needed to determine A and C at imaginary frequencies, so these accurate expressions for the shorter-range dispersion energies are of little practical use. Fortunately, there is a useful simplification that can be made which leads to an approximate formula for the dispersion energy in terms of the *static* polarizabilities. A relationship of this type was originally proposed by London,[28] and it rests upon an average-energy, or Unsöld,[73] approximation. On inserting the multipole expansion (1.46) into (1.29) for the dispersion energy, a general expansion in which the first term is (1.119) is obtained for $u_{\text{dispersion}}$. Separation into single-centre contributions can be achieved by replacing the energy denominators in (1.29) as follows:

$$\frac{1}{(W_{p_a}-W_{m_a})+(W_{p_b}-W_{m_b})}=\frac{U_a U_b(1+\Delta)}{(U_a+U_b)(W_{p_a}-W_{m_a})(W_{p_b}-W_{m_b})} \tag{1.127}$$

$$\Delta=\frac{U_a^{-1}-(W_{p_a}-W_{m_a})^{-1}+U_b^{-1}-(W_{p_b}-W_{m_b})^{-1}}{(W_{p_a}-W_{m_a})^{-1}+(W_{p_b}-W_{m_b})^{-1}} \tag{1.128}$$

The simplification is made by omitting Δ from (1.127) and by setting U_a and U_b equal to fixed energies associated with molecules a and b in the states m_a and m_b. Then the multipole expansion of the dispersion energy reduces to[24]

$$\begin{aligned}u_{\text{dispersion}}=-\frac{U_aU_b}{4(U_a+U_b)}&[T_{\alpha\beta}T_{\gamma\delta}\alpha^{(m_a)}_{\alpha\gamma}\alpha^{(m_b)}_{\beta\delta}\\&+\tfrac{2}{3}T_{\alpha\beta}T_{\gamma\delta\epsilon}(\alpha^{(m_a)}_{\alpha\gamma}A^{(m_b)}_{\beta,\delta\epsilon}-A^{(m_a)}_{\beta,\delta\epsilon}\alpha^{(m_b)}_{\alpha\gamma})\\&+\tfrac{1}{3}T_{\alpha\beta\gamma}T_{\delta\epsilon\phi}(\alpha^{(m_a)}_{\alpha\delta}C^{(m_b)}_{\beta\gamma,\epsilon\phi}+C^{(m_a)}_{\beta\gamma,\epsilon\phi}\alpha^{(m_b)}_{\alpha\delta}-\tfrac{2}{3}A^{(m_a)}_{\alpha,\epsilon\phi}A^{(m_b)}_{\delta,\beta\gamma})\\&-\tfrac{2}{9}T_{\alpha\beta}T_{\gamma\delta\epsilon\phi}A^{(m_a)}_{\alpha,\gamma\delta}A^{(m_b)}_{\beta,\epsilon\phi}+\ldots]\end{aligned} \tag{1.129}$$

where the polarizabilities $\boldsymbol{\alpha}$, $\mathbf{A}$, $\mathbf{C}$, ... are the static polarizabilities of molecules a and b in the states m_a and m_b and are given by (1.50), (1.61) and (1.62). If m_a and m_b are the ground electronic states, U_a and U_b may be set equal to the first ionization potentials, or to the lowest allowed excitation energies; in this latter case, equation (1.129) gives a lower bound to the magnitude of the leading term in the dispersion energy.

If the molecules depart from spherical symmetry the polarizabilities are functions of orientation, and $u_{\text{induction}}$ and $u_{\text{dispersion}}$ are angle-dependent. This may be crucial to some experiments, such as the rotational relaxation of molecules,[74] but for some purposes it is of interest to average over the angular variables and to obtain an effective central potential which depends only on the separation R. These average energies of interaction are obtained from (1.116) and (1.129) by replacing the appropriate tensors by their isotropic parts,

$$\overline{u^{(b)}_{\text{induction}}}=-\tfrac{1}{2}\alpha^{(m_b)}F^{(b)}_{\alpha}F^{(b)}_{\alpha}-\tfrac{1}{6}C^{(m_b)}F^{(b)}_{\alpha\beta}F^{(b)}_{\alpha\beta}-\ldots \tag{1.130}$$

and $F_\alpha^{(b)}F_\alpha^{(b)}$ and $F_{\alpha\beta}^{(b)}F_{\alpha\beta}^{(b)}$ can be written in terms of the permanent moments of molecule a by means of (1.45) (see also (1.4) and (1.6),

$$\overline{u_{\text{induction}}^{(b)}} = -(4\pi\epsilon_0)^{-2}\Bigg[\tfrac{1}{2}\alpha^{(m_b)}\Big\{q^{(a)2}R^{-4}+2\mu_\alpha\mu_\alpha R^{-6}+2\Theta_{\alpha\beta}\Theta_{\alpha\beta}R^{-8}+\dots + \frac{2^n n!(n+1)!}{(2n)!}\xi^{(n)2}R^{-(2n+4)}+\dots\Big\} + \tfrac{1}{6}C^{(m_b)}\Big\{6q^{(a)2}R^{-6}+30\mu_\alpha\mu_\alpha R^{-8}+56\Theta_{\alpha\beta}\Theta_{\alpha\beta}R^{-10}+\dots + \frac{2^n n!(n+2)!(2n+3)}{(2n)!}\xi^{(n)2}R^{-(2n+6)}+\dots\Big\}+\dots\Bigg] \quad (1.131)$$

where

$$\xi^{(n)2} = \xi^{(n)}_{\alpha\beta\gamma\dots\nu}\xi^{(n)}_{\alpha\beta\gamma\dots\nu}$$

$$\mu_\alpha\mu_\alpha = \mu_x^2+\mu_y^2+\mu_z^2 \quad \text{and} \quad \Theta_{\alpha\beta}\Theta_{\alpha\beta} = \Theta_{xx}^2+\Theta_{yy}^2+\Theta_{zz}^2$$

if x, y, z are the principal axes of $\mathbf{\Theta}$. The mean dispersion energy is

$$\overline{u_{\text{dispersion}}} = -(4\pi\epsilon_0)^{-2}\frac{3U_aU_b}{2(U_a+U_b)}R^{-6}[\alpha^{(m_a)}\alpha^{(m_b)} + 5R^{-2}(\alpha^{(m_a)}C^{(m_b)}+C^{(m_a)}\alpha^{(m_b)})+\dots] \quad (1.132)$$

For two like molecules in interaction in their ground states this simplifies to

$$\overline{u_{\text{dispersion}}} = -(4\pi\epsilon_0)^{-2}\frac{3U\alpha^2}{4R^6}\left[1+\frac{10C}{\alpha R^2}+\dots\right] \quad (1.133)$$

The ratio of the R^{-8} to the R^{-6} term is therefore $10\,C/\alpha R^2$ in this approximation. For the H atom,

$$(4\pi\epsilon_0)^{-1}\alpha = \tfrac{9}{2}a_0^3 \quad \text{and} \quad (4\pi\epsilon_0)^{-1}C = \tfrac{15}{2}a_0^5$$

where a_0 is the Bohr radius,[21] so the ratio is (50/3) $(a_0/R)^2$; the accurate value is $19{\cdot}14(a_0/R)^2$.[30]

If one or both of the interacting molecules is in an excited state, the long-range interaction energy may be very different from its value for the ground state. The polarizability $\alpha^{(m)}$ may even be negative, as for the sodium atom in the $1s^22s^22p^64p^1$ excited state[75], making the induction energy positive. The approximate equation (1.129) for the dispersion energy would normally be less reliable for molecules in excited electronic states, since the average-energy approximation is likely to fail when some transitions have positive and others negative energy changes $\hbar\omega_{pm}$. However, for the lowest triplet level the equation should be useful since the interaction Hamiltonian $\mathcal{H}'$ does not mix states of different multiplicity.

The longest-range dispersion energy results from an intermolecular correlation of the fluctuating dipole moments and varies as R^{-6}. Similarly, the R^{-2} energy of attraction between dissolved polyelectrolyte molecules at their isoelectric point (where their time-average charge is zero), discovered in 1952 by Kirkwood and Shumaker,[76] can be considered to be due to fluctuations in the charges of the molecules, although in this case the presence of the medium is essential, for it provides the source and the sink for the protons.

IV. SHORT-RANGE INTERMOLECULAR FORCES

When the overlap of the electron clouds of two molecules is significant, it is essential that the total wavefunction be antisymmetric with respect to exchange of all pairs of electrons. If we do not impose this condition, then the wavefunction does not represent observable states of the system, for it does not conform to the postulate of the indistinguishability of electrons. This point has been considered by Claverie.[77]

One technique for ensuring that the total wavefunction in antisymmetric is to choose a basis set for a variational calculation that is antisymmetric and such a basis set could be the antisymmetrized product functions $\mathcal{A}p_a p_b$ where p_a and p_b are eigenfunctions of the free-molecule Hamiltonians $\mathcal{H}^{(a)}$ and $\mathcal{H}^{(b)}$ of equation (1.22) and $\mathcal{A}$ is the operator which antisymmetrizes with respect to intermolecular exchange of electrons. The set of functions $\mathcal{A}p_a p_b$ are not orthogonal at separations at which overlap is significant and cannot therefore be eigenfunctions of a Hamiltonian. The normal quantum-mechanical perturbation theory is therefore not applicable and because the basis is overcomplete there is no unique transformation to an orthogonal set.

However it is possible to perform a variational calculation with a trial wavefunction which is a sum of a finite number of terms of the set $\mathcal{A}p_a p_b$

$$\psi = \sum_{p_a,\, p_b} c_{p_a p_b} \mathcal{A} p_a p_b \tag{1.134}$$

The coefficients $c_{p_a p_b}$ could be determined through the variation principle.

A simpler technique is to use the usual self-consistent-field theory[78] to obtain the best one-electron wavefunctions for the interacting molecular pair and to deduce the interaction energy by subtracting the energy computed for the separate molecules. This approach can give useful results for the short-range repulsive energy but if only a single configuration is employed there can be no electron correlation and hence no dispersion energy. Since the dispersion energy is the only source of attraction between inert gas atoms, it is to be expected that the Hartree–Fock potential curve for these systems should have no minimum. Minima have been obtained for He_2[79] but these result from the use of an incomplete basis set[80,81] and can be eliminated by the addition of 'ghost orbitals' in the calculations on the separate molecules to compensate for the extension of the basis set for the pair;[82] one uses the basis set for the pair but has a total of only n_a or n_b electrons, and it is necessary to compute the energy of the separate molecules for each R and for each orientation of the molecules.

Unfortunately in these variational calculations the interaction energy is not bound, for it is the difference between the total energy for a fixed configuration and the energy of the isolated molecules.

Numerous variational calculations, involving extensive configuration interaction, have recently been performed on He_2.[83–88] The minimum in the potential has a depth of approximately 15×10^{-23} J = 11 K at a separation of $3{\cdot}0\times10^{-10}$ m, and for $R>4=10^{-10}$ m, the results agree with the energy C_6R^{-6} deduced from semi-empirical calculations of C_6.[69]

A variational calculation which includes intermolecular correlation has been performed on Ne_2,[89] yielding a well depth of 54×10^{-23} J = 39·2 K at $3{\cdot}08\times10^{-10}$ m and long-range behaviour closely following the semi-empirical value for C_6R^{-6}.[69] However, if there is no intra-atomic correlation in the calculation one might expect the computed long-range behaviour to correspond to a 'Hartree–Fock' C_6 which would differ from the true C_6 by the factor $(\alpha_{\text{Hartree–Fock}}/\alpha_{\text{true}})^2$. Since

$$(4\pi\epsilon_0)^{-1}\alpha_{\text{Hartree–Fock}}=0{\cdot}349\times10^{-30}\ \text{m}^3$$ [90]

and

$$(4\pi\epsilon_0)^{-1}\alpha_{\text{Ne}}=0{\cdot}397\times10^{-30}\ \text{m}^3,$$ [91]

this factor is 0·77.

Accurate variational calculations including electron correlation are difficult to perform on all but the simplest atoms and molecules. Boys and Bernardi[92] proposed employing the bivariational techniques of Boys and Handy[93,94] together with 'counterpoise' procedures to minimize errors in small energy differences, but useful calculations of intermolecular energies have not yet been performed.

In the region of small overlap, the interaction energy is small compared to the total energy and it is tempting to seek a perturbation scheme for evaluating intermolecular forces. Unfortunately there are difficulties stemming from the need to antisymmetrize the total wavefunction with respect to exchange of electrons between the molecules; these difficulties have been considered by Hirschfelder.[95] Several perturbation schemes have been suggested for the short and intermediate ranges,[96–101] and their efficacy in describing the energy of the ground ($^1\Sigma_g^+$) and first-excited ($^3\Sigma_u^+$) states of the hydrogen molecule were tested by Certain *et al.*,[102] who found that while some gave satisfactory results, none offered any computational advantage over the standard variational calculation of Kołos and Wolneiwicz.[103] The perturbation approach to the calculation of intermolecular forces in the intermediate region has been well reviewed by Certain and Bruch.[104]

If the unperturbed wavefunction is taken to be the antisymmetrized product of the wavefunctions of the free-molecules, then the first-order energy $\langle\Psi_0|\mathcal{H}-W_{m_a}-W_{m_b}|\Psi_0\rangle/\langle\Psi_0|\Psi_0\rangle$ separates into a *Coulomb* and an *exchange* contribution. The first is similar to the classical electrostatic interaction of the overlapping charge distributions and may be of either sign; for inert gas atoms it is negative and vanishes when the overlap is zero. The second is quantum-mechanical in origin and is due to the antisymmetrization of Ψ_0. There is a contribution to the exchange energy for every exchange between an orbital of a and an orbital of b. For the inert gas atoms the exchange energy is larger than the Coulomb energy and is positive.

For a pair of H atoms with parallel electron spins, the wavefunction Ψ_0 may be taken[10] to be the Heitler–London wavefunction

$$\Psi_0 = 2^{-1/2}(1s_a(1)1s_b(2) - 1s_a(2)1s_b(1)) \tag{1.135}$$

The Coulomb interaction energy is then

$$u_{\text{Coulomb}} = \frac{\langle 1s_a(1)1s_b(2)| - R_{b1}^{-1} - R_{a2}^{-1} + r_{12}^{-1} + R^{-1} |1s_a(1)1s_b(2)\rangle}{1-S^2} \tag{1.136}$$

$$= \frac{-2\langle 1s_a(1)|\, R_{b1}^{-1}\, |1s_a(1)\rangle + \langle 1s_a(1)1s_b(2)|\, r_{12}^{-1}\, |1s_a(1)1s_b(2)\rangle + R^{-1}}{1-S^2} \tag{1.137}$$

where the overlap integral

$$S = \langle 1s_a(1)|1s_b(1)\rangle \tag{1.138}$$

The exchange energy is

$$u_{\text{exchange}} = \frac{\langle 1s_a(1)1s_b(2)| - R_{b1}^{-1} - R_{a2}^{-1} + r_{12}^{-1} + R^{-1} |1s_a(2)1s_b(1)\rangle}{1-S^2} \tag{1.139}$$

$$= \frac{-2S\langle 1s_a(1)|\, R_{b1}^{-1}\, |1s_b(1)\rangle + \langle 1s_a(1)1s_b(2)|\, r_{12}^{-1}\, |1s_a(2)1s_b(1)\rangle + S^2R^{-1}}{1-S^2} \tag{1.140}$$

and is proportional to S^2 for small S. Following Mulliken[105] it is sometimes assumed that $u_{\text{exchange}} = kS^2R^{-1}$ where k is a constant; for two H atoms k has the value 1·23 atomic units.[12]

In going to the second-order of perturbation, there appears to be no natural way to proceed. Murrell and Shaw[98] use the perturbed wavefunction

$$\Psi = \Psi_0 + \sum_{p_a,\, p_b}{}' c_{p_a p_b} p_a p_b \tag{1.141}$$

in which the perturbation to Ψ_0 is a simple product function. This enables them to introduce various second-order energies such as those due to induction, dispersion and exchange-polarization.

The limitations of the multipolar expansion of $\mathcal{H}'$ in the overlap region have been investigated by Murrell and Shaw[98] and by Meath *et al.*[106,107] Using the Unsöld average-energy approximation,[107] the second-order energy u was evaluated without expanding $\mathcal{H}'$,

$$u = -U^{-1}[\langle m_a m_b|\, \mathcal{H}'^2\, |m_a m_b\rangle - \langle m_a m_b|\, \mathcal{H}'\, |m_a m_b\rangle^2] \tag{1.142}$$

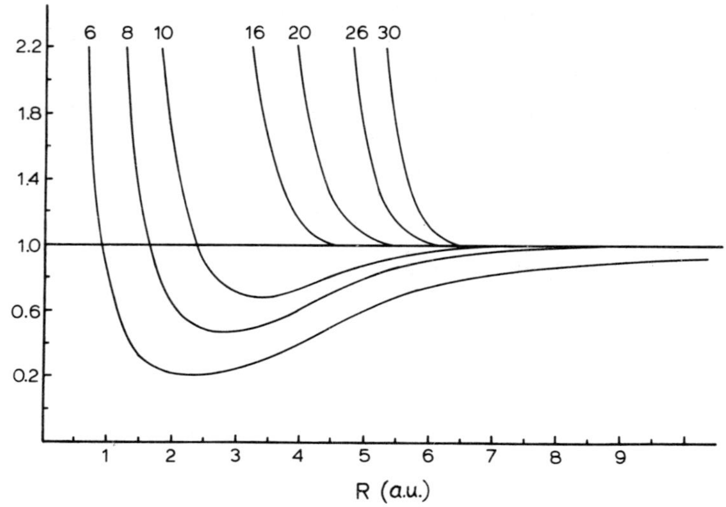

Figure 1.13. A comparison for He_2 of the expansion

$$\sum_{n=3}^{N} C_{2n}R^{-2n}$$

for N from 3 to 15 with the second-order interaction energy u. The value of $2N$ is indicated on each curve. The asymptotic nature of the inverse-power series expansion for the dispersion energy may be clearly seen—outside the region of validity of the expansion, inclusion of more terms leads to a *poorer* approximation. The curves are those of Kreek, Pan and Meath[107]

and compared with expansions of the dispersion energy

$$u_{\text{dispersion}} = \sum_{n=3}^{N} C_{2n}R^{-2n} \tag{1.143}$$

for various values of N. The coefficients C_{2n} were evaluated for the same average energy U, using the multipolar expansion (1.47) for $\mathcal{H}'$. The comparison for He_2 is illustrated in Figure 1.13 for N from 3 to 15. At large R, the expansion agrees well with the exact result, but for small R the series overestimate the energy and diverge rapidly as R decreases. Actually the dispersion series

$$\sum_{n=3} C_{2n}R^{-2n}$$

asymptotically diverges at all finite R;[108,109] for large R successive terms initially decrease rapidly and diverge only for large values of N; if the series is terminated when the terms become negligible, an accurate result is obtained. However, at small R, the divergence starts at the early terms in the expansion, and, as shown in Figure 1.13, including more terms in the series yields a much poorer approximation to the accurate result. Meath *et al.*[107] also studied the He–H(1*s*), H(1*s*)–H(1*s*) and H(1*s*)–H(2*s*) systems and found that the divergence occurs at rapidly increasing R values as one proceeds along this series from He–He to H(1s)–H(2s).

The *ab initio* compution of short-range intermolecular forces can be a lengthy procedure and is limited to systems of rather few electrons. There is a need therefore for efficient approximate methods. One useful method[110–112] involves the use of the statistical model first applied to atoms by Thomas and Fermi (this technique has been reviewed by Gombas[113]) with the additional assumption that the electron density is additive. The effects of distortion of the electron clouds by the intermolecular forces are therefore excluded. In the statistical model the kinetic and exchange energy densities are determined by the electron density.[115,116] Using this model, Gordon and Kim[112] gave a useful description of the intermolecular potentials in pairs of inert gas atoms and in alkali halide molecules. Their results are good at separations out to the potential minimum. However, the model excludes electron correlation in the absence of overlap, and since it is precisely this correlation which yields the long-range dispersion energy, the accuracy of the Gordon–Kim potentials at the minimum may be fortuitous; certainly the attractive force decreases far too quickly as R increases from the minimum. In a subsequent paper[116] they combined their electron-gas theory of short-range interactions with a description of the long-range induction and dispersion energies that is based on a Drude model[117] which treates the molecules as if they were three-dimensional harmonic oscillators.

As a conclusion to this section, it can be noted that although there are formidable difficulties associated with the accurate evaluation of short-range forces, this need not lead to an impasse in our progress in understanding and calculating molecular interaction energies. The short-range repulsive energies rise extremely rapidly as R decreases and reflect the shape of the interacting molecules; the effect of an error in the magnitude of the repulsive force may therefore do little more harm than to cause a slight expansion or contraction in the position and depth of the attractive well.

Much effort has been devoted to the development of a perturbation theory of short-range forces.[104] This may have been in vain since electron exchange causes the orbitals in the pair to be qualitatively different from those in the free molecules. It may not be reasonable to seek a theory which expresses, as in the long-range case, the interaction energy in terms of the properties of the non-interacting molecules. In the absence of such a theory, the evaluation of short-range intermolecular forces must be performed separately for each system at various positions and relative orientations.

V. THE HYDROGEN BOND AND DONOR–ACCEPTOR INTERACTIONS

The intermolecular or intramolecular interaction between a hydrogen atom bonded to an electronegative atom and another electronegative atom or group of atoms, known as the hydrogen bond, has long been of interest to physicists, chemists and biologists. The hydrogen bond has a dissociation energy of the order of 10 kJ mol^{-1} and is therefore intermediate in strength between that of a typical chemical bond (~500 kJ mol^{-1}) and a 'van der Waals' interaction energy[8] (~1 kJ mol^{-1}). A hydrogen bond may be detected in a number of ways:

(i) through molecular association in liquids and solutions leading to abnormal molecular weights and transport properties;
(ii) through crystal structure determinations, particularly by means of neutron diffraction, in which a proton is found between two electronegative atoms;
(iii) through abnormally high heats of vaporization of solids and liquids;
(iv) through large red-shifts in the vibrational frequencies and a broadening of the vibrational spectral lines;
(v) through a blue shift in the electronic spectra associated with $\pi^* \leftarrow n$ transitions;[119]
(vi) through a large downfield shift in the proton magnetic resonance spectrum.[120]

In some instances, as in ice,[121] in carboxylic acid dimers,[122] and in the bifluoride ion FHF^-[123] the hydrogen bonding is strong enough to hold the molecule in a well-defined structure. But it one examines carefully the motion of the protons and deuterons in the hydrogen fluoride dimer[124] or in carboxylic acid dimers,[122] one sees evidence for the tunnelling motion of the light nuclei through the barrier in the double-minimum potential surface. And because the potential well is shallow and very anharmonic in relation to the usual molecular potential surfaces, there are normally many thermally populated states that differ significantly in their structures and properties.[125]

The hydrogen bond results from interatomic forces that probably should not be divided into components, although no doubt electrostatic and overlap interactions are the principal ingredients. It is not easy to study experimentally the energy and electronic properties of hydrogen-bonded molecules as a function of the important distances, namely X–H and H–Y in –X–H . . . Y, where X and Y are electronegative atoms such as F, O, N, Cl, S and P. However, theoretically it is feasible to examine the effect of changes in these and other variables and thereby to expose the hydrogen bond. Progress in this direction has recently been made. Thus Clementi[126] used self-consistent-field techniques to obtain a potential surface for NH_3 + HCl. The dimer $(HF)_2$ has been investigated by Kollman and Allen,[127] Kraemers and Diercksen,[128] Del Bene and Pople,[129] and Schaefer *et al.*[130] using the SCF method, and Lischka[131] looked at correlation effects. A near linear F–H . . . F hydrogen bond is expected with an F–F distance of $2{\cdot}8 \times 10^{-10}$ m, an angle H . . . F–N of 128° and a dissociation energy of 19 kJ mol^{-1}. The structure deduced from molecular beam spectroscopy[124] is very similar.

It is of interest to note that the SCF description of the hydrogen bond can be quite accurate. This is because electron pairs are conserved in dissociating an H-bond, unlike a covalent bond, so that the correct products are obtained, and secondly because the usual slight overestimation of the electrostatic energy is partially compensated by an underestimation of the induction energy.[132]

The water dimer, $(H_2O)_2$, has been extensively studied both experimentally and theoretically. It is predicted that the most stable dimer has an approximately linear hydrogen bond with an O–O distance greater than that in liquid water and in ice; the bifurcated and cyclic structures (see Figure 1.14) are significantly less favoured,[133–137] and may not even be associated with minima in the potential surface.[134,137,138] The microwave spectrum of the water dimer has been studied in a supersonic nozzle beam by Dyke and Muenter[139,140] and they have shown that it has a linear H-bond with an O–O distance of $2{\cdot}98 \pm 0{\cdot}04 \times 10^{-10}$ m. The dissociation energy has not yet been accurately determined; its value is probably in the range 16 to 36 kJ mol^{-1}.[132]

The water dimer has also been studied experimentally by infrared spectroscopy. Tursi and Nixon[141] concluded from their observation of four dimeric O–H stretching frequencies in a nitrogen matrix at 20 K that the dimer is linear. In earlier work, van Thiel, Becker and Pimentel[142] saw only

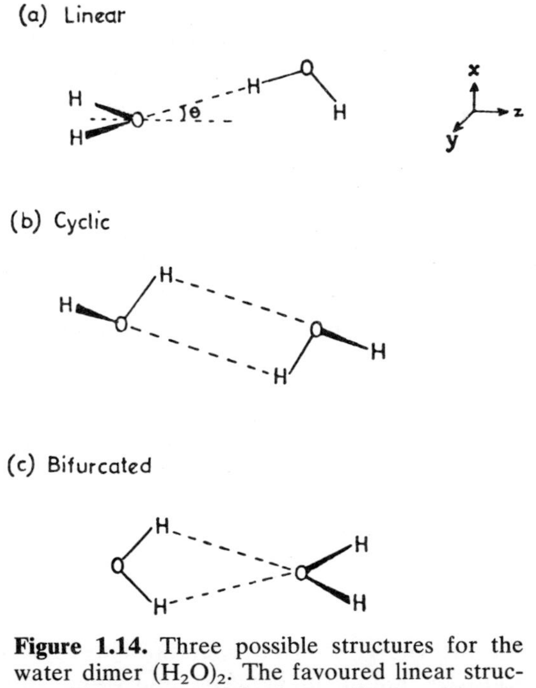

Figure 1.14. Three possible structures for the water dimer $(H_2O)_2$. The favoured linear structure (a) is symmetric in the *xz* plane and has an approximately linear O–H --- O hydrogen bond

two OH stretching frequencies in a nitrogen matrix and concluded that the stable dimer was centrosymmetric and therefore cyclic. Magnusson[143] also reported just two frequencies in CCl_4 solution at 298 K and deduced that the dimer is cyclic in this solvent. Alternative explanations of Magnusson's observations have been proposed.[143,144]

There is evidence for an important non-additive and cooperative interaction involving hydrogen-bonded species. The presence of a negative charge near the proton in an H–X bond enhances the charge separation H^+–X^- in the bond and therefore makes the X^- more attractive to protons in other H–X bonds. This cooperative effect has been investigated by Del Bene and Pople[134,146] and Hankins *et al.*[135] who found a negative three–body energy of $-5{\cdot}7\ \mathrm{kJ\,mol^{-1}}$ for the sequential trimer $(H_2O)_3$. Similar non-additive contributions to the binding of clusters of HF molecules have been calculated.[147,148]

The question of what it is, if anything, that is special about a hydrogen bond has been considered by a number of authors;[132,149,150] it is probable that the great strength of some hydrogen bonds compared to other intermolecular energies is due to the weakness of the short-range overlap repulsive forces. The proton in an H–X bond is in a region of low electron density; it has no 'inner shell' of electrons and can therefore approach abnormally close to electro-negative atoms, thereby increasing the negative energy resulting from the electrostatic and other forces of attraction. Thus we do not often need to speak of a 'lithium bond', and one explanation for this would be that the electron core of the lithium atom prevents the nucleus from approaching as closely to an electronegative atom as can the proton, in spite of stronger electrostatic forces.

A topic closely related to the hydrogen bond is that of charge-transfer or donor–acceptor interactions. The charge-transfer model was introduced by Mulliken[151] to describe the ground and excited states of molecules, one of which has a low ionization potential and the other a high electron affinity. The distortion that results from interaction of the donor D with the acceptor A can be approximately accommodated by the ground state wavefunction

$$\Psi_0 = \Psi_{DA} + \lambda \Psi_{D^+A^-} \tag{1.144}$$

where $\Psi_{D^+A^-}$ is an ionic representation and Ψ_{DA} a more typical pair wavefunction; the parater λ is normally small in the ground state. There is an excited state orthogonal to Ψ_0 and of the form

$$\Psi_{\text{excited}} = \Psi_{D^+A^-} - \lambda \Psi_{DA} \tag{1.145}$$

which is largely ionic in character, and transitions between Ψ_0 and Ψ_{excited} are responsible for the charge-transfer spectra. These broad spectra were observed in solutions of iodine in benzene and other solvents by Hildebrand[152] and have been reviewed by Mulliken and Person,[153] Murrell[154] and Foster.[155,156]

There have been a number of attempts[157,158,159] to partition the interaction energy of hydrogen bonded and other complexes into various contributions, such as those due to electrostatic, induction, dispersion, charge transfer and exchange repulsion.

Molecular complex formation, and hydrogen bonding in particular, are topics of great importance in many branches of science. For example, the structure and function of biological macromolecules[160] is dependent upon these interactions, as are the properties of many liquids, solids and ionic solutions. But a full appreciation of complex formation in these situations is dependent upon an understanding of the role of the medium.

VI. THE EFFECT OF A MEDIUM

A medium of relative permittivity, or dielectric constant, ϵ_r, reduces the electrostatic energy of interaction of two molecules immersed in it by a factor of ϵ_r. The effects of the medium on dispersion energy have been examined.[161–164,59] McLachlan[162] has distinguished five effects relating to the interaction of molecules in a liquid:

(i) the force depends on the thermodynamic free energy $A(\mathbf{R}) = U(\mathbf{R}) - TS(\mathbf{R})$ and therefore involves an average over the configurations of other molecules;
(ii) the fixed molecules disturb the local order of the liquid, producing a 'solution force' between them;
(iii) the electromagnetic forces which would exist between the molecules in free space are modified by the medium;
(iv) Archimedes' principle is involved, because a dissolved molecule can only move by displacing the liquid from its path;
(v) the electron clouds overlap with those of the solvent and change the properties of the molecules.

It is convenient to introduce an 'effective' or 'excess' polarizability α^* which may be used to give the effective intermolecular energy.[162,163] The dispersion force between any two similar spherical systems is always attractive, regardless of the nature of the medium, so that two bubbles or two colloidal particles attract one another. However, if the interacting bodies a and b have different permittivities ϵ_a and ϵ_b such that $\epsilon_a > \epsilon_r > \epsilon_b$, where ϵ_r is the permittivity of the medium, the energy of interaction is positive.[165]

The presence of polarizable matter between interacting molecules may *increase* their mutual potential energy. For example, if a sphere of polarizability α is at the point midway between a pair of charges $+q$ and $-q$ at a separation R (see Figure 1.15), the interaction energy is

$$u = -(4\pi\epsilon_0)^{-1}q^2R^{-1}[1 + 32\alpha(4\pi\epsilon_0R^3)^{-1}] \qquad (1.146)$$

However, if the two charges were of the same sign, the induced dipole on the sphere would be zero (although it would possess an induced quadrupole

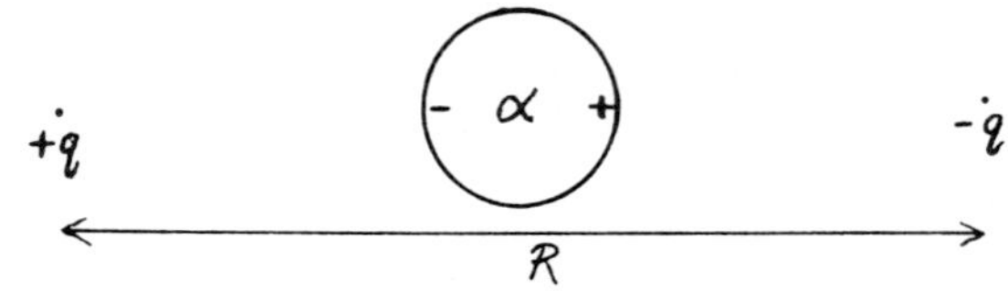

Figure 1.15. A polarizable sphere at the point mid-way between opposite charges q and $-q$. The sphere acquires a dipole of the sense shown and this clearly *increases* the force of attraction between the charges (see equation (1.147))

proportional to CR^{-3} at large R), and the repulsive force would not be affected by the polarizable sphere.

If two spheres each of polarizability α are at a constant distance d beyond each of the charges (see Figure 1.16) the magnitude of the force between the charges is *reduced* and takes the value

$$-\left(\frac{\partial u}{\partial R}\right)_d = -(4\pi\epsilon_0)^{-1}q^2R^{-2}\left[1-\frac{4\alpha}{(4\pi\epsilon_0)Rd^2}\left(1+\frac{2d}{R}\right)\left(1+\frac{d}{R}\right)^{-5}\right] \quad (1.147)$$

If the two charges have the same sign, the force of repulsion is reduced to

$$-\left(\frac{\partial u}{\partial R}\right)_d = (4\pi\epsilon_0)^{-1}q^2R^{-2}\left[1-\frac{4\alpha}{(4\pi\epsilon_0)Rd^2}\left(1+\frac{2d}{R}+\frac{2d^2}{R^2}\right)\left(1+\frac{d}{R}\right)^{-5}\right] \quad (1.148)$$

Much effort is being devoted to the quantum-mechanical evaluation of solvent effects. The approaches vary from those employing the traditional reaction field of Onsager[166] to the 'supermolecule' approach of Pullman.[167] In the former, the effects of the environment on interacting molecules are simulated by those of a continuum having the bulk properties of the medium. This approach has been explored by Linder.[168] In the latter, more fundamental, approach, the interacting molecules are treated as a single large quantum-mechanical system and the energy evaluated using conventional SCF techniques. An intermediate approach in which the reaction

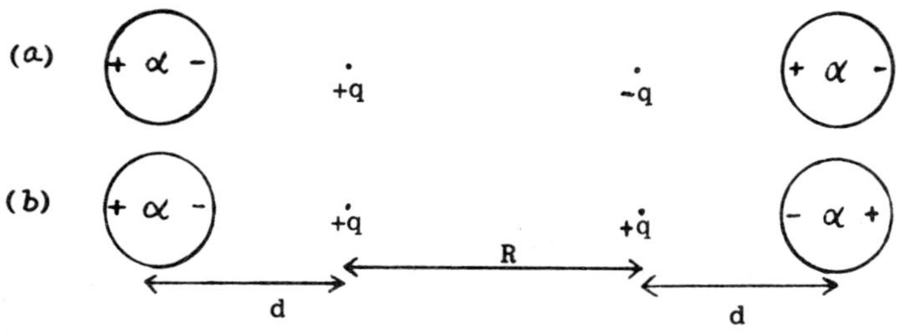

Figure 1.16. (a) Two polarizable spheres at a fixed distance d beyond the point charges q and $-q$ which are distant R apart. The induced dipoles have the signs shown and reduce the attraction between the charges (see equation (1.147)). (b) The situation is similar to (a) except that the charges are of the *same* sign. The induced dipoles then lead to a reduced repulsive force

field is incorporated into a non-linear effective Hamiltonian has been explored by Tapia and Goscinski.[169]

All methods may be expected to have their strong points and their weak points. It may be that the soundest way forward is to exploit the strengths of each approach—that is, to use the reaction field to simulate the effects of the long-range electrostatic and induction interactions and the supermolecule route to the study of shorter-range effects for which it is the only sound technique. The complexity of the problem means that we may have to be content with empirical or semi-empirical procedures for some time to come.

In addition to affecting the forces between molecules, the medium can have a pronounced influence on optical, electric and magnetic properties of molecules.[170] For example, Figure 1.17 shows how the mean polarizability of a pair of helium atoms varies with the internuclear distance *R*. Three facts concerning the mean polarizability require the existence of the positive maximum and the negative minimum in the curve—these are:

(i) the *increase* proportional to R^{-6} due to the classical dipole–dipole coupling and the distortion associated with the long-range dipole dispersion force;[170,171]
(ii) the observed *negative* second dielectric virial coefficient,[172,173] implying that the statistical average of the polarizability of the pair is *less* than that of two isolated atoms;[174]
(iii) the united atom corresponding to $R = 0$ is beryllium ($1s^2 2s^2$) and its polarizability is approximately sixteen times that of two He atoms.[175]

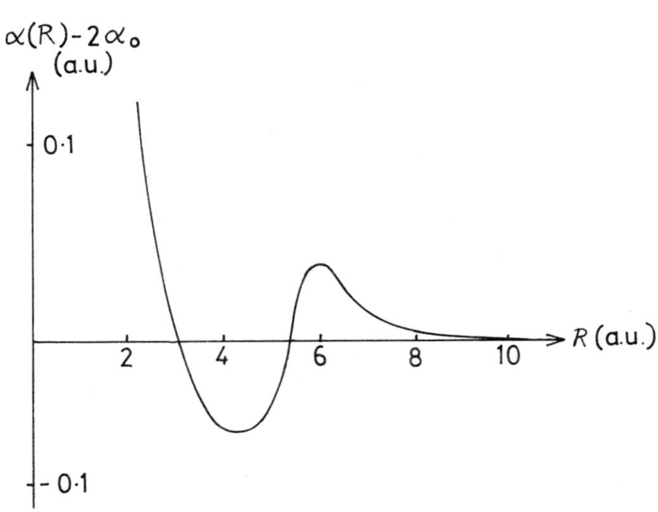

Figure 1.17. A schematic representation of the change in the mean polarizability of two helium atoms as a function of their separation *R*. The value of the polarizability of a free atom is $1{\cdot}3832$ a.u. $= 0{\cdot}20497 \times 10^{-30}$ m^3. The minimum in the potential energy occurs at $R = 5{\cdot}7$ a.u.

VII. SOME PARTICULAR MOLECULAR INTERACTIONS

To illustrate some of the formulae given in earlier sections, some particular cases are now treated.

A. A point charge and a linear molecule

If θ is the angle between the axis of the linear molecule (which might be HF or OCS) and the vector R from the origin on the axis of the molecule to the point charge q, equation (1.47) becomes

$$u_{\text{electrostatic}} = q \sum_{n=1}^{\infty} (-1)^n \frac{2^n n!}{(2n)!} T_{\alpha\beta\gamma\ldots\nu} \xi^{(n)}_{\alpha\beta\gamma\ldots\nu} \quad (1.149)$$

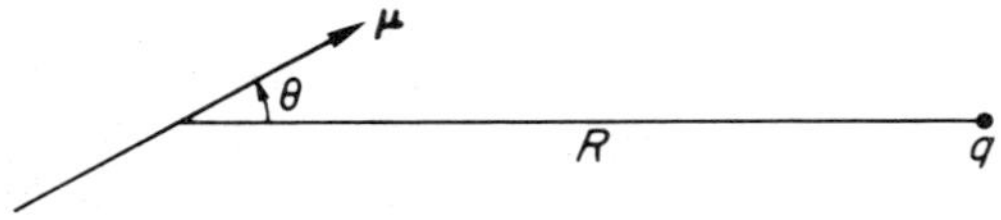

and with (1.68)–(1.70) reduces to

$$u_{\text{electrostatic}} = (4\pi\epsilon_0)^{-1} q[\mu R^{-2} \cos\theta + \Theta R^{-3}(\tfrac{3}{2}\cos^2\theta - \tfrac{1}{2}) + \Omega R^{-4}(\tfrac{5}{2}\cos^3\theta - \tfrac{3}{2}\cos\theta) - \ldots - \xi^{(n)} R^{-(n+1)} P_n(\cos\theta) - \ldots] \quad (1.150)$$

where $\xi^{(n)} = \xi^{(n)}_{333\ldots3}$ (see equations (1.68)–(1.71) and $P_n(\cos\theta)$ is a Legendre polynomial. The induction energy of equation (1.116) reduces to

$$u_{\text{induction}} = -\tfrac{1}{2}(4\pi\epsilon_0)^{-2} q^2 R^{-4}[\alpha + \tfrac{1}{3}(\alpha_\parallel - \alpha_\perp)(3\cos^2\theta - 1) + A_\parallel R^{-1}(3\cos^3\theta - \cos\theta) + 4A_\perp R^{-1}\cos\theta\sin^2\theta + 2CR^{-2} + \ldots] \quad (1.151)$$

and $u_{\text{dispersion}}$ is zero in this case.

There are also short-range interactions arising from the non-vanishing electron density at the point charge.

The energy must be independent of the choice of origin on the axis of the molecule. A movement of this origin changes R and θ as well as Θ, Ω, $A_\parallel$, $A_\perp$ and C. However, both the electrostatic and the induction energies are invariant.[24]

B. A point charge and a tetrahedral molecule

The first non-vanishing electric moment of the molecule is its octopole $\boldsymbol{\Omega}$, and (1.47) together with (1.75) and (1.76) yield

$$u_{\text{electrostatic}} = q \sum_{n=3}^{\infty} (-1)^n \frac{2^n n!}{(2n)!} T_{\alpha\beta\gamma\ldots\nu} \xi^{(n)}_{\alpha\beta\gamma\ldots\nu} = (4\pi\epsilon_0)^{-1} q R^{-4}[6\Omega\cos\theta_x\cos\theta_y\cos\theta_z + \tfrac{5}{2}\Phi R^{-1}(\cos^4\theta_x + \cos^4\theta_y + \cos^4\theta_z - \tfrac{3}{5}) + \ldots] \quad (1.152)$$

And (1.116), (1.73), (1.74), (1.76) and (1.77) give

$$u_{\text{induction}} = -\tfrac{1}{2}(4\pi\epsilon_0)^{-2}q^2R^{-4}[\alpha + 12AR^{-1}\cos\theta_x\cos\theta_y\cos\theta_z + 2CR^{-2}$$
$$+3\Delta CR^{-2}(\cos^4\theta_x + \cos^4\theta_y + \cos^4\theta_z - \tfrac{3}{5})$$
$$-2q\beta R^{-2}(4\pi\epsilon_0)^{-1}\cos\theta_x\cos\theta_y\cos\theta_z + \ldots] \tag{1.153}$$

where θ_α is the angle between the α-axis of the cube enclosing the tetrahedron (see Figure 1.11 above) and the vector from its centre to the other molecule (the point charge).

The order of magnitude of the dipole–quadrupole polarizability A may be obtained from a model in which the tetrahedron is comprised of four atoms at a distance R_0 from the centre and having polarizabilities $\alpha_\parallel$ and $\alpha_\perp$ along and at right angles to their bonds to the central atom. For this system,

$$A = \frac{4}{\sqrt{3}}(\alpha_\parallel - \alpha_\perp)R_0 \tag{1.154}$$

so that for a molecule such as CH_4 or CF_4 $(4\pi\epsilon_0)^{-1}A$ may be about $10^{-40}\,m^4$. The dipole–quadrupole polarizability A thus depends on the *distribution of polarizability* in the molecule—rather as the quadrupole moment depends upon the distribution of dipole in the molecule.

C. A spherical and a tetrahedral molecule

If both the sphere b and the tetrahedron a are uncharged,

$$u_{\text{electrostatic}} = 0 \tag{1.155}$$

and $u_{\text{induction}}$ is proportional to R^{-10} and is negligible compared to $u_{\text{dispersion}}$. From equation (1.129)

$$u_{\text{dispersion}} = -(4\pi\epsilon_0)^{-2}\frac{3U_aU_b}{2(U_a+U_b)R^6}[\alpha^{(a)}\alpha^{(b)}$$
$$+8A^{(a)}\alpha^{(b)}R^{-1}\cos\theta_x\cos\theta_y\cos\theta_z$$
$$+5\alpha^{(a)}C^{(b)}R^{-2} + 5\alpha^{(b)}C^{(a)}R^{-2}$$
$$+\tfrac{5}{2}\alpha^{(b)}\Delta C^{(a)}R^{-2}(\cos^4\theta_x + \cos^4\theta_y + \cos^4\theta_z - \tfrac{3}{5}) + \ldots] \tag{1.156}$$

Thus the $A^{(a)}$ contribution to the dispersion energy in R^{-7} is the longest-range angle-dependent energy and it could be important in rotational-translational energy conversion in collisions of say a CH_4 molecule and an argon atom. It could also be a significant contributor to the binding energy of solids comprised of tetrahedral molecules but since it vanishes on averaging over all orientations, it plays only a minor role in determining the equilibrium properties of gases such as the second virial coefficient in the equation of state.

The overlap energy is of short range and will have both a central and angle-dependent part, like (1.156). One useful empirical model for the overlap energy might involve the sum of atom–atom potentials in which each atom is considered to be a centre of interaction with every atom in the other molecule.[176–178]

D. Two linear polar molecules

For two linear molecules a and b in the configuration shown

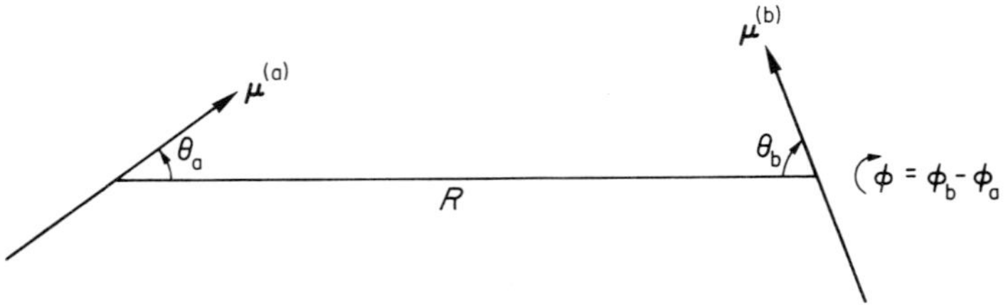

$$
\begin{aligned}
u_{\text{electrostatic}} = {} & (4\pi\epsilon_0)^{-1}[\mu^{(a)}\mu^{(b)}R^{-3}\{2\cos\theta_a\cos\theta_b + \sin\theta_a\sin\theta_b\cos\phi\} \\
& + \tfrac{3}{2}\mu^{(a)}\Theta^{(b)}R^{-4}\{\cos\theta_a(3\cos^2\theta_a - 1) + 2\sin\theta_a\sin\theta_b\cos\theta_b\cos\phi\} \\
& + \tfrac{3}{2}\mu^{(b)}\Theta^{(a)}R^{-4}\{\cos\theta_b(3\cos^2\theta_a - 1) \\
& \qquad + 2\sin\theta_b\sin\theta_a\cos\theta_a\cos\theta\} \\
& + \tfrac{3}{4}\Theta^{(a)}\Theta^{(b)}R^{-5}\{1 - 5\cos^2\theta_a - 5\cos^2\theta_b + 17\cos^2\theta_a\cos^2\theta_b \\
& + 2\sin^2\theta_a\sin^2\theta_b\cos^2\phi + 16\sin\theta_a\cos\theta_a\sin\theta_b\cos\theta_b\cos\phi\} \\
& + \ldots]
\end{aligned}
\qquad (1.157)
$$

$$
\begin{aligned}
u^{(b)}_{\text{induction}} = {} & -\tfrac{1}{2}(4\pi\epsilon_0)^{-2}[\alpha^{(b)}\{\mu^{(a)2}R^{-6}(3\cos^2\theta_a + 1) \\
& + 12\mu^{(a)}\Theta^{(a)}R^{-7}\cos^3\theta_a + \ldots\} \\
& + \tfrac{1}{3}(\alpha_\parallel - \alpha_\perp)^{(b)}\{\mu^{(a)2}R^{-6}(12\cos^2\theta_a\cos^2\theta_b \\
& \qquad + 3\sin^2\theta_a\sin^2\theta_b\cos^2\phi \\
& - 3\cos^2\theta_a - 1 + 12\sin\theta_a\cos\theta_a\sin\theta_b\cos\theta_b\cos\phi) + \ldots\} \\
& + 3A^{(b)}_\parallel\{\mu^{(a)2}R^{-7}(6\cos^2\theta_a\cos^3\theta_b - 2\cos^2\theta_a\cos\theta_b \\
& + 7\sin\theta_a\cos\theta_a\sin\theta_b\cos^2\theta_b\cos\phi \\
& - \sin\theta_a\cos\theta_a\sin\theta_b\cos\phi \\
& + 2\sin^2\theta_a\sin^2\theta_b\cos\theta_2\cos^2\phi) + \ldots\} \\
& - 4A^{(b)}_\perp\{\mu^{(a)2}R^{-7}(6\cos^2\theta_a\cos^3\theta_b - 5\cos^2\theta_a\cos\theta_b \\
& + 7\sin\theta_a\cos\theta_a\sin\theta_b\cos^2\theta_b\cos\phi \\
& - 2\sin\theta_a\cos\theta_a\sin\theta_b\cos\phi \\
& + 2\sin^2\theta_a\sin^2\theta_b\cos\theta_b\cos^2\phi - \cos\theta_b) + \ldots\} + \ldots]
\end{aligned}
\qquad (1.158)
$$

$$u_{\text{dispersion}} = -(4\pi\epsilon_0)^{-2}\frac{3U_aU_b}{2(U_a+U_b)R^6}[\alpha^{(a)}\alpha^{(b)}+\tfrac{1}{3}\alpha^{(a)}(\alpha_\parallel-\alpha_\perp)^{(b)}$$
$$\times(\tfrac{3}{2}\cos^2\theta_b-\tfrac{1}{2})$$
$$+\tfrac{1}{3}\alpha^{(b)}(\alpha_\parallel-\alpha_\perp)^{(a)}(\tfrac{3}{2}\cos^2\theta_a-\tfrac{1}{2})$$
$$+2\alpha^{(a)}A_\parallel^{(b)}R^{-1}\cos^3\theta_b+2\alpha^{(b)}A_\parallel^{(a)}R^{-1}\cos^3\theta_a$$
$$+\tfrac{4}{3}\alpha^{(a)}A_\perp^{(b)}R^{-1}(3\cos\theta_b-2\cos^3\theta_b)$$
$$+\tfrac{4}{3}\alpha^{(b)}A_\perp^{(a)}R^{-1}(3\cos\theta_a-2\cos^3\theta_a)$$
$$+5\alpha^{(a)}C^{(b)}R^{-2}+5\alpha^{(b)}C^{(a)}R^{-2}+\ldots] \qquad (1.159)$$

The short-range overlap energy, as in Part C of Section VII, must reflect both the size and shape of the molecule.

E. A sphere and a linear polar molecule

The interaction energy of a spherical atom b, such as argon, with a dipolar molecule a may be obtained from the expressions in the last Section D by setting

$$\mu^{(b)}=0$$
$$\Theta^{(b)}=0$$
$$(\alpha_\parallel-\alpha_\perp)^{(b)}=0$$
$$A_\parallel^{(b)}=0$$
$$A_\perp^{(b)}=0$$

The long-range electrostatic energy vanishes and

$$u=u_{\text{overlap}}-\tfrac{1}{2}(4\pi\epsilon_0)^{-2}\alpha^{(b)}\{\mu^{(a)2}R^{-6}(3\cos^2\theta_a+1)+12\mu^{(a)}\Theta^{(a)}R^{-7}\cos^3\theta_a+\ldots\}$$
$$-(4\pi\epsilon_0)^{-2}\frac{3U_aU_b}{2(U_a+U_b)R^6}\{\alpha^{(a)}\alpha^{(b)}+\tfrac{1}{3}\alpha^{(b)}(\alpha_\parallel-\alpha_\perp)^{(a)}(\tfrac{3}{2}\cos^2\theta_a-\tfrac{1}{2})$$
$$+2\alpha^{(b)}A_\parallel^{(a)}R^{-1}\cos^3\theta_a$$
$$+\tfrac{4}{3}\alpha^{(b)}A_\perp^{(a)}R^{-1}(3\cos\theta_a-2\cos^3\theta_a)$$
$$+5\alpha^{(a)}C^{(b)}R^{-2}+5\alpha^{(b)}C^{(a)}R^{-2}+\ldots\} \qquad (1.160)$$

The dipole–quadrupole induction energy in $\alpha^{(b)}\mu^{(a)}\Theta^{(a)}R^{-7}\cos^3\theta_a$ was first considered by Anderson[179] in an explanation of the pressure broadening of the ammonia inversion microwave spectrum by spherical atoms. The additional angle-dependent terms in (1.160) in $A_\parallel^{(a)}R^{-7}$ and $A_\perp^{(a)}R^{-7}$ are likely also to be important in these and other collision experiments; the ratio of the $A_\parallel^{(a)}$ term to the $\mu^{(a)}\Theta^{(a)}$ induction energy is

$$\frac{U_aU_b}{2(U_a+U_b)}\frac{A_\parallel^{(a)}}{\mu^{(a)}\Theta^{(a)}}$$

which is generally not small. The ratio may be large for the alkali halide molecules because of their large polarizabilities and small quadrupole moments.

Since $A_\parallel$ and $A_\perp$ depend on the choice of origin (see (1.67), the R^{-7} contribution to the dispersion energy can be varied at will, although the total dispersion energy is unaffected by this arbitrary choice. There is in general no unique 'centre of action' for the dispersion force relative to which the R^{-7} contribution to the energy is zero for all θ. However, it is possible to define a point for which the term in $R^{-7}\cos\theta$ vanishes and this was termed the 'centre of the dispersion force' by Herman;[180] this is the point at which $A_\parallel = -2A_\perp$ and according to (1.160) it is independent of the nature of the perturbing spherical atom. However, there remains a term in

$$(3A_\parallel - 4A_\perp)R^{-7}(\tfrac{5}{2}\cos^3\theta - \tfrac{3}{2}\cos\theta)$$

with a $P_3(\cos\theta)$ angular dependence. Herman[180] calculated that the centre of the dispersion force in HCl is about $0{\cdot}11\times10^{-10}$ m from the centre of mass in the direction of the proton; however, he did not consider the two independent quantities $A_\parallel$ and $A_\perp$. Reliable values of $A_\parallel$ and $A_\perp$ are not yet available.

A similar R^{-7} contribution to the dispersion energy of a hydrogen atom in the first excited state and an inert gas atom has been evaluated by Bukta and Meath.[181]

F. Two linear centrosymmetric molecules

If the linear molecules are centrosymmetric, like H_2, CO_2, HC≡CH, etc, the odd multipoles $\boldsymbol{\xi}^{(2n+1)}$ vanish if the molecular centre of symmetry is chosen as the origin. The polarizabilities $\mathbf{A}$ and $\boldsymbol{\beta}$ are also zero there. The long-range energies are

$$\begin{aligned}
u_{\text{electrostatic}} = {} & \frac{3\Theta^{(a)}\Theta^{(b)}}{4(4\pi\epsilon_0)R^5}[1-5\cos^2\theta_a-5\cos^2\theta_b+17\cos^2\theta_a\cos^2\theta_b \\
& +2\sin^2\theta_a\sin^2\theta_b\cos^2\phi+16\cos\theta_a\sin\theta_a\cos\theta_b\sin\theta_b\cos\phi] \\
& -\frac{15\Theta^{(a)}\Phi^{(b)}}{16(4\pi\epsilon_0)R^7}[1-7\cos^2\theta_a-14\cos^2\theta_b+21\cos^4\theta_b \\
& \qquad +74\cos^2\theta_a\cos^2\theta_b \\
& -91\cos^2\theta_a\cos^4\theta_b+4\sin^2\theta_a\sin^2\theta_b\cos^2\phi(1-7\cos^2\theta_b) \\
& +16\cos\theta_a\sin\theta_a\cos\theta_b\sin\theta_b\cos\phi(3-7\cos^2\theta_b)] \\
& -\frac{15\Phi^{(b)}\Phi^{(a)}}{16(4\pi\epsilon_0)R^7}[1-7\cos^2\theta_b-14\cos^2\theta_a+21\cos^4\theta_a \\
& \qquad +74\cos^2\theta_a\cos^2\theta_b \\
& -91\cos^2\theta_b\cos^4\theta_a+4\sin^2\theta_a\sin^2\theta_b\cos^2\phi(1-7\cos^2\theta_a) \\
& +16\cos\theta_a\sin\theta_a\cos\theta_b\sin\theta_b\cos\phi(3-7\cos^2\theta_b)]
\end{aligned} \tag{1.161}$$

$$u_{\text{induction}} = -(4\pi\epsilon_0)^{-2}[\tfrac{9}{8}\alpha^{(b)}\Theta^{(a)^2}R^{-8}(1-2\cos^2\theta_a+5\cos^4\theta_a)+\ldots] \quad (1.162)$$

$$u_{\text{dispersion}} = -(4\pi\epsilon_0)^{-2}\frac{3U_aU_b}{2(U_a+U_b)R^6}[\alpha^{(a)}\alpha^{(b)}+\tfrac{1}{3}\alpha^{(a)}(\alpha_\parallel-\alpha_\perp)^{(b)}(\tfrac{3}{2}\cos^2\theta_b-\tfrac{1}{2})$$
$$+\tfrac{1}{3}\alpha^{(b)}(\alpha_\parallel-\alpha_\perp)^{(a)}(\tfrac{3}{2}\cos^2\theta_a-\tfrac{1}{2})$$
$$+5\alpha^{(a)}C^{(b)}R^{-2}+5\alpha^{(b)}C^{(a)}R^{-2}+\ldots] \quad (1.163)$$

The dipole induced in a hydrogen molecule b by the quadrupole $\mathbf{\Theta}$ of a neighbouring molecule a rotates twice for each rotation of a and therefore leads to pressure-induced transitions in the infrared in which the change ΔJ in the angular momentum of a is ± 2.[182,183] The hexadecapole rotates four times for each rotation of a, so it can induce transitions in which $\Delta J = \pm 4$; the pure rotational transition $J=1$ to $J=5$ has been seen in compressed hydrogen at 195 K and at pressures in the range 180–300 amagat.[184]

VIII. MANIFESTATIONS OF INTERMOLECULAR FORCES

The forces between molecules influence the properties of matter at a microscopic level and in the bulk phases. In this section, various manifestations of intermolecular forces are mentioned and their importance for revealing the nature and magnitude of the force is briefly discussed.

A. Equilibrium properties of imperfect gases

Early studies, including those of van der Waals, concentrated on the pressure–volume–temperature relationships of gases. This is still a useful source of information about molecular interactions.

The initial deviation of an equilibrium property of a gas from ideal-gas behaviour is proportional to the difference between the average contribution of a pair of molecules and that of two non-interacting molecules. If the appropriate bulk equilibrium property Q is expressed as a power series in the gas density ρ

$$Q = A_Q + B_Q\rho + C_Q\rho^2 + \ldots \quad (1.164)$$

$$A_Q = \lim_{\rho\to 0} Q \quad (1.165)$$

is the value Q would have in the ideal gas, when each molecule contributes to the bulk property as if it alone were present.

$$B_Q = \lim_{\rho\to 0}[(Q-A_Q)\rho^{-1}] \quad (1.166)$$

and B_Q, C_Q, ... are called the second, third, ... Q virial coefficients.[174] B_Q describes the initial deviation from ideal-gas behaviour. For one mole of gas

$$A_Q = N_A q_1 \quad (1.167)$$

where $N_A = 6{\cdot}022 \times 10^{23}$ is Avogadro's number and q_1 is the mean contribution of molecule 1 to the bulk property Q. The classical statistical-mechanical expression for B_Q is[174]

$$B_Q = N_A^2 \int (\tfrac{1}{2}q_{12} - q_1) \exp(-u_{12}/kT)\, dv_2 \tag{1.168}$$

where q_{12} is the contribution of the pair of molecules 12 whose intermolecular potential energy is u_{12}; q_{12} and u_{12} are functions of the relative positions of molecules 1 and 2 and the integral in (1.168) is over all positions and orientations of molecule 2 relative to 1 and $\int dv_2 = V_m$, the volume occupied by the mole of gas. The quantum corrections to equation (1.168) are only significant for light gases such as hydrogen and helium at low temperature, and can be evaluated if required.[8] The quantum contributions to the second dielectric virial coefficient of ^{3}He and ^{4}He were evaluated by Bruch, Fortune and Berman.[185]

A critical compilation of experimental second virial coefficients $B(T)$ in the equation of state of gases,

$$\frac{pV_m}{RT} = 1 + B(T)\rho + C(T)\rho^2 + \ldots \tag{1.169}$$

has been made by Dymond and Smith.[186] $B(T)$ is not very sensitive to angle-dependent intermolecular forces.[187] However the second dielectric virial coefficient B_ϵ is extremely sensitive to angle-dependence in the short-range repulsive forces, that is, to molecular shape,

$$\frac{\epsilon_r - 1}{\epsilon_r + 2} V_m = A_\epsilon + \frac{B_\epsilon}{V_m} + \frac{C_\epsilon}{V_m^2} + \ldots \tag{1.170}$$

The Debye equation[188] for A_ϵ is

$$A_\epsilon = \frac{4\pi N_A}{3(4\pi\epsilon_0)} \left(\alpha + \frac{\mu^2}{3kT}\right) \tag{1.171}$$

where α is the mean polarizability and μ^2 the mean-square 'permanent' dipole moment of the molecule. And[189]

$$B_\epsilon = \frac{4\pi N_A^2}{3(4\pi\epsilon_0)} \int [\tfrac{1}{2}\alpha_{12} - \alpha + \frac{1}{3kT}(\tfrac{1}{2}\mu_{12}^2 - \mu^2) \exp(-u_{12}/kT)]\, dv_2 \tag{1.172}$$

where the integral is to be taken over a spherical volume. For dipolar gases, one-half the mean-square dipole moment of the pair may be either smaller or larger than μ^2, leading to negative and positive values for B_ϵ. The sign reflects the shape of the molecule and B_ϵ is negative for cigar-like molecules with dipoles along the axis of the cigar and positive for disc-shaped molecules[189] (see Figure 1.18). The magnitude of B_ϵ generally decreases as the temperature T increases.

The mean-square dipole moment of a pair of non-polar molecules may

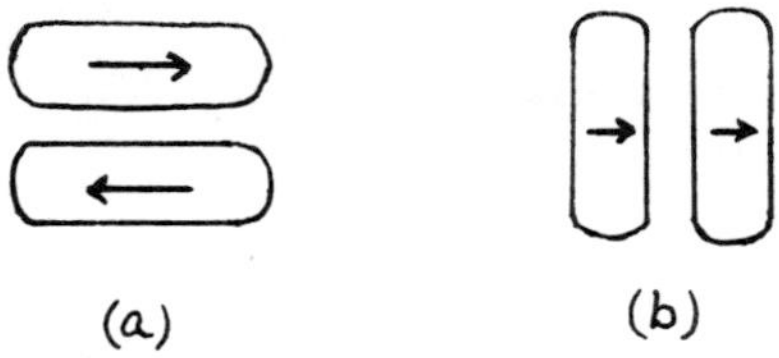

Figure 1.18. (a) Favourable configurations for cigar-shaped molecules with dipoles along their axes, and (b) disc-shaped molecules with dipoles at right-angles to the disc. The mean-square dipole of the pair is reduced by the interaction in (a) and enhanced by that in (b), leading to negative and positive second dielectric virial coefficients for the gases

not vanish. The quadrupole and higher moments of a molecule produce an electric field in its vicinity in accord with equation (1.45) and this field F_β induces a dipole $\alpha_{\alpha\beta}^{(b)} F_\beta$ in a nearby molecule b. This induced dipole leads to pressure-induced absorption of electromagnetic waves and may dominate the second dielectric virial coefficient.[189] In a pair of identical inert gas atoms, μ_{12} is zero and the polarizability change $(\frac{1}{2}\alpha_{12}-\alpha)$ is the only contributor to B_ϵ. The second refractivity virial coefficient B_R is also proportional to the average value of $(\frac{1}{2}\alpha_{12}-\alpha)$ but at optical rather than static frequencies. A differential technique has recently been applied to the measurement of B_R in simple gases.[91]

The subject of second dielectric virial coefficients was recently reviewed by Sutter.[190]

B. Non-equilibrium properties of fluids

Since the dynamics of molecular collisions need to be considered, the theory of transport properties is much more difficult than that of equilibrium properties. Useful information about intermolecular forces may nevertheless be extracted from experimental transport data. The early theoretical work of Enskog,[191] Chapman and Cowling,[192] and Hirschfelder, Curtiss and Bird[8] provided expressions relating the transport coefficients of a gas to the intermolecular potential. Non-equilibrium properties are now usually interpreted in terms of auto-correlation functions in the system at equilibrium.[193] This development is surprisingly recent and is based on the fluctuation-dissipation theorem.[194]

The computational technique known as *molecular dynamics* is proving to be an important tool in the study of both the equilibrium and non-equilibrium properties of fluids.[195] The motion of a finite number of interacting molecules is followed for a period of time by solving the classical equations of motion for various initial conditions. Various intermolecular potentials may be used and the computed equilibrium and non-equilibrium properties compared to experimental values.

C. The structure and properties of crystals and the conformations of molecules

The structure of a molecular crystal and its properties such as the compressibility, heat of sublimation and lattice vibrations, are determined by the intermolecular forces. The relationship between crystal structure and intermolecular forces has been considered for hydrocarbons by Williams,[177] Kitaigorodsky[178] and others,[196] for the halogens,[197] hydrogen cyanide[198] and hydrogen.[199] Similarly the conformations of molecules, and the tertiary and quaternary structure of macromolecules, is determined by the interaction of non-bonded atoms.[200]

Deviation from the Cauchy relation for the elastic constants of cubic crystals are determined by many-body interactions.[201] Non-additivity of the long-range dispersion energy and of the short-range interactions contribute to this deviation[202,203] and to other properties of the crystal and dense fluid. The three-body triple–dipole dispersion energy[204] of three molecules 1, 2, 3 may be obtained from the third-order perturbed energy (see equation (1.25) and expressed in the form

$$u_{123}-u_{12}-u_{23}-u_{31}=c(1+3\cos\theta_1\cos\theta_2\cos\theta_3)R_1^{-3}R_2^{-3}R_3^{-3} \tag{1.173}$$

where c is a constant and R_1, R_2, R_3 are the sides and θ_1, θ_2, θ_3 the interior angles of the triangle formed by the atoms. For three H atoms, $c=21{\cdot}6425$ a.u.[205]

D. Spectroscopic properties

Perturbations arising from molecular interactions affect spectra in various ways. Changes in the energy levels lead to shifts in absorption and emission maxima; distortion of the molecules induces changes in intensities (including the appearance of forbidden lines), and simultaneous transitions may occur in interacting molecules, leading to completely new spectral frequencies; and there may be changes in the line shape and width, reflecting lifetimes of the relevant states and changes of phase. The various forms of spectroscopy permit different aspects of the interaction to be studied; for example, perturbation of rotational states is caused by angle-dependent interactions and vibrational spectra reflect the dependence of the interaction on the nuclear coordinates.

The transient dipole moments arising from interacting pairs of dissimilar inert gas atoms lead to broad 'translational' absorption bands first found by Kiss and Welsh.[206–208] The induced dipole consists of a short-range overlap contribution and a term in R^{-7} at long range due to the distortion that results from the dispersion forces[18,209] (see Figure 1.5). Pressure-induced spectra are generally very broad and are attributable to the dipole moments resulting from the distortion of the electron distributions and nuclear configurations of molecules interacting in binary, ternary and higher-order collisions. These dipoles change with the translational, rotational and vibrational motion, leading to pressure-induced absorption spectra at far-infrared and infrared frequencies.[182] The dipole moment of an interacting pair may be considered to come from four distinct sources:

(i) sum of the permanent dipole moments;
(ii) the dipoles induced in each molecule by the non-uniform electric field of the neighbour (see equation (1.58));
(iii) distortion of the charge distribution resulting from dispersion-type forces (see Figure 1.5);
(iv) distortion of the electron distribution and nuclear configuration by short-range forces.

Effects (iii) and (iv) are solely responsible for the absorption in mixtures of the inert gases but are generally much smaller than effect (ii).[210]

Double resonance spectroscopy provides a powerful new tool for studying the role of intermolecular forces in causing a transition between particular stationary states.[74,211] Measurement of the effect on the spectral line due to transitions $\psi_3 \leftarrow \psi_2$ of an intense pulse of radiation at the resonant frequency of the transition $\psi_4 \leftarrow \psi_1$ as a function of gas pressure gives the probability a collision causes the quantum transition from ψ_4 to ψ_3 (see Figure 1.19).

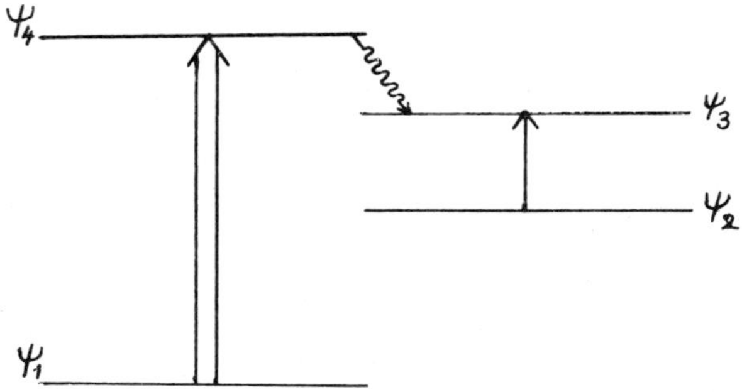

Figure 1.19. In a double-resonance experiment an intense pulse of radiation at the frequency ω_{41} causes transitions from ψ_1 to ψ_4 and their effect on the intensity of the transition $\psi_3 \leftarrow \psi_2$ at ω_{32} is monitored. The wavy line connecting ψ_4 to ψ_3 represents a collision-induced transition

E. Molecular beam scattering

This is a very powerful and direct method of studying intermolecular forces between small molecules. The collisions may be elastic, inelastic or reactive, and the technique is now well-developed and useful.[212–214] When coupled with other experimental information, it is capable of yielding very accurate intermolecular potentials for the inert gases.[13]

F. Forces between macroscopic bodies

The force between two close macroscopic bodies depends upon the long-range interaction energies between the constituent molecules. If the separation R between the molecules a and b is large compared to the reduced wavelength $\lambda\!\!\!^{-} = c\omega^{-1}$ of the transitions that are responsible for the fluctuations in the charge density that give the attractive interaction, the usual R^{-6} dispersion energy is 'retarded' and becomes

$$u(R) = -(4\pi\epsilon_0)^{-2}\frac{23\hbar c}{4\pi}\frac{\alpha^{(a)}\alpha^{(b)}}{R^7}, \qquad R \gg \lambda \tag{1.174}$$

where $\alpha^{(a)}$ and $\alpha^{(b)}$ are the *static* polarizabilities.[67,59,163,215] These macroscopic forces may be significant in colloidal suspensions, but they are extremely weak because of the large separations R.

G. Chemical effects

The environment of a molecule may have a profound influence on its chemical reactivity, as in surface chemistry or through 'solvent effects'. This is one of the main reasons the topic of intermolecular forces is so important and fascinating at the present time.

IX. REFERENCES

1. C. Tanford, *The Hydrophobic Effect*, Wiley, New York, 1973.
1a. 'A discussion on rubber elasticity', *Proc. Roy. Soc. A*, **351,** 295–406 (1976).
2. M. Born and J. R. Oppenheimer, *Ann. Phys.*, **84,** 457 (1927).
3. M. Born and H. Huang, *Dynamical Theory of Crystal Lattices*, Oxford University Press, London, 1954, pp. 166 and 402.
4. K. W. Ford and J. A. Wheeler, *Ann. Phys., New York*, **7,** 259, 287 (1959).
5. W. H. Miller, *Adv. Chem. Phys.*, **25,** 69 (1974).
6. R. G. Newton, *Scattering Theory of Waves and Particles*, McGraw-Hill, New York, 1966.
7. R. B. Bernstein, *Adv. Chem. Phys.*, **10,** 75, (1966).
8. J. O. Hirschfelder, C. F. Curtiss and R. B. Bird, *Molecular Theory of Gases and Liquids*, Wiley, New York, 1964.
9. J. O. Hirschfelder, Ed., *Adv. Chem. Phys.*, **12** (1967).
10. H. Margenau and N. R. Kestner, *Intermolecular Forces*, 2nd Edition, Pergamon Press, London 1971.

11. R. J. Bell and I. J. Zucker in *Rare Gas Solids*, M. L. Klein and J. A. Venables, Ed., Academic Press, New York, 1976, Ch. 2.
12. J. N. Murrell in *Rare Gas Solids*, M. L. Klein and J. A. Venables, Ed., Academic Press, New York, 1976, Ch. 3.
13. J. A. Barker in *Rare Gas Solids*, M. L. Klein and J. A. Venables, Ed., Academic Press, New York, 1976, Ch. 4.
14. P. R. Certain, J. O. Hirschfelder, W. Kołos, and L. Wolniewicz, *J. Chem. Phys.*, **49,** 24 (1968).
15. W. Kołos and L. Wolniewicz, *J. Chem. Phys.*, **41,** 3663 (1964); **43,** 2429 (1965); **49,** 404 (1968).
16. H. Hellmann, *Quantenchemie*, Demticke, Leipzig, 1937, p. 285.
17. R. P. Feynman, *Phys. Rev.*, **56,** 340 (1939).
18. A. D. Buckingham, *CNRS Colloques*, **77,** 57 (1959).
19. J. O. Hirschfelder and M. A. Eliason, *J. Chem. Phys.*, **47,** 1164 (1967).
20. J. O. Hirschfelder and W. J. Meath, *Adv. Chem. Phys.*, **12,** 3 (1967).
21. A. D. Buckingham, *Quarterly Reviews* (*Chem. Soc., London*), **13,** 189 (1959).
22. C. G. Gray, *Canad. J. Phys.*, **54,** 505 (1976).
23. R. Bonaccorsi, E. Scrocco and J. Tomasi, *J. Chem. Phys.*, **52,** 5270 (1970).
23a. R. Bonaccorsi, A. Pullman, E. Scrocco, and J. Tomasi, *Theoret. Chim. Acta*, **24,** 51 (1972).
24. A. D. Buckingham, *Adv. Chem. Phys.*, **12,** 107 (1967).
25. W. H. Keesom, *Physik Z.*, **22,** 129 (1921).
26. P. Debye, *Physik Z.*, **22,** 302 (1921).
27. S. C. Wang, *Physik Z.*, **28,** 663 (1927).
28. F. London, *Z. Physik*, **63,** 245 (1930); *Z. Physik. Chem. B.*, **11,** 221 (1930).
29. F. London, *Trans. Faraday Soc.*, **33,** 8 (1937).
30. L. Pauling and J. Y. Beach, *Phys. Rev.*, **47,** 686 (1935).
31. J. O. Hirschfelder and P.-O. Löwdin, *Mol. Phys.*, **2,** 229 (1959); **9,** 491 (1965).
32. W. J. Deal Jr., *Intern. J. Quantum Chem.*, **6,** 593 (1972).
33. H. C. Longuet-Higgins, *Discussions Faraday Soc.*, **40,** 7 (1965).
34. G. Zundel, *Hydration and Intermolecular Interaction*, Academic Press, New York, 1969.
35. G. Zundel and E. C. Weidemann, *Trans. Faraday Soc.*, **66,** 1941 (1970).
36. W. S. Benedict and R. Herman, *J. Quant. Spect. Radiative Transfer*, **3,** 265 (1963); R. G. Breene, Jr, *The Shift and Shape of Spectral Lines*, Pergamon Press, Oxford, 1961.
37. J. E. Griffiths, M. Clerc, and P. M. Rentzepis, *J. Chem. Phys.*, **60,** 3824 (1974).
38. A. S. Davydov, *Theory of Molecular Excitons*, McGraw-Hill, New York, 1962.
39. D. P. Craig and S. H. Walmsley, *Excitons in Molecular Crystals*, Benjamin, New York, 1968.
40. J. B. Birks, *Photophysics of Aromatic Molecules*, Wiley, New York, 1970; *Reports Prog. Phys.*, **38,** 903 (1975).
41. C. Mavroyannis and M. J. Stephen. *Molec. Phys.*, **5,** 262 (1962).
42. D. P. Craig, E. A. Power and T. Thirunamachandran, *Proc. Roy. Soc. A*, **322,** 165 (1971).
43. J. S. Dahler and J. O. Hirschfelder, *J. Chem. Phys.*, **25,** 986 (1956).
44. H. Jeffreys, *Cartesian Tensors*, Cambridge University Press, 1931.
45. G. Temple, *Cartesian Tensors*, Methuen, London, 1960.
46. L. Pauling and E. B. Wilson, *Introduction to Quantum Mechanics*, McGraw-Hill, New York, 1935.
47. E. U. Condon and G. H. Shortley, *The Theory of Atomic Spectra*, Cambridge University Press, 1935.
48. L. I. Schiff, *Quantum Mechanics*, McGraw-Hill, New York, 3rd Edn., 1968.

49. S. Bhagavantam, *Crystal Symmetry and Physical Properties*, Academic Press, 1966.
50. H. A. Jahn, *Acta Cryst.*, **2,** 30 (1949).
51. S. Bhagavantam and D. Suryanarayana, *Acta Cryst.* **2,** 21 (1949).
52. R. R. Birss, *Reports Progress Physics*, **26,** 307 (1963).
53. A. D. Buckingham, in *Physical Chemistry an Advanced Treatise*, Vol. 4, Ch. 8, H. Eyring, W. Jost and D. Henderson, Eds, Academic Press, New York, 1970.
54. A. D. Buckingham in MTP *International Review of Science, Physical Chemistry Series* 1, Vol. 3, Chap. 3, D. A. Ramsay, Ed., Butterworths, London, 1972.
55. A. D. Buckingham and J. A. Pople, *Proc. Phys. Soc. A*, **68,** 905 (1955).
56. A. D. McLean and M. Yoshimine, *J. Chem. Phys.*, **47,** 1927 (1967).
57. A. Dalgarno, *Adv. Chem. Phys.*, **12,** 143 (1967).
58. L. Jansen, *Physica*, **23,** 599 (1957).
59. J. Mahanty and B. W. Ninham, *Dispersion Forces*, Academic Press, New York, 1976.
60. B. I. Bleaney and B. Bleaney, *Electricity and Magnetism*, Oxford University Press, London, 2nd Edn., 1965.
61. A. D. Buckingham and R. E. Raab, *Proc. Roy. Soc. A*, **345,** 365 (1975).
62. R. E. Raab, *Molec. Phys.*, **29,** 1323 (1975).
63. A. D. Buckingham and P. J. Stephens, *Ann. Rev. Phys. Chem.*, **17,** 399 (1966).
64. A. D. Buckingham and M. B. Dunn, *J. Chem. Soc. A*, 1988 (1971).
65. J. A. Barker, *Proc. Roy. Soc. A*, **219,** 367 (1953).
66. A. D. Buckingham and J. A. Pople, *Trans. Faraday Soc.*, **51,** 1173 and 1179 (1955).
67. H. B. G. Casimir and D. Polder, *Phys. Rev.*, **73,** 360 (1948).
68. A. Dalgarno and W. D. Davison, *Advan. Atom Molec. Phys.*, **2,** 1 (1966).
69. G. Starkschall and R. G. Gordon, *J. Chem. Phys.*, **54,** 663 (1971); **56,** 2801 (1972).
70. P. W. Langhoff, R. G. Gordon and M. Karplus, *J. Chem. Phys.*, **55,** 2126 (1971).
71. G. Starkschall and R. G. Gordon, *J. Chem. Phys.*, **56,** 2102 (1972).
72. W. D. Davison, *Proc. Phys. Soc.*, **87,** 133 (1966).
73. A. Unsöld, *Z. Physik*, **43,** 563 (1927).
74. T. Oka, *Advan. Atom Molec. Phys.*, **9,** 127 (1973).
75. R. W. Schmieder, A. Lurio and W. Happer, *Phys. Rev. A*, **3,** 1209 (1971).
76. J. G. Kirkwood and J. B. Shumaker, *Proc. Nat. Acad. Sci.*, **38,** 863 (1952).
77. P. Claverie, *Int. J. Quant. Chem.*, **5,** 273 (1971).
78. C. C. J. Roothaan, *Rev. Mod. Phys.*, **23,** 69 (1951).
79. P. E. Phillipson, *Phys. Rev.*, **125,** 1981 (1962).
80. T. L. Gilbert and A. C. Wahl, *J. Chem. Phys.*, **47,** 3425 (1967).
81. N. R. Kestner, *J. Chem. Phys.*, **48,** 252 (1968).
82. N. S. Ostlund and D. L. Merrifield, *Chem. Phys. Letters*, **39,** 612 (1976).
83. H. F. Schaefer, D. R. McLaughlin, F. E. Harris, and B. J. Alder, *Phys. Rev. Lett.*, **25,** 988 (1970).
84. P. Bertoncini and A. C. Wahl, *Phys. Rev. Lett.*, **25,** 991 (1970).
85. D. R. McLaughlin and H. F. Schaefer, *Chem. Phys. Lett.*, **12,** 244 (1971).
86. P. Bertoncini and A. C. Wahl, *J. Chem. Phys.*, **58,** 1259 (1973).
87. B. Liu and A. D. McLean, *J. Chem. Phys.*, **59,** 4557 (1974).
88. B. Liu and A. D. McLean, see Ch. 4, Fig. 23 in *Rare Gas Solids*, M. L. Klein and J. A. Venables, Ed., Academic Press, New York, (1976).
89. W. J. Stevens, A. C. Wahl, M. A. Gardner and A. M. Karo, *J. Chem. Phys.*, **60,** 2195 (1974).
90. R. R. Teachout and R. T. Pack, *Atomic Data*, **3,** 195 (1971).

91. A. D. Buckingham and C. Graham, *Proc. Roy. Soc. A*, **336,** 275 (1974).
92. S. F. Boys and F. Bernardi, *Molec. Phys.*, **19,** 553 (1971).
93. S. F. Boys, *Proc. Roy. Soc. A*, **309,** 195 (1969).
94. S. F. Boys and N. C. Handy, *Proc. Roy. Soc. A*, **309,** 209; **310,** 43, 63; **311,** 309 (1969).
95. J. O. Hirschfelder, *Chem. Phys. Lett.*, **1,** 325 (1967).
96. R. Eisenschitz and F. London, *Z. Phys.*, **45,** 2188 (1966).
97. J. O. Hirschfelder and R. Silbey, *J. Chem. Phys.*, **45,** 2188 (1966).
98. J. N. Murrell and G. Shaw, *J. Chem. Phys.*, **46,** 1768 (1967); **49,** 4731 (1968).
99. A. T. Amos and J. I. Musher, *Chem. Phys. Lett.*, **1,** 149 (1967).
100. A van der Avoird, *Chem. Phys. Lett.*, **1,** 411 (1967); *J. Chem. Phys.*, **47,** 3649 (1967).
101. J. O. Hirschfelder, *Chem. Phys. Lett.*, **1,** 363 (1967).
102. P. R. Certain, J. O. Hirschfelder, W. Kołos and L. Wolniewicz, *J. Chem. Phys.*, **49,** 24 (1968).
103. W. Kołos and L. Wohniewicz, *J. Chem. Phys.*, **41,** 3663 (1964); **43,** 2429 (1965).
104. P. R. Certain and L. W. Bruch, *International Review of Science*, Phys. Chem. Series 1, Vol. 1, Chap. 4, W. Byers Brown, Ed., Butterworths, London, 1972.
105. R. S. Mulliken, *J. Chem. Phys.*, **46,** 497, 675 (1949).
106. H. Kreek and W. J. Meath, *J. Chem. Phys.*, **50,** 2289 (1969).
107. H. Kreek, Y. H. Pan and W. J. Meath, *Molec. Phys.*, **19,** (1970).
108. F. C. Brooks, *Phys. Rev.*, **86,** 92 (1952).
109. A. Dalgarno and J. T. Lewis, *Proc. Phys. Soc. A*, **69,** 57 (1956).
110. P. T. Wedepohl, *Proc. Phys. Soc.*, **92,** 79 (1967).
111. W. D. Wilson and C. L. Bisson, *Phys. Rev.*, **B3,** 3984 (1971).
112. R. G. Gordon and Y. S. Kim, *J. Chem. Phys.*, **56,** 3122 (1972).
113. P. Gombas, *Handbuch der Physik*, S. Flugge, Ed., Vol. 36, Springer-Verlag, Berlin, 1956.
114. J. C. Slater, *Quantum Theory of Atomic Structure*, Vol. 2, Appendix 22, McGraw-HIll, New York, 1960.
115. H. A. Bethe and R. W. Jackiw, *Intermediate Quantum Mechanics*, 2nd Edn., Ch. 5, Benjamin, New York, 1968.
116. Y. S. Kim and R. G. Gordon, *J. Chem. Phys.*, **61,** 1 (1974).
117. P. Drude, *The Theory of Optics*, Longmans, London, 1902.
118. A. G. Gaydon, *Dissociation Energies*, 3rd edn., Chapman and Hall, London, 1968.
119. G. J. Brealey and M. Kasha, *J. Amer. Chem. Soc.*, **77,** 4462 (1955).
120. J. A. Pople, W. G. Schneider and H. J. Bernstein, *High-Resolution Nuclear Magnetic Resonance*, Ch. 15, McGraw-Hill, New York, 1959.
121. N. H. Fletcher, *The Chemical Physics of Ice*, Cambridge University Press, 1970.
122. C. C. Costain and G. P. Srivastava, *J. Chem. Phys.*, **41,** 1620 (1964).
123. J. A. Ibers, *J. Chem. Phys.*, **41,** 25 (1964).
124. T. R. Dyke, B. J. Howard and W. Klemperer, *J. Chem. Phys.*, **56,** 2442 (1972).
125. B. J. Howard, *Int. Rev. Science, Physical Chemistry Series* 2, Vol. 2, A. D. Buckingham, Ed., Ch. 3, Butterworths, London, 1975.
126. E. Clementi, *J. Chem. Phys.*, **46,** 3851 (1967).
127. P. A. Kollman and L. C. Allen, *J. Chem. Phys.*, **52,** 5085 (1970).
128. G. H. F. Diercksen and W. P. Kraemer, *Chem. Phys. Lett.*, **6,** 419 (1970).
129. J. Del Bene and J. A. Pople, *J. Chem. Phys.*, **55,** 2296 (1971).
130. D. R. Yarkony, S. V. O'Neil, H. F. Schaefer, C. P. Baskin and C. F. Bender, *J. Chem. Phys.*, **60,** 855 (1974).
131. H. Lischka, *J. Amer. Chem. Soc.*, **96,** 4761 (1974).
132. P. A. Kollman, *Hydrogen Bonding and Donor–Acceptor Interactions*, (to be

published in *Modern Theoretical Chemistry*, H. F. Schaefer, Ed., Plenum, New York, 1977).
133. K. Morokuma and L. Pederson, *J. Chem. Phys.*, **48,** 3275 (1968).
134. J. Del Bene and J. A. Pople, *J. Chem. Phys.*, **52,** 4858 (1970).
135. D. Hankins, J. W. Moskowitz and F. H. Stillinger, *J. Chem. Phys.*, **53,** 4544 (1970).
136. G. H. F. Diercksen, *Theor. Chem. Acta*, **21,** 35 (1971).
137. H. Popkie, H. Kistenmacher and E. Clementi, *J. Chem. Phys.*, **59,** 1325 (1973).
138. L. Shipman and H. A. Scheraga, *J. Phys. Chem.*, **78,** 2055 (1974).
139. T. R. Dyke and J. S. Muenter, *J. Chem. Phys.*, **60,** 2929 (1974).
140. T. R. Dyke and J. S. Muenter, *Int. Rev. Sci. Phys. Chem. Series* 2, Vol. 2, Ch. 2, A. D. Buckingham, Ed., 1975.
141. A. J. Tursi and E. R. Nixon, *J. Chem. Phys.*, **52,** 1521 (1970).
142. M. Van Thiel, E. D. Becker and G. C. Pimentel, *J. Chem. Phys.*, **27,** 486 (1957).
143. L. B. Magnusson, *J. Phys. Chem.*, **74,** 4221 (1970); *Molec. Phys.*, **21,** 571 (1971).
144. P. A. Kollman and A. D. Buckingham, *Molec. Phys.*, **21,** 567 (1971).
145. P. W. Atkins and M. C. R. Symons, *Molec. Phys.* **23,** 831 (1972).
146. J. Del Bene and J. A. Pople, *J. Chem. Phys.*, **58,** 3605 (1973).
147. P. A. Kollman and L. C. Allen, *J. Amer. Chem. Soc.*, **92,** 4108 (1970).
148. J. Del Bene and J. A. Pople, *J. Chem. Phys.*, **55,** 2296 (1971).
149. C. A. Coulson, *Research*, **10,** 149 (1957).
150. J. N. Murrell, *Chem. in Britain*, **5,** 107 (1969).
151. R. S. Mulliken, *J. Amer. Chem. Soc.*, **72,** 600 (1950); **74,** 811 (1952); *J. Phys. Chem.*, **56,** 801 (1952).
152. H. A. Benesi and J. H. Hildebrand, *J. Amer. Chem. Soc.* **71,** 2703 (1949).
153. R. S. Mulliken and W. B. Person, *Molecular Complexes*, Wiley, New York, 1969.
154. J. N. Murrell, *Quart. Rev. Chem. Soc.*, **15,** 191 (1961).
155. R. Foster, *Organic Charge Transfer Complexes*, Academic Press, London, 1969.
156. R. Foster, Ed., *Molecular Association*, Academic Press, London, 1975.
157. K. Morokuma, *J. Chem. Phys.*, **55,** 1236 (1971).
158. A. Pullman, *Chem. Phys. Letters*, **20,** 29 (1973).
159. A. Pullman and P. Schuster, *Chem. Phys. Letters*, **24,** 472 (1974).
160. *A Discussion on the Physics and Chemistry of Biological Recognition, Phil. Trans. Roy. Soc. B*, **272,** 1–198 (1975).
161. N. R. Kestner and O. Sinanoğlu, *J. Chem. Phys.*, **38,** 1730 (1963).
162. A. D. McLachlan, *Discussions Faraday Soc.*, **40,** 239 (1965).
163. J. N. Israelachvili and D. Tabor, *Prog. Surface Membrane Sci.*, **7,** 1 (1973).
164. D. Langbein, *Theory of van der Waals Attraction*, Springer, Berlin, 1974.
165. D. Gingell, *Phil. Trans. Roy. Soc. B*, **272,** 11 (1975).
166. L. Onsager, *J. Amer. Chem. Soc.*, **58,** 1486 (1936).
167. A. Pullman in *The World of Quantum Chemistry*, R. Daudel and B. Pullman, Eds., p. 239, Reidel, Dordrecht, 1974.
168. B. Linder, *Adv. Chem. Phys.*, **12,** 225.
169. O. Tapia and O. Goscinski, *Molec. Phys.*, **29,** 1653 (1975).
170. A. D. Buckingham in *The World of Quantum Chemistry*, R. Daudel and B. Pullman, Eds., p. 253, Reidel, Dordrecht, 1974.
171. P. R. Certain and P. J. Fortune, *J. Chem. Phys.*, **55,** 5818 (1971).
172. R. H. Orcutt and R. H. Cole, *J. Chem. Phys.*, **46,** 697 (1967).
173. D. Vidal and M. Lallemand, *J. Chem. Phys.*, **64,** 4293 (1976).
174. A. D. Buckingham and J. A. Pople, *Discussions Faraday Soc.*, **22,** 17 (1956).
175. P. W. Langhoff, M. Karplus and R. P. Hurst, *J. Chem. Phys.* **44,** 505 (1966).

176. H. A. Scheraga, S. J. Leach, R. A. Scott and G. Némethy, *Discussions Faraday Soc.*, **40,** 268 (1965).
177. D. E. Williams, *J. Chem. Phys.*, **45,** 3770 (1966); **47,** 4680 (1967).
178. A. I. Kitaigorodsky, *Molecular Crystals and Molecules*, Academic Press, New York, 1973.
179. P. W. Anderson, *Phys. Rev.*, **80,** 511 (1950).
180. R. M. Herman, *J. Chem. Phys.*, **44,** 1346 (1966).
181. J. F. Bukta and W. J. Meath, *Molec. Phys.*, **29,** 1409 (1975).
182. H. L. Welsh, *International Rev. Sci.*, *Physical Chemistry Series* 1, Vol. 3, D. A. Ramsay, Ed., Chap. 2, Butterworths, London, 1972.
183. J. Van Kranendonk, *Physica*, **24,** 347 (1958); **73,** 156 (1974).
184. P. W. Gibbs, C. G. Gray, J. L. Hunt, S. Paddi Reddy, R. H. Tipping and K. S. Chang, *Phys. Rev. Letters*, **33,** 256 (1974).
185. L. W. Bruch, P. J. Fortune, and D. H. Berman, *J. Chem. Phys.*, **61,** 2626 (1974).
186. J. H. Dymond and E. B. Smith, *The Virial Coefficients of Gases*, Oxford University Press, 1969.
187. J. A. Pople, *Proc. Roy. Soc. A*, **221,** 508 (1954).
188. P. Debye, *Polar Molecules*, Chemical Catalog Company, New York, 1929.
189. A. D. Buckingham and J. A. Pople, *Trans. Faraday Soc.*, **51,** 1029; **51,** 1179 (1955).
190. H. G. Sutter, in *Dielectric and Related Molecular Processes*, Vol. 1, M. Davies, Ed., p. 65, Specialist Periodical Reports, Chem. Soc. London, 1972.
191. D. Enskog, *Ark. Mat. Astron. Fys.*, **16,** 16 (1922).
192. S. Chapman and T. G. Cowling, *The Mathematical Theory of Non-Uniform Gases*, Cambridge University Press, 1952.
193. P. C. Martin, *Measurements and Correlation Functions*, Gordon and Breach, New York, 1970.
194. R. Kubo, *Rep. Prog. Phys.*, **29,** 255 (1966).
195. W. G. Hoover and W. T. Ashurst in *Theoretical Chemistry, Advances and Perspectives*, Vol. 1, 1, H. Eyring and D. Henderson, 1975.
196. D. P. Craig, P. A. Dobosh, R. Mason and D. P. Santry, *Discuss. Faraday Soc.*, **40,** 110 (1965).
197. S. C. Nyburg, *J. Chem. Phys.*, **40,** 2493 (1964).
198. A. I. M. Rae, *Molec. Phys.*, **16,** 257 (1969).
199. H. M. James, *Phys. Rev.*, **167,** 862 (1968).
200. A. J. Hopfinger, *Conformational Properties of Macromolecules*, Academic Press, New York, 1973.
201. P.-O. Löwdin, *Advan. Phys.*, **5,** 1 (1956).
202. G. G. Chell, *J. Phys. C*, **1,** 266 (1968).
203. E. Lombardi, L. Jansen and R. Ritter, *Phys. Rev.*, **185,** 1150, 1158 (1969).
204. B. M. Axilrod and E. Teller, *J. Chem. Phys.*, **11,** 299 (1943).
205. A. Dalgarno and G. A. Victor, *Proc. Phys. Soc.*, **90,** 605 (1967).
206. Z. J. Kiss and H. L. Welsh, *Phys. Rev. Lett.*, **2,** 166 (1959).
207. A. Watanabe and H. L. Welsh, *Can. J. Phys.*, **43,** 818 (1965).
208. D. R. Bosomworth and H. P. Gush, *Can. J. Phys.*, **43,** 751 (1965).
209. S. Weiss, *Can. J. Phys.*, **54,** 584 (1976).
210. A. D. Buckingham and A. J. C. Ladd, *Can. J. Phys.*, **54,** 611 (1976).
211. A. M. Ronn and E. B. Wilson, *J. Chem. Phys.*, **46,** 3262 (1967).
212. J. Ross, Ed., *Adv. Chem. Phys.*, **10,** 8 (1966).
213. *Discussions Faraday Soc.*, **55,** 41 (1973).
214. K. P. Lawley, Ed., *Adv. Chem. Phys.*, **30,** 109 (1975).
215. E. A. Power, *Adv. Chem. Phys.*, **12,** 167 (1967).

CHAPTER 2

Elaboration of approximate formulas for the interactions between large molecules: Applications in organic chemistry

P. Claverie

Institut de Biologie Physico-Chimique, Laboratoire de Chimie Quantique. Paris, France

I. INTRODUCTION

The very kind of phenomena whose study involves the calculation of intermolecular interactions constitutes a strong incentive to derive formulas as simple as possible to achieve such calculations.

Indeed, if we consider, by contrast, the study of an isolated molecule, the so-called 'chemical' or 'valence' interactions between the atoms result in the existence of *well-defined* and *transferable* geometrical configurations of the nuclei. This transferability from one molecule to other ones corresponds to the existence of characteristic bond lengths and valency angles (related to a small number of hybridization schemes), and it allows the calculation of molecular properties by using a standard geometry. Even if higher accuracy is desired, such a standard geometry at least gives a convenient starting point for the search of a better one. Thus, a number of problems of quantum chemistry concerning the isolated molecule could be treated by considering only a small number of nuclear configurations, and consequently the computational procedure (for a single configuration) could be complex and time-consuming (*ab initio* SCF or perturbation methods, large basis sets of atomic orbitals, Configuration Interaction) without making prohibitive the whole treatment of the problem. Of course, as soon as *numerous* conformations become involved (as it happens in problems of conformational analysis or chemical reactivity), the situation changes, and simpler, faster methods for evaluating at least the energy of each configuration become strongly necessary.

Now, it is just this kind of situation which almost always holds in the field of intermolecular interactions. There is in general no strongly preferred geometry (or at least it is not easy to foresee *a priori*), and consequently the interaction energy must be calculated in principle for a large number of relative positions of the molecules, in order to scan the whole configuration space. Indeed, the physical problems involving several molecules quite generally lead either to the direct study of the energy hypersurface as a function of the configuration (especially the knowledge of minima and maxima), either to integrations involving the interaction potential if some statistical mechanical treatment is to be performed. In order to fulfil such purposes, the fastest methods for computing the interaction potential will be welcome. Moreover, if we want to have a chance of expressing analytically (at least partially) such a statistical treatment, it is of course necessary to express the interaction energy itself by an explicit analytical formula, as simple as possible. As a striking example of such a situation, we would mention the recent developments of the distribution function method for polyatomic liquids, namely the 'Reference Interaction Site Model' (RISM) (Andersen and Chandler, 1972: Lowden and Chandler, 1973, 1974; Chandler, 1973, 1976; Ladanyi and Chandler, 1975; H. C. Andersen, 1975): in this method, the intermolecular potential is assumed to be a sum of atom–atom terms:

$$U^{(1,2)} = \sum_{\alpha}^{(1)} \sum_{\gamma}^{(2)} u_{\alpha\gamma}(r_{\alpha\gamma}) \tag{2.1}$$

each of these terms depending only on the *distance* $r_{\alpha\gamma} = |\vec{r}_\gamma - \vec{r}_\alpha|$ of the atoms α and γ (the superscripts 1, 2 label the molecules). Then, for such an intermolecular potential, all interesting thermodynamic functions may be deduced from various site–site (atom–atom) functions (direct correlation functions $c_{\alpha\gamma}(r_{\alpha\gamma})$, distribution functions $g_{\alpha\gamma}(r_{\alpha\gamma})$). In the framework of the RIS Model, these functions are first calculated for hard-sphere interatomic potentials $u_{\alpha\gamma}^{(HS)}$, then a perturbation procedure is used for getting the functions corresponding to the 'soft' potentials $u_{\alpha\gamma}$ of equation (2.1). Such a procedure was very successful in the case of simple liquids, i.e. monoatomic liquids interacting through an isotropic potential $U^{(1,2)}(r)$ (Barker and Henderson, 1967; Weeks *et al.* 1971; Andersen *et al.* 1971): see the books edited by Eyring *et al.* (1971) and the review papers by Henderson and Davidson (1967), Barker and Henderson (1971, 1972), Fitts (1972), Chandler (1974), and H. C. Andersen (1975). This success is a very strong incentive to develop the RIS Model for liquids of polyatomic molecules, but it is then necessary to use simplified formulas of the type (2.1). Thus we have here a good example of the importance of developing simplified formulas in view of applications in Statistical Mechanics.

Even in the case of a type of interaction for which the geometry of the energy minimum is better defined (when compared to complexes involving mainly dispersion interaction) the possibility of using simplified formulae would considerably extend the field of applications.

Owing to these grounds, it is not astonishing that the search for interaction potentials, more or less empirical, and with the simplest possible form, has been pursued for a long time. Thus the well-known Lennard-Jones potential (1924)

$$U(r) = 4\varepsilon\left(\left(\frac{\sigma}{r}\right)^{12} - \left(\frac{\sigma}{r}\right)^{6}\right) \tag{2.2}$$

was proposed even before Eisenschitz and London (1930) gave some theoretical basis for the existence of an attractive energy varying like r^{-6} between any pair of atoms or molecules. At the time of its proposal, the Lennard–Jones potential should rather appear as a semi-empirical representation for the interaction potential between rare gas atoms, and actually the future development of the theory confirmed the r^{-6} attractive part (as dispersion term) but not the r^{-12} repulsive one (for which an exponential like variation Be^{-br} (Born–Mayer) is theoretically founded in the region of the Van der Waals minimum: see e.g. Cohen and Pack (1974); Murrell and Varandas (1975)). Already before the Lennard–Jones potential, the electrostatic (dipole–dipole) and polarization (dipole-induced dipole) potentials had been proposed on the basis of classical electromagnetic concepts (Keesom, Debye).

Later on, other potentials were proposed for the interaction between rare gas atoms (Buckingham, Kihara); the essential improvement was the replacement of the $(\sigma/r)^{12}$ term by an exponential one Be^{-br} for representing the short-range repulsion. In that case, at some short distance, the formula $-A/r^6 + Be^{-br}$ must be replaced by some other one, in order to avoid the spurious decrease which occurs at very short distance due to the $-A/r^6$ term which goes to $-\infty$ while the other one remains finite (e.g. see Egelstaff

(1967, Ch. 3, p. 28), Hirschfelder *et al.* (1954, Ch. 3) and the Section VI.A.5 of the present chapter). Moreover, the dipole-dipole term (Stockmayer potential) was added for dealing with polar small molecules, and more generally non-isotropic potentials, based upon spherical harmonics expansion (around *one* centre in each molecule) were also used for the dispersion and repulsion terms between small (non-spherical) molecules (e.g. see Cook and Rowlinson (1953), Rowlinson (1969, Ch. 7)). But all these improvements remained essentially limited to the interaction between atoms or very small molecules, since a *single* centre was used for each of the interacting subsystems (concerning the kind of semi-empirical potentials considered in this paragraph, see Hirschfelder, Curtiss et Bird (1954, Ch. 3); Kihara (1958, 1967); Fitts (1966); Rowlinson (1969, Ch. 7); Mason and Spurling (1969, Ch. 4); Buckingham and Utting (1970); Barker and Henderson (1972).

It is clear that, for large molecules at intermediate or short distances (i.e. distances not large with respect to the size of the molecules themselves), any realistic intermolecular potential must take into account in a rather explicit way the size and shape of the molecules. Two kinds of semi-empirical potentials obeying this requirement may be quoted: the convex-core potentials introduced by Kihara (1953, 1955, 1967, 1970) (for applications see Kihara and Jhon (1970), Kihara *et al.* (1971), Kihara and Yamazaki (1971), Jhon *et al.* (1972)), and the 'many-centre' potentials, essentially sums of atom–atom interaction terms as mentioned above (equation (2.1)). Such 'atom–atom' potentials were actually used not only for intermolecular but also for intramolecular problems (conformational studies of large molecules). These potentials will be considered below in Section VI (let us quote here Cignitti and Allen (1965); Kitaigorodskii (1961a, 1965); Kitaigorodskii and Mirskaya (1964a); Kitaigorodskii *et al.* (1969); Liquori (1966, 1969); Giglio (1969, 1970 and references therein); Coiro *et al.* (1972); Scheraga (1968); Momany *et al.* (1974a,*b, 1975), Shipman *et al.* (1975, 1976); Burgess *et al.* (1975, 1976); Hagler *et al.* (1974)).

Besides this 'semi-empirical' line of development, there was, of course, the more 'theoretical' line of the works whose authors tried to derive from a quantum-theoretical basis convenient formulas expressing the intermolecular potential. Let us quote London (1930a,b; 1937; 1942), Margenau (1939a,b), Slater (1928), Rosen (1950), Longuet-Higgins (1956), Longuet-Higgins and Salem (1961), Salem (1960, 1961, 1962a), Hirschfelder (1965), the collection of papers edited by Hirschfelder (1967c), the book by Margenau and Kestner (1969)† and the reviews by Kihara (1958, 1967, 1970); during the past few years, several reviews were published, most of them dealing essentially with the theory of the interactions between atoms or small molecules (Buckingham and Utting, 1970; Certain and Bruch,

* This paper contains numerous references concerning semi-empirical formulas.

† This book displays an extensive bibliography of papers dealing with the theory of intermolecular interactions.

1972; Murrell, 1974), but the recent reviews by Rein (1973) and by Amos and Crispin (1976c) also consider the evaluation of the interaction energy between large molecules (essentially as a sum of interaction terms between small sub-units in each molecule), thus following the same line of thought as in the present chapter and in the author's thesis (Claverie, 1973).

Besides these general references, it appears worthwhile to mention some specific topics:

—the series of works about exchange perturbation theory: Hirschfelder (1967a,b), Chipman (1972), Murrell and Shaw (1967a,b; 1968b), Murrell and Varandas (1975), and references in these papers (a more entensive bibliography will be given in Section II below);

—the series of works about the study of dispersion energy in terms of the dynamic polarizability of the subsystems: Mavroyannis and Stephen (1962), McLachlan (1963a,b) Dzyaloshinskii *et al.* (1961), Dalgarno and Davison (1966), Goscinski (1968); Langhoff and Karplus (1970a,b), Langhoff *et al.* (1971), Tang and Karplus (1968), Tang (1968, 1969, 1970, 1971), Starkschall and Gordon (1971, 1972a,b); the extension of this type of method to many-molecular systems (so-called 'reaction-field' techniques) is described in review papers by Jehle (1965, 1969) (who used the name 'charge fluctuation forces'), Linder (1967), Linder and Rabenold (1972), Parsegian (1973), Israelachvili (1974); see also Langbein (1973) and references therein.

—the application of the 'electron gas model' has led during the last years to important practical advances in the evaluation of the short-range interactions (see Section IV.D below), and such a simplified evaluation opens the way to the systematic determination of atom–atom short-range potentials of the Born-Mayer type (Be^{-br}) obtained by fitting the numerical values computed according to the simplified procedure: Gaydaenko and Nikulin (1970); Gordon and Kim (1972); Rae (1973), Cohen and Pack (1974).

All these theoretical works concerned essentially the interactions between atoms or at most very small molecules. Consequently, it appears necessary to establish a connection between this theoretical basis and the 'semi-empirical' formulas developed for dealing with the interaction between large molecules. It is the purpose of the present chapter to contribute to this program (see also Rein, 1973; Amos and Crispin, 1976b,c).

The theoretical basis for the derivation of simplified formulas expressing the interaction energy—whether for atoms or small or large molecules—is the concept of perturbation with respect to the isolated subsystems. Indeed, the complexity of a quantum-mechanical *ab initio* calculation is considerable, so that a simplified evaluation of the interaction energy may be expected only by using the knowledge of the separate subsystems and then evaluating directly the energy *change* when these subsystems are put at finite distance (the smallness of this change justifying the recourse to perturbation methods), instead of computing directly the *total energy* of the supersystem and then subtracting the sum of the energies of the subsystems computed previously. It is customary nowadays to refer to this last procedure as the 'supermolecule approach'; actually, there is a somewhat continuous transition between the usual supersystem approach (namely that using the *ab initio* SCF or some other quantum-chemical method for computing the energy of the subsystems and of the supersystem) and the simplified perturbation approach, namely the possibility of performing *ab initio* or eventually less refined *perturbation* computations: for example, see Certain *et al.* (1968); Chipman (1972); Chipman *et al.* (1973); Chipman and Hirschfelder (1973); Conway and Murrell (1972, 1974); Murrell and Varandas (1975); Daudey *et al.* (1974a,b), Daudey (1974), Kochanski (1974, 1975); Kochanski and Gouyet (1975a,b); Kolos (1974, 1975); Jeziorski and Van Hemert (1976) and more generally the works concerning the exchange perturbation theories quoted in Section II.B.1 below. Let us also mention here the theoretical analysis of the SCF perturbation method, by Basilevsky and Berenfeld (1972b) and the semi-empirical SCF-perturbation treatments by Devaquet (1970), Sustmann and Binsch (1971a,b) and Magnasco (1971), Magnasco and Musso (1971a,b).

Besides this first fundamental feature of the perturbation approach, namely the direct evaluation of the energy difference, the second fundamental feature is the attempt to split this interaction energy into several terms, each of these terms having a simple variation law as a function of distance, so that it becomes possible to find for each of these terms an appropriate simplified formula, and thus to recover an evaluation of the total interaction energy. We shall thus encounter the long-range terms (electrostatic, polarization, dispersion, ...) with an r^{-n}-like variation law, and the short-range terms (first-order exchange repulsion, second-order exchange terms (involving mainly the charge transfer term), ...) with a decreasing exponential variation law.

Besides its practical interest (namely the derivation of simplified formulas), such a decomposition allows a better understanding of the behaviour of the total interaction energy, and this conceptual interest explains its use for analysing the results of variational calculations in the 'supermolecule approach': see, for example, Dreyfus (1970), Dreyfus and Pullman (1970), Kollman and Allen (1970), Kutzelnigg and Gelus (1970), Kutzelnigg *et al.* (1972), Morokuma (1971), Kolos (1967), Kolos and Wolniewicz (1974).

Such analyses also make easier to avoid the difficulties involved in variational supermolecule calculations, which are due to the smallness of the interaction energy (with respect to the total energy) and to the almost unavoidable approximations (finite dimension of the orbital basis, lack of correlation energy). A good example of this situation is given by the analysis of the so-called 'basis extension effect' between the supermolecule and the separated subsystems (Clementi, 1967; Morokuma and Pedersen, 1968; Liu and McLean, 1973; Daudey *et al.*, 1974a,b; Daudey, 1974). Conversely, such elaborate *ab initio* calculations, when performed with appropriate cautions, may give reference values, which allow to check the simplified approximate formulas for the interaction energy (concerning such *ab initio* calculations, see, for example, the review by Buckingham and Utting (1970); Gilbert and Wahl (1967); the calculations of the He . . . He interactions by Bertoncini and Wahl (1970), Schaeffer III *et al.* (1970). McLaughlin and Schaeffer III (1971), Liu and McLean (1973), Salahub (1974), Daudey *et al.* (1974a), Wormer *et al.* (1975); the calculations on various other small systems by Das and Wahl (1971), Wagner *et al.* (1974), Kochanski (1974, 1975), Wormer and Van der Avoird (1975), Geurts *et al.* (1975), and the thesis of Wormer (1975). As concerns the *ab initio* studies of the hydrogen bond, some recent reviews are: Bratoz (1967), Dreyfus (1970), Diercksen (1971), Kollman and Allen (1972), Joesten and Schaad (1974), Diercksen *et al.* (1975), Allen (1975a,b) and the chapter by Schuster in the present book (an *ab initio* perturbation treatment of the H_2O . . . H_2O interaction has also been made by Daudey (1974b)).

The third important point concerning the derivation of simplified formulas for the calculation of the interaction energy between *large* molecules is the possibility of expressing this energy in terms of contributions from small subunits (into which each molecule is decomposed). Note that the properties attached to these subunits may be transferable from one molecule to another (e.g. bond polarizabilities), or not directly transferable (e.g. atomic net charges). A theoretical basis for performing such an analysis in terms of subunits is displayed by the possibility of representing with reasonable accuracy the ground state of many closed-shell molecules by a single Slater determinant built from *localized* orbitals (e.g. see Diner *et al.* (1970), England *et al.* (1971), and Millié *et al.* (1975), Daudey *et al.* (1975) in the two volumes edited by Chalvet *et al.* (1975, 1976)). This remarkable property may be used first for analysing various molecular quantities (e.g. polarizability or magnetic susceptibility) in terms of local (e.g. bond) contributions (Amos and Musher, 1968; Amos and Yoffe, 1975d; Amos and Crispin, 1975, 1976a), and these decompositions may then be used for deriving simple expressions of the interaction energy (e.g. expressing the polarization and dispersion contributions in terms of bond polarizabilities) (Claverie (1973 Ch. III section 4, 1976), Amos and Crispin (1976a,b,c) and references therein). Similarly, the representation of the total charge distribution of a molecule by a set of local multipole moments (especially charges) has been investigated by Pollak and Rein (1967), Cohan (1969), Rein *et al.* (1972), Rabinowitz and Rein (1972), Rein (1973), Claverie (1973, Ch. III, Section 4.B; 1976), Hall (1973), Tait and Hall (1973), Shipman (1975), Amos *et al.* (1975), Amos and Yoffe (1975c), Amos and Crispin (1976b,c).

The content of the present chapter will be as follows:

In Section II, some basic theoretical developments necessary for the derivation of simplified formulae will be given: the inadequacy at medium- and short-range of the usual Rayleigh–Schrödinger perturbation treatment (which leads to the well-known long-range forces) will be analysed, and the design of 'exchange perturbation' treatments in order to overcome this deficiency will be examined. The expressions obtained by such a treatment (up to second order) will give a basis for the derivation of approximate simplified formulae.

This derivation will be the purpose of Sections III to V. Four terms will be dealt with in detail: the short-range first-order exchange (repulsive) term (both the exact expressions and the 'electron gas' model will be considered) and the three long-range terms; electrostatic (first-order); polarization and dispersion (second-order). Arguments will be presented in favour of expressions of these terms as sums of *local* contributions (atom–atom, atom–bond, bond–bond). Such local expressions will involve not only local but also transferable properties like the bond polarizabilities, but local and non-transferable properties (e.g. atomic net charges) will also occur. It must be emphasized that the simplified formulas suggested by the theoretically exact expressions may display an unsuspected accuracy, which appears rather difficult to prove in a direct deductive way: the representation of the first-order short-range repulsion between atoms by an exponential formula Be^{-br} will be a striking example. The other terms not considered in detail (among them essentially the charge-transfer term) will be briefly discussed (the problem of representing the hydrogen bond in some simple way will be considered in Section VI.A.4). Finally, for an assembly of many molecules, the pair-wise additivity of the interaction energy and the possible deviations from it are of crucial importance, and a brief discussion of this problem will also be presented.

Finally, in Section VI, some applications of the simplified formulas will be described. A short review of the semi-empirical formulas proposed by various authors will first be made. Then, the following domains of applications will be considered: interactions between two molecules, interaction between a molecule and a solid surface, solute–solvent interaction (using either a 'discrete' or a 'continuum' model for the solvent), interactions in crystals. All these applications will concern essentially organic medium-size molecules: biologically oriented applications will be treated in other chapters of this book. Only a short mention will be made of the application of intermolecular-type atom–atom potentials to the intramolecular interactions in macromolecules (e.g. see Liquori (1966, 1969); J. E. Williams *et al.* (1968); Scheraga (1968, 1971); Ramachandran and Sasisekharan (1968); Popov (1975) and references therein; Burgess and Sheraga (1975); Hagler *et al.* (1974)).

We shall not review here the experimental methods giving information about the intermolecular interactions (essentially between atoms or small molecules). We refer the reader interested in this subject to the reviews by Schlier (1969), Leonas (1972), the book edited by Lawley (1975) (these works deal mainly with molecular beam scattering), the series of papers in the part II of the book edited by Hirschfelder (1967c), the review by Ewing (1972) (who considers the study of Van der Waals complexes by electron diffraction, spectroscopy and mass spectrometry) and the review by Certain and Bruch (1972) (in their Section 4.3 they consider the determination of intermolecular forces from thermodynamic, spectroscopic, and molecular scattering data).

Except for Sections IV.D and V.6, the theoretical developments presented in Parts II to V are based, to a large extent, on the content of author's doctoral dissertation (Claverie, 1973). A shortened version concerning essentially the decomposition of the interaction as a sum of local contributions was published recently (Claverie, 1976). Only interactions between ground-state closed-shell atoms or molecules will be considered.

II. BASIC THEORY

The purpose of this section is only to lay out the basic theoretical developments necessary for the derivation of simplified formulas in the next section. More exhaustive reports may be found in the collection of papers edited by Hirschfelder (1967c; see especially Hirschfelder and Meath, p. 3), in the book by Margenau and Kestner (1969); see also the chapter by Buckingham in this book. Nevertheless, these general references do not cover with great detail the 'exchange perturbation' theory (e.g. see Chipman, 1972), which will be considered in the present section. In Section II.A, the usual Rayleigh–Schrödinger perturbation treatment for the long-range interaction will be recalled, and the reason for its inadequacy at medium and short-range (region of the Van der Waals Minimum) will be stated. In Section II.B, we shall consider a possible remedy to this inadequacy: the elaboration of so-called 'exchange' or 'symmetry-adapted' perturbation treatments. In Section II.C, such a treatment (Claverie (1973) Ch. II, Section 4) will be described, and a small number of perturbation terms (altogether common to most 'symmetrized' perturbation treatments) will be retained as a suitable starting point for the derivation of simplified formulas in Section III. By the way, the *ab initio* treatments (Daudey *et al.*, 1974a,b, Daudey, 1974) connected with this perturbation scheme will be briefly mentioned.

A. The Rayleigh–Schrödinger (RS) long-range treatment ('polarization approximation') and its inadequacy at short distance

1. *Outline of the treatment*

This theory has been known for a long time (London, 1937; Margenau, 1939a; Longuet-Higgins, 1956) and has been reviewed in a detailed manner quite recently (Hirschfelder and Meath, p. 3, Buckingham, p. 107, and Dalgarno, p. 143, in Hirschfelder (1967c); Dalgarno and Davison, 1966; Margenau and Kestner 1969). Following Hirschfelder (1967a), the name 'polarization approximation'* is often used nowadays, especially in the papers concerned with the numerous 'exchange' perturbation treatments, to be considered later (Section B). The formulation given by Longuet-Higgins (1956) is used here, with some changes in the notation (Claverie, 1968a).

A bracketed superscript (i) or (m) will label the molecule: thus the Hamiltonian of molecule i will be denoted by $H^{(i)}$, its eigenfunctions by:

$$W_0^{(i)}, W_1^{(i)}, \ldots, W_a^{(i)}, \ldots, W_b^{(i)}, \ldots \tag{2.3}$$

and the corresponding eigenvalues (ordered with increasing values) by:

$$E_0^{(i)}, E_1^{(i)}, \ldots, E_a^{(i)}, \ldots, E_b^{(i)}, \ldots \tag{2.4}$$

The number of electrons of molecule i will be denoted by n_i (we shall restrict ourselves almost always to the interaction between two molecules only, but the extension of the developments to a larger number of molecules does not involve any difficulty of a fundamental nature).

We now introduce as unperturbed Hamiltonian denoted H_0 the sum of the Hamiltonians $H^{(i)}$ of the isolated molecules:

$$\begin{aligned} H_0(1, \ldots, n_1, n_1+1, \ldots, n_1+n_2) \\ = H^{(i)}(1, \ldots, n_1) + H^{(2)}(n_1+1, \ldots, n_1+n_2) \end{aligned} \tag{2.5}$$

(according to the language of linear algebra, $H_0 = H^{(1)} + H^{(2)}$ is a 'direct sum' ($\oplus$), and not an ordinary sum ($+$), since the operators $H^{(1)}$ and $H^{(2)}$ act on functions of *different* subsets of variables: $(1, \ldots, n_1)$ and $(n_1+1, \ldots, n_1+n_2)$. But is is unnecessary for us to use this refinement of notation).

This choice of H_0 as the unperturbed Hamiltonian completely defines the Rayleigh–Schrödinger expansion nowadays considered as the convenient one for the treatment of long-range forces. By the way, it is interesting to note that the first work of Eisenschitz and London (1930) did *not* introduce this RS treatment, but an 'exchange' or 'symmetrized' perturbation treatment, historically the first one of that kind of treatments (which will be studied in the Section B below). But the further papers by London (1930a, b; 1937; 1942) actually dealt with the theory and applications of the long-range interaction energy.

The eigenfunctions of H_0 are simply the products of an eigenfunction of $H^{(1)}$ and an eigenfunction of $H^{(2)}$:

$$|W_0^{(1)}W_0^{(2)}\rangle, \ldots, |W_0^{(1)}W_b^{(2)}\rangle, \ldots, |W_a^{(1)}W_0^{(2)}\rangle, \ldots, |W_a^{(1)}W_b^{(2)}\rangle, \ldots \tag{2.6}$$

and the eigenvalues are just the sums of the corresponding eigenvalues of $H^{(1)}$ and $H^{(2)}$:

$$E_0^{(1)}+E_0^{(2)}, \ldots, E_0^{(1)}+E_b^{(2)}, \ldots, E_a^{(1)}+E_0^{(2)}, \ldots, E_a^{(1)}+E_b^{(2)}, \ldots \tag{2.7}$$

* Confusion should be carefully avoided between this denomination of the long-range RS treatment, and the polarization (also called induction) term, which is only a part of the second-order perturbation term derived in the framework of this RS treatment.

The simplicity of the connection between the spectra of $H^{(1)}$, $H^{(2)}$ and the spectrum of the unperturbed operator H_0 is the essential pleasant feature of the present RS treatment, a feature which will be lost in the 'exchange perturbation' treatments of the next section.

We denote $H(1, \ldots, n_1, n_1+1, \ldots, n_1+n_2)$ the total Hamiltonian of the complex, and we introduce the interaction Hamiltonian V:

$$V = H - H_0 = H - (H^{(1)} + H^{(2)}) \tag{2.8}$$

whose explicit expression is:

$$V = e^2 \left[\sum_{\mu^{(1)}} \sum_{\mu^{(2)}} \frac{z_{\mu^{(1)}} z_{\mu^{(2)}}}{r_{\mu^{(1)}\mu^{(2)}}} - \sum_{\mu^{(1)}} \sum_{j^{(2)}} \frac{z_{\mu^{(1)}}}{r_{\mu^{(1)}j^{(2)}}} - \sum_{j^{(1)}} \sum_{\mu^{(2)}} \frac{z_{\mu^{(2)}}}{r_{j^{(1)}\mu^{(2)}}} + \sum_{j^{(1)}} \sum_{j^{(2)}} \frac{1}{r_{j^{(1)}j^{(2)}}} \right] \tag{2.9}$$

where r denotes the distance between two particles (nuclei or electrons), μ denotes the nuclei, j the electrons and z the nuclear charges; e denotes the elementary charge (1 in atomic units).

Using V as the perturbation operator and applying the Rayleigh–Schrödinger perturbation method to the 'ground state' eigenfunction of H_0, namely the product $W_0^{(1)} W_0^{(2)}$ of the ground state eigenfunctions of $H^{(1)}$ and $H^{(2)}$, we get the following expansion for the perturbation energy:

$$\begin{aligned}
\Delta E^{(RS)} &= \langle W_0^{(1)} W_0^{(2)} | V | W_0^{(1)} W_0^{(2)} \rangle && \text{(electrostatic energy)} \quad \text{first order} \\
&- \sum_a{}'^{(1)} \frac{|\langle W_0^{(1)} W_0^{(2)} | V | W_a^{(1)} W_0^{(2)} \rangle|^2}{E_a^{(1)} - E_0^{(1)}} && \text{(polarization energy of (1) by (2))} \\
&- \sum_b{}'^{(2)} \frac{|\langle W_0^{(1)} W_0^{(2)} | V | W_0^{(1)} W_b^{(2)} \rangle|^2}{E_b^{(2)} - E_0^{(2)}} && \text{(polarization energy of (2) by (1))} \quad \text{second order} \\
&- \sum_a{}'^{(1)} \sum_b{}'^{(2)} \frac{|\langle W_0^{(1)} W_0^{(2)} | V | W_a^{(1)} W_b^{(2)} \rangle|^2}{(E_a^{(1)} + E_b^{(2)}) - (E_0^{(1)} + E_0^{(2)})} && \text{(dispersion energy)} \\
&+ \cdots \text{ higher order terms}
\end{aligned} \tag{2.10}$$

(in the symbol $\sum'^{(m)}$, the superscript (m) refers to the molecule and the means that the label 0 of the ground state is excluded from the summation).

As concerns the physical meaning of the terms of this expansion (which is recalled in the formula (2.10)), we refer to the general texts quoted at the beginning of this Subsection A.1, and to the analysis presented in Sections III to V below in order to derive simplified formulas.

As already indicated, we use here the name 'standard' or 'usual' or 'long-range' Rayleigh–Schrödinger (RS) perturbation treatment for the procedure described above; there is no exact agreement between various authors as concerns the terminology: thus Hirschfelder (1976a) used the name 'polarization approximation'. Anyway, it is necessary to avoid any confusion between this procedure and other applications of the Rayleigh–Schrödinger method, which use unperturbed Hamiltonians *different* from $(H^{(1)}+H^{(2)})$ (see Certain *et al.* (1968) and references therein: Sternheimer, 1954; Corinaldesi, 1962, 1963; Corinaldesi and Lin, 1963; Hirschfelder *et al.* 1964; Jansen; 1967); the 'symmetrized' perturbation treatment in Section B below will also make use of the RS method, with a different zero-order starting point.

2. Inadequacy at intermediate and short distances

We only summarize here the results of the detailed analysis given in previous works (Claverie, 1971a, 1973, Ch. I).

At long distance (more than 8 Å between atoms, say), the above defined treatment has always been used with success for evaluating the interaction energy.

The terms of the expansion (2.10) may be expressed approximately as sums of R^{-n} terms (asymptotic multipole expansion). For example, between neutral dipolar molecules, the electrostatic term varies at long distance like R^{-3}, while the second order terms vary like R^{-6} and are *negative* (attraction). Between atoms (rare gases), the R^{-6} dispersion term dominates, and even at medium-range (region of the Van der Waals minimum) it is much larger than the third-order term (see the example of the interactions between two helium atoms according to Stogryn (1971) at the beginning of Section III). Therefore no minimum would be obtained at the correct distance by using the standard RS procedure (at least limited at second-order), since only one dominant attractive term appears. But this is only the ascertaining of a practical inadequacy, and neither and explanation nor a proof (it could be argued that the necessary repulsive terms are embedded in numerous high-order terms of the perturbation expansion; this would still imply the rejection of the usual RS procedure at medium-range, but for a purely *practical* and not theoretical reason).

The theoretical argument commonly given against the use of the standard RS treatment at medium-range was the fact that the basis (2.6) of the eigenfunctions of $(H^{(1)}+H^{(2)})$ is made from functions which do not satisfy the Pauli principle for the (n_1+n_2) electrons of the complex. Indeed the functions $W_a^{(1)}(1,\ldots,n_1)$ are antisymmetric with respect to the exchange of electron labels $(1,\ldots,n_1)$, and similarly the functions $W_b^{(2)}(n_1+1,\ldots,n_1+n_2)$ with respect to the permutations of electron labels $(n_1+1,\ldots,n_1+n_2)$; their products $W_a^{(1)}W_b^{(2)}$ have the same properties, but have no special symmetry property with respect the the 'inter-system' exchanges (i.e. those exchanging electron labels between the sets $(1,\ldots,n_1)$ and $(n_1+1,\ldots,n_1+n_2)$). Then it is argued that, 'at short enough distances', the Pauli principle must be satisfied for all electron exchanges, and therefore the set of *antisymmetrized* products $A\ (W_a^{(1)}W_b^{(2)})$ (or at least the antisymmetrized product $A(W_0^{(1)}W_0^{(2)})$ of the ground states, see Murrell and Shaw (1967a)) should be used for an appropriate treatment of the complex at such 'short distances'.

But this argument against the RS long-range treatment is not convincing. Indeed, for getting a final state of given permutational symmetry (i.e. belonging to some irreducible representation of the permutation group $\mathscr{S}_n$ for n objects), it is certainly *sufficient* to use from the beginning a basis of symmetrized functions, but this is *not at all necessary.* It is only necessary that the basis spans a vector space *containing* the symmetric functions. Now, this condition is precisely fulfilled in our present case: it may easily be shown that the basis of *all* simple products $\{W_a^{(1)}W_b^{(2)}\}$ (where the bases $\{W_a^{(1)}\}$ and $\{W_b^{(2)}\}$ are complete for the antisymmetric subspaces of functions corresponding to the subsystems) actually spans a subspace (of the space of n-particle functions) which is an invariant subspace of $(H_0+\lambda V)$ whatever λ, and which *contains* the 'smaller' subspace of the antisymmetric n-particle functions (Claverie, 1973, Ch. I, Section 1.B and Appendix 1): see Appendix B of the present chapter.

A somewhat refined version of the argument (Musher, 1967, Section IV-B; Murrel and Shaw 1967a) considers the spins of the functions $W_a^{(1)}$, $W_b^{(2)}$ and their relationship with the spin desired for the final exact function, but this argument also was proved to be non-convincing. The essential point here is the 'exchange degeneracy'. i.e. the fact that the irreducible representations of the symmetric group $\mathscr{S}_n$ have a dimension greater than 1 (except the totally symmetric and antisymmetric ones). As a consequence, the RS perturbation treatment could conceivably give a correct energy (eigenvalue) with a perturbed function belonging to the correct 'exchange degenerate' subspace (and the usual eigenfunction, chosen in that subspace as being totally antisymmetric with respect to simultaneous *space and spin* exchanges, would simply be recovered by applying the total anti-symmetrizer to the function resulting from the perturbation treatment).

To sum up, the arguments based on the 'deficiency' of the basis of simple products must be rejected, and Hirshfelder (1966, Section XIII, pp. 28–29) was right in stating that the inadequacy of the usual RS treatment at short distances was not properly understood at that time.

As we have seen, the molecular interaction problem exhibits a rather special feature (as emphasized by Hirschfelder (1967b)): the symmetry of the unperturbed Hamiltonian $H^{(1)}+H^{(2)}$ (with respect to electron permutations) is *lower* than the symmetry of the perturbed Hamiltonian H (the symmetry group for $H^{(1)}+H^{(2)}$ is the direct product $\mathscr{S}_{n_1}\otimes\mathscr{S}_{n_2}$, and this is a subgroup of $\mathscr{S}_{(n_1+n_2)}$, which is the symmetry group for H). This situation is just the opposite of the more familiar one, where the symmetry of the *unperturbed* Hamiltonian is *lowered* (and not increased) by the perturbation V.

This 'accidental' increase of symmetry of $H_0+\lambda V$ for $\lambda=1$ was considered as a possible cause for the limitation of the convergence of the perturbation expansion, so that the following type of argument was proposed for explaining the inadequacy of the standard RS treatment: it would not be convergent, but only asymptotically convergent (Hirschfelder, 1967a), the best possible result (of the asymptotic series) becoming worse and worse as the intermolecular distance decreases. This argument was based upon results of Robinson (1961) for the one-dimensional delta-function model of H_2^+, according to which the convergence radius of the perturbation expansion of $E(\lambda)$ (the eigenvalue of $H_0+\lambda V$) would be smaller than 1, except for very small internuclear distance R: such a result would actually imply the inadequacy of the RS treatment for getting the energy $E(1)$ of H_0+V. But is was later proved that Robinson used a too poor lower bound, and that a better lower bound was *larger* than 1 for any R (Claverie, 1969); actually, this bound was later proved to be the convergence radius itself (Certain and Byers Brown, 1972), while Ahlrichs and Claverie (1972) studied the difference between the convergence radii for the eigenvalue and the eigenfunction respectively. Later on, other examples were given by Ahlrichs (1973a) where an accidental increase of symmetry does not limit the convergence of the perturbation expansion (the increased symmetry occurs for $\lambda=1$ while the convergence radius is larger than 1). Thus the argument against the convergence of the RS treatment also loses its foundation.

Let us therefore reexamine the problem from the beginning. It is necessary to distinguish carefully *two* questions when faced with any perturbation treatment. When the so-called perturbation parameter λ varies from $-\infty$ to $+\infty$, every eigenvalue of the operator $H(\lambda)$ (in the RS treatment $H(\lambda)=H_0+\lambda V$) varies, as represented on Figure 2.1, and we are interested

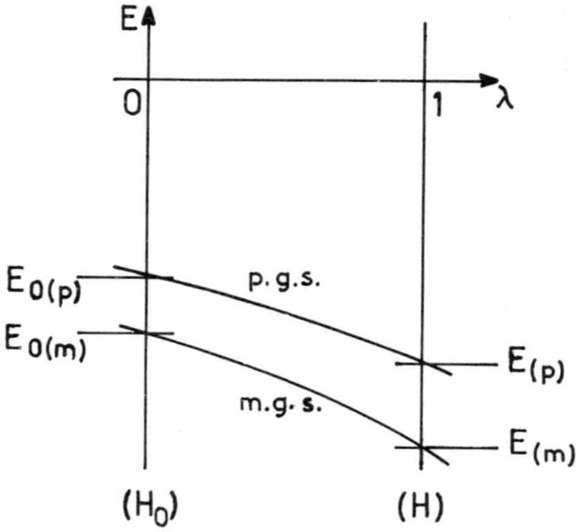

Figure 2.1. The connections between the eigenvalues of $H(0)(\equiv H_0)$ and $H(1)(\equiv H)$ which are established by the curves $E_i(\lambda)$ (eigenvalues of $H(\lambda)=H_0+\lambda V$)

in obtaining some eigenvalue of $H(1)$, starting from an appropriate eigenvalue of $H(0)$. The RS perturbation expansion tries to solve this problem by using the Taylor expansion of the functions $E_i(\lambda)$ around the value $\lambda = 0$. the *two* relevant questions now appear clearly:

(1) How are the eigenvalues of $H(1)$ connected with those of $H(0)$ through the curves $E_i(\lambda)$? It is actually trivial that a given eigenvalue of $H(1)$, say $E_b^{(1)}$, may be obtained only if we start from some eigenvalue $E_a^{(0)}$ of $H(0)$ connected with $E_b^{(1)}$ through someone among the curves $E_i(\lambda)$ (i.e. $E_i(0) = E_a^{(0)}$ and $E_i(1) = E_b^{(1)}$). Note that, owing to the possible degeneracies of $H(1)$, we may have several curves passing through the same $E_b^{(1)}$, and correspondingly several possible starting points among the eigenvalues of $H(0)$.

(2) What is the value of the convergence radius of the Taylor expansion around $\lambda = 0$ for the chosen branch $E_i(\lambda)$, or, more precisely, is this convergence radius larger than 1?

It must be emphasized that the first question is of more general importance than the second one, because this second one may be circumvented, at least in principle, by resorting to some method having a range of validity less limited than the Taylor expansion (let us quote the perturbation theory in closed form described by Löwdin (1962), the Fredholm perturbation formula (Morse et Feshbach, 1953, Vol. II, Ch. 9, Section 9.1), and the Padé approximants method (Baker, 1965; Baker and Gammel; 1970; Goscinski and Brändas, 1971, Longman 1971). On the contrary, the first question always remains perfectly relevant, independently of the method used for obtaining the functions $E_i(\lambda)$. It must be emphasized that this question of connection between H_0 and H when λ varies from 0 to 1 is conceptually *independent* from the variation of the intermolecular distance R (or, more generally, of the relative positions of the two molecules), since the variation of λ is a mathematical procedure performed for fixed relative positions of the molecules. The *practical* (not conceptual) differences occurring between the case of large and short intermolecular distances will be discussed *afterwards* (see Figure 2.3 below).

This question of connection between the eigenvalues of H_0 and H is so easily solved in most applications of perturbation theory, that it is often not asked at all. This problem has sometimes been considered in many-body theory: thus in the theory of superconductivity, it turned out that the usual unperturbed ground state used in many-body problems was not connected with the exact (perturbed) ground state (March, Young and Sampanthar, 1967, Ch. 4, Section 4.1, p. 68). Now, the problem of connection is also non-trivial in the theory of intermolecular interaction, and here lies the deep *conceptual* ground for the failure of the RS treatment at short distances (Claverie, 1971a; 1973, Ch. I). A *practical* reason, namely the slowness of convergence (as advocated by Hirschfelder, 1967a) may also exist, and will be discussed afterwards.

We may now summarize the results obtained previously (Claverie, 1971a, 1973, Ch. I). The RS treatment works in the space of all simple products $W_a^{(1)}W_b^{(2)}$, which *contains* the subspace of the totally antisymmetric (with respect to space–spin exchange) functions, and this inclusion is a *strict* one, except for $n_1 = n_2 = 1$ (i.e. there is in the simple product space a non-empty subspace of functions orthogonal to the subspace of totally antisymmetric functions). The lowest eigenvalue in the totally antisymmetric subspace gives the *physical ground state* (p.g.s.) of the complex, since, according to the Pauli principle, the other functions are not physically acceptable. But, the Hamiltonian H has other eigenvalues in the orthogonal 'non-physical' subspace, and the lowest of all (physical and non-physical) eigenvalues *just belongs to these non-physical ones*; we shall refer to it as the '*mathematical ground state*' (m.g.s.). This is seen by considering explicitly the space parts of the functions. For $n_1 + n_2 = 3$ or 4 electrons, the property is easily obtained, because the lowest of all eigenvalues of H corresponds to a totally symmetric space function (which, up to 4 electrons, belongs to the space of the simple products of subsystem functions), and for more than two electrons, this function is forbidden by the Pauli principle (indeed, no spin function may be combined with such a function in order to give a totally antisymmetric space–spin function). When at least one of the subsystems has more than two electrons, the argument becomes much more involved, because the functions $\{W_a\}$ of this subsystem are themselves restricted by the Pauli principle, so that the space of simple products $\{W_a^{(1)}W_b^{(2)}\}$ does no more contain the totally symmetric space functions for $(n_1 + n_2)$ electrons; but, roughly speaking, it still contains space functions which are more symmetrical than those of the (space–spin) antisymmetric subspace, and the main idea of the argument then consists in proving that the lowest eigenvalue in the simple product space corresponds to one of these more symmetrical (and therefore 'physically' forbidden) functions.

Now, the unperturbed function of the RS treatment, namely the product $W_0^{(1)}W_0^{(2)}$ is the lowest eigenstate (m.g.s.) of the operator $H_0 = H^{(1)} + H^{(2)}$, and it is easily proven that, then λ varies, this m.g.s. of H_0 is connected with the m.g.s. of H in the simple product space spanned by $\{W_a^{(1)}W_b^{(2)}\}$: up to 4 electrons, the non-degeneracy of the m.g.s. may be used for the proof, and in the general case the non-crossing rule (see its recent thorough proof by Razi Naqvi and Byers Brown (1972)) is used. Since the m.g.s. of H (in the simple product space) is *not* the p.g.s. (except for $n_1 + n_2 = 2$), the standard RS perturbation treatment clearly appears inadequate, independently of any convergence question. But this inadequacy is a matter of unsatisfactory connection between the eigenstates of H_0 and those of H: there exist some curve(s) $E(\lambda)$ passing through the p.g.s. of H for $\lambda = 1$ (see curve p.g.s. on Figure 2.1), and these curves pass, for $\lambda = 0$, through some state(s) of H_0, which would be theoretically convenient unperturbed states for getting the p.g.s. of H (despite the fact that these unperturbed functions are simple products as well as $W_0^{(1)}W_0^{(2)}$ and do not satisfy the antisymmetry principle for (n_1+n_2) particles). But the perturbation treatment starting from $E_{0(p)}$ (see Figure 2.1) would give as perturbation energy $E-E_{0(p)}$, and not the difference which is interesting for the intermolecular interaction problem, namely $E-E_0$. It is impossible to get directly this difference by using $H_0+\lambda V$ as 'interpolation Hamiltonian', owing to the lack of connection between the two levels of interest.

The convergence of the usual RS treatment ('polarization approximation') towards the lowest possible state has actually been verified for the H_2^+ case (interaction $H\ldots H^+$) by Schulman (1972) for a model problem (a basis restricted to the two 1s orbitals centred at the two protons), and by Chalasinski, Jeziorski and Szalewicz (1977) who performed a highly accurate calculation (this important work will be discussed below in Sections II.B.1 and III.A).

It is instructive to consider the cases of two 1-electron systems (e.g. hydrogen atoms) and two 2-electron systems (e.g. helium atoms) in the framework of the molecular orbital approximation, as depicted in Figure 2.2. In the first case, the two depicted states are physically allowed (because we have only two electrons); the lowest one (two electrons in the lowest orbital) is space-symmetric and spin-antisymmetric (hence the name: singlet), and the second (one electron in each of the lowest two orbitals) is space-antisymmetric (and spin–symmetric: triplet; there would also be a singlet state built with the same two orbitals, but it is of no interest here). In the case of the two 2-electron systems, the lowest state (m.g.s.) corresponds to the four electrons in the lowest orbital, and is no more physically allowed, while the higher state (p.g.s.) corresponds to two electrons in the lowest orbital and two in the next one. In both cases, the standard RS treatment would connect the lowest state of $H^{(1)}+H^{(2)}$ to the lowest state of H, namely the singlet state of the two-electron complex, and the m.g.s. of the four-electron complex respectively. The problem of getting the p.g.s. of the four-electron complex appears equivalent to getting the lowest triplet state (instead of the lowest singlet) of the two-electron complex.

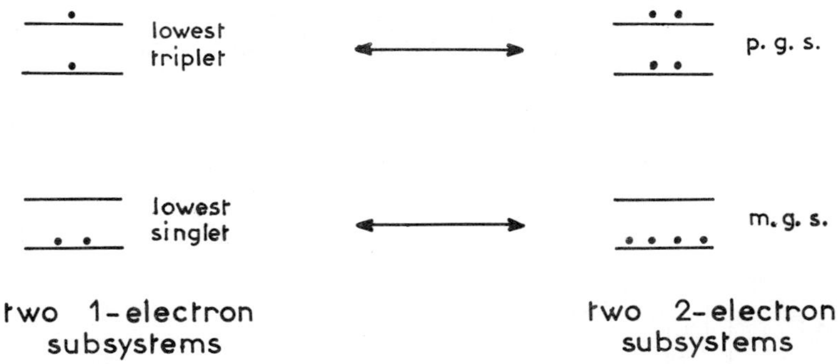

Figure 2.2. The analogy between the interaction of two 1-electron subsystems and that of two 2-electron subsystems (m.g.s.: mathematical ground state; p.g.s.: physical ground state)

3. Behaviour of the interaction energy as a function of distance: the practical usefulness of the RS treatment at long distances

The non-connection between the ground state $W_0^{(1)}W_0^{(2)}$ of $H^{(1)}+H^{(2)}$ and the p.g.s. of H is a *qualitative* property, and as such independent of the relative positions of the molecules. But let us now consider the variation of the energies of the m.g.s. and the p.g.s. of the Hamiltonian H as functions of the intermolecular distance R. The m.g.s. energy curve, of course, always remains the lowest one, and, as $R\to\infty$, both the m.g.s. and p.g.s. energy curves tend to $E_0=E_0^{(1)}+E_0^{(2)}$ (see Figure 2.3). The crucial point now is to compare the 'physical' interaction energy

$$\Delta E = E - E_0 \tag{2.11}$$

and the 'mathematical' one

$$\Delta E_{\rm m} = E_{\rm m} - E_0 \tag{2.12}$$

For the interaction between two hydrogen atoms, the analytical asymptotic behaviour of the difference $\Delta E-\Delta E_{\rm m}$ (namely the singlet–triplet separation) is known (Herring and Flicker (1964) use the method of Gor'kov and Pitaevskii (1964) and correct a numerical factor obtained by these authors): this difference is an exponentially decreasing function of R.

$$\Delta E\,(\text{Triplet}) - \Delta E(\text{Singlet}) = 1{\cdot}641\; R^{5/2}{\rm e}^{-2R} + O(R^2{\rm e}^{-2R}) \tag{2.13}$$

The validity for arbitrary systems of this property (exponential decrease of the energy difference between the mathematical and the physical ground states) has been proved quite recently by Ahlrichs (1976).

This exponential decrease of the energy difference between p.g.s. and m.g.s. is related with the exponential decrease of the eigenfunctions $W_0^{(1)}$

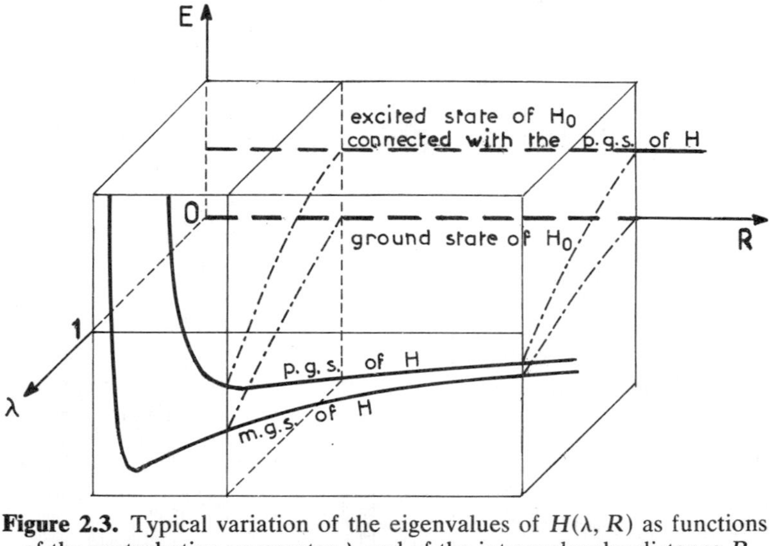

Figure 2.3. Typical variation of the eigenvalues of $H(\lambda, R)$ as functions of the perturbation parameter λ and of the intermolecular distance R

and $W_0^{(2)}$ of the isolated subsystems at large distances, a property whose general validity was also proven rather recently (Ahlrichs, 1972b, 1973a, 1973b; O'Connor, 1973). But if we remember that the 'mathematical' interaction energy ΔE_m (given by the usual RS treatment) decreases much more slowly with distance, namely at least like the R^{-6} dispersion term, we can understand the *practical* usefulness of this RS treatment at large distances: the difference $\Delta E - \Delta E_m$ indeed becomes negligible with respect to ΔE_m at large R, and it therefore does not matter that the exact perturbation energy would be ΔE_m instead of ΔE (the perturbation terms beyond second order, which are almost always neglected, become themselves much larger than an exponentially decreasing difference $\Delta E - \Delta E_m$, since they themselves vary like inverse powers of R). In summary, the asymptotic $1/R$ expansion is valid for the mathematical ground state and the physical ground state as well (see Ahlrichs, 1976).

On the contrary, when R decreases, the separation between m.g.s. and p.g.s. increases and becomes non negligible with respect to the interaction energies ΔE_m and ΔE. The p.g.s. energy reaches the Van der Waals minimum at a distance where the m.g.s. energy still steadily decreases, in perfect agreement with the lack of a convenient repulsive term in the usual RS treatment. The m.g.s. energy would reach a deep minimum for a short distance, of the order of magnitude of a chemical bond distance, as exemplified by the hydrogen–hydrogen interaction (the m.g.s. curve corresponds to the singlet state, i.e. the bonding state of the H_2 molecule, whose minimum corresponds to the equilibrium bond distance of this molecule). It would be of no use to consider the terms of the usual RS treatment beyond second order, since they would at most bring the interaction energy closer to the 'unphysical' too large value ΔE (see also the accurate calculations concerning H_2^+ by Chalasinski *et al.*, 1977). It may be mentioned that this analysis of the RS treatment (theoretical inadequacy and practical adequacy at large distance) perfectly holds for the treatment of a many-molecule system as a gas of 'composite particles' proposed by Girardeau (1963, 1970): indeed, as shown by Claverie (1973, Appendix 5 of Ch. I), this treatment is essentially equivalent to the usual (long-range) RS treatment applied to the many-molecule system.

After having recalled the fundamental ground for the inadequacy of the standard RS treatment (with $H^{(1)} + H^{(2)}$ as unperturbed Hamiltonian) and the reason for its practical usefulness at large distance, we shall consider in the next sections B and C the 'symmetrized' or 'exchange' perturbation treatments as a possible method for getting the physical interaction energy ΔE.

B. The 'symmetrized' or 'exchange' perturbation treatments for the medium-range interactions

This section B is based essentially on the chapter II of author's doctoral dissertation (Claverie, 1973).

1. Introduction

As we have seen in the previous section, the long-range Rayleigh–Schrödinger perturbation treatment would give (if it converges for the value $\lambda = 1$ of the perturbation parameter λ), instead of the 'physical' interaction energy ΔE (corresponding to the physical ground state of the complex), some 'mathematical' interaction energy ΔE_{m} corresponding to some mathematical ground state of the complex). We could therefore write

$$\Delta E = \Delta E_{\mathrm{m}} + (\Delta E - \Delta E_{\mathrm{m}}) \tag{2.14}$$

and try to evaluate, besides ΔE_{m}, the difference $(\Delta E - \Delta E_{\mathrm{m}})$ (for which the name 'complete repulsion energy' could be proposed) in order to add it to ΔE_{m} calculated according to the standard RS treatment. We use the word 'complete' in order to distinguish $(\Delta E - \Delta E_{\mathrm{m}})$ from the difference $(\Delta E - \Delta E_{(2)})$ introduced below as an 'effective' repulsion ($\Delta E_{(2)}$ being the approximation of ΔE_{m} obtained at second order of the RS standard treatment.) But it may be feared that ΔE_{m} would be obtained with high accuracy only at a high order of perturbation. Thus let us consider the interaction between two hydrogen atoms with eigenfunctions $\{a_i(1)\}$ and $\{b_j(2)\}$ respectively; the unperturbed function of the complex in the RS treatment is $a_0(1)b_0(2)$, while the perturbed function to be obtained would be space-symmetric (see Section A above), i.e. having the symmetry of $a_0(1)b_0(2) + b_0(1)a_0(2)$. Now, the expansion of the term $b_0(1)a_0(2)$ in the basis of simple products could converge rather slowly: this term is indeed the product of the expansions

$$b_0(1) = \sum_i \langle b_0 \mid a_i \rangle a_i(1)$$

and

$$a_0(2) = \sum_j \langle a_0 \mid b_j \rangle b_j(2)$$

which at intermolecular equilibrium distances involve a large number of individually rather small terms.

Moreover it would still remain to evaluate the difference $\Delta E - \Delta E_{\mathrm{m}}$ (the singlet–triplet separation in the case of two hydrogen atoms).

Thus, in order to get simplified analytical formulae, we shall rather limit ourselves to the second-order RS energy, thus using $\Delta E_{(2)} = \varepsilon_{(1)} + \varepsilon_{(2)}$, and we shall try to evaluate the difference $\Delta E - \Delta E_{(2)}$, which could be named 'effective repulsion energy'. The recent calculations on H_2 (Kolos, 1974; Kolos and Wolniewicz, 1974) and especially on H_2^+ (Chalasinski, Jeziorski and Szalewicz, 1977) indeed show that, at medium- and long-range (i.e. distances of the order of magnitude of the equilibrium distance or larger), the first terms of the RS expansion give quite accurately $(\Delta E + \Delta E_{\mathrm{m}})/2$ (which is traditionally called the Coulomb part of the interaction energy in these systems): thus, for H_2^+, at $R = 12{\cdot}5$ a.u., already $\Delta E_{(2)}$ differs from $(\Delta E + \Delta E_m)/2$ by -1 per cent only (and $\Delta E_{(4)}$ by only $0{\cdot}002$ per cent); but the remaining terms of the RS expansion, which should therefore give $(\Delta E_m - \Delta E)/2$ in order to finally reach ΔE_m, converge extremely slowly. It is only at quite short distances (e.g. 3 a.u. for H_2^+) that the RS expansion would converge better to ΔE_m. This behaviour is in agreement with the analytical results obtained for the δ-function model of H_2^+, namely a convergence radius (for the energy) decreasing from $+\infty$ to 1 as R increases from 0 to $+\infty$, so that, for the value $\lambda = 1$ of the perturbation parameter, the convergence of the series must be good when R is small and become worse and worse as R increases (Claverie, 1969; Certain and Byers Brown, 1972; Ahlrichs and Claverie, 1972).

To sum up: in the region of interest for us (namely medium- and long-range), there appears to be (according to the analysis of Chalasinski *et al.* (1977) for the H_2^+ case) a kind of '*intermediate*' convergence of the RS expansion towards the mean value $(\Delta E + \Delta E_m)/2$ of the mathematical and physical ground states. By intermediate convergence, we mean that terms ε_n^{RS}, of the RS expansion at first decrease rapidly (beyond ε_2^{RS}), thus seeming to converge towards some *intermediate sum*, and afterwards the higher-order ε_n^{RS} remain very small with respect to the intermediate sum, but decrease *extremely slowly*, so that the total sum including these remaining terms would converge extremely slowly towards a *complete sum* (ΔE_m in our case) significantly different from the intermediate sum.

A simple example of this behaviour is displayed by the sum of two geometric series with very different convergence rates but with equal sums: we take as the general term of our series $u_n(\lambda) = A_1(\lambda/\rho_1)^n + A_2(\lambda/\rho_2)^n$, with $\rho_1 > 1$, $\rho_2 = 1 + \eta(\eta \ll 1)$; for $\lambda = 1$, the sums of the two subseries A_1/ρ_1^n and A_2/ρ_2^n are respectively $S_1 = A_1/(1 - 1/\rho_1)$ and $S_2 = A_2/(1 - 1/\rho_2) = A_2\rho_2/(\rho_2 - 1) = A_2(1+\eta)/\eta$. By choosing $A_2/\eta = A_1\rho_1/(\rho_1 - 1)$, we shall have the desired behaviour: the first subseries A_1/ρ_1^n practically converges towards S_1 with a small number of terms, while the corresponding terms A_2/ρ_2^n $A_2 \sim \eta$ of the second subseries remain very small with respect to S_1, and a very large number of terms of this second subseries will be necessary for getting the corresponding sum S_2 with a sufficient accuracy.

Returning to the RS expansion ('polarization approximation') for the intermolecular interaction, it appears possible, even by taking $\Delta E_{(2)}$ only, to get the *intermediate* sum $(\Delta E + \Delta E_m)/2$ with an accuracy of about 3 per cent for H . . . H: $-8{\cdot}421\ \text{cm}^{-1}$, instead of $-8{\cdot}274\ \text{cm}^{-1}$ at $R = 8$ a.u. (Kolos, 1974). Consequently, at such distances the 'effective' repulsion energy which must be added to this value is $(\Delta E - \Delta E_m)/2$, i.e. half the 'complete' separation $\Delta E - \Delta E_m$.

Instead of trying to evaluate separately this difference $\Delta E - \Delta E_m$, it appears desirable to devise some method for getting directly ΔE by means of a perturbation expansion, which must, at large distance, give the same asymptotic expansion according to the powers of $1/R$ as ΔE_m (or its low order approximations) does, since $\Delta E - \Delta E_m = O(e^{-R})$ as $R \to \infty$. These treatments are based essentially on the use of functions having the correct symmetry associated to the complex, instead of the simple product functions (in our intermolecular problem: use of the totally antisymmetrized functions $A(W_a^{(1)}W_b^{(2)})$ instead of the products $W_a^{(1)}W_b^{(2)}$). These treatments have often been referred to as 'exchange perturbation', or more explicitly as perturbation calculation of the 'exchange interaction'. This is only a way of speaking, since, as emphasized by many authors (e.g. Dushman and Seitz, 1937, p. 238; Coulson, 1953, Ch. V, Section 2, p. 113) there are no peculiar physical 'exchange forces' or 'exchange energy' but only a rearrangement of the eigenfunction either in a way which lowers the energy (m.g.s. of the complex, corresponding to the singlet of H . . . H) or in a way which raises it (p.g.s. of the complex, corresponding to the triplet of H . . . H). This appears clearly from the study of the standard RS treatment, which was more or less considered as not involving 'exchange effects', while the study of Section II.A shows that this treatment is theoretically able to the m.g.s. energy (and actually does so practically at short enough distances according to Chalasinski *et al.* (1977)). Actually, the common basic feature to these exchange treatments, as mentioned above, is to introduce from the beginning (i.e. already for the unperturbed function) the symmetry requirement which holds for the exact function, namely in our case the antisymmetry with respect to all electron permutations. In the H_2^+ problem, this requirement is the geometrical symmetry of the (electronic) function with respect to the symmetry plane (perpendicular to the segment joining the nuclei in its middle). Thus we prefer to speak about the 'treatment of symmetry properties in perturbation theory' (Löwdin, 1968a,b), or 'symmetry–adapted perturbation theory' (Klein, 1971), or more briefly 'symmetrized perturbation'.

It is rather remarkable that the early paper by Eisenschitz and London (1930), always quoted as the original work on the interaction between atoms or molecules (especially as concerns the long-range dispersion term) was actually the first of the symmetrized perturbation treatments: the Part I (pp. 491–508) of the paper is devoted to this problem; at that time, Eisenschitz and London used neither Slater determinants nor antisymmetrization operators, so that they resorted to explicit group-theoretical techniques for taking into account the requirement of total (space–spin) antisymmetry. Thus, Van der Avoird (1967a) rewrote their treatment in a modern form, much more compact and easy to read. Under this form, it may be shown (Chipman, 1972, Section 5.2, p. 58; Claverie, 1973, Ch. II, Section 3, p. 106) that the Eisenschitz–London (EL) treatment is identical with the HAV treatment of Hirschfelder (1967b), which is itself equivalent to the Van der Avoird (1967b) treatment, as shown by Johnson and Epstein (1968). This EL–HAV treatment gives an underestimated long-range second order interaction (and certainly also underestimated higher-order terms). This fact was observed in computations (Certain, Hirshfelder, Kolos and Wolniewicz, 1968), proved for the case of finite bases (Claverie, 1973 Ch. II, Section 3–C, p. 114), and appears true also for exact calculations (Certain, 1968; Chipman, 1972, p. 126 and Appendix D). The inadequacy of the EL-HAV treatment was also studied by Basilevsky and Berenfeld (1972a). It is only in Part II of their paper (pp. 509–524) that Eisenschitz and London propose an 'asymptotic form' for the long-range interaction energy, by replacing their treatment of Part I by the treatment nowadays known as the standard RS treatment (or polarization approximation: Hirschfelder, 1967a) studied in Section II.A, apparently without being aware of the non-equivalence at long range of these two treatments (non-equivalence which was recognized only recently, as just emphasized). The other works of London (1930a,b; 1937; 1942) were devoted to the long-range interaction only. Other authors tried to complete this interaction in the equilibrium range by so-called exchange terms, essentially the first-order Heitler–London term (Slater, 1928; Gentile, 1930; N. Rosen, 1931; Slater and Kirkwood, 1931; Bleick and Mayer, 1934; P. Rosen, 1950), and Margenau (1939b) evaluated a second-order exchange term. The Heitler–London energy is defined as the expectation value:

$$E_1^{\mathrm{HL}} = \frac{\langle AW_0^{(1)}W_0^{(2)}|\, H\, |AW_0^{(1)}W_0^{(2)}\rangle}{\langle AW_0^{(1)}W_0^{(2)}\,|\, AW_0^{(1)}W_0^{(2)}\rangle}$$

The corresponding first-order interaction energy is: $\varepsilon_1^{\mathrm{HL}} \equiv \Delta E_1^{\mathrm{HL}} = E_1^{\mathrm{HL}} - (E_0^{(1)} + E_0^{(2)})$, and the exchange part of it is: $\Delta E_1^{\mathrm{exch}} = \Delta E_1^{\mathrm{HL}} - \Delta E_1^{\mathrm{RS}}$ (see formulas (2.95, 2.96, 2.97) in Section II.B.3.b below).

But the early attempt of Eisenschitz and London, namely deriving a full symmetrized perturbation treatment, was not undertaken again before the years 1965. At that time, renewed interest in this subject appeared, resulting in a sudden increase in the number of publications: besides the works previous to 1960 quoted by Hirschfelder (1967a; references (1) to (7)), we may quote Carr (1963), Corinaldesi (1962, 1963, 1964a,b,c), Corinaldesi and Lin (1963), Corinaldesi *et al.* (1963), Murrell, Randić and Williams (1965), Murrell and Shaw (1967a), Salem (1965), Musher and Salem (1966), Hirschfelder and Silbey (1966), Musher (1967), Musher and Amos (1967), Amos and Musher (1967), Van der Avoird (1967a,b,c; 1968), Jansen (1967), Jansen and Lombardi (1967), Ritchie (1968a,b), Byers Brown (1968a), Amos and Musher (1969) and finally Micha (1968) (who considered the case where the non-interacting (unperturbed) system is degenerate). Comparative studies of these methods were given by Hirschfelder (1967a,b), Johnson and Epstein (1968), Epstein and Johnson (1968), Certain, Hirschfelder, Kolos, and Wolniewicz (1968), Hirschfelder and Certain (1968a,b), Certain, Hirschfelder, and Epstein (1969), Certain (1968) and Damburg and Propin (1971). More general formalisms including several previous treatments as special cases were also proposed: Lekkerkerker and Laidlaw (1970a,b); Laidlaw *et al.* (1971); Amos (1970); Löwdin (1968a,b; 1971). New developments followed: Sanders (1969) (following Hirschfelder and Silbey, 1966); Certain and Hirschfelder (1970a,b), Certain, Dion and Hirschfelder (1970); Certain, Hirschfelder and Dion (1970);

Hirschfelder (1971); Laughlin and Amos (1971); the spin-free formulation by Matsen and Junker (1971); Klein (1971; 1972); Klein and Seitz (1973); Goscinski and Brändas (1968), Brändas and Goscinki (1970). The works of Byers Brown (1968b); Byers Brown and Power (1970) (where references to previous works, especially those of Bingel, may be found) Ahlrichs (1972a); Certain and Merkel (1975a,b) deal with very short-range interatomic interactions ($R < 1$ a.u.), where the unperturbed state is taken to be the united atom (see also the Section 4.2.3 in the review by Certain and Bruch (1972)). This subject is outside the scope of the present work. An important review of several methods together with their numerical application to H_2^+ was made by Chipman (1972); see also Certain and Bruch (1972, Section 4.2.2), Chipman *et al.* (1973); Chipman and Hirschfelder (1973). Methods of the many-body theory were applied to the interaction problem by Dutta *et al.* (1970, 1971), Boehm and Yaris (1971) and Kvasnička *et al.* (1974); in this respect, it may be mentioned that the many-electron theory of Sinanoglu (1962a,b; 1964; 1965) was applied essentially to the He ... He problem (Kestner and Sinanoglu, 1966; Kestner, 1966). Piela (1970, 1971), Jeziorski and Piela (1972) and Piela and Jeziorski (1972), following Jansen (1967) and Byers Brown (1968a), also derived new treatments and applied them to the H ... H interaction (lowest singlet and lowest triplet states); their results, together with those of Certain, Hirschfelder, Kolos and Wolniewicz (1968), and those of Chipman (1972) show that some (but not all) of the numerous symmetrized perturbation treatments proposed thus far appear indeed able to give satisfactory results when compared with the accurate energies available for the small systems involved (H_2^+ and H_2). Other symmetrized perturbation treatments were proposed by Basilevsky and Berenfeld (1972a), Farberov *et al.* (1972), Rumyantsev (1973), Berrondo (1973), Peierls (1973), Atalay *et al.* (1973), Gouyet (1973, 1974), Kochanski and Gouyet (1975a,b) who used biorthogonalized orbitals, Daudey *et al.* (1974a,b), Daudey (1974) whose work is partly based upon the theoretical analysis of Claverie (1973, Ch. II, Section 4-D), Amos and Yoffe (1975a,b), Snook and Spurling (1975) who essentially apply the Murrell and Shaw (1967a) formulas up to second-order for the He ... He interaction and evaluate the coefficients of the long-range expansion of the dispersion term $C_6/R^6 + C_8/R^8 + C_{10}/R_{10}$ (it is quite interesting that they found their second-order exchange term completely negligible: see the discussion at the beginning of Part III below); Magnasco *et al.* (1976) propose a modification of the MS–MA procedure (MS: Murrell and Shaw (1967a); MA: Musher and Amos (1967)) and apply it to the H_2^+ case.

That such a number of symmetrized perturbation treatments have been proposed clearly illustrates the indeterminacy of the problem, which was discussed by Certain, Hirschfelder and Epstein (1969), Lekkerkerker and Laidlaw (1970a) and also Amos (1970). The problem is indeed to treat by the perturbation method an 'interpolation' eigenvalue equation (Certain *et al.*, 1969):

$$F[\lambda, E(\lambda)]\Phi(\lambda) = 0 \qquad (2.15)$$

where $F(\lambda, E)$ is an 'interpolation operator' connecting the *appropriate* eigenvalues of H_0 and H, i.e. an operator such that for $\lambda = 0$, $E(0)$ is the eigenvalue $E_0 = E_0^{(1)} + E_0^{(2)}$ of $(H_0 - E_0)\Psi_0 = 0$ (where $\Psi_0 = W_0^{(1)} W_0^{(2)}$), while for $\lambda = 1$, $E(1)$ is the eigenvalue E of the physical ground state of H, i.e. $(H - E)\Psi = 0$ (with Ψ satisfying the Pauli principle). The long-range RS treatment corresponds to the obvious interpolation operator

$$F(\lambda, E) \equiv H_0 + \lambda V - E \qquad (2.16)$$

but unfortunately, as we have seen in Section A, this operator does not realize the desired connection between the eigenvalues of interest. It is therefore necessary to search other interpolation operators, and there is obviously an infinity of possible choices, corresponding to the infinity of continuous curves $E(\lambda)$ which connect two prescribed values $E(0) = E_0$ and $E(1) = E$. The perturbation method consists in representing the curve $E(\lambda)$ by its Taylor expansion around the value $\lambda = 0$, and from that point of view, an interpolation operator will be 'good' if the Taylor expansion of the corresponding $E(\lambda)$ still converges rapidly for $\lambda = 1$ (the practice of any perturbation method indeed shows that it becomes rapidly prohibitive to go beyond the second order, except for very simple systems). It is of course very difficult to evaluate *a priori* by purely theoretical arguments the quality of a given interpolation operator, and applications on simple systems have been up to now necessary for assessing this quality (nevertheless, some results could also be obtained by a theoretical analysis, e.g. the inadequacy of the long-range behaviour of the EL–HAV treatment mentioned above).

This non-uniqueness of the symmetrized perturbation treatments has been considered as a rather negative feature, e.g. by Certain and Bruch (1972, Section 4.4.2) and Whitton and Byers Brown (1976). Moreover, these authors emphasize that the Heitler–London exchange interaction (see just above) has not the correct asymptotic (long-range) behaviour (as shown by Gorkov and Pitaevsky (1964) and Herring and Flicker (1964)), and they assert that it would therefore be undesirable for a perturbation treatment to have as its first-order term the Heitler–London interaction energy; now, this situation actually holds for almost all exchange perturbation treatments, which would thus be subject to this criticism. But, if this criticism is actually valid from the theoretical standpoint, it is not decisive from the practical standpoint (let us recall here the result of the Section II.A above: despite its theoretical inadequacy the long-range RS treatment is perfectly useful at large distances from the practical point of view).

We are indeed essentially interested in the behaviour of the 'exchange' terms at the *intermediate* (or short) distances, and not so much at large distances, where the exchange terms are anyway exponentially small and therefore negligible with respect to the usual long-range terms (electrostatic, polarization, dispersion). If, at intermediate distances, one or several exchange perturbation treatments appear able to give satisfactory evaluations of the interaction energy by using the Heitler–London exchange energy, the usual long-range terms (and eventually some second-order exchange term such as a charge-transfer term), this is all we need, especially for large molecules, for which only approximate evaluations by simplified formulas are conceivable in actual practice. Now, as it will be discussed more thoroughly in Part III, the situation just described seems rather plausible, as supported, for example, by the recent works of Chalasinski *et al.* (1977) and Jeziorski and Van Hemert (1976).

In the next Section (2) we shall present a general method for the problems involving symmetry constraints, and especially the symmetry-adapted perturbation treatments (this subject was also considered, in a different way, in the important work by Löwdin (1968a,b)). Then, in Section 3, we shall present on this basis a symmetry-adapted perturbation expansion, every order of which may be easily decomposed into the long range term (of the standard RS treatment) plus an 'exchange' short-range term, so that this expansion truncated at low order (the second-order in most cases) will be a suitable basis for the derivation of simplified approximate formulas in Section III.

2. *The 'ideal space' method for dealing with the symmetry constraints and its application in perturbation theory*

a. The eigenvalue problem in a non-orthogonal and overcomplete 'basis' set. As already mentioned, we are led to use, instead of the basis of simple products $W_a^{(1)}W_b^{(2)}$, the set of antisymmetrized products $A(W_a^{(1)}W_b^{(2)})$: in that way we shall indeed avoid the occurrence of undesired states with wrong symmetry. The various antisymmetrization operators used in the present work are described in Appendix A; we note

$$\mathscr{A}=\sum_P(-1)^P P \tag{2.17}$$

$$\mathfrak{A}=\mathscr{A}/\sqrt{n!} \tag{2.18}$$

$$\mathrm{A}=\mathscr{A}/n! \tag{2.19}$$

A is the projection antisymmetrizer and $\mathfrak{A}$, when applied to a product of n orthonormal one-particle functions, generates the normalized Slater determinant, while $\mathscr{A}$ generates the unnormalized determinant. But if $\{W_a^{(1)}W_b^{(2)}\}$ is the (orthogonal and complete) set of all products, the projected set $\{A(W_a^{(1)}W_b^{(2)})\}$ is then necessarily overcomplete and not orthogonal since it spans only the antisymmetric *subspace* contained (in the strict sense) in the product space (Claverie, 1973, Ch. I, Section 1-B and Appendix 1): see Appendix B. If the sets $\{W_a^{(1)}\}$ and $\{W_b^{(2)}\}$, and therefore $\{W_a^{(1)}W_b^{(2)}\}$ are not complete, the set $\{A(W_a^{(1)}W_b^{(2)})\}$ is still non-orthogonal but may be linearly independent (and would become increasingly almost-dependent as the set $\{W_a^{(1)}W_b^{(2)}\}$ grows toward completeness).

We therefore must solve the eigenvalue problem of the total Hamiltonian in a non-orthogonal and possibly overcomplete (or at least almost linearly dependent) basis set. A usual procedure is to extract from this set an orthogonal basis, but it is not very convenient in the present problem, especially for a general theoretical analysis. For solving directly the eigenvalue problem in an arbitrary basis set (even a linearly dependent one), we shall follow the work by Löwdin (1967), especially the Section 2 of this paper, which corresponds to the general concept of using a so-called 'ideal' space (Dyson, 1956; Fleckinger, Gomes, and Soulet, 1976; Fleckinger and Soulet, 1976); Amos (1971) also proposed to apply this kind of method to the symmetry-adapted perturbation problems.

Let us denote by $\{\Psi_i\}$ the basis (eventually non-orthogonal and overcomplete); in our intermolecular problem, we have $\Psi_i = A\phi_i$, where $\{\phi_i\}$ denote for brevity the orthonormal basis of the simple products $\{W_a^{(1)} W_b^{(2)}\}$. We introduce the matrix H with elements

$$H_{ij} = \langle \Psi_i | \, H \, | \Psi_j \rangle \tag{2.20}$$

and the metric or overlap matrix S with elements

$$S_{ij} = \langle \Psi_i \mid \Psi_j \rangle \tag{2.21}$$

Every function Ψ of the 'physical' space spanned by $\{\Psi_i\}$ may be expressed as a linear expansion

$$\Psi = \sum_k c_k \Psi_k \tag{2.22}$$

and there are an infinity of such decompositions if the set $\{\Psi_k\}$ is linearly dependent. To each such expansion we associate the column vector

$$\mathbf{c} = \begin{bmatrix} c_0 \\ \cdot \\ \cdot \\ \cdot \\ c_k \\ \cdot \\ \cdot \\ \cdot \end{bmatrix} \tag{2.23}$$

It is now possible to deal directly with the matrices **H**, **S**, **c** without further consideration of the basis $\{\Psi_i\}$ and the 'physical' operators (H and the identity operator I) which generated them. These matrices may be considered as expressing operators and vectors in the 'natural' basis:

$$\mathbf{e}_0 = \begin{bmatrix} 1 \\ 0 \\ 0 \\ \cdot \\ \cdot \\ \cdot \end{bmatrix}, \quad \mathbf{e}_1 = \begin{bmatrix} 0 \\ 1 \\ 0 \\ \cdot \\ \cdot \\ \cdot \end{bmatrix}, \quad \ldots \tag{2.24}$$

These abstract basis vectors are in one-to-one correspondence with the functions Ψ_i. The abstract linear space spanned by this basis will be called the 'ideal' space, and it will be made a metric space by considering the basis $\{\mathbf{e}_0, \mathbf{e}_1, \ldots, \mathbf{e}_i, \ldots\}$ as orthonormal by definition. Thus, the scalar product in the ideal space, denoted $(\mid)$ (while the scalar product in the physical space is denoted $\langle \mid \rangle$) is expressed as:

$$(c \mid d) = \sum_i c_i^* d_i \tag{2.25}$$

Dyson (1956), Fleckinger *et al.* (1976) directly introduced an ideal space $\mathscr{E}^{\text{ideal}}$ spanned by a basis $\{\Psi_i^{\text{ideal}}\}$, orthonormal by definition, whose elements are in a one-to-one correspondence with the physical basis functions Ψ_i: this is obviously equivalent with our less abstract introduction through the matrices (2.20, 2.21, 2.23). An essential concept of the ideal space formulation is therefore the following: after the physical problem has been translated in mathematical form (here in matrix form), the physical meaning may be somewhat forgotten in order to find the best mathematical solution.

Löwdin (1967) then shows that the eigenvalues of H in the space $\{\Psi_i\}$ (or their best approximations if the set $\{\Psi_i\}$ is not complete in the physical space) may be obtained as the solutions of the generalized eigenvalue problem:

$$(\mathbf{H} - E\mathbf{S})\mathbf{c} = 0 \tag{2.26}$$

which satisfy the condition ($\mathbf{c}^\dagger$ means the hermitian conjugate of $\mathbf{c}$):

$$\mathbf{c}^\dagger \mathbf{S} \mathbf{c} = (\mathbf{c} \mid \mathbf{S}\mathbf{c}) \neq 0 \quad \text{or} \quad \mathbf{S}\mathbf{c} \neq 0 \tag{2.27}$$

This supplementary condition excludes the solutions of (2.26) which would correspond to the zero function in the physical space. It is of course fulfilled if $\{\Psi_i\}$ is a linearly independent set, because $\mathbf{S}$ is then *strictly* positive definite and non-singular. But if $\{\Psi_i\}$ is linearly dependent, there are combinations

$$\sum_i a_i \Psi_i = 0$$

with the a_i's not all equal zero. Denoting $\mathbf{a}$ the vector of such coefficients, it is easily seen that $\mathbf{Ha} = 0$, $\mathbf{Sa} = 0$, and more generally $\mathbf{Ba} = 0$, where $\mathbf{B}$ is the matrix with elements $B_{ij} = \langle \Psi_i | B | \Psi_j \rangle$ corresponding to any physical operator B:

$$\begin{aligned}(\mathbf{Ba})_i &= \sum_j B_{ij} a_j = \sum_j \langle \Psi_i | B | \Psi_j \rangle a_j \\ &= \langle \Psi_i | B \left| \sum_j a_j \Psi_j \right\rangle = \langle \Psi_i | B | 0 \rangle = 0\end{aligned}$$

Thus, we have $(\mathbf{H} - E\mathbf{S})\,\mathbf{a} = 0$ whatever E, hence the necessity of excluding these physically spurious solutions.

It is evident that these vectors $\mathbf{a}$ such that $\sum_i a_i \Psi_i = 0$ form a linear subspace $\mathscr{V}_0$ of $\mathscr{E}^{\text{ideal}}$; we shall denote $\mathscr{V}_1$ the orthogonal subspace. Actually, $\mathscr{V}_0$ is the eigenspace of $\mathbf{S}$ corresponding to the eigenvalue 0: we already saw that $\mathbf{a} \in \mathscr{V}_0$ implies $\mathbf{Sa} = 0$, and conversely $\mathbf{Sa} = 0$ means $(\mathbf{Sa})_i = 0$ whatever i, but

$$(\mathbf{Sa})_i = \sum_j \langle \Psi_i \mid \Psi_j \rangle a_j = \left\langle \Psi_i \middle| \sum_j a_j \Psi_j \right\rangle$$

and

$$\left\langle \Psi_i \middle| \sum_j a_j \Psi_j \right\rangle = 0$$

for every i means

$$\sum_j a_j \Psi_j = 0,$$

i.e. $\mathbf{a} \in \mathscr{V}_0$.

Since $\mathscr{V}_0$ is the eigenspace of $\mathbf{S}$ for the eigenvalue 0, and since $\mathbf{S}$ is hermitian, the orthogonal subspace $\mathscr{V}_1$ is invariant with respect to $\mathbf{S}$, and the restriction of $\mathbf{S}$ to $\mathscr{V}_1$ is strictly positive definite ($\mathscr{V}_1$ is indeed the direct sum of all eigenspaces of $\mathbf{S}$ with strictly positive eigenvalues). Similarly, for any physical operator B, $\mathscr{V}_0$ is an eigenspace with the (spurious) eigenvalue 0 for the corresponding matrix $\mathbf{B}$, so that $\mathscr{V}_1$ is the sum of the eigenspaces corresponding to the physical eigenvalues. Actually, $\mathscr{V}_1$ is in one-to-one correspondence with the physical space spanned by the set $\{\Psi_i\}$, because among all vectors of coefficients $\mathbf{c}$ corresponding to any given physical Ψ, there is one and only one which belongs to $\mathscr{V}_1$. Let us indeed introduce the projection operator $\mathbf{\Pi}_1$ on the subspace $\mathscr{V}_1$, and, for a given $\mathbf{c}$ corresponding to Ψ, we introduce $\mathbf{c}_1 = \mathbf{\Pi}_1 \mathbf{c}$. Then $\mathbf{c}_1$ is also a set of coefficients for Ψ, because the difference $\mathbf{c} - \mathbf{c}_1 = (1 - \mathbf{\Pi}_1)\mathbf{c}$ belongs to $\mathscr{V}_0$, and is therefore a set of coefficients for the zero function. Moreover $\mathbf{c}_1$ is the same whatever the initial choice of $\mathbf{c}$, because for any other $\mathbf{c}'$ corresponding to Ψ, $\mathbf{c}' - \mathbf{c}$ corresponds to the zero function, therefore $\mathbf{c}' - \mathbf{c}$ belongs to $\mathscr{V}_0$ and $\mathbf{\Pi}_1(\mathbf{c}' - \mathbf{c}) = 0$, i.e. $\mathbf{\Pi}_1 \mathbf{c}' = \mathbf{\Pi}_1 \mathbf{c} = \mathbf{c}_1$.

Using the modern algebraic terminology, we would say that the relation '$\mathbf{c}$ and $\mathbf{c}'$ are sets of coefficients for the same physical Ψ' or as well '$\mathbf{c} - \mathbf{c}'$ belongs to the subspace $\mathscr{V}_0$' is an *equivalence relation* in $\mathscr{E}^{\text{ideal}}$, and $\mathscr{V}_1$ is the quotient set of $\mathscr{E}^{\text{ideal}}$ defined by this equivalence relation. $\mathscr{V}_1$ is therefore the subspace of $\mathscr{E}^{\text{ideal}}$ isomorphic with the original physical subspace $\{\Psi_i\}$ (cf. Fleckinger *et al.*, 1976).

A possible way for solving the eigenvalue problem (2.26, 2.27) would therefore consist in considering directly the problem in the subspace $\mathscr{V}_1$, and this means nothing but using the orthogonalization method of Rowe (1969) for obtaining an orthogonal basis in the physical subspace (this method consists in diagonalizing the matrix S and retaining as basis vectors the eigenvectors corresponding to non-zero eigenvalues, which are just an orthogonal basis of our subspace $\mathscr{V}_1$). Thus, projecting directly the problem (2.26, 2.27) into the invariant subspace $\mathscr{V}_1$ is strictly equivalent with introducing an orthonormal basis in the physical space and then solving the usual eigenvalue problem.

The general problem (2.26, 2.27) has the same eigenvalues, but the eigenvectors are indeterminate: indeed, if we consider the eigenvector $\mathbf{c}_1 \in \mathscr{V}_1$ corresponding to some eigenvalue E, all vectors $\mathbf{c} = \mathbf{c}_1 + \mathbf{a}$, where $\mathbf{a}$ is an arbitrary vector of $\mathscr{V}_0$, are also solutions of (2.26, 2.27) for the same eigenvalue E, since $\mathbf{Ha} = 0$ and $\mathbf{Sa} = 0$.

The projection onto the subspace $\mathscr{V}_1$ is an obvious method for removing this indeterminacy (and would seem quite convenient for usual problems in a finite basis), but it would not have been necessary to introduce the ideal space concept for defining it. Actually, for a general discussion, including the case of infinite bases and the derivation of perturbation treatments, the broader framework of the ideal space, with its *a priori* indeterminate eigenvectors, will offer interesting new possibilities (e.g. see the method used by Dyson (1956) for the so-called spin waves, and the discussion by Fleckinger *et al.* (1976) showing that a number of quantum-theoretical methods, including the use of pseudo-potentials in solid-state physics as well as several symmetry-adapted perturbation treatments, may be embedded in the general framework of the ideal space concept).

b. The perturbation treatments in the ideal space. In order to solve by the perturbation method the problem (2.26, 2.27), we split the matrix $\mathbf{H}$ into an 'unperturbed' matrix $\mathbf{H}^0$ and a 'perturbation' matrix $\mathbf{V} = \mathbf{H} - \mathbf{H}^0$. Presently, $\mathbf{H}^0$ is not precisely defined, and appropriate definitions will appear as a result of the development of the method (thus there is no *a priori* connection between the unperturbed matrix $\mathbf{H}^0$ and the operator $H_0 = H^{(1)} + H^{(2)}$ introduced previously).

We now consider the 'interpolation' equation (which appears as a generalized eigenvalue problem: Laughlin and Amos (1971) and Laughlin *et al.* (1974) also studied generalized eigenvalue problems by perturbation methods, and considered the special case of the symmetry-adapted perturbation treatments):

$$[\mathbf{H}^0 + \lambda \mathrm{V} - E(\lambda)\mathbf{S}]\mathbf{c}(\lambda) = 0 \tag{2.28}$$

where the perturbation parameter λ will vary from 0 to 1. Expressing $E(\lambda)$ and $\mathbf{c}(\lambda)$ by their series expansions:

$$E(\lambda) = \sum_{n=0}^{\infty} \varepsilon_n \lambda^n \tag{2.29}$$

$$\mathbf{c}(\lambda) = \sum_{n=0}^{\infty} \mathbf{c}_n \lambda^n \tag{2.30}$$

putting these expressions in (2.28), and equating to zero the coefficients of the successive powers of λ, we get the set of perturbation equations:

$$\left.\begin{array}{lr}(\mathbf{H}^0-\varepsilon_0\mathbf{S})\mathbf{c}_0=0 & (0)\\ (\mathbf{H}^0-\varepsilon_0\mathbf{S})\mathbf{C}_1+(\mathbf{V}-\varepsilon_1\mathbf{S})\mathbf{c}_0=0 & (1)\\ \cdots\cdots\cdots & \\ (\mathbf{H}^0-\varepsilon_0\mathbf{S})\mathbf{C}_n+(\mathbf{V}-\varepsilon_1\mathbf{S})\mathbf{c}_{n-1}-\varepsilon_2\mathbf{S}\mathbf{c}_{n-2}-\ldots-\varepsilon_n\mathbf{S}\mathbf{c}_0=0 & (n)\\ \cdots\cdots\cdots & \end{array}\right\}(2.31)$$

Except for the appearance of the matrix **S** instead of the identity matrix, these are identical with usual Rayleigh–Schrödinger perturbation equations. It is convenient, at this point, to assume that $\mathbf{H}^0$ is an hermitian matrix (**H** and **S** already are, by their very definition (2.20, 2.21)). Then the solution of this sequence of equations (2.31) may be obtained in the usual way: we introduce the projection matrix $\mathbf{O}_0=\mathbf{c}_0\mathbf{c}_0^\dagger$ on $\mathbf{c}_0$, its complementary $\mathbf{P}_0=1-\mathbf{O}_0$ and the matrix

$$\mathbf{R}_0=\mathbf{P}_0[\alpha\mathbf{O}_0+\mathbf{P}_0(\varepsilon_0\mathbf{S}-\mathbf{H}^0)\mathbf{P}_0]^{-1}\mathbf{P}_0 \qquad (2.32)$$

which corresponds to the so-called 'reduced resolvent' of the usual perturbation treatments (Löwdin, 1962, 1966, 1968a,b); $\mathbf{R}_0$ is essentially the inverse of the restriction of $(\varepsilon_0\mathbf{S}-\mathbf{H}^0)$ to the subspace orthogonal to $\mathbf{c}_0$ (whose eigenprojector is $\mathbf{P}_0$). This inverse exists provided $\mathbf{c}_0$ is the *unique* eigenvector corresponding to the eigenvalue zero of $(\varepsilon_0\mathbf{S}-\mathbf{H}^0)$, and we shall therefore impose to our $\mathbf{H}^0$ the fulfilment of this condition (the term $\alpha\mathbf{O}_0$, with α non-zero, is introduced only for allowing the inversion in the whole space, and $\mathbf{R}_0$ is independent on the precise value of α). Then we may proceed in the usual way:

(a) we multiply on the left the equations (2.31n) by $\mathbf{c}_0^\dagger$ (or, otherwise stated, we perform the scalar product with $(c_0|$ on the left) in order to get ε_n:

$$\mathbf{c}_0^\dagger(\mathbf{H}^0-\varepsilon_0\mathbf{S})\mathbf{c}_n+\mathbf{c}_0^\dagger\mathbf{V}\mathbf{c}_{n-1}-\sum_{r=1}^{n}\varepsilon_r\mathbf{c}_0^\dagger\mathbf{S}\mathbf{c}_{n-r}=0$$

Since $\mathbf{H}^0$ is hermitian, $\mathbf{c}_0^\dagger(\mathbf{H}^0-\varepsilon_0\mathbf{S})=[(\mathbf{H}^0-\varepsilon_0\mathbf{S})\mathbf{c}_0]^\dagger=0$ (according to (2.31) (0)); moreover, $\mathbf{c}_0^\dagger\mathbf{S}\mathbf{c}_0\neq 0$ provided that $\mathbf{c}_0$ does not belong to $\mathscr{V}_0$, i.e. the corresponding function $\sum_i c_{0,i}\Psi_i$ in the physical space is not identically zero. Since $\mathbf{H}^0$ is not fixed *a priori*, it is quite possible to fulfil this condition. Then we get:

$$\varepsilon_n=\frac{1}{\mathbf{c}_0^\dagger\mathbf{S}\mathbf{c}}\left[\mathbf{c}_0^\dagger\mathbf{V}\mathbf{c}_{n-1}-\sum_{r=1}^{n-1}\varepsilon_r\mathbf{c}_0^\dagger\mathbf{S}\mathbf{c}_{n-r}\right] \qquad (2.33)$$

(b) From the very definition of $\mathbf{R}_0$, we have $\mathbf{R}_0(\varepsilon_0\mathbf{S}-\mathbf{H}^0)=\mathbf{P}_0$, hence, multiplying the equations (2.33) on the left by $\mathbf{R}_0$, we get:

$$\mathbf{P}_0\mathbf{c}_n=\mathbf{R}_0\left[\mathbf{V}\mathbf{c}_{n-1}-\sum_{r=1}^{n}\varepsilon_r\mathbf{S}\mathbf{c}_{n-r}\right] \qquad (2.34)$$

As in any Rayleigh-Schrödinger treatment, the component of $\mathbf{c}_n$ on the unperturbed vector $\mathbf{c}_0$ remains indeterminate, and may be chosen arbitrarily without modifying the values ε_r and the direction of the total perturbed vector $\mathbf{c}(\lambda)$ (e.g. see Hirschfelder, Byers–Brown and Epstein (1964) Section II, and Epstein (1968a)). We shall chose the value zero for this component:

$$\mathbf{c}_0^\dagger \mathbf{c}_n = 0 \quad \text{for every} \quad n > 1, \quad \text{hence } \mathbf{c}_n = \mathbf{P}_0 \mathbf{c}_n \tag{2.35}$$

and the solution of (2.31) is now completed through (2.33, 2.34, 2.35). Thus, we obtain with $\mathbf{c}(\lambda = 1)$ a solution of (2.26, 2.27) (provided that the condition $\mathbf{Sc}(1) \neq 0$ is fulfilled, which will actually be proved below) which is perfectly determined, once $\mathbf{H}^0$ has been chosen. At this point, the reader may wonder how we manage to escape the original indeterminacy of the problem (2.26, 2.27). The answer, of course, is that we have introduced the new matrix $\mathbf{H}^0$, with three definite requirements up to now:

(1) $\mathbf{H}^0$ must be hermitian (2.36)
(2) the eigenvector $\mathbf{c}_0$ of $(\mathbf{H}^0 - \varepsilon_0 \mathbf{S})\mathbf{c}_0$ must be unique (non-degeneracy) (2.37)
(3) $\mathbf{c}_0$ must not belong to $\mathscr{V}_0$, or $\mathbf{Sc}_0 \neq 0$, or $\mathbf{c}_0^\dagger \mathbf{Sc}_0 \neq 0$ (i.e. the corresponding function

$$\Psi = \sum_i c_{0,i} \Psi_i \tag{2.38}$$

in the physical space must not be identically zero)

Now, the unicity requirement (2.37) was just not fulfilled by the total matrix $\mathbf{H}$, owing to the fact that $\mathbf{Ha} = 0$ for every $\mathbf{a} \in \mathscr{V}_0$, and (2.37) implies that $\mathbf{H}^0$ just must *not* have this last property. Then, this property will also *not* hold for $\mathbf{V} = \mathbf{H} - \mathbf{H}^0$, and therefore also *not* for $\mathbf{H}(\lambda) = \mathbf{H}^0 + \lambda \mathbf{V}$, except of course for $\lambda = 1$. Then, the solution $\mathbf{c}(\lambda)$ of (2.28) such that $\mathbf{c}(0) = \mathbf{c}_0$ will also be unique, and $\mathbf{c}(1)$ will then be uniquely defined as $\lim_{\lambda \to 1} \mathbf{c}(\lambda)$.

Of course, if we change $\mathbf{H}^0$, we expect in general a different $\mathbf{c}(1)$, so that the indeterminacy of the solution $\mathbf{c}$ of the original problem (II–20) has now been displaced on the choice of $\mathbf{H}^0$. But this is quite interesting because, as we shall see in Section C below, quite natural choices of $\mathbf{H}^0$ can be made, while no analogous preferable way of removing the indeterminacy directly appeared in the original statement (2.26, 2.27) of the problem.

Let us now verify that the requirement (2.27) $\mathbf{Sc} \neq 0$ may actually be fulfilled. For that purpose, we shall prove that the basic equation (2.28) of our perturbation treatment may be replaced by an equivalent equation in the subspace $\mathscr{V}_1$ (whose all vectors fulfil (2.27). Let us indeed use a change of basis, defined by a unitary matrix $\mathbf{U}$ ($\mathbf{c}' = \mathbf{U}^{-1}\mathbf{c} = \mathbf{U}^\dagger \mathbf{c}$; $\mathbf{B}$ becomes $\mathbf{B}' = \mathbf{U}^{-1}\mathbf{BU} = \mathbf{U}^\dagger \mathbf{BU}$) such that the new basis is adapted to the decomposition of the ideal space into $\mathscr{V}_1 \oplus \mathscr{V}_0$ (i.e. the new basis is the union of a basis of $\mathscr{V}_1$ and a basis of $\mathscr{V}_0$; we could for example take a matrix $\mathbf{U}$ which diagonalizes $\mathbf{S}$, since $\mathscr{V}_0$ and $\mathscr{V}_1$ are invariant subspaces of $\mathbf{S}$). Then the matrices $\mathbf{H}' = \mathbf{U}\dagger\mathbf{HU}$ and $\mathbf{S}' = \mathbf{U}\dagger\mathbf{SU}$ transformed of $\mathbf{H}$ and $\mathbf{S}$ respectively are block-diagonal:

$$\mathbf{H}' = \begin{array}{c} \begin{matrix} \mathscr{V}_1 & \mathscr{V}_0 \end{matrix} \\ \left[\begin{array}{c|c} \mathbf{H}'_{11} & 0 \\ \hline 0 & 0 \end{array}\right] \end{array} \begin{matrix} \mathscr{V}_1 \\ \mathscr{V}_0 \end{matrix} \qquad \mathbf{S}' = \begin{array}{c} \begin{matrix} \mathscr{V}_1 & \mathscr{V}_0 \end{matrix} \\ \left[\begin{array}{c|c} \mathbf{S}'_{11} & 0 \\ \hline 0 & 0 \end{array}\right] \end{array} \begin{matrix} \mathscr{V}_1 \\ \mathscr{V}_0 \end{matrix} \tag{2.39}$$

where the block corresponding to $\mathscr{V}_0$ is zero since $\mathscr{V}_0$ is eigenspace of $\mathbf{H}$ and $\mathbf{S}$ with the eigenvalue 0, while no such properties hold for $\mathbf{H}^{0\prime} = \mathbf{U}^\dagger \mathbf{H}^0 \mathbf{U}$, nor for $\mathbf{V}' = \mathbf{U}^\dagger \mathbf{V} \mathbf{U}$:

$$\mathbf{H}^{0\prime} = \left[\begin{array}{c|c} \mathbf{H}^{0\prime}_{11} & \mathbf{H}^{0\prime}_{10} \\ \hline \mathbf{H}^{0\prime}_{01} & \mathbf{H}^{0\prime}_{00} \end{array}\right] \qquad \mathbf{V}' = \left[\begin{array}{c|c} \mathbf{V}'_{11} & \mathbf{V}'_{10} \\ \hline \mathbf{V}'_{01} & \mathbf{V}'_{00} \end{array}\right] \tag{2.40}$$

and, owing to $\mathbf{V}' = \mathbf{H}' - \mathbf{H}^{0\prime}$, we have

$$\mathbf{V}'_{11} = \mathbf{H}'_{11} - \mathbf{H}^{0\prime}_{11}; \qquad \mathbf{V}'_{10} = -\mathbf{H}^{0\prime}_{10}; \qquad \mathbf{V}_{01} = -\mathbf{H}^{0\prime}_{01}; \qquad \mathbf{V}'_{00} = -\mathbf{H}^{0\prime}_{00} \tag{2.41}$$

and the special block-diagonal form (2.39) will hold for $\mathbf{H}'(\lambda) = \mathbf{H}^{0\prime} + \lambda \mathbf{V}'$ only for $\lambda = 1$.

Now the express the transformed problem (2.28), namely $(\mathbf{H}'(\lambda) - E\mathbf{S}')\mathbf{c}' = 0$, by using the partitioning method (Löwdin, 1963, 1966):

$$\left[\begin{array}{c|c} \mathbf{H}'_{11}(\lambda) - E\mathbf{S}'_{11} & \mathbf{H}'_{10}(\lambda) \\ \hline \mathbf{H}'_{01}(\lambda) & \mathbf{H}'_{00}(\lambda) \end{array}\right] \left[\begin{array}{c} \mathbf{c}'_1 \\ \hline \mathbf{c}'_0 \end{array}\right] = 0 \tag{2.42}$$

or

$$\begin{cases} (\mathbf{H}'_{11}(\lambda) - E\mathbf{S}'_{11})\mathbf{c}'_1 + \mathbf{H}'_{10}(\lambda)\mathbf{c}'_0 = 0 & (2.43) \\ \mathbf{H}'_{01}(\lambda)\mathbf{c}'_1 + \mathbf{H}'_{00}(\lambda)\mathbf{c}'_0 = 0 & (2.44) \end{cases}$$

or, using the relations (2.41):

$$\begin{cases} (\lambda \mathbf{H}'_{11} + (1-\lambda)\mathbf{H}^{0\prime}_{11} - E\mathbf{S}'_{11})\mathbf{c}'_1 + (1-\lambda)\mathbf{H}^{0\prime}_{10}\mathbf{c}'_0 = 0 & (2.45) \\ (1-\lambda)\mathbf{H}^{0\prime}_{01}\mathbf{c}'_1 + (1-\lambda)\mathbf{H}^{0}_{00}\mathbf{c}'_0 = 0 & (2.46) \end{cases}$$

(2.46) implies

$$\mathbf{H}^{0\prime}_{01}\mathbf{c}'_1 + \mathbf{H}^{0\prime}_{00}\mathbf{c}'_0 = 0 \tag{2.47}$$

for every $\lambda \neq 1$, and since we search the solution $\mathbf{c}(1)$ as the continuous limit of $\mathbf{c}(\lambda)$ when $\lambda \to 1$, we must still use (2.47) for $\lambda = 1$.

The requirements (2.36, 2.37, 2.38) hold of course for the transformed matrix $\mathbf{H}^{0\prime}$, and therefore the matrix

$$\mathbf{H}^{0\prime} - \varepsilon_0 \mathbf{S}' = \begin{bmatrix} \mathbf{H}^{0\prime}_{11} - \varepsilon_0 \mathbf{S}' & \mathbf{H}^{0\prime}_{10} \\ \mathbf{H}^{0}_{01} & \mathbf{H}^{0\prime}_{00} \end{bmatrix} \tag{2.48}$$

has a *single* eigenvector $\mathbf{c}'_0$ corresponding to the eigenvalue 0. Now, for pursuing the solution of (2.45, 2.46), it is necessary to assume a fourth requirement concerning $\mathbf{H}^0$, namely:

$$\mathbf{H}^{0\prime}_{00} \text{ must be regular (i.e. to have an inverse)} \tag{2.49}$$

An important special case where this property is a consequence of the previous requirements (2.36, 2.37, 2.38) is obtained when $\mathbf{H}^0 - \varepsilon_0\mathbf{S}$ is *positive semi-definite* (let us recall that a hermitian matrix $\mathbf{B}$ is said positive semi-definite, or non-negative, if $\mathbf{c}^\dagger\mathbf{B}\mathbf{c} \geqslant 0$ for every $\mathbf{c}$; this is equivalent to say that all eigenvalues are non-negative). Indeed $\mathbf{H}^{0\prime} - (\varepsilon_0 - \alpha)\mathbf{S}' = (\mathbf{H}^{0\prime} - \varepsilon_0\mathbf{S}') + \alpha\mathbf{S}'$, where α is any positive number, is then strictly positive definite: $\mathbf{c}^\dagger(\mathbf{H}^{0\prime} - \varepsilon_0\mathbf{S}')\mathbf{c}$ is positive for every $\mathbf{c}$ different from $\mathbf{c}_0$, and is zero for $\mathbf{c} = \mathbf{c}_0$, but then $\mathbf{c}_0^\dagger\mathbf{S}\mathbf{c}_0 > 0$ (because it is non-zero according to (2.38) and $\mathbf{S}$ is non-negative), therefore $\mathbf{c}^\dagger[\mathbf{H}^{0\prime} - (\varepsilon_0 - \alpha)\mathbf{S}']\mathbf{c} > 0$ for any $\mathbf{c}$. But $\mathbf{H}^{0\prime}_{00}$ is the restriction of $\mathbf{H}^{0\prime} - (\varepsilon_0 - \alpha)\mathbf{S}'$ to $\mathcal{V}_0$ whatever α, according to (2.48) (it is only necessary to replace there ε_0 by $\varepsilon_0 - \alpha$), and the eigenvalues of a restricted matrix always lie between the extreme eigenvalues of the complete matrix: this is an obvious consequence of the 'separation theorem' which asserts that the eigenvalues of a restricted matrix of order $(n-1)$ 'separate' the eigenvalues of the matrix of order n, i.e. the rth eigenvalue of the restricted matrix belongs to the segment defined by the rth and $(r+1)$th eigenvalues of the matrix (e.g. see McDonald (1933) and Löwdin (1963, 1965)). Consequently, any restricted matrix from a strictly positive matrix is itself strictly positive, thus $\mathbf{H}^{0\prime}_{00}$ is strictly positive and therefore regular (no zero eigenvalue).

We shall just encounter this case in Section C below for treating the ground state of the intermolecular complexes: actually, we shall introduce a matrix $\mathbf{H}^0 - \varepsilon_0\mathbf{S}$ diagonal with the first diagonal element equal to zero and following ones positive.

$\mathbf{H}^{0\prime}_{00}$ being regular, we get from (2.47):

$$\mathbf{c}'_0 = -(\mathbf{H}^{0\prime}_{00})^{-1}\mathbf{H}^{0\prime}_{01}\mathbf{c}'_1 \tag{2.50}$$

hence, inserting this result in (2.45):

$$[\lambda\mathbf{H}'_{11} + (1-\lambda)\mathbf{H}^{0\prime}_{11} - (1-\lambda)\mathbf{H}^{0\prime}_{10}(\mathbf{H}^{0\prime}_{00})^{-1}\mathbf{H}^{0\prime}_{01} - E\mathbf{S}'_{11}]\mathbf{c}'_1 = 0 \tag{2.51}$$

Our initial perturbation problem (2.28) is therefore reduced to the problem (2.51) which is entirely defined in subspace $\mathcal{V}_1$ (isomorphic to the physical space). In particular, for $\lambda = 1$, (2.51) reduces to

$$(\mathbf{H}'_{11} - E\mathbf{S}'_{11})\mathbf{c}'_1 = 0 \tag{2.52}$$

and this equation is nothing but the reduction of the eigenvalue problem to the subspace $\mathcal{V}_1$; now, as we saw at the end of the previous section (a), (2.52) is rigorously equivalent to the eigenvalue problem in the physical space. Since, according to (2.38), the unperturbed vector $\mathbf{c}_0$ has a non-zero projection $\mathbf{c}_{0,1}$ in $\mathcal{V}_1$, we may be sure that the solution of (2.51) with the corresponding non-zero unperturbed vector $\mathbf{c}'_{0.1} = (\mathbf{U}^\dagger\mathbf{c}_0)_1$ for $\lambda = 0$ will give us for $\lambda = 1$ some non-zero solution belonging to $\mathcal{V}_1$ (and therefore physical).

It only remains to check whether the connection between the unperturbed and the perturbed eigenvalues is a correct one (in Section A above we indeed saw that this connection was not necessarily a trivial matter). Here, the situation is more favourable than for the RS treatment in the simple product basis, since, according to the previous analysis, we are sure to get a physical solution. We want go get the lowest of these physical eigenvalues. Now, the equation (2.51) may be written as:

$$[\mathbf{B}(\lambda)-E\mathbf{S}'_{11}]\mathbf{c}'_1=0 \tag{2.53}$$

where $\mathbf{B}(\lambda)=\lambda\mathbf{H}'_{11}+(1-\lambda)[\mathbf{H}^{0\prime}_{10}-\mathbf{H}^{0\prime}_{10}(\mathbf{H}^{0\prime}_{00})^{-1}\mathbf{H}^{0\prime}_{01}]$ is a hermitian matrix, while $\mathbf{S}'_{11}$ is regular, according to the very definition of $\mathcal{V}_1$. Consequently, the generalized eigenvalue problem (2.53) may be reduced to a usual one by the $(\mathbf{S}'_{11})^{-1/2}$ orthogonalization (which is equivalent to multiplying (2.53) on the left by this matrix and factorizing $(\mathbf{S}'_{11})^{1/2}\mathbf{c}'_1$ as the eigenvector):

$$[(\mathbf{S}'_{11})^{-1/2}\mathbf{B}(\lambda)(\mathbf{S}'_{11})^{-1/2}-E](\mathbf{S}'_{11})^{1/2}\mathbf{c}'_1=0 \tag{2.54}$$

Now, the (hermitian) matrix $(\mathbf{S}'_{11})^{-1/2}\mathbf{B}(\lambda)(\mathbf{S}'_{11})^{-1/2}$ depends, as $\mathbf{B}(\lambda)$, on the *single* perturbation parameter λ, and the non-crossing rule (Razi Naqvi and Byers Brown, 1972) therefore applies to the eigenvalues $E(\lambda)$ of (2.54). Thus, the lowest eigenvalue of (2.52), namely the physical ground state, will be connected with the lowest eigenvalue of the unperturbed (generalized) eigenvalue problem $(\mathbf{H}^0-E\mathbf{S})\mathbf{c}=0$. We therefore need only to choose this lowest eigenvalue as the ε_0 in (2.31), with $\mathbf{c}_0$ being the corresponding eigenvector; in actual practice, $\varepsilon_0=\mathbf{E}_0^{(1)}+\mathbf{E}_0^{(2)}$ and $c_0=(1,0,\ldots 0)$ corresponding to $\Psi_0=A(W_0^{(1)}W_0^{(2)})$ are fixed *a priori* and then $\mathbf{H}^0$ will be built in such a way as to fulfil the requirement that ε_0 is the lowest eigenvalue of $(\mathbf{H}^0-E\mathbf{S})\mathbf{c}=0$. Note also that, if $\mathbf{H}^0$ is so chosen that the subspaces corresponding to functions with a given total spin are invariant with respect to it, the above reasoning holds not only for the whole physical space but also for each of these invariant subspaces separately.

c. A special case: the physical basis is obtained by a projection from an initial orthonormal basis. This case corresponds to the symmetry-adapted perturbation treatments: the original basis is an orthonormal one, and a symmetry-adapted 'basis' is obtained by applying to this basis some suitable symmetrization operator. Thus, in the intermolecular interaction problem, the orthonormal basis is made from the simple products $\{\phi_i=W_a^{(1)}W_b^{(2)}\}$, while the symmetry-adapted 'basis' $\{\Psi_i\}$ is made from the totally antisymmetrized functions

$$\Psi_i=A\phi_i \tag{2.55}$$

(for practical purposes, the functions $\mathfrak{A}\phi_i$, which are normalized in the limit of infinite intermolecular distance R, may be preferable).

In such a case, the abstract ideal space introduced previously may be realized in a simple way by considering the linear functional space generated by the *non-symmetrized* orthonormal set $\{\phi_i\}$: these basis vectors are indeed in a one-to-one correspondence with the physical basis functions $\{\Psi_i\}$, and they are orthonormal as required for the abstract ideal basis $\{\mathbf{e}_i\}$ introduced at the beginning of Section (a). In actual practice, the set $\{\phi_i\}$ is almost always incomplete in the functional Hilbert space (for example $\{\phi_i$ is a finite set), and it will be appropriate to introduce the projection operator

$$P = \sum_i |\phi_i\rangle\langle\phi_i| \tag{2.56}$$

corresponding to the linear subspace spanned by the set $\{\phi_i\}$.

The matrices **H** and **S** will now appear as expressing the operators AHA and A^2 respectively, restricted to the subspace $\{\phi_i\}$, or, otherwise stated, the matrices of the operators $PAHAP$ and PA^2P:

$$(\mathbf{H})_{ij} = \langle\Psi_i|\,H\,|\Psi_j\rangle = \langle A\phi_i|\,H\,|A\phi_j\rangle = \langle\phi_i|\,AH\,|\phi_j\rangle \tag{2.57}$$

$$(\mathbf{S})_{ij} = \langle\Psi_i\,|\,\Psi_j\rangle = \langle A\phi_i\,|\,A\phi_j\rangle = \langle\phi_i|\,A^2\,|\phi_j\rangle \tag{2.58}$$

Thus, the problem (2.26, 2.27) may be written

$$(PAHAP - EPA^2P)\Phi = 0 \tag{2.59}$$

with

$$PA\Phi \neq 0 \tag{2.60}$$

where $\Phi = \sum_i c_i\phi_i$. Note that Φ, which replaces the column vector of coefficients $\mathbf{c}$, is *not* the physical solution $\Psi = \sum_i c_i\Psi_i$; we only have, according to (2.55): $\Psi = A\Phi$.

If A is the projection operator itself, the vector $\mathbf{Sc}$ has as elements the scalar products $\langle\phi_i\,|\,\Psi\rangle$: indeed,

$$(\mathbf{Sc})_i = \sum_j S_{ij}c_j = \sum_j \langle\phi_i\,|\,A^2\phi_j\rangle c_j = \sum_j \langle\phi_i\,|\,A\phi_j\rangle c_j$$

$$= \sum_j \langle\phi_i\,|\,\Psi_j\rangle c_j = \langle\phi_i\,|\,\sum_j c_j\Psi_j\rangle = \langle\phi_i\,|\,\Psi\rangle$$

(with $\mathfrak{A} = \sqrt{Q}A$ or $\mathscr{A} = QA$, we would find instead $(\mathbf{Sc})_i = \sqrt{Q}\langle\phi_i\,|\Psi\rangle$ or $(\mathbf{Sc})_i = Q\langle\phi_i\,|\,\Psi\rangle$ respectively).* Thus the vector $\mathbf{Sc}$ corresponds to the function

$$\sum_i (\mathbf{Sc})_i\phi_i = \sum_i \langle\phi_i\,|\,\Psi\rangle\,|\phi_i\rangle = \Big(\sum_i |\phi_i\rangle\langle\phi_i|\Big)\Psi = P\Psi = PA\Phi \tag{2.61}$$

We shall consider also:

$$\sum_i \langle\phi_i\,|\,\Psi\rangle|\Psi_i = \sum_i \langle\phi_i\,|\,\Psi\rangle|A\phi_i\rangle = A\sum_i \langle\phi_i\,|\,\Psi\rangle|\phi_i\rangle = AP\Psi \tag{2.62}$$

It is clear that, in the general case where the basis $\{\phi_i\}$ is incomplete A is not a sub-projection of P (i.e. the antisymmetric subspace is not contained in the subspace spanned by the basis $\{\phi_i\}$), the operators AP, PA and A are different, and the two functions above (2.61, 2.62) differ from Ψ.

* see Appendix A for the definition of the various antisymmetrization operators, and especially the formulas (2.397, 2.398) where $Q = N_{\text{int}}$.

But if we consider the complete basis ϕ_i, a lot of simplifications appear, owing to the fact that we then have

$$AP = PA = A \tag{2.63}$$

since the antisymmetric subspace is contained in the space spanned by *all* simple products (see Section A2 and Appendix B). Then (2.59, 2.60) become

$$(AHA - EA)\Phi = 0 \tag{2.64}$$

with

$$A\Phi \neq 0 \tag{2.65}$$

and from $A^2 = A$, it results that $\mathbf{S}^2 = \mathbf{S}$ and also $\mathbf{SH} = \mathbf{HS} = \mathbf{H}$. The expansion (2.61) now gives Ψ itself and not only some projection $P\Psi$ (indeed $P\Psi = P(A\Psi) = A\Psi$ (see (2.63)) $= \Psi$). Moreover, the coefficients of this expansion, namely the scalar products $b_i = \langle \phi_i \mid \Psi \rangle$, are *also* suitable coefficients for the (non-unique) expansion of Ψ in the overcomplete set $\{\Psi_i\}$: indeed, according to (2.62),

$$\sum_i b_i \Psi_i = AP\Psi = A\Psi = \Psi.$$

This special choice of the coefficients for the expansion of a function Ψ in the set $\{\Psi_i\}$ was explicitly stated by Van der Avoird (1967a) in his reformulation of the treatment by Eisenschitz and London (1930, Part I, Section 2), who introduced the same choice in a more complicated way: among all possible expansions

$$\Psi = \sum_i c_i \Psi_i$$

Eisenschitz and London chose the one for which

$$\sum_i |c_i|^2$$

is the smallest possible. Now, all these expansions may be generated by the functions Φ such that $A\Phi = \Psi$: then, from

$$\Phi = \sum_i c_i \phi_i \text{ (with } c_i = \langle \phi_i \mid \Phi \rangle)$$

we get

$$\Psi = A\Phi = \sum_i c_i (A\phi_i) = \sum_i c_i \Psi_i$$

Then,

$$\sum_i |c_i^2| = \|\Phi\|^2, \quad \text{and} \quad \|\Phi\|^2 > \|A\Phi\|^2 = \|\Psi\|^2$$

since A is a projection operator: the minimum of

$$\sum_i |c_i|^2 = \|\Phi\|^2$$

is clearly obtained for $\Phi = \Psi$, hence $c_i = \langle \phi_i \mid \Psi \rangle = b_i$.

These special properties were used by Van der Avoird (1967b,c) (the treatment HAV of Hirschfelder (1967b) was proved to be equivalent by Johnson and Epstein (1968), and these equivalent methods are themselves equivalent to the original Eisenschitz–London treatment (Chipman, 1972; Claverie, 1973) as mentioned in Section B.1 above). But the uncautious use of these properties leads to formulae which are no more justified if an incomplete basis $\{\phi_i\}$ must be used, as occurs in practice. The shortcomings of this EL–HAV treatment have been studied in detail by Basilevsky and Berenfeld (1972, Appendix), Chipman (1972, p. 126 and Appendix D), Claverie (1973, Ch. II, Section 3).

3. Application to the intermolecular interaction problem

a. Definition of the perturbation treatment. According to the previous section, the basic idea of the treatment is to define an 'unperturbed' matrix $\mathbf{H}^0$ which satisfies the three requirements (2.36–38) from which appropiate consequences have been deduced (essentially the correct connection of the desired physical ground state with a suitable 'unperturbed' lowest eigenvalue). We shall see that, together with (2.36, 2.37, 2.38), the requirement that $(\mathbf{H}^0 - \varepsilon_0\mathbf{S})$ is a diagonal matrix (in order to easily solve the perturbation equations (2.31)) determines almost completely the matrix $\mathbf{H}^0$.

Let us first define completely the sets of functions used. We start from a (complete or incomplete) orthonormal basis of simple products $\{W_a^{(1)}W_b^{(2)}\}$, denoted $\{\phi_i\}$ for brevity. Then we introduce the set of antisymmetrized functions

$$\Psi_i = \mathfrak{A}_{\text{int}}\phi_i \equiv \mathfrak{A}_{\text{int}}(W_a^{(1)}W_b^{(2)}) \tag{2.66}$$

$$\mathfrak{A}_{\text{int}} = \sqrt{N_{\text{int}}}A_{\text{int}} = \frac{1}{\sqrt{N_{\text{int}}}}\mathscr{A}_{\text{int}} \tag{2.67}$$

is the *intersystem* antisymmetrizer which insures that the functions Ψ_i are normalized in the limit of infinite inter-system distance (see Appendix A).

The use of these Ψ_i instead of the projections $A_{\text{int}}\phi_i$ will make easier the comparison with the long-range RS treatment.

Let us consider again the basic perturbation problem (2.28, 2.31). As unperturbed vector $\mathbf{c}_0$, we choose $\mathbf{e}_0 = (1, 0, 0, \ldots)$, which corresponds to the function $\Psi_0 = \mathfrak{A}_{\text{int}}\phi_0 = \mathfrak{A}_{\text{int}}[W_0^{(1)}W_0^{(2)}]$ (this antisymmetrized product of the ground state eigenfunctions of the subsystems almost always appears as the unperturbed function in the 'exchange' perturbation treatments). Since Ψ_0 is a non-zero physical function, the condition (2.38) is satisfied. The first equation (2.31(0)) then becomes $(\mathbf{H}^0 = \varepsilon_0\mathbf{S})\mathbf{e}_0 = 0$, i.e *the first column (and line) of* $\mathbf{H}^0 - \varepsilon_0\mathbf{S}$ *must be zero.* Moreover, in order to make easy the calculation of the reduced resolvent $\mathbf{R}_0$ defined by (2.32), i.e. the inverse of $(\varepsilon_0\mathbf{S} - \mathbf{H}^0)$ in the subspace $\{\mathbf{e}_1, \ldots, \mathbf{e}_k, \ldots\}$ orthogonal to $\mathbf{c}_0 = \mathbf{e}_0$, an obvious way is to require that $\mathbf{H}^0 - \varepsilon_0\mathbf{S}$ is a *real diagonal* matrix (it is therefore hermitian: (2.36) is satisfied):

$$\mathbf{H}^0 - \varepsilon_0\mathbf{S} = \mathbf{D} \tag{2.68}$$

whose diagonal elements $(\mathbf{D})_{ii}$ will be denoted D_i for brevity. Then, the requirement that the first line and column of $\mathbf{H}^0-\varepsilon_0\mathbf{S}$ are zero reduces to

$$D_0=0 \tag{2.69}$$

and $\mathbf{c}_0=\mathbf{e}_0$ will obviously be the unique vector for which $(\mathbf{H}^0-\varepsilon_0\mathbf{S})\mathbf{c}=0$ provided *all other diagonal elements D_i are non-zero.*

At this stage, the three conditions (2.36, 2.37, 2.38) are therefore fulfilled. Moreover, if the D_i's are positive for $i>0$, $\mathbf{H}^0-\varepsilon_0\mathbf{S}$ is non-negative, and this implies the fulfilment of (2.49), i.e. the insurance of getting a non-zero physical solution (see Section 2.c above).

This physical eigenstate will be the lowest one if ε_0 is the lowest eigenvalue of the (generalized) eigenvalue problem $(\mathbf{H}^0-E\mathbf{S})\mathbf{c}=0$, according to the analysis of Section 2.c. This property is also implied by the positiveness of the D_i's. Let us indeed consider such a solution $E\neq\varepsilon_0$, $\mathbf{c}\neq\mathbf{c}_0$:

$$(\mathbf{H}^0-E\mathbf{S})\mathbf{c}=0 \tag{2.70}$$

hence $\mathbf{c}^\dagger(\mathbf{H}^0-E\mathbf{S})\mathbf{c}=0$. Since $\mathbf{H}^0=\mathbf{D}+\varepsilon_0\mathbf{S}$ according to (2.68), we get:

$$\mathbf{c}^\dagger\mathbf{D}\mathbf{c}+(\varepsilon_0-E)\mathbf{c}^\dagger\mathbf{S}\mathbf{c}=0 \tag{2.71}$$

but $\mathbf{S}$ is non-negative, therefore $\mathbf{c}^\dagger\mathbf{S}\mathbf{c}\geqslant 0$, and $\mathbf{c}^\dagger\mathbf{D}\mathbf{c}>0$, because $\mathbf{D}$ is also non-negative, and $\mathbf{c}^\dagger\mathbf{D}\mathbf{c}$ vanishes only for $\mathbf{c}=\mathbf{c}_0$, which is not the case: therefore (2.71) implies $\mathbf{c}^\dagger\mathbf{S}\mathbf{c}>0$ (strictly) and $\varepsilon_0-E<0$, i.e. $E>\varepsilon_0$, which is the desired property. Thus the correct connection of the unperturbed 'state' $(\varepsilon_0, \mathbf{c}_0)$ with the physical ground state is insured.

We still have some freedom of choice for the diagonal elements D_i, which play the role of the energy differences appearing in the denominators of usual perturbation expansions (see (2.10)). Two rather natural choices appear:

(1) the difference between the unperturbed eigenvalues

$$D_i=E_i^0-E_0^0 \tag{2.72}$$

where $H_0\phi_i=E_i^0\phi_i$ (i.e. if $\phi_i=W_a^{(1)}W_b^{(2)}$, $E_i^0=E_a^{(1)}+E_b^{(2)}$). With this choice, the denominators appearing in the perturbation expansion will be identical with those of the long-range RS treatment (2.10), and this will allow an easy comparison of the two treatments at large distances. This choice of the D_i's is therefore appropriate for theoretical analysis. For the same reason it is convenient to choose $\varepsilon_0=E_0^0=E_0^{(1)}+E_0^{(2)}$, and the definition of $\mathbf{H}^0=\mathbf{D}+E_0^0\mathbf{S}$ is thus complete.

(2) the D_i's, and ε_0, are determined by the requirement that the diagonal part of the perturbation matrix $\mathbf{V}=\mathbf{H}-\mathbf{H}^0$ vanishes. This requirement therefore takes advantage of the freedom of choice for ε_0 and $\mathbf{D}$ in order to reduce the 'magnitude' of the perturbation $\mathbf{V}$, and the corresponding choice would therefore appear practically interesting, e.g. for actual *ab initio* calculations.

$$(\mathbf{V})_{ii}=0 \quad \text{means} \quad (\mathbf{H}^0)_{ii}=(\mathbf{H})_{ii} \tag{2.73}$$

hence

$$D_i = (\mathbf{H}^0)_{ii} - \varepsilon_0(\mathbf{S})_{ii} = (\mathbf{H})_{ii} - \varepsilon_0(\mathbf{S})_{ii} \tag{2.74}$$

Then (2.69) determines ε_0:

$$\varepsilon_0 = \frac{(\mathbf{H})_{00}}{(\mathbf{S})_{00}} = \frac{H_{00}}{S_{00}} = \frac{\langle \Psi_0 | H | \Psi_0 \rangle}{\langle \Psi_0 | \Psi_0 \rangle} = E_0 \tag{2.75}$$

This is the well-known 'Heitler–London' energy which appears as the first-order total energy in almost all 'exchange' perturbation treatments. Here we denote it by E_0, which must be carefully distinguished from the unperturbed physical energy $E_0^0 = E_0^{(1)} + E_0^{(2)}$ introduced above. Quite generally, we shall denote

$$E_i = \frac{\langle \Psi_i | H | \Psi_i \rangle}{\langle \Psi_i | \Psi_i \rangle} = \frac{H_{ii}}{S_{ii}} = \frac{(\mathbf{H})_{ii}}{(\mathbf{S})_{ii}} \tag{2.76}$$

the expectation value of the total Hamiltonian H with respect to the antisymmetrized produce $\Psi_i = \mathfrak{A}_{\text{int}}\phi_i$. Then (2.74) gives, for $i > 0$:

$$D_i = (\mathbf{H})_{ii} - E_0(\mathbf{S})_{ii} = (\mathbf{S})_{ii}(E_i - E_0) = S_{ii}(E_i - E_0) \tag{2.77}$$

Thus this second choice essentially uses as denominators the differences between the expectation values of the total Hamiltonian (taken with the antisymmetrized products Ψ_i), while the first choice used the differences between the expectation values of the unperturbed Hamiltonian $H_0 = H^{(1)} + H^{(2)}$ taken with respect to the simple products ϕ_i (which are its eigenfunctions). This second choice of H^0 and V (with zero diagonal elements for V) corresponds to the so-called 'Epstein–Nesbet' partition of the total Hamiltonian used for perturbation treatment of Configuration Interaction by Claverie, Diner and Malrieu (1967). The change from denominators $(E_i^0 - E_0^0)$ to $(E_i - E_0)$ may be considered as a reorganization of the perturbation expansion, i.e. for each order, a partial summation of higher-order terms resulting in geometric series whose sums generate the new denominators (Claverie *et al.*, 1967; Rubinstejn and Yaris, 1971). Note that, for large intermolecular distances, $S_{ii} \to 1$ and $E_i \to E_i^0$, $E_0 \to E_0^0$ so that the two perturbation expansions tend to become identical order by order.

Before considering in more detail the properties of the perturbation expansions corresponding to these two choices, we shall write the perturbation formulas under their general form, valid for any choice of $\mathbf{H}^0$.

According to its definition (2.32) and to (2.68), the reduced resolvent $\mathbf{R}_0$ is the diagonal matrtix with elements

$$(\mathbf{R}_0)_{00} = 0 \qquad (\mathbf{R}_0)_{ii} = -\frac{1}{D_i} \tag{2.78}$$

Since $\mathbf{c}_0^\dagger = \mathbf{e}_0^\dagger = [1\ 0\ 0 \ldots]$, the product $\mathbf{c}_0^\dagger \mathbf{b}$ for any vector $\mathbf{b}$ is simply the first component $(\mathbf{b})_0$. Thus the formulas (2.33, 2.34, 2.35) give

$$\varepsilon_n = \frac{1}{(\mathbf{S})_{00}} \left[(\mathbf{V}\mathbf{c}_{n-1})_0 - \sum_{r=1}^{n-1} \varepsilon_r (\mathbf{S}\mathbf{c}_{n-r})_0 \right] \tag{2.79}$$

$$(\mathbf{c}_n)_0 = 0, \qquad (\mathbf{c}_n)_k = -\frac{1}{D_k} \left[(\mathbf{V}\mathbf{c}_{n-1})_k - \sum_{r=1}^{n} \varepsilon_r (\mathbf{S}\mathbf{c}_{n-r})_k \right] \quad (\text{for } k > 0) \tag{2.80}$$

For example, the lowest-order corrections are:

$$\varepsilon_1 = \frac{(\mathbf{V}\mathbf{c}_0)_0}{(\mathbf{S})_{00}} = \frac{(\mathbf{V}\mathbf{e}_0)_0}{(\mathbf{S})_{00}} = \frac{(\mathbf{V})_{00}}{(\mathbf{S})_{00}} \tag{2.81}$$

$$(\mathbf{c}_1)_0 = 0, \qquad (\mathbf{c}_1)_k = -\frac{(\mathbf{V}\mathbf{c}_0)_k}{D_k} = -\frac{(\mathbf{V}\mathbf{e}_0)_k}{D_k} = -\frac{(\mathbf{V})_{k0}}{D_k} \tag{2.82}$$

$$\begin{aligned} \varepsilon_2 &= \frac{1}{(\mathbf{S})_{00}} [(\mathbf{V}\mathbf{c}_1)_0 - \varepsilon_1 (\mathbf{S}\mathbf{c}_1)_0] = \frac{[(\mathbf{V} - \varepsilon_1 \mathbf{S})\mathbf{c}_1]_0}{(\mathbf{S})_{00}} = \frac{1}{(\mathbf{S})_{00}} \sum_k (\mathbf{V} - \varepsilon_1 \mathbf{S})_{0k} (\mathbf{c}_1)_k \\ &= -\frac{1}{(\mathbf{S})_{00}} \sum_k{}' \frac{[(\mathbf{V})_{0k} - \varepsilon_1 (\mathbf{S})_{0k}](\mathbf{V})_{k0}}{D_k} \end{aligned} \tag{2.83}$$

(Let us recall that $\sum_k'$ means that the value $k = 0$ is excluded from the summation.)

We shall now specialize these expressions for each of the two choices defined above, and especially examine their relationship with the long-range RS treatment defined in section *A*.

b. A first choice for the unperturbed matrix.

$$D_i = E_i^0 - E_0^0, \qquad \varepsilon_0 = E_0^0$$

From $\mathbf{H}^0 = \mathbf{D} + E_0^0 \mathbf{S}$, we get

$$\mathbf{V} = \mathbf{H} - (\mathbf{D} + E_0^0 \mathbf{S})$$

or

$$(\mathbf{V})_{ij} = (\mathbf{H})_{ij} - E_0^0 (\mathbf{S})_{ij} = \mathbf{H}_{ij} - E_0^0 S_{ij} = \langle \Psi_i | H - E_0^0 | \Psi_j \rangle \qquad (i \neq j) \tag{2.84}$$

$$\begin{aligned} (\mathbf{V})_{ii} &= (\mathbf{H})_{ii} - E_0^0 (\mathbf{S})_{ii} - (E_i^0 - E_0^0) \\ &= H_{ii} - E_0^0 S_{ii} - (E_i^0 - E_0^0) \\ &= \langle \Psi_i | H - E_0^0 | \Psi_i \rangle - (E_i^0 - E_0^0) \\ &= S_{ii}(E_i - E_0^0) - (E_i^0 - E_0^0) \\ &= S_{ii} E_i - E_i^0 + (1 - S_{ii}) E_0^0 \end{aligned} \tag{2.85}$$

A matrix element $\langle\Psi_i|H|\Psi_j\rangle$ may be transformed as follows (see Murrell and Shaw, 1967a):

$$\begin{aligned}\langle\Psi_i|H|\Psi_j\rangle &= \langle\mathfrak{A}_{\text{int}}\phi_i|H|\mathfrak{A}_{\text{int}}\phi_j\rangle \\ &= \langle\phi_i|H|(\mathfrak{A}_{\text{int}})^2\phi_j\rangle = \langle\phi_i|H|\mathcal{A}_{\text{int}}\phi_j\rangle \\ &= \langle(\mathfrak{A}_{\text{int}})^2\phi_i|H|\phi_j\rangle = \langle\mathcal{A}_{\text{int}}\phi_i|H|\phi_j\rangle \end{aligned} \tag{2.86}$$

since $(\mathfrak{A}_{\text{int}})^2[W^{(1)}W^{(2)}] = \mathcal{A}_{\text{int}}[W^{(1)}W^{(2)}]$ (see Appendix A, formula (2.411)). Then

$$\begin{aligned}\langle\phi_i|H|\mathcal{A}_{\text{int}}\phi_j\rangle &= \langle\phi_i|H_0+V|\mathcal{A}_{\text{int}}\phi_j\rangle \\ &= E_i^0\langle\phi_i|\mathcal{A}_{\text{int}}\phi_j\rangle + \langle\phi_i|V|\mathcal{A}_{\text{int}}\phi_j\rangle\end{aligned}$$

and similarly for $\langle\mathcal{A}_{\text{int}}\phi_i|H|\phi_j\rangle$, hence the two alternative expressions (which reduce to a single one for $i=j$):

$$\langle\Psi_i|H|\Psi_j\rangle = \langle\phi_i|V\mathcal{A}_{\text{int}}|\phi_j\rangle + E_i^0\langle\phi_i|\mathcal{A}_{\text{int}}|\phi_j\rangle \tag{2.87}$$

$$= \langle\phi_i|\mathcal{A}_{\text{int}}V|\phi_j\rangle + E_j^0\langle\phi_i|\mathcal{A}_{\text{int}}|\phi_j\rangle \tag{2.88}$$

note that

$$\langle\phi_i|\mathcal{A}_{\text{int}}|\phi_j\rangle = \langle\phi_i|(\mathfrak{A}_{\text{int}})^2|\phi_j\rangle = \langle\Psi_i|\Psi_j\rangle = S_{ij} \tag{2.89}$$

Thus (2.86, 2.87) may be rewritten as:

$$H_{ij} = \langle\Psi_i|H|\Psi_j\rangle = \langle\phi_i|V\mathcal{A}_{\text{int}}|\phi_j\rangle + E_i^0 S_{ij} \tag{2.90}$$

$$= \langle\phi_i|\mathcal{A}_{\text{int}}V|\phi_j\rangle + E_j^0 S_{ij} \tag{2.91}$$

and therefore

$$(\mathbf{V})_{ij} = \langle\phi_i|V\mathcal{A}_{\text{int}}|\phi_j\rangle + (E_i^0 - E_0^0)S_{ij} \tag{2.92}$$

$$= \langle\phi_i|\mathcal{A}_{\text{int}}V|\phi_j\rangle + (E_j^0 - E_0^0)S_{ij} \qquad (i\neq j) \tag{2.93}$$

$$(\mathbf{V})_{ii} = \langle\phi_i|V\mathcal{A}_{\text{int}}|\phi_i\rangle + (E_i^0 - E_0^0)(S_{ii}-1) \tag{2.94}$$

Then the first-order perturbation energy (2.81) becomes, according to (2.85):

$$\varepsilon_1 = \frac{\langle\Psi_0|H-E_0^0|\Psi_0\rangle}{\langle\Psi_0|\Psi_0\rangle} = \frac{\langle\Psi_0|H|\Psi_0\rangle}{\langle\Psi_0|\Psi_0\rangle} - E_0^0 = E_0 - E_0^0 \tag{2.95}$$

Thus ε_1 is nothing but the 'Heitler–London' interaction energy (that we shall denote $\varepsilon_1^{\text{HL}}$), namely the difference between the total Heitler–London energy and the unperturbed energy of the separated subsystems $E_0^0 = E_0^{(1)} + E_0^{(2)}$. An explicit form of ε_1 may be obtained by using (2.87, 2.88, 2.89).

$$\varepsilon_1^{\text{HL}} = \frac{\langle\phi_0|V\mathcal{A}_{\text{int}}|\phi_0\rangle + E_0^0\langle\Psi_0|\Psi_0\rangle}{\langle\Psi_0|\Psi_0\rangle} - E_0^0 = \frac{\langle\phi_0|V\mathcal{A}_{\text{int}}|\phi_0\rangle}{\langle\Psi_0|\Psi_0\rangle} = \frac{\langle\phi_0|V\mathcal{A}_{\text{int}}|\phi_0\rangle}{\langle\phi_0|\mathcal{A}_{\text{int}}|\phi_0\rangle} \tag{2.96}$$

The comparison with $\varepsilon_1^{\text{RS}} = \langle\phi_0|V|\phi_0\rangle$ (see (2.10) and remember that $\phi_0 = W_0^{(1)}W_0^{(2)}$) can now easily be made by using the expression $\mathcal{A}_{\text{int}} = 1 - \mathcal{A}'_{\text{int}}$ (see Appendix A, equation (2.380)), where $\mathcal{A}'_{\text{int}}$ is the proper exchange part of $\mathscr{A}_{\text{int}}$:

$$\varepsilon_1^{\text{HL}} = \frac{\langle\phi_0|\, V\,|\phi_0\rangle - \langle\phi_0|\, V\mathscr{A}'_{\text{int}}\,|\phi_0\rangle}{1-\langle\phi_0|\,\mathscr{A}'_{\text{int}}\,|\phi_0\rangle}$$

since $\langle\phi_0\,|\,\phi_0\rangle = 1$, hence:

$$\begin{aligned}\varepsilon_1^{\text{HL}} &= \varepsilon_1^{\text{RS}}\left[1+\frac{\langle\phi_0|\,\mathscr{A}'_{\text{int}}\,|\phi_0\rangle}{1-\langle\phi_0|\,\mathscr{A}'_{\text{int}}\,|\phi_0\rangle}\right]-\frac{\langle\phi_0|\, V\mathscr{A}'_{\text{int}}\,|\phi_0\rangle}{1-\langle\phi_0|\,\mathscr{A}'_{\text{int}}\,|\phi_0\rangle}\\ &= \varepsilon_1^{\text{RS}} + \frac{\langle\phi_0|\, V\,\phi_0\rangle\langle\phi_0|\,\mathscr{A}'_{\text{int}}\,|\phi_0\rangle - \langle\phi_0|\, V\mathscr{A}'_{\text{int}}\,|\phi_0\rangle}{1-\langle\phi_0|\,\mathscr{A}'_{\text{int}}\,|\phi_0\rangle}\end{aligned} \quad (2.97)$$

When the intermolecular distance R grows to infinity, all integrals involving the exchange operator $\mathscr{A}'_{\text{int}}$ decrease exponentially to zero (this is a consequence of the exponential decrease at large distance of the bound state functions of atoms and molecules, a behaviour whose general validity was recently proved by Ahlrichs (1972a,b,; 1973b,c). Thus $\varepsilon_1^{\text{HL}} - \varepsilon_1^{\text{RS}}$ exponentially decreases to zero as $R \to \infty$.

Similarly, the second-order term (2.83) becomes by inserting (2.84):

$$\begin{aligned}\varepsilon_2 &= -\frac{1}{S_{00}}\sum_k{}' \frac{(H_{0k} - E_0^0 S_{0k} - \varepsilon_1 S_{0k})(H_{k0} - E_0^0 S_{k0})}{E_k^0 - E_0^0}\\ &= -\frac{1}{S_{00}}\sum_k{}' \frac{(H_{0k} - E_0 S_{0k})(H_{k0} - E_0^0 S_{k0})}{E_k^0 - E_0^0}\end{aligned} \quad (2.98)$$

since $E_0^0 + \varepsilon_1 = E_0$ according to (2.95), and the comparison with

$$\varepsilon_2^{\text{RS}} = -\sum_k{}' \frac{|\langle\phi_0|\, V\,|\phi_k\rangle|^2}{E_k^0 - E_0^0}$$

is easily made by using (2.86) and (2.89):

$$\begin{aligned}S_{0k} &= \langle\phi_0|\,\mathscr{A}_{\text{int}}\,|\phi_k\rangle\\ H_{0k} &= \langle\phi_0|\, V\mathscr{A}_{\text{int}}\,|\phi_k\rangle + E_0^0 S_{0k}\end{aligned}$$

so that (2.98) becomes:

$$\varepsilon_2 = -\frac{1}{\langle\phi_0|\,\mathscr{A}_{\text{int}}\,|\phi_0\rangle}\sum_k{}' \frac{\langle\phi_0|\,(V-\varepsilon_1)\mathscr{A}_{\text{int}}\,|\phi_k\rangle\langle\phi_k|\,\mathscr{A}_{\text{int}} V\,|\phi_0\rangle}{E_k^0 - E_0^0} \quad (2.99)$$

$$\begin{aligned}&= -\frac{1}{1-\langle\phi_0|\,\mathscr{A}'_{\text{int}}\,|\phi_0\rangle}\sum_k{}' [\langle\phi_0|\, V\,|\phi_k\rangle - \langle\phi_0|\,(V-\varepsilon_1)\mathscr{A}'_{\text{int}}\,|\phi_k\rangle]\\ &\quad\times[\langle\phi_k|\, V\,|\phi_0\rangle - \langle\phi_k|\,\mathscr{A}'_{\text{int}} V\,|\phi_0\rangle]/(E_k^0 - E_0^0)\end{aligned} \quad (2.100)$$

As $R \to \infty$, $\langle\phi_0|\,\mathscr{A}'_{\text{int}}\,|\phi_0\rangle$ and $\langle\phi_0|\, V\mathscr{A}'_{\text{int}}\,|\phi_k\rangle$ decreases exponentially to zero, thus the difference between ε_2 and

$$-\sum_k{}' \frac{|\langle\phi_0|\, V\,|\phi_k\rangle|^2}{E_k^0 - E_0^0} = \varepsilon_2^{\text{RS}}$$

should become exponentially small. This argument is rather informal, a more rigorous one should correctly take into account the fact that the summation $\sum_k'$ must cover the continuum, and that the exponential decrease of the exchange matrix elements is not uniform with respect to k. Thus, the exponential decrease of $\varepsilon_2-\varepsilon_2^{\mathrm{RS}}$ is ensured if we use a *finite* basis only (see the related problem concerning the EL–HAV treatment in Certain, 1968 and Chipman, 1972). Note that ε_2 may be expressed in closed form by using the RS reduced resolvent

$$R_0=\sum_k{}' \frac{|\phi_k\rangle\langle\phi_k|}{E_0^0-E_k^0} \tag{2.101}$$

which is immediately introduced in (2.99), thus giving:

$$\varepsilon_2=\frac{1}{S_{00}}\langle\phi_0|\,(V-\varepsilon_1)\mathscr{A}_{\mathrm{int}}R_0\mathscr{A}_{\mathrm{int}}V\,|\phi_0\rangle \tag{2.102}$$

The *informal* argument may be extended to further orders for suggesting that $\varepsilon_n-\varepsilon_n^{\mathrm{RS}}=O(\mathrm{e}^{-R})$. Indeed, according to (2.92, 2.93, 2.94),

$$\begin{aligned}(\mathbf{V})_{ij}&=\langle\phi_i|\,V\mathscr{A}_{\mathrm{int}}\,|\phi_j\rangle+(E_i^0-E_0^0)(S_{ij}-\delta_{ij})\\&=\langle\phi_i|\,V\,|\phi_j\rangle+\langle\phi_i|\,V\mathscr{A}'_{\mathrm{int}}\,|\phi_j\rangle+(E_i^0-E_0^0)(S_{ij}-\delta_{ij}) &(2.103)\\&=\langle\phi_i|\,V\,|\phi_j\rangle+O(\mathrm{e}^{-R})=V_{ij}^{\mathrm{RS}}+O(\mathrm{e}^{-R}) &(2.104)\end{aligned}$$

Indeed

$$S_{ij}=\langle\psi_i\mid\psi_j\rangle=\langle\phi_i|\,\mathscr{A}_{\mathrm{int}}\,|\phi_j\rangle=\langle\phi_i\mid\phi_i\rangle-\langle\phi_i|\,\mathscr{A}'_{\mathrm{int}}\,|\phi_j\rangle=\delta_{ij}+O(\mathrm{e}^{-R});$$

therefore we also have for large R

$$\mathbf{S}=\mathbf{1}+O(\mathrm{e}^{-R}) \tag{2.105}$$

Finally, the reduced resolvent matrix $\mathbf{R}_0$ is exactly identical with the reduced resolvent matrix of the RS treatment in the $\{\phi_i\}$ basis (namely it is diagonal with $(\mathbf{R}_0)_{00}=0$ and $(\mathbf{R}_0)_{ii}=1/E_0^0-E_i^0)$). Then, since the matrix perturbation equations (2.31) differ from those of the long-range RS treatment only by the presence of $\mathbf{V}$ instead of $\mathbf{V}^{\mathrm{RS}}$ (with elements $(\mathbf{V}^{\mathrm{RS}})_{ij}=V_{ij}^{\mathrm{RS}}=\langle\phi_i|\,V\,|\phi_j\rangle$) and of $\mathbf{S}$ instead of $\mathbf{1}$, and since ε_0 and $\mathbf{c}_0$ are the same in both cases ($\varepsilon_0=E_0^0$, and $\mathbf{c}_0=(1,0,\ldots)$ which represents ψ_0 in the 'exchange' treatment and ϕ_0 in the long-range RS treatment) it is clear from (2.104) and (2.105) that we shall have, at any order, and when using a *finite* basis:

$$\varepsilon_n-\varepsilon_n^{\mathrm{RS}}=O(\mathrm{e}^{-R})\qquad c_n-c_n^{\mathrm{RS}}=O(\mathrm{e}^{-R}) \tag{2.106}$$

c. A second choice for the unperturbed matrix.

$$D_i=(E_i-E_0)/S_{ii},\qquad \varepsilon_0=E_0$$

Then $\mathbf{V}=\mathbf{H}-E_0\mathbf{S}-\mathbf{D}$, i.e. for $i\neq j$ (let us recall that $(\mathbf{V})_{ii}=0$ by the very definition of this second choice):

$$\begin{aligned}(\mathbf{V})_{ij} &= (\mathbf{H})_{ij} - E_0(\mathbf{S})_{ij} \\ &= H_{ij} - E_0 S_{ij} \\ &= \langle\psi_i|\, H - E_0\, |\psi_j\rangle\end{aligned} \tag{2.107}$$

and, using (2.90, 2.91)

$$(\mathbf{V})_{ij} = \langle\phi_i|\, V\mathscr{A}_{\text{int}}\, |\phi_j\rangle + (E_i^0 - E_0)S_{ij} \tag{2.108}$$

or the alternative expression

$$(\mathbf{V})_{ij} = \langle\phi_i|\, \mathscr{A}_{\text{int}} V\, |\phi_j\rangle + (E_j^0 - E_0)S_{ij} \tag{2.109}$$

The first-order correction $\varepsilon_1 = (\mathbf{V})_{00}/(\mathbf{S})_{00}$ is zero, so that the total energy up to first order is

$$\varepsilon_0 = E_0 = \frac{\langle\psi_0|\, H\, |\psi_0\rangle}{\langle\psi_0 \mid \psi_0\rangle}$$

i.e. the Heitler–London energy (2.75) as in the previous choice (we already noticed after equation (2.75) that the first-order total energy is common to most exchange perturbation treatments).

The second-order energy is:

$$\varepsilon_2 = -\frac{1}{S_{00}} \sum_k{}' \frac{|H_{0k} - E_0 S_{0k}|^2}{(E_k - E_0)S_{kk}} \tag{2.110}$$

$$= -\frac{1}{S_{00}} \sum_k{}' \frac{|H_{0k} - E_0 S_{0k}|^2}{H_{kk} - E_0 S_{kk}} \tag{2.111}$$

These formulae are very similar to those obtained by Hirschfelder (1976a) by using the Brillouin perturbation treatment: in the second-order term, the only difference is the replacement of E_0 by the exact (unknown) energy E in the Brillouin-type formula. Replacing E by its expansion $\sum_r \varepsilon_r \lambda^r$ and identifying term by term the various powers of λ would reproduce our formula (2.111). Our explicit treatment of course presents several advantages: the validity has been insured even for an 'overcomplete' basis, the correct connection between the perturbed and unperturbed eigenstates may be ascertained and the convergence properties would be easier to study.

The formula (2.111) for ε_2 differs from the previous one (2.98) by the replacement of E_0^0, E_k^0 by the corresponding mean values E_o, E_k. Upon insertion of (2.108, 2.109), we get:

$$\varepsilon_2 = -\frac{1}{S_{00}} \sum_k{}' \frac{|\langle\phi_0|\, V\mathscr{A}_{\text{int}}\, |\phi_k\rangle + (E_0^0 - E_0)S_{0k}|^2}{E_k - E_0} \tag{2.112}$$

Now, the differences $E_k - E_k^0$ are of the order of magnitude of V; the calculation is the same as was done previously for $k=0$ in order to get the Heitler–London interaction energy (2.96, 2.97) by using (2.90, 2.91):

$$
\begin{aligned}
E_k - E_k^0 &= \frac{H_{kk}}{S_{kk}} - E_k^0 = \frac{\langle \phi_k | V\mathscr{A}_{\text{int}} | \phi_k \rangle + E_k^0 S_{kk}}{S_{kk}} - E_k^0 \\
&= \frac{\langle \phi_k | V\mathscr{A}_{\text{int}} | \phi_k \rangle}{S_{kk}} = \frac{\langle \phi_k | V\mathscr{A}_{\text{int}} | \phi_k \rangle}{\langle \phi_k | \mathscr{A}_{\text{int}} | \phi_k \rangle} \\
&= \frac{\langle \phi_k | V | \phi_k \rangle - \langle \phi_k | V\mathscr{A}'_{\text{int}} | \phi_k \rangle}{1 - \langle \phi_k | \mathscr{A}'_{\text{int}} | \phi_k \rangle} \\
&= \langle \phi_k | V | \phi_k \rangle + \frac{\langle \phi_k | V | \phi_k \rangle \langle \phi_k | \mathscr{A}'_{\text{int}} | \phi_k \rangle - \langle \phi_k | V\mathscr{A}'_{\text{int}} | \phi_k \rangle}{1 - \langle \phi_k | \mathscr{A}'_{\text{int}} | \phi_k \rangle}
\end{aligned}
\tag{2.113}
$$

Thus, the principal term of ε_2, which is of order of magnitude V^2, is still the RS long-range term $\varepsilon_2^{\text{RS}} = -\sum_k' |\langle \phi_0 | V | \phi_k \rangle|^2 / (E_k^0 - E_0^0)$. The supplementary term differs from the one in the previous choice (2.98, 2.99, 2.100) not only by terms of higher order in V due to the different denominators, but also by the so-called exchange terms since, in the numerator, the second matrix element is here $\langle \phi_k \mid \mathscr{A}_{\text{int}}(V - \varepsilon_1^{\text{HL}})\phi_0 \rangle$ (formula (2.112), with $\varepsilon_1^{\text{HL}} = E_0 - E_0^0$) instead of $\langle \phi_k | \mathscr{A}_{\text{int}} V | \phi_0 \rangle$ (formula (2.99)). Since $S_{0k} = \langle \phi_0 | \mathscr{A}_{\text{int}} | \phi_k \rangle$, we may rewrite (2.112) as

$$
\varepsilon_2 = -\frac{1}{S_{00}} \sum_k{}' \frac{|\langle \phi_0 | (V - \varepsilon_1^{\text{HL}}) \mathscr{A}_{\text{int}} | \phi_k \rangle|^2}{E_k - E_0} \tag{2.114}
$$

$$
= \frac{\langle \phi_0 | (V - \varepsilon_1^{\text{HL}}) \mathscr{A}_{\text{int}} R \mathscr{A}_{\text{int}} (V - \varepsilon_1^{\text{HL}}) | \phi_0 \rangle}{\langle \phi_0 | \mathscr{A}_{\text{int}} | \phi_0 \rangle} \tag{2.115}
$$

where we have introduced the new reduced resolvent

$$
R = \sum_k{}' \frac{|\phi_k\rangle\langle\phi_k|}{E_o - E_k} \tag{2.116}
$$

Under this form (2.115), we see that our second choice ε_2 resembles the EL–HAV ε_2 (Van der Avoird, 1967b; Chipman, 1972, Table 5.3, p. 85) but differs from it by the presence of R instead of R_0 (2.101) and of $\mathscr{A}_{\text{int}} = N_{\text{int}} A_{\text{int}}$ instead of the projection operator A itself (remember that for the functions $\phi_i = W_a^{(1)} W_b^{(2)}$, we have $A_{\text{int}} \phi_i = A\phi_i$ (Appendix A, formula (2.407)). This second difference will obviously avoid the appearance of the wrong factor $1/N_{\text{int}}$ which makes the EL–HAV treatment clearly incorrect at large distances (see the discussion in Section 1 above). Similarly our first-choice ε_2 (2.112) involves R_0 itself but contains V instead of $(V - \varepsilon_1^{\text{HL}})$ and $\mathscr{A}_{\text{int}}$ instead of A.

Our second choice (c), together with the super-secular equation method (Löwdin, 1967), the perturbation theory for projected states (Peierls, 1973; Atalay *et al.*, 1973) and the primitive function methods (Chipman *et al.*,

1973) are discussed by Fleckinger *et al.* (1976) in the general framework of the 'ideal space' concept. On another hand, it is easily seen that none of the perturbation treatments reviewed by Chipman (1972, see especially Table 5.3, p. 84) gives for ε_2 the same result as one of our two choices. Indeed, as we mentioned in Section I above, the 'exchange' perturbation treatments begin to differ essentially at the second-order exchange contribution. But this lack of uniqueness (it would make no sense of speaking of 'ambiguity', since each treatment is perfectly defined for itself) is not as serious a defect as has been feared by some authors (e.g. see the Introduction of Gordon and Kim (1972), or Certain and Bruch (1972) Section 4.2.2). Indeed:

(1) Actual calculations on simple model systems (H_2: Certain *et al.* (1968); H_2^+: Chipman (1972), Chipman *et al.* (1973), Chipman and Hirschfelder (1973); Chalasinski *et al.* (1977) allow a comparison of the various exchange treatments, and already show that some of them are quite satisfactory, at least in the regions of interest here, namely medium- and long-range (e.g. see the Hirschfelder–Silbey (HS) and the Murrell–Shaw–Musher–Amos (MS–MA) treatments in Chipman's (1972) thesis, and the HS treatment in the study of H_2^+ by Chalasinski *et al.*, 1977).

(2) The satisfactory treatments give very close numerical results in the region of interest, so that the choice of either one becomes relatively immaterial.

(3) As already mentioned, several treatments begin to differ only at the second-order exchange term (for example this is the case for the HS, MS–MA and the two expansions proposed in the present work). Now, this term may be much smaller than the common terms ε_1^{HL} and ε_2^{RS} (e.g. see the evaluation for He . . . He by Snook and Spurling (1975)), and in such cases the expression $\varepsilon_1^{HL}+\varepsilon_2^{RS}$ will give a suitable unique approximation to the interaction energy (this proposal was already made by Murrell and Shaw (1967b), and it will be discussed more thoroughly at the beginning of the next Section III; this approximation has been used recently with success by Jeziorski and Van Hemert, 1976).

d. Remark concerning the 'charge-transfer' term. Thus far we considered the functions $\psi_i = \mathfrak{A}_{int}[W_a^{(1)}W_b^{(2)}]$. If $W_a^{(1)}$ and $W_b^{(2)}$ are Slater determinants for the subsystems, the functions ψ_i are themselves Slater determinants built from n_1 spin-orbitals of subsystem (2). By contrast, the 'charge-transfer' functions are Slater determinants with (n_1+n_2) electrons built from (n_1+p) spin-orbitals of (1) and (n_2-p) spin-orbitals of (2) (with $p=\pm 1, \pm 2, \ldots$). If the *complete* bases $\{W_a^{(1)}, W_b^{(2)}\}$ were used for the subsystems, the (overcomplete) set $\{\mathfrak{A}_{int}(W_a^{(1)}W_b^{(2)})\}$ would be perfectly sufficient for getting the exact energy of the complex; and the charge-transfer functions may actually be expressed in terms of this overcomplete basis. As a consequence, no clear identification of a separate charge-transfer contribution to the interaction energy could be made, and it is actually *not* made in the accurate calculations by Kolos (1974) and Kolos and Wolniewicz (1974) concerning the

H...H interactions. The situation becomes different when, for practical reasons, it is necessary to use a rather limited finite basis of orbitals for each subsystem. Then, the charge-transfer functions are almost orthogonal to the limited *incomplete* set of the usual (non-transfer) functions at large and medium distances (indeed, for representing a charge-transfer state, an infinity of usual states, especially highly excited and even continuum states, would be necessary). Thus, adding the charge-transfer functions to the ordinary ones realizes an improvement towards completeness of the limited set of totally antisymmetrized functions which are used for the perturbation treatment, as was already stressed by Murrell, Randic, and Williams (1965). The formulae obtained in Section (c) are more appropriate since their denominators involve only mean values of H (and not mean values of the hamiltonians $H^{(m)}$ of the neutral subsystems). Thus, denoting by k the ordinary (non-transfer) functions and by k' the transfer ones, the second-order term may be written as the sum of an ordinary contribution and a charge-transfer one: (2.111) becomes

$$\varepsilon_2 = -\frac{1}{S_{00}}\left[\sum_k{}' \frac{|H_{0k}-E_0S_{0k}|^2}{H_{kk}-E_0S_{kk}} + \sum_{k'} \frac{|H_{0k'}-E_0S_{0k'}|^2}{H_{k'k'}-E_0S_{k'k'}}\right] \tag{2.117}$$

At higher orders, we could distinguish ordinary terms, pure charge-transfer terms and mixed terms. But it must be once more emphasized that these decompositions do *not* have an intrinsic character and depend on the *finite* basis used.

We noticed previously that the numerator $H_{0k}-E_0S_{0k} = \langle\psi_0|\,H-E_0\,|\psi_k\rangle = \langle\phi_0|\,V\mathscr{A}_{\text{int}}\,|\phi_k\rangle$ (see (2.108)) had (for a non-transfer function) a long range part $\langle\phi_0|\,V\,|\phi_k\rangle$ of order V and a short-range 'exchange' one $-\langle\phi_0|\,V\mathscr{A}'_{\text{int}}|\phi_k\rangle$ of order VS^2 (where S denotes the order of magnitude of an overlap integral between an orbital of system (1) and an orbital of system (2)). No such difference appears for a charge-transfer numerator $H_{0k'}-E_0S_{0k'} = \langle\phi_0|\,V\,|\phi_{k'}\rangle - \langle\phi_0|\,V\mathscr{A}'_{\text{int}}\,|\phi_{k'}\rangle$ (here $\phi_{k'}$ denotes the product of an ordinary state of one system by a charge-transfer excited state of the other, i.e. a state where a spin-orbital of the system has been replaced by an (empty) spin-orbital of the other). Then, it is easily seen that both terms have the same order of magnitude, namely VS (indeed, the first term contains electronic and nuclear hybrid integrals, of order VS, and the second term contains integrals of the same order, corresponding to the exchange involving the receiving orbital of the acceptor and any other orbital of the acceptor). Consequently, the charge-transfer term cannot be split into a 'polarization' (long-range RS) part of order V^2 and an 'exchange' part of order $V(VS^2) = V^2S^2$: it appears as a homogeneous short-range contribution of order $(VS)^2 = V^2S^2$. A further discussion will be found in Section III.A below, concerning the kinds of complexes where the evaluation of a part of the second-order term ε_2 as a charge-transfer contribution appears desirable.

Let us finally mention that the thorough analysis of the interaction problem performed in the framework of the 'exchange' perturbation treatments may be applied in the practical case of computation involving non-exact ground state functions of the subsystems and finite bases, thus allowing to devise better *ab initio* perturbational or variational procedures (e.g. by making a clear analysis of the 'basis extension effect' pointed out by Clementi (1967) and Morokuma and Pedersen (1968): see Claverie (1973, Ch. II, Section 4-D), Daudey *et al.* (1974a,b), Daudey (1974).

III. GENERAL CONCEPTS AND RESULTS CONCERNING THE DERIVATION OF SIMPLIFIED FORMULAS

A. Search for approximate expressions of the interaction energy

1. Proposal of an approximation and discussion of the neglected terms

In the parts III, IV, V of this Chapter we shall be concerned with the derivation of simplified formulae from the theoretical expressions discussed in the previous section. More specifically, we shall use our 'first choice' of the unperturbed matrix $\mathbf{H}^0$, because its relationship with the RS long-range treatment is more direct (the denominators $E_i^0 - E_j^0$ are the same). Thus the interaction energy is

$$\Delta E = \varepsilon_1^{\mathrm{HL}} + \sum_{r=2}^{\infty} \varepsilon_r \tag{2.118}$$

where $\varepsilon_1^{\mathrm{HL}} = \varepsilon_1$ and ε_2 have been explicitly written in the section II.B.3.c above (see formulae (2.95, 2.96, 2.97, 2.98, 2.99, 2.100, 2.102)). Each term ε_r may be split into a long-range part $\varepsilon_r^{\mathrm{RS}}$ (corresponding to the RS long-range treatment or 'polarization approximation' recalled in Section II.A) and a short-range supplementary 'exchange' term, say $\varepsilon_r^{\mathrm{ex}}$, which involves integrals of the overlap, hybrid and exchange type, so that this term decreases exponentially as a function of the intermolecular distance R. It must be emphasized that each term $\varepsilon_r^{\mathrm{RS}}$ itself contains, on one hand the multipolar long-range parts (which vary as R^{-n}, and are obtained from multipole expansions of nuclear and Coulomb integrals appearing in $\varepsilon_r^{\mathrm{RS}}$), and on the other hand the short-range (exponentially decreasing) parts brought about by the so-called 'penetration' part of the intermolecular nuclear and Coulomb integrals (e.g. see Roothaan's, 1951, formulae).

In the present section we shall be concerned essentially with the simplified evaluation of the Heitler–London (first-order) interaction $\varepsilon_1^{\mathrm{HL}}$ and the multipolar part ε_2^m of the second-order RS term $\varepsilon_2^{\mathrm{RS}} = \varepsilon_2^m + \varepsilon_2^{\mathrm{pen}}$. By retaining only these terms Murrell and Shaw (1967b, 1968b) indeed found values of the interaction energy for He . . . He in good agreement with experimental results. Several authors (Margenau, 1939a; Margenau and Rosen, 1953; Kreek and Meath, 1969; Kolos, 1974) also came to the conclusion that a practical evaluation of the interaction energy could often be obtained without a detailed evaluation of the second-order penetration and exchange parts (sometimes called 'charge overlap effects'). The validity of this approximation is certainly not completely general, and for example Certain (1971) mentions that it does not apply well to the He . . . H^+ interaction. But the fact that it appears convenient for He . . . He is encouraging, since this system is just a case of interaction between neutral closed-shell systems, Now, in order to discuss this approximation more thoroughly, let us list the neglected terms:

(a) *The penetration part* $\varepsilon_2^{\text{pen}}$ *of* $\varepsilon_2^{\text{RS}}$. This non-multipolar part of $\varepsilon_2^{\text{RS}}$ has been studied under the names 'overlap effect' by Murrell and Shaw (1968c) or 'charge overlap effect' by Kreek and Meath (1969); Singh, Kreek and Meath (1970); Kreek, Pan and Meath (1970); Pan and Meath (1971); Bukta and Meath (1972), and is also discussed, together with the exchange part $\varepsilon_2^{\text{ex}}$ by Kreek and Meath (1971), Chalasinski and Jeziorski (1973, 1974). According to Murrell and Shaw (1968c), Kreek and Meath (1969, 1971), the penetration part $\varepsilon_2^{\text{pen}}$ is still small at the equilibrium distance for He . . . He (about 5·75 a.u.), and reduces the absolute value of the dispersion energy, because it reduces the absolute value of the integrals $\langle W_0^{(1)} W_0^{(2)} | V | W_a^{(1)} W_b^{(2)} \rangle$ which appear in the numerators of the perturbation expansion (2.10). Kolos (1974) confirms this behaviour for the H . . . H interaction. The situation seems less simple as concerns the polarization term. The reduction of the values of the integrals involving not too high excited states (with respect to their multipolar approximation) and the resulting reduction of the polarization terms certainly exists as for the dispersion terms; but for the integrals involving very high excited states (and continuum states), the multipolar approximation loses its meaning at medium distances, and, as we saw in the Section II.B.3.d above, these terms correspond to the so-called 'charge-transfer' terms which replace them in practical computations involving necessarily a finite number of basis functions (or of atomic orbitals). Since, as pointed out previously, these charge-transfer terms cannot be conveniently split into penetration parts of RS long-range terms and 'exchange' parts, we defer their further discussion to point (c) below.

At sufficiently short distances, the penetration part compensates for the increase to infinity of the multipolar part and insures that the integrals involved in the second-order terms (polarization and dispersion) of formula (2.10), and therefore these terms themselves, remain finite as the distance $R \rightarrow 0$. In the simplified formulae based upon multipolar terms, it is therefore necessary to introduce appropriate 'cut-offs' at short distance in order to avoid the spurious divergence to infinity of these terms when the distance decreases to zero (see Section VI.A.5).

(b) *The second-order 'exchange' contribution* $\varepsilon_2^{\text{ex}} = \varepsilon_2 - \varepsilon_2^{\text{RS}}$. This contribution differs according to the particular exchange treatments used, but, fortunately, the available numerical experience indicates that it is quite small for the best exchange treatment used so far. Thus, for H_2^+ (interaction $H \ldots H^+$),* Chalasinski *et al.* (1977) find, at $R = 12{\cdot}5$ a.u. (for the ungerade state, which is mathematically analogous to the physical ground state, while the gerade state corresponds to the mathematical ground state): $\varepsilon_2^{RS} = -9{\cdot}425 \times 10^{-5}$ a.u. and $\varepsilon_2^{HS} = -9{\cdot}116 \times 10^{-5}$ a.u. (where the superscript HS refers to the Hirschfelder–Silbey (1966) treatment), hence $\varepsilon_2^{ex(HS)} = 0{\cdot}31 \times 10^{-5}$ a.u., i.e. -3.2 per cent of ε_2^{RS} and $-5{\cdot}26$ per cent of the exact interaction energy $\Delta E = -6{\cdot}077 \times 10^{-5}$ a.u. ($R = 12{\cdot}5$ a.u. corresponds to the minimum region for the ungerade state). For the H... H interaction, in the triplet state, at $R = 8{\cdot}0$ a.u. (which corresponds to the Van der Waals minimum region), with Kolos and Wolniewicz's (1974) conversion factor: 1 a.u. (energy) $= 219\,474{\cdot}62$ cm^{-1} (see also Kauzmann (1957), Appendix II, p. 730), Certain *et al.* (1968) evaluated $\varepsilon_2^{ex} = 0{\cdot}3$ cm^{-1} for three 'exchange' treatments (EL–HAV (Eisenschitz and London, 1930); Van der Avoird, 1967a,b,c); Hirschfelder, 1967b); HS (Hirschfelder and Silbey, 1966); MS–MA (Murrell and Shaw, 1967a; Musher and Amos, 1967; Amos and Musher, 1967)); now, they find $\varepsilon_2^{RS} = -8{\cdot}0$ cm^{-1} (hence $\varepsilon_2^{ex} = -3{\cdot}75$ per cent of ε_2^{RS}) and a total interaction energy ΔE of about $-4{\cdot}3$ cm^{-1} ($-4{\cdot}42$ cm^{-1} in Kolos (1974), $-4{\cdot}43$ cm^{-1} in Kolos and Wolniewicz (1974)), i.e. $\varepsilon_2^{ex} = -6{\cdot}8$ per cent of ΔE. For the He... He interaction, Snook and Spurling (1975) found a completely negligible ε_2^{ex} for the MS–MA treatment: see their Table 3: at $R = 5{\cdot}5$ a.u., the largest exchange term amounts to only $0{\cdot}0003 \times 10^{-5}$ a.u., while the dipole–dipole dispersion energy amounts to $-5{\cdot}27 \times 10^{-5}$ a.u. and the total second-order (dispersion) energy to $-7{\cdot}295$ a.u. (let us recall that Murrell and Shaw (1968b) gave $-6{\cdot}998$ a.u. with a two-term formula $-C_6/R^6 - C_8/R^8$, while Snook and Spurling also include the two $1/R^{10}$ terms). It is quite interesting that, among the three simple systems just considered, it is the He... He system, namely the case of two closed shell systems, which gives by far the most negligible second-order exchange contribution: this situation actually gives good hope to get sufficiently reliable formulas for the interaction energy between closed-shell systems, even when neglecting this contribution ε_2^{ex}. At this point, it may be mentioned that Kreek and Meath (1971) studied the He... He interaction energy up to second-order by applying the Unsöld (1927) approximation (i.e. replacing the various denominators by an average one) to a symmetrized perturbation treatment of the Brillouin–Wigner type. These authors therefore include in their treatment both the penetration and exchange parts discussed above, and actually conclude that, in the equilibrium region and at larger distance, these contributions may be practically neglected with respect to the second-order (suitably truncated) multipole part ε_2^m.

* As already mentioned, the calling 'exchange' treatment shows here its inadequacy: we actually have a 'symmetrized' treatment, where the symmetry is not a permutational one (between electrons) but rather a geometrical one (symmetry of the electronic function with respect to the nuclear positions).

(c) *The charge-transfer* (*C.T.*) *contribution.* In a previous work (Claverie, 1973, p. 160) it was stated that the 'exchange' contribution to the interaction would include the charge-transfer (C.T.) contribution. Actually, the status of this C.T. contribution is less simple: according to the analysis of Section II.B.3.d, the C.T. states could be expanded in terms of an infinity of highly excited (and continuum) ordinary states, and the C.T. terms would then appear as the sum of a *part* of $\varepsilon_2^{\text{pen}}$ and a *part* of $\varepsilon_2^{\text{ex}}$, both parts being of the same order of magnitude. The use of the C.T. states then appears as convenient practical method for evaluating a part of the sum of $\varepsilon_2^{\text{pen}}$ and $\varepsilon_2^{\text{ex}}$, a part which would dominate $\varepsilon_2^{\text{pen}}+\varepsilon_2^{\text{ex}}$ (and be non-negligible with respect to the total interaction energy) for some types of complexes.

This exponentially varying contribution is an attractive one, and its magnitude of course changes considerably from a system to another: while being negligible for the He . . . He or H . . . H interactions, it becomes of the same order of magnitude as the other terms (at equilibrium distance) in the case of the hydrogen bond (Daudey, 1974). As concerns He . . . He (equilibrium distance: about 5·75 a.u.), Daudey *et al.* (1974b) find at $R=5{\cdot}5$ a.u. a proper charge-transfer energy of $-0{\cdot}08\times10^{-5}$ a.u., to be compared with a 'polarization' energy (contribution of the local excitations) of $-0{\cdot}14\times 10^{-5}$ a.u. and a multipolar dispersion energy of $-7{\cdot}0\times10^{-5}$ a.u. (Murrell and Shaw, 1968b). As concerns H . . . H, Kolos (1974) finds $|\varepsilon_2^{\text{RS}}|<|\varepsilon_2^{m}|$ at $R=8$ a.u., i.e. $\varepsilon_2^{\text{pen}}>0$, while the reverse would hold if the attractive charge-transfer contribution were important. Actually, Murrell and Teixeira–Dias (1971, Table 2) evaluate for the H . . . H interaction at $R=8$ a.u.: E (dispersion) $=-2{\cdot}3962\times10^{-5}$ a.u., E (charge-transfer) $=-0{\cdot}0795\times10^{-5}$ a.u. $\simeq3{\cdot}32$ per cent of E (dispersion). However, it remains less important in complexes having equilibrium distances larger than those for the hydrogen bond, even for the so-called charge-transfer complexes. It is now widely recognized that, in such complexes, *the charge transfer occurs in the excited state function, but remains rather small in the ground state of the complex and accordingly brings only a relatively small contribution to the interaction energy* (Dewar and Thompson, 1966; Malrieu and Claverie, 1968; Mantione, 1968a,b, 1969a,b, 1971; Cook and Schug, 1970; Lippert *et al.*, 1969; Hanna and Lippert, 1973; Soos and Klein, 1975). As a rule, it could be said that the charge-transfer contribution becomes important only in those complexes with fairly short equilibrium distances, e.g. the halogen–lone pair complexed (Malrieu and Claverie, 1968, Section II.4), or the already quoted hydrogen bonded complexes.

In the present chapter, we shall not try to derive simplified formulae for the charge transfer contribution. The literature devoted to the so-called charge transfer complexes is indeed very abundant; e.g. see besides the works quoted above: Mulliken (1950, 1952, 1964), Brieglieb (1961), Andrews and Keefer (1964), Rose (1967), Herndon and Feuer (1968), Mulliken and Person (1962, 1969), Foster (1969, 1973), Slifkin (1971), Soos (1974) and Johansen (1975). We shall therefore consider the other contributions. It must also be emphasized that, by using directly the multipolar part

ε_2^m of ε_2^{RS}, the omission of the reduction which would be brought (for the low-lying excited states) by the penetration part may cancel to some extent the omission of an explicit charge transfer contribution: thus, by using only $\varepsilon_1^{HL}+\varepsilon_2^m$ (as we intend to do), Alvarez-Rizzati and Mason (1975) obtained a quite good evaluation of the interaction energy between rare-gas atoms and closed-shell ions, although an independent evaluation of the charge transfer contribution alone indicated that it was not negligible.

(d) *The remainder* $\sum_{r=3}^{\infty}\varepsilon_r$ *of the series* (2.118) *beyond second order.* A few evaluations of the higher-order terms or of accurate total interaction energies are available, and actually suggest that the neglect of the terms beyond second-order is a possible approximation from the practical point of view.

The most complete results are those of Chalasinski *et al.* (1977) for H_2^+: for the ungerade state, at $R = 12{\cdot}5$ a.u., the Hirschfelder–Silbey (1966) symmetrized perturbation treatment already gives at second-order a relative error of only $-0{\cdot}63$ per cent, and even the simple approximation $\Delta E = \varepsilon_1^{HL}+\varepsilon_2^{RS}$ gives a relative error of only $-4{\cdot}5$ per cent (${}^uE_{exact} = -6{\cdot}077\times10^{-5}$ a.u., $\varepsilon_1^{HL} = (1-1{\cdot}50621)\,{}^uE_{exact} = 3{\cdot}08\times10^{-5}$ a.u. and $\varepsilon_2^{RS} = -9{\cdot}425 = 10^{-5}$ a.u.). The next long-range RS higher-order terms beyond the second one actually decrease quite rapidly: $\varepsilon_3^{RS}/\varepsilon_2^{RS} = 0{\cdot}0134$, $\varepsilon_4^{RS}/\varepsilon_3^{RS} = 0{\cdot}125$.

Let us recall that, as explained previously in Section II.B.1, there is no interest in going too far in the RS expansion: the first terms (up to fourth order) give *practically* the mean value $({}^uE+{}^gE)/2$ (g denotes the gerade state, which is here the mathematical ground state), and adding further terms would make the sum of the RS series converge *extremely slowly* towards ${}^gE$; the interesting procedure is obviously to stop the RS series at the first few terms, and then to add an 'effective repulsion energy' $({}^uE-{}^gE)/2$, which is, to a convenient accuracy, simply given by $\varepsilon_1^{HL}-\varepsilon_1^{RS}$.

As concerns the H . . . H interaction, a comparison of the multipole (long-range) parts ε_2^m and ε_3^m (of ε_2^{RS} and ε_3^{RS} respectively) has been made by Arrighini, Biondi and Guidotti (1973), who give the asymptotic expansions:

$$\varepsilon_2^m(R) = -\left[\frac{6{\cdot}492}{R^6}+\frac{123{\cdot}9}{R^8}+\frac{3273{\cdot}0}{R^{10}}+\frac{121000{\cdot}0}{R^{12}}+\dots\right] \qquad (2.119)$$

$$\varepsilon_3^m(R) = \frac{3475{\cdot}0}{R^{11}}+\frac{2914\times10^2}{R^{13}}+\frac{2305\times10^4}{R^{15}}+\dots \qquad (2.120)$$

where R and ε are expressed in atomic units. At $R = 8{\cdot}0$ a.u. (which corresponds to the equilibrium region for the triplet state: according to the accurate calculations of Kolos and Wolniewicz (1974) the minimum occurs between 7·8 and 7·85 a.u.), we therefore get

$$\varepsilon_2^m(8{\cdot}0) = -\frac{1}{(8{\cdot}0)^6}[6{\cdot}492+1{\cdot}935+0{\cdot}8+0{\cdot}46+\dots]$$

$$\varepsilon_3^m(8{\cdot}0) = \frac{1}{(8{\cdot}0)^6}[0{\cdot}106+0{\cdot}111+0{\cdot}1715+\dots]$$

Thus, the asymptotic expansion for ε_3^m cannot give an accurate estimate of $\varepsilon_3^{\mathrm{RS}}$ at $R=8$ a.u., but if we consider only the order of magnitude, we see that $\varepsilon_3^{\mathrm{RS}}(8\cdot0)/\varepsilon_2^{\mathrm{RS}}(8\cdot0) \simeq (0\cdot388/9\cdot687) = -0\cdot04$. On another hand, the direct calculation by Kolos (1974) gives $\varepsilon_1^{\mathrm{RS}} = -0\cdot393\ \mathrm{cm}^{-1} = -0\cdot1791\times10^{-5}$ a.u., $\varepsilon_1^{\mathrm{RS}}+\varepsilon_2^{\mathrm{RS}} = -8\cdot421\ \mathrm{cm}^{-1} = -3\cdot84\times10^{-5}$ a.u. and $\varepsilon_1^{\mathrm{RS}}+\varepsilon_2^{\mathrm{RS}}+\varepsilon_3^{\mathrm{RS}} = -8\cdot123$ $\mathrm{cm}^{-1} = -3\cdot7\times10^{-5}$ a.u., hence $\varepsilon_3^{\mathrm{RS}}(8\cdot0)/\varepsilon_2^{\mathrm{RS}}(8\cdot0) = -(0\cdot14/3\cdot661) = -0\cdot0382$. The agreement between the two evaluations is quite remarkable, and if we note that the further terms would partly cancel the third-order term, it may be concluded also for the H . . . H interaction that the neglect of the RS terms beyond second-order appears quite justified since we already intend to get only *simplified approximate* formulae for the $\varepsilon_1^{\mathrm{RS}}$ and $\varepsilon_2^{\mathrm{RS}}$. Kolos (1974) gives for $(\Delta E(\text{triplet})+\Delta E(\text{singlet}))/2$ (considered, as explained previously, as an evaluation of the first few terms of the RS series) the value $-8\cdot274\ \mathrm{cm}^{-1} = -3\cdot77\times10^{-5}$ a.u., and the difference with $\varepsilon_1^{\mathrm{RS}}+\varepsilon_2^{\mathrm{RS}}$ amounts only to $0\cdot067\times10^{-5}$ a.u., i.e. $-(0\cdot067/3\cdot658) = 0\cdot0183 = 1\cdot83$ per cent of $\varepsilon_2^{\mathrm{RS}}$.

Finally, Stogryn (1971) studied the asymptotic multipolar expansions of $\varepsilon_2^{\mathrm{RS}}$ and $\varepsilon_3^{\mathrm{RS}}$ for the general case (subsystems eventually possessing non-zero permanent multipoles) and gave the explicit results for the He . . . He interaction:

$$\varepsilon_2^m(R) = -\left(\frac{1\cdot471}{R^6}+\frac{14\cdot1}{R^8}+\frac{182\cdot0}{R^{10}}+\ldots\right) \qquad (2.121)$$

$$\varepsilon_3^m(R) = \frac{24\cdot4}{R^{11}}+\ldots \qquad (2.122)$$

(ε and R in a.u.). Note that the R^{-9} term of ε_3^m has a zero coefficient for the interaction between spherically symmetric systems (compare (2.120)). At the distance $R = 5\cdot75$ a.u. (van der Waals minimum) we therefore get:

$$\varepsilon_2^m(5\cdot75) = -(4\cdot46+1\cdot292+0\cdot505+\ldots)10^{-5}\ \text{a.u.}$$

$$\varepsilon_3^m(5\cdot75) = (0\cdot0118+\ldots)10^{-5}\ \text{a.u.}$$

Thus, the first term of the $1/R$ expansion of ε_3^m amounts to only $0\cdot2$ per cent of ε_2^m at the equilibrium distance (for the H . . . H interaction, the corresponding ratio was $0\cdot106/9\cdot687 = 1\cdot1$ per cent). The first term of the fourth-order contribution, of the form C/R^{12}, could have a coefficient larger than that of R^{11} in (2.122), but the result obtained with ε_3^m nevertheless suggests that the neglect of the perturbations terms beyond the second-order one may be a reasonable approximation for He . . . He at large and medium distances (up to the equilibrium region). It is interesting that the accuracy of this approximation is better for He . . . He (which is an interaction between closed-shell systems) than for H . . . H. By the way, it must be stressed that, even for H . . . H, the error resulting from the neglect of the terms beyond second-order remains of the order of some per cent: according to the Table III of Kolos (1974), we have $\varepsilon_1^{\mathrm{RS}}+\varepsilon_2^{\mathrm{RS}} = E_{\mathrm{pol}}^{(2)} = -8\cdot421\ \mathrm{cm}^{-1}$ and $E_{\mathrm{pol}} =$ $-8\cdot274\ \mathrm{cm}^{-1}$, i.e. an accuracy of $0\cdot147/8\cdot274 = 1\cdot78$ per cent (here E_{pol} is defined as $(\Delta E(\text{singlet})+\Delta E(\text{triplet}))/2$, and, as we have seen previously (Section II.B.1), the work of Chalasinski *et al.* (1977) actually suggests that there is a kind of 'intermediate convergence' of the RS series towards this mean value). The accuracy is less good for the total interaction energy, because its absolute value is smaller than that of E_{pol} (see Table IV of Kolos (1974)). But this accuracy may be considered as sufficient from our point of view, namely the derivation of simplified formulae (although it was considered as rather low by Kolos (1974) whose purpose was a highly accurate analysis).

More generally, summations of long-range RS terms beyond the second order performed by using many-body diagrammatic techniques (Malrieu, 1971a,b) support the conclusion that neglecting these terms may be a reasonable approximation in a number of usual cases.

At the end of this discussion of the remainder $\sum_{r=3}^{\infty}\varepsilon_r$, it must be emphasized that, up to now, the exact eigenfunctions and eigenvalues of the subsystems were supposed to be used. If this is not the case, serious difficulties may occur: the calculation of the energy of the complex indeed involves some spurious terms which correspond to a lowering of the *intra*molecular energies: this is the so-called 'basis extension effect' (Clementi, 1967; Morokuma and Pedersen, 1968; Liu and McLean, 1973; Daudey *et al.*, 1974a,b; Daudey, 1974). Instead of suppressing directly these spurious terms in the series for the interaction energy, Daudey *et al.* (1974b) compute (also by the perturbation method) the energy lowering of the subsystems in a suitably extended base (they add to the basis of each subsystems the virtual molecular orbitals of the other), and they subtract the sum of these lowerings from the (too large) interaction energy computed by its own series. But the cancellation of the spurious terms does not occur exactly order by order, so that the series obtained by the difference (of the series for the complex and the subsystems) converges only very slowly in the He . . . He case (the convergence is better but still not very fast in the $H_2O \ldots H_2O$ case (Daudey, 1974)). This slow convergence is only due to the use of a difference of two series, instead of a directly defined series for the interaction energy. Such results therefore do not contradict our previous conclusion concerning the practically negligible magnitude of the remainder already at low order in a directly defined perturbation series for the interaction energy: they stress the kind of difficulty which may occur, even in *ab initio* computations, when a small quantity is obtained as the difference of two larger quantities instead of being evaluated directly.

2. Conclusion

We shall therefore conclude, from our discussion of points (a)–(d) above, that the approximate expression for the interaction energy

$$\Delta E \simeq \varepsilon_1^{\text{HL}} + \varepsilon_2^{m} \tag{2.123}$$

may be a suitable starting point for the derivation of simplified formula. Of course, the multipolar expansion ε_2^m of ε_2^{RS} must be suitably truncated since it is only an asymptotic expansion (e.g. see Erdelyi (1956) for the definition of such expansions) and not a convergent series (see Young (1975) and references therein, Dalgarno and Lewis (1956), and also Deal and Young (1976)).

For rare gas atoms, the proposal (2.123) becomes equivalent with the suggestion made by Das and Wahl (1971), namely to add the Hartree–Fock interaction energy ε_{HF} and the long-range Van der Waals interaction (which reduces to the dispersion in the present case): the latter interaction is indeed our ε_2^m, and the former is practically equal to our ε_1^{HL} for rare gas atoms: for two He atoms, $E_{coul}+E_{exch}$ (see Note 1, p. 305) (we note $\varepsilon_1^{pen}=\varepsilon_1^{RS}$ for E_{coul} and $\varepsilon_1^{ex(1)}$ for E_{exch}, see Section IV.C below). Now, as recalled recently by Cohen and Pack (1974, p. 2735), it was shown by Toennies (1973) that this suggestion of Das and Wahl works well for all like pairs of rare gas atoms. For systems having a non-zero electrostatic long-range interaction, the situation is less simple, because ε_{HF} then includes a polarization contribution (which is a part of ε_2^m); thus, only the dispersion part of ε_2^m should be added to ε_{HF}. But this procedure is still not quite sure, because it is possible that the polarization part of ε_{HF} is overestimated with respect to that of the long-range RS perturbation treatment: indeed the Hartree–Fock procedure cannot take into account the dispersion term correctly (see formula (2.10): the dispersion term involves doubly excited functions $W_a^{(1)}W_b^{(2)}$ with respect to the ground state function $W_0^{(1)}W_0^{(2)}$, while the Hartree–Fock method includes properly only the single excited functions $W_0^{(1)}W_b^{(2)}$ and $W_a^{(1)}W_0^{(2)}$), but the minimization of the energy may then result in an overestimation of the only available contribution, namely the polarization term.

Let us recall once more that we are interested here only in the interactions between closed-shell ground-state systems. Among these interactions, some cases probably cannot be represented by (2.123) with sufficient accuracy, and appropriate supplementary terms or modifications should then be used: we think to the complexes with rather short intermolecular equilibrium distances, namely the hydrogen-bonded complexes (see Section VI.A.4 below) and some strong charge-transfer complexes (halogen-lone pair complexes: see Malrieu and Claverie (1968), Section II.4). As concerns the hydrogen-bonded complexes, it must be mentioned that Jeziorski and Van Hemert (1976) obtained satisfactory results for the water–water complex by performing an *ab initio* calculation of $\Delta E=\varepsilon_1^{HL}+\varepsilon_2^{RS}$: the difference with (2.123) is the use of the *complete* second-order RS perturbation term $\varepsilon_2^{RS}=\varepsilon_2^m+\varepsilon_2^{pen}$, instead of its multipole approximation ε_2^m (note that the basis used for the calculation of this ε_2^{RS} included so-called charge-transfer states). On another hand, the terms beyond second-order may also play a non-negligible role when the interactions between more than two molecules are considered (differences with respect to pair-wise additivity, or more briefly 'non-additivity effects'): e.g. see Margenau and Stamper (1966),

Margenau and Kestner (1969, Ch. 5), Stogryn (1970, 1971), and Sections IV.E, V.D.3, V.E.4.a, V.F.2. Let us now consider more closely the formula (2.123). According to the formula (2.97) of Section II.B.3.b above, $\varepsilon_1^{\mathrm{HL}}$ may be decomposed into $\varepsilon_1^{\mathrm{RS}}$ (electrostatic energy) and the first-order exchange contribution $\varepsilon_1^{\mathrm{ex}}$. Moreover, $\varepsilon_1^{\mathrm{RS}} = \varepsilon_1^m + \varepsilon_1^{\mathrm{pen}}$ is itself the sum of a multipolar part ε_1^m (which vanishes in interactions between spherically symmetrical systems like the rare gas atoms) and a penetration part $\varepsilon_1^{\mathrm{pen}}$, which varies exponentially with distance in the medium and long-range regions. It appears therefore interesting to represent in the simplest possible way the sum $\varepsilon_1^{\mathrm{ex}} + \varepsilon_1^{\mathrm{pen}}$ of the exponentially-varying contributions, and in the Section IV below (devoted to the short-range part of the energy) we shall actually deal with these two contributions. In order to stress this decomposition of $\varepsilon_1^{\mathrm{HL}}$, we may rewrite (2.123) as follows:

$$\Delta E \simeq (\varepsilon_1^{\mathrm{ex}} + \varepsilon_1^{\mathrm{pen}}) + \varepsilon_1^m + \varepsilon_2^m \tag{2.124}$$

It is now easy to establish the relationship between this theoretically-founded approximate formula and the so-called semi-empirical potentials (Lennard-Jones, Buckingham, Kihara . . .) quoted in the introduction of this chapter (Part I). The attractive term $(-1/R^6)$ of these potentials corresponds to ε_2^m, since for rare-gas atoms, ε_2^m reduces to the dispersion term, while the repulsive short-range term ($1/R^{12}$ or e^{-bR}) corresponds to $(\varepsilon_1^{\mathrm{ex}} + \varepsilon_1^{\mathrm{pen}})$; the multipolar electrostatic energy ε_1^m vanishes for rare-gas atoms, but the dipole–dipole energy (Stockmayer potential) was actually added for dealing with quasi-spherical polar molecules. (Of course, more elaborate evaluations of the electrostatic energy, quite similar to those studied below in Section V.B have also been considered: e.g. see the study of anthracene dimers by Morris (1974), where the necessity of a correct evaluation of the electrostatic term is emphasized).

The next Section III.B and Parts IV, V will be devoted to the derivation of simplified formulas for the four contributions appearing in (2.124): the short-range (exponential-like) repulsion energy $(\varepsilon_1^{\mathrm{ex}} + \varepsilon_1^{\mathrm{pen}})$, the multipolar electrostatic energy (ε_1^m), and the multipolar polarization and dispersion energies (ε_2^m). It must be emphasized that the word 'multipolar' *does not* take here its restricted sense of 'one-centre multipolar expansion': on the contrary, an essential purpose of our developments will consist in introducing *many-centred* multipole expansions, in order to obtain formulas suited for large molecules with arbitrary shapes (see Section V.A). Indeed, it should be rather clear that the only general way for taking systematically into account the size and shapes of molecules when calculating their interaction is to introduce several 'centres of force', conveniently distributed in the molecules. Following this line of thought, systematic attempts will be developed for expressing the various interaction terms as sums of local contributions (atom–atom, atom–bond, bond–bond) (Claverie, 1973, 1976; Rein 1973; Amos and Crispin 1976b,c). For that purpose, an important role is played by the possibility of representing, with a good accuracy, the ground state function of many molecules by a single Slater determinant built from *localized* orbitals (e.g. see Diner *et al.* (1970), England *et al.* (1971), and the book edited by Chalvet *et al.* (1975), especially the paper by Millié *et al.*, p. 59). Thus, it will appear that the calculation of the interaction energy between large molecules as a sum of local contributions, although being an approximate procedure, certainly has some theoretical foundation. Some works challenging this kind of approximation will be discussed in Section VI.A.6, and it will be seen that they do not invalidate the previous conclusion, essentially because these works are concerned with highly refined properties of very small systems, while we are concerned with a not highly accurate (at most 1 per cent accuracy, say) of the interaction energy between medium- or large-size molecules; moreover, these works often reduce the atom–atom formulas to the dispersion and short-range repulsion term, even when polar molecules are involved: in such cases, electrostatic and polarization contributions (atom–atom or atom–bond) should be included, and their neglect invalidates any conclusion about the accuracy of local-type formulas.

As a final remark, it must never be forgotten that a rather simple analytical formula, having only an approximate or qualitative theoretical justification, may work surprisingly well in practice (i.e. when compared with the results obtained by an *a priori* accurate but much more cumbersome method): a striking example of such a situation will be given by the representation of the He . . . He short-range energy ($\varepsilon_1^{\text{ex}}+\varepsilon_1^{\text{pen}}=\varepsilon_1^{\text{HL}}$ in that case) in terms of a simple exponential (Section IV.C). Such situations simply show that it may be very difficult to *prove* rigorously that a simple analytical formula is an accurate approximation for some complicated expression, but the lack (at the present time) of a suitable proof does not eliminate either the existence of such a simple formula or the right to use it for practical calculations.

B. Expression of the matrix elements involving the interaction Hamiltonian

The long-range RS perturbation series (2.10) involves matrix elements of the general form $\langle W_a^{(1)}W_b^{(2)}|\, V\, |W_c^{(1)}W_d^{(2)}\rangle$, where V is the interaction Hamiltonian (2.9). The symmetrized perturbation treatments (Section II.B.3) involve analogous matrix elements with the operator V replaced by the product $V\mathscr{A}_{\text{int}}$ (see formulas (2.87, 2.88, 2.89, 2.90, 2.91, 2.92, 2.93, 2.94)). We shall derive new useful expressions successively for these two kinds of matrix elements.

1. *The matrix elements of V*

The expression of these matrix elements proposed by Longuet-Higgins (1956) will be very useful for our purpose: they may be considered as electrostatic interaction energies between appropriate charge distributions defined independently in molecules (1) and (2). This expression is obtained by introducing the 'local charge density operators' associated with the two molecules ($m=1, 2$):

$$\boldsymbol{\rho}^{(m)}(\vec{r})=\boldsymbol{\rho}_{\text{nuc}}^{(m)}(\vec{r})-\boldsymbol{\rho}_{\text{el}}^{(m)}(\vec{r}) \tag{2.125}$$

with

$$\boldsymbol{\rho}_{\text{nuc}}^{(m)}(\vec{r})=\sum_{\mu}^{(m)} Z_\mu \delta(\vec{r}-\vec{r}_\mu) \tag{2.126}$$

$$\boldsymbol{\rho}_{\text{el}}^{(m)}(\vec{r})=\sum_{i}^{(m)} \delta(\vec{r}-\vec{r}_i) \tag{2.127}$$

These operators are simply functions of the nuclear coordinates $\vec{r}_\mu$ and electronic coordinates $\vec{r}_i$, and they simply act by scalar multiplication on the functions of these coordinates. In practice, we shall use functions of the electronic coordinates only (Born–Oppenheimer approximation) with the nuclear coordinates kept fixed. The dummy space variable $\vec{r}$ must be carefully distinguished from the electronic and nuclear coordinates: it is the arbitrary point of space at which the charge density is considered. Let us indeed consider the mean value of $\boldsymbol{\rho}(\vec{r})$ with respect to some electronic state function W_a (we drop the superscript m for simplicity):

$$\begin{aligned}\langle W_a|\boldsymbol{\rho}(\vec{r})|W_a\rangle &= \langle W_a \mid W_a\rangle \sum_\mu Z_\mu\delta(\vec{r}-\vec{r}_\mu)-\langle W_a|\sum_{i=1}^{n}\delta(\vec{r}-\vec{r}_i)|W_a\rangle \\ &= \sum_\mu Z_\mu\delta(\vec{r}-\vec{r}_\mu)-n\langle W_a|\delta(\vec{r}-\vec{r}_1)|W_a\rangle \\ &= \sum_\mu Z_\mu\delta(\vec{r}-\vec{r}_\mu)-n\int\ldots\int W_a^*(\vec{r},\vec{r}_2,\ldots,\vec{r}_n) \\ &\quad\times W_a(\vec{r},\vec{r}_2,\ldots,\vec{r}_n)\,\mathrm{d}\vec{r}_2\ldots\mathrm{d}\vec{r}_n \\ &= \boldsymbol{\rho}_{\text{nuc}}(\vec{r})-d_{1,a}(\vec{r},\vec{r})\end{aligned} \tag{2.128}$$

where $\boldsymbol{\rho}_{\text{nuc}}(\vec{r})$ is the charge density associated to the point *nuclear charges* Z_μ located at the points $\vec{r}_\mu$, and $d_{1,a}(\vec{r},\vec{r})$ is the 'diagonal' part of the one-particle reduced 'density matrix' (the kernel of the 'density operator') associated with the state W_a, i.e. the total *electronic charge* density (at the point $\vec{r}$) corresponding to this state (for the definition and use of the complete and reduced density matrices see, for example, McWeeny (1960); McWeeny and Yonezawa (1965); McWeeny and Sutcliffe (1969). We thus see that any confusion must be carefully avoided between the 'local charge density operator' $\boldsymbol{\rho}(\vec{r})$, which is independent of any particular state, and the 'density matrix' associated with a peculiar state: the relationship between them is given only be (2.128). We also see that the mean value $\langle W_a|\boldsymbol{\rho}(\vec{r})|W_a\rangle$ is actually nothing but the *total* charge density at the point $\vec{r}$ of the system described by the function W_a (hence the name of the operator

$\boldsymbol{\rho}(\vec{r})$). Now, it is easily seen that the operator V may be rewritten as

$$V = \iint \frac{\boldsymbol{\rho}^{(1)}(\vec{r}^1)\boldsymbol{\rho}^{(2)}(\vec{r}^2)}{|\vec{r}^1 - \vec{r}^2|} \mathrm{d}\vec{r}^1\,\mathrm{d}\vec{r}^2 \tag{2.129}$$

Indeed $\boldsymbol{\rho}(\vec{r})$ is nothing but a usual function of $\vec{r}$ and the nuclear and electron coordinates, so that the integration in (2.129) is an ordinary one, and, upon introduction of the definition (2.125, 2.126, 2.127) for $\boldsymbol{\rho}^{(1)}$ and $\boldsymbol{\rho}^{(2)}$, it readily gives the expression (2.9) for V.

The advantage of (2.129) with respect to the original expression (2.9) is that each operator $\boldsymbol{\rho}^{(m)}(r)$ depends only on the coordinates of the subsystem (m), so that the separation of these variables for the two subsystems is realized in the integrand, and a general matrix element of V therefore becomes:

$$\begin{aligned}
&\langle W_a^{(1)} W_b^{(2)} |\, V \,| W_c^{(1)} W_d^{(2)}\rangle \\
&\quad = \iint \frac{\langle W_a^{(1)} W_b^{(2)} |\, \boldsymbol{\rho}^{(1)}(\vec{r}^1)\boldsymbol{\rho}^{(2)}(\vec{r}^2) \,| W_c^{(1)} W_d^{(2)}\rangle}{|\vec{r}^1 - \vec{r}^2|} \mathrm{d}\vec{r}^1\,\mathrm{d}\vec{r}^2 \\
&\quad = \iint \frac{\langle W_a^{(1)} |\, \boldsymbol{\rho}^{(1)}(\vec{r}^1) \,| W_c^{(1)}\rangle\langle W_b^{(2)} |\, \boldsymbol{\rho}^{(2)}(\vec{r}^2) \,| W_d^{(2)}\rangle}{|\vec{r}^1 - \vec{r}^2|} \mathrm{d}\vec{r}^1\,\mathrm{d}\vec{r}^2 \\
&\quad = \iint \frac{f_{ac}^{(1)}(\vec{r}^1) f_{bd}^{(2)}(\vec{r}^2)}{|\vec{r}^1 - \vec{r}^2|} \mathrm{d}\vec{r}^1\,\mathrm{d}\vec{r}^2
\end{aligned} \tag{2.130}$$

where

$$f_{ac}^{(1)}(\vec{r}^1) = \langle W_a^{(1)} |\, \boldsymbol{\rho}^{(1)}(\vec{r}^1)\, \mathbf{g} W_c^{(1)}\rangle \tag{2.131}$$

$$f_{bd}^{(2)}(\vec{r}^2) = \langle W_b^{(2)} |\, \boldsymbol{\rho}^{(2)}(\vec{r}^2) \,| W_d^{(2)}\rangle \tag{2.132}$$

Thus the matrix element $\langle W_a^{(1)} W_b^{(2)} |\, V \,| W_c^{(1)} W_d^{(2)}\rangle$ actually appears as the 'electrostatic interaction' between the two 'charge densities' $f_{ac}^{(1)}(\vec{r}^1)$ and $f_{bd}^{(2)}(\vec{r}^2)$.

For the sake of simplicity we shall now omit the superscripts ($m = 1$ or 2) in the discussion of these 'charge densities'.

We may distinguish the *state* charge distributions $f_{aa}(\vec{r})$ already discussed (equations 2.125, 2.126, 2.127, 2.128) which are the total electronic and nuclear) charge densities corresponding to the state functions W_a, and the *transition* charge distributions $f_{ac}(\vec{r})$ for $a \neq c$. Since W_a and W_c are purely electronic functions, we get, owing to the orthogonality relation $\langle W_a \mid W_c\rangle = 0$:

$$\begin{aligned}
f_{ac}(\vec{r}) &= \langle W_a |\, \boldsymbol{\rho}_{\mathrm{nuc}}(\vec{\mathbf{r}}) \,| W_c\rangle - \langle W_a |\, \boldsymbol{\rho}_{\mathrm{el}}(\vec{\mathbf{r}}) \,| W_c\rangle \\
&= \langle W_a \mid W_c\rangle \boldsymbol{\rho}_{\mathrm{nuc}}(\vec{\mathbf{r}}) - \langle W_a |\, \boldsymbol{\rho}_{\mathrm{el}}(\vec{r}) \,| W_c\rangle \\
&= -\langle W_a |\, \boldsymbol{\rho}_{\mathrm{e1}}(\vec{r}) \mid W_c\rangle = -\langle W_a |\, \sum_{i=1}^{n} \delta(\vec{r} - \vec{r}_i) \mid W_c\rangle
\end{aligned} \tag{2.133}$$

The advantage of the expression (2.130) of the matrix elements is twofold: first, the charge distributions involved are defined for each molecule independently of the other according to (2.131, 2.132); second, at moderate (and of course large) intermolecular distances, the electrostatic interaction between these distributions will be obtained with sufficient accuracy by replacing the continuous (electronic) part of the charge distributions by appropriate multipole expansions, and this procedure allows considerable conceptual and computational simplifications.

We shall finally give explicit forms of the distributions (2.131, 2.132) for the special case where the molecular functions W_a are Slater determinants (for the sake of simplicity, we drop the label m of the molecule). Since $\boldsymbol{\rho}(\vec{r})$ is a mono-electronic operator (i.e. a sum of one-electron terms), its matrix elements will be non-zero only if the two determinants W_a and W_c differ by at most one spin-orbital. Hence the two cases:

a. State charge distributions $f_{aa}(r)$. As $\sum_i \delta(\vec{r}-\vec{r}_i)$ is a symmetric sum of one-electron operators, the matrix element $\langle W_a | \boldsymbol{\rho}_{\mathrm{el}}(\vec{r}) | W_a \rangle$ reduces to the element involving only the diagonal terms of the determinants W_a:

$$\langle W_a | \sum_{i=1}^{n} \delta(\vec{r}-\vec{r}_i) | W_a \rangle$$

$$= \langle b_1(\vec{r}_1)\sigma_1(1) \ldots b_n(\vec{r}_n)\sigma_n(n) | \sum_{i=1}^{n} \delta(\vec{r}-\vec{r}_i) | b_1(\vec{r}_1)\sigma_1(1) \ldots b_n(\vec{r}_n)\sigma_n(n) \rangle$$

where we have explicitly written the spin-orbitals $a_i(i) = b_i(r_i)\sigma_i(i)$ in terms of the orbitals b_i and the spin parts σ_i)

$$= \sum_{i=1}^{n} \langle \sigma_i(i) | \sigma_i(i) \rangle \int b_i^*(\vec{r}_i)\delta(\vec{r}-\vec{r}_i) b_i(\vec{r}) \, \mathrm{d}\vec{r}$$

$$= \sum_{i=1}^{n} |b_i(\vec{r})|^2$$

and the total charge distribution thus writes:

$$f_{aa}(\vec{r}) = \sum_{\mu} Z_\mu \delta(\vec{r}-\vec{r}_\mu) - \sum_{i=1}^{n} |b_i(\vec{r})|^2 \tag{2.134}$$

Of course, if we have N orbitals b_I with occupation numbers $n_I^{\mathrm{occ}}(n_I^{\mathrm{occ}} = 1$ or $2)$, the sum $\sum_{i=1}^{n}$ over the spin-orbitals becomes a sum over the orbitals, namely

$$\sum_{i=1}^{n} |b_i(\vec{r})|^2 = \sum_{I=1}^{N} n_I^{\mathrm{occ}} |b_I(\vec{r})|^2 \tag{2.135}$$

b. Transitition charge distribution $f_{ac}(\vec{r})$. The functions W_a and W_c differ by one spin orbital, say $a_i = b_i\sigma_i$ and $a_j = b_j\sigma_j$. Then the matrix element (2.133) reduces to the matrix element of the one-electron operator taken with respect to these different spin-orbitals:

$$\begin{aligned}
&\langle a_1(1)\ldots a_i(l)\ldots|\sum_{k=1}^{n}\delta(\vec{r}-\vec{r}_k)\,|a_1(1)\ldots a_j(l)\ldots\rangle \\
&=\langle a_i(l)|\,\delta(\vec{r}-\vec{r}_l)\,|a_j(l)\rangle \qquad (2.136)\\
&=\langle b_i(\vec{r}_l)|\,\delta(\vec{r}-\vec{r}_l)\,|b_j(\vec{r}_l)\rangle\langle\sigma_i(l)\mid\sigma_j(l)\rangle \\
&=b_i^*(\vec{r})b_j(\vec{r})\delta_{ij}^{\text{spin}}
\end{aligned}$$

where $\delta_{ij}^{\text{spin}}=\langle\sigma_i\mid\sigma_j\rangle=1$ if $\sigma_i=\sigma_j$ (same spin function α or β) and 0 if $\sigma_i\neq\sigma_j$. Hence (2.133) finally gives

$$f_{ac}(\vec{r})=\left\langle W_a\middle|\boldsymbol{\rho}(\vec{r})\middle|W_a\left(\frac{j}{i}\right)\right\rangle=-b_i^*(\vec{r})b_j(\vec{r})\delta_{ij}^{\text{spin}} \qquad (2.137)$$

(here $W_a(j/i)$ is an obvious notation indicating the replacement of the spin-orbital i by the spin-orbital j).

Let us remark that these 'transition charge distributions' are only mathematical aids for the computation of the matrix elements of V: *no direct physical* meaning should be ascribed to them. As concerns the state charge distributions, their physical meaning depends on the life-time of the state under consideration: there is no problem for the ground state, but, as concerns the excited states, some direct physical meaning may be attributed to their charge distribution only if they have sufficiently long life-time. It must never be forgotten that the perturbation treatment uses the excited state functions only as a mathematically convenient basis, without any concern about their degree of physical meaning.

2. *The matrix elements involving $\mathscr{A}_{\text{int}}$*

The inter-system antisymmetrizer $\mathscr{A}_{\text{int}}$ may be written (see Appendix A, equations (2.380, 2.381)):

$$\mathscr{A}_{\text{int}}=1-\mathscr{A}'_{\text{int}}=1-P_{(1)}+P_{(2)}+\ldots+(-1)^kP_{(k)}+\ldots+(-1)^{n_0}P_{(n_0)} \qquad (2.138)$$

where $n_0=\text{Min}\,(n_1,n_2)$ and $P_{(k)}$ denotes the sum of all permutations which exchange exactly k electrons labels between the two subsystems. In the matrix elements involving $\mathscr{A}_{\text{int}}$ (equations (2.86) to (2.94)), $P_{(k)}$ generates terms which vary like the $(2k)$th power of the overlap $\mathbf{S}$ between the molecular orbitals of the subsystems (we suppose that the state functions $W_a^{(i)}$ are expressed in terms of one-electron spin-orbitals). We shall speak of 'terms of order $\mathbf{S}^{2k}$', as is now usual in works dealing with 'exchange' perturbation treatments (Murrell *et al.* 1965; Murrell and Shaw, 1967a).

Now, the intermolecular orbital overlap S is quite small for most complexes (even at equilibrium distance), especially between the orbitals occupied in the ground state function. S may be of the order of magnitude of 1/100: in the He . . . He case (see Table 2 in Section IV.C below), $\mathbf{S}=$

$0{\cdot}766\times10^{-2}$ at $R=5{\cdot}5$ a.u. and $0{\cdot}425\times10^{-2}$ at $R=6{\cdot}0$ a.u., the energy minimum occurring at about $R=5{\cdot}75$ a.u.; the same conclusion holds for the neon . . . neon interaction according to the calculations of Dacre and McWeeny (1970). Since the first-order (Heitler–London) interaction energy (equation (2.96)) involves just these ground-state orbitals, a convenient practical approximation will be obtained by neglecting the $P_{(k)}$'s beyond the one-exchange term $-P_{(1)}$ in (2.138). Thus, we shall consider matrix elements of $VP_{(1)}$ instead of matrix elements of $V\mathscr{A}'$. For the sake of simplicity, we shall also treat explicitly only the case where the subsystem functions are Slater determinants (a more general treatment using the reduced density matrices could be developed for arbitrary functions: see McWeeny and Yonezawa (1965); Claverie (1973, Ch. III, Section 3.C1.c, p. 209)).

We therefore consider the matrix elements

$$\langle W_a^{(1)}W_b^{(2)}|\,VP_{(1)}\,|W_c^{(1)}W_d^{(2)}\rangle = \langle \mathfrak{A}^{(1)}(w_a^{(1)})\mathfrak{A}^{(2)}(w_b^{(2)})|\,VP_{(1)}\,|\mathfrak{A}^{(1)}(w_c^{(1)})\mathfrak{A}^{(2)}(w_d^{(2)})\rangle \tag{2.139}$$

where

$$\mathfrak{A}^{(s)}=\frac{1}{\sqrt{n_s!}}\sum_{P^{(s)}}(-1)^{P^{(s)}}P^{(s)}$$

denotes the antisymmetrizer for the subsystem (s), and the w's denote the Hartree products of spin-orbitals:

$$\begin{aligned} w_a^{(1)} &= a_1^{(1)}(1)\ldots a_{n_1}^{(1)}(n_1)\\ w_c^{(1)} &= c_1^{(1)}(1)\ldots c_{n_1}^{(1)}(n_1)\\ w_b^{(2)} &= b_1^{(2)}(n_1+1)\ldots b_{n_2}^{(2)}(n_1+n_2)\\ w_d^{(2)} &= d_1^{(2)}(n_1+1)\ldots d_{n_2}^{(2)}(n_1+n_2) \end{aligned} \tag{2.140}$$

Now,

$$P_{(1)}=\sum_{i_1}^{(1)}\sum_{j_2}^{(2)}P_{i_1j_2}\equiv\sum_{i=1}^{n_1}\sum_{j=1}^{n_2}P_{i,j+n_1} \tag{2.141}$$

is obviously symmetric with respect to the subsets $\{i_1=1,\ldots,n_1\}$ and $\{j_2=n_1+1,\ldots,n_1+n_2\}$, and consequently it commutes with any intra-system permutation $Q^{(s)}$ relative to subsystem (s) (an explicit proof is given in Appendix A: see equation (2.389)). Thus $P_{(1)}$ also commutes with $\mathfrak{A}^{(1)}$ and $\mathfrak{A}^{(2)}$, so that we may transform the function $P_{(1)}|W_c^{(1)}W_d^{(2)}\rangle$ appearing in (2.138) as follows:

$$\begin{aligned} &P_{(1)}[\mathfrak{A}^{(1)}(w_c^{(1)})\mathfrak{A}^{(2)}(w_d^{(2)})]\\ &\quad=P_{(1)}\mathfrak{A}^{(1)}\mathfrak{A}^{(2)}[w_c^{(1)}w_d^{(2)}]\\ &\quad=\mathfrak{A}^{(1)}\mathfrak{A}^{(2)}P_{(1)}[w_c^{(1)}w_d^{(2)}]\\ &\quad=\sum_i^{(1)}\sum_j^{(2)}\mathfrak{A}^{(1)}\mathfrak{A}^{(2)}[P_{i,j+n_1}(w_c^{(1)}w_d^{(2)})] \end{aligned} \tag{2.142}$$

now

$$
\begin{aligned}
P_{i,j+n_1}(w_c^{(1)}w_d^{(2)}) &= P_{i,j+n_1}[c_1^{(1)}(1)\ldots c_i^{(1)}(i)\ldots c_{n_1}^{(1)}(n_1) \\
&\quad \times d_1^{(2)}(n_1+1)\ldots d_j^{(2)}(n_1+j)\ldots d_{n_2}^{(2)}(n_1+n_2)] \\
&= c_1^{(1)}(1)\ldots c_i^{(1)}(n_1+j)\ldots c_{n_1}^{(1)}(n_1) \\
&\quad \times d_1^{(2)}(n_1+1)\ldots d_j^{(2)}(i)\ldots d_{n_2}^{(2)}(n_1+n_2) \\
&= [c_1^{(1)}(1)\ldots d_j^{(2)}(i)\ldots c_{n_1}^{(1)}(n_1)] \\
&\quad \times [d_1^{(2)}(n_1+1)\ldots c_i^{(1)}(n_1+j)\ldots d_{n_2}^{(2)}(n_1+n_2)] \\
&= \left(w_c^{(1)} \,\middle|\, \frac{j}{i}\right)\left(w_d^{(2)} \,\middle|\, \frac{i}{j}\right)
\end{aligned}
\tag{2.143}
$$

where $(w_c^{(1)} \mid i/j)$ denotes the first bracketed expression (replacement of the spin-orbital $c_i^{(1)}$ by $d_j^{(2)}$) and $(w_d^{(2)} \mid i/j)$ denotes the second one (replacement of $d_j^{(2)}$ by $c_i^{(1)}$). Inserting (2.143) in (2.142), we obtain the Slater determinants $W^{(s)} = \mathfrak{A}^{(s)}w^{(s)}$ corresponding respectively to the various Hartree products $w^{(s)}$:

$$
P_{(1)}[W_c^{(1)}W_d^{(2)}] = \sum_{i=1}^{n_1}\sum_{j=1}^{n_2}\left(W_c^{(1)} \,\middle|\, \frac{j}{i}\right)\left(W_d^{(2)} \,\middle|\, \frac{i}{j}\right) \tag{2.144}
$$

Thus, we have reduced $P_{(1)}[W_c^{(1)}W_d^{(2)}]$ to a sum of simple products of determinants pertaining to the subsystems; these are formally 'charge-transfer' determinants corresponding to opposite transfers between all possible pairs of orbitals $c_i^{(1)}$, $d_j^{(2)}$. In this way, all integrals involving functions of the type $P_{(1)}[W_c^{(1)}W_d^{(2)}]$ may be reduced to sums of integrals involving simple products $W^{(1)}W^{(2)}$ of 'opposite transfer' determinants. Note that the generalization to the k-exchange terms involving $P_{(k)}$ would be obvious: we would have products of determinants resulting from k opposite transfers instead of 1. Note also that this transformation of $P_{(1)}[W_c^{(1)}W_d^{(2)}]$ (or more generally $\mathscr{A}'_{\text{int}}[W_c^{(1)}W_d^{(2)}]$) is independent of the remaining part of the matrix element and could be applied with any operator replacing V, e.g. the identity operator (see below the derivations following the formula (2.150)).

We are therefore brought again to the evaluation of matrix elements between simple products of subsystem functions that we treated in the previous Subsection B.1. In order to use the formulae (2.134, 2.136) derived above, it is, however, necessary to use Slater determinants built from orthogonal spin-orbitals. For that purpose, in a 'transfer' determinant like $(W_c^{(1)} \mid j/i)$, we split the spin-orbital $d_j^{(2)}$ into its component in the subspace spanned by the spin-orbitals $c_i^{(1)}$, namely $\sum_{k=1}^{n_1} S_{kj}c_k^{(1)}$ (where we note $S_{kj} = \langle c_k^{(1)} \mid d_j^{(2)}\rangle$ the overlap integral), and its component orthogonal to that subspace, that we note

$$d_j'^{(2)} = d_j^{(2)} - \sum_{k=1}^{n_1} S_{kj} c_k^{(1)} \tag{2.145}$$

Then $(W_c^{(1)} \mid j/i)$ may be written

$$\left(W_c^{(1)} \middle| \frac{j}{i}\right) = \left(W_c^{(1)} \middle| \frac{j'}{i}\right) + \sum_{k=1}^{n_1} S_{kj} \left(W_c^{(1)} \middle| \frac{k}{i}\right) \tag{2.146}$$

where j' denotes $d_j'^{(2)}$. Now $(W_c^{(1)} \mid k/i)$ is identically zero if $k \neq i$ since the determinant then contains twice the spin-orbital $c_k^{(1)}$, while $(W_c^{(1)} \mid i/i) \equiv W_c^{(1)}$. Therefore (2.146) reduces to

$$\left(W_c^{(1)} \middle| \frac{j}{i}\right) = \left(W_c^{(1)} \middle| \frac{j'}{i}\right) + S_{ij} W_c^{(1)} \tag{2.147}$$

and similarly we have

$$\left(W_d^{(2)} \middle| \frac{i}{j}\right) = \left(W_d^{(2)} \middle| \frac{i'}{j}\right) + S_{ji} W_d^{(2)} \tag{2.148}$$

where i' denotes $c_i'^{(1)} = c_i^{(1)} - \sum_{l=1}^{n_2} S_{il} d_l^{(2)}$. The right-hand sides of (2.147, 2.148) now contain only Slater determinants built from orthogonal spin-orbitals. Thus (2.144) becomes, upon insertion of (2.147, 2.148):

$$P_{(1)}[W_c^{(1)} W_d^{(2)}] = \sum_{i=1}^{n_1} \sum_{j=1}^{n_2} \left\{ \left(W_c^{(1)} \middle| \frac{j'}{i}\right)\left(W_d^{(2)} \middle| \frac{i'}{j}\right) + S_{ij}\left[W_c^{(1)} \left(W_d^{(2)} \middle| \frac{i'}{j}\right) \right.\right.$$
$$\left.\left. + \left(W_c^{(1)} \middle| \frac{j'}{i}\right) W_d^{(2)} \right] + S_{ij}^2 W_c^{(1)} W_d^{(2)} \right\} \tag{2.149}$$

Then the matrix element (2.139) becomes:

$$\langle W_a^{(1)} W_b^{(2)} | \, VP_{(1)} \, | W_c^{(1)} W_d^{(2)} \rangle = \sum_{i=1}^{n_1} \sum_{j=1}^{n_2} \left\{ \left\langle W_a^{(1)} W_b^{(2)} \middle| V \middle| \left(W_c^{(1)} \middle| \frac{j'}{i}\right)\left(W_d^{(2)} \middle| \frac{i'}{j}\right) \right\rangle \right.$$
$$+ S_{ij}\left[\left\langle W_a^{(1)} W_b^{(2)} \middle| V \middle| W_c^{(1)} \left(W_d^{(2)} \middle| \frac{i'}{j}\right) \right\rangle + \left\langle W_a^{(1)} W_b^{(2)} \middle| V \middle| \left(W_c^{(1)} \middle| \frac{j'}{i}\right) W_d^{(2)} \right\rangle \right]$$
$$\left. + S_{ij}^2 \langle W_a^{(1)} W_b^{(2)} | \, V \, | W_c^{(1)} W_d^{(2)} \rangle \right\} \tag{2.150}$$

Similarly, the matrix elements involving $P_{(1)}$ alone (which correspond to the scalar products $\langle \psi_I \mid \psi_J \rangle = \langle \phi_I | \, \mathcal{A}_{\text{int}} \, | \phi_J \rangle = \langle W_a^{(1)} W_b^{(2)} | \, \mathcal{A}_{\text{int}} \, | W_c^{(1)} W_d^{(2)} \rangle$ encountered in the symmetrized perturbation treatments, see Section II.B.3 above) become:

$$\begin{aligned}
&\langle W_a^{(1)} W_b^{(2)} | P_{(1)} | W_c^{(1)} W_d^{(2)} \rangle \\
&\quad = \sum_{i=1}^{n_1} \sum_{j=2}^{n_2} \left\{ \left\langle W_a^{(1)} W_b^{(2)} \,\middle|\, \left(W_c^{(1)} \,\middle|\, \frac{j'}{i} \right) \left(W^{(2)} \,\middle|\, \frac{i'}{j} \right) \right\rangle \right. \\
&\qquad + S_{ij} \left[\left\langle W_a^{(1)} W_b^{(2)} \,\middle|\, W_c^{(1)} \left(W_d^{(2)} \,\middle|\, \frac{i'}{j} \right) \right\rangle + \left\langle W_a^{(1)} W_b^{(2)} \,\middle|\, \left(W_c^{(1)} \,\middle|\, \frac{j'}{i} \right) W_d^{(2)} \right\rangle \right] \\
&\qquad \left. + S_{ij}^2 \langle W_a^{(1)} W_b^{(2)} \mid W_c^{(1)} W_d^{(2)} \rangle \right\} \\
&\quad = \sum_{j=1}^{n_1} \sum_{j=2}^{n_2} \left\{ \left\langle W_a^{(1)} \,\middle|\, \left(W_c^{(1)} \,\middle|\, \frac{j'}{i} \right) \right\rangle \left\langle W_b^{(2)} \,\middle|\, \left(W_d^{(2)} \,\middle|\, \frac{i'}{j} \right) \right\rangle \right. \qquad (2.151) \\
&\qquad + S_{ij} \left[\langle W_a^{(1)} \mid W_c^{(1)} \rangle \left\langle W_b^{(2)} \,\middle|\, \left(W_d^{(2)} \,\middle|\, \frac{i'}{j} \right) \right\rangle + \left\langle W_a^{(1)} \,\middle|\, \left(W_c^{(1)} \,\middle|\, \frac{j'}{i} \right) \right\rangle \right. \\
&\qquad\qquad \left. \times \langle W_b^{(2)} \mid W_d^{(2)} \rangle \right] \\
&\qquad \left. + S_{ij}^2 \langle W_a^{(1)} \mid W_c^{(1)} \rangle \langle W_b^{(2)} \mid W_d^{(2)} \rangle \right\} \\
&\quad = \sum_{i=1}^{n_1} \sum_{j=1}^{n_2} \left[\mathcal{S}_{ac}^{(1)} \left(\frac{j'}{i} \right) + S_{ij} \delta_{ac} \right] \left[\mathcal{S}_{bd}^{(2)} \left(\frac{i'}{j} \right) + S_{ij} \delta_{bd} \right]
\end{aligned}$$

where we used the orthonormality relations $\langle W_a^{(1)} \mid W_c^{(1)} \rangle = \delta_{ac}$, $\langle W_b^{(2)} \mid W_d^{(2)} \rangle = \delta_{bd}$ and we introduced the notation

$$\mathcal{S}_{ac}^{(1)} \left(\frac{j'}{i} \right) = \left\langle W_a^{(1)} \,\middle|\, \left(W_c^{(1)} \,\middle|\, \frac{j'}{i} \right) \right\rangle \qquad \mathcal{S}_{bd}^{(2)} \left(\frac{i'}{j} \right) = \left\langle W_b^{(2)} \,\middle|\, \left(W_d^{(2)} \,\middle|\, \frac{i'}{j} \right) \right\rangle \tag{2.152}$$

but it must be emphasized that the scalar product $\mathcal{S}_{ac}^{(1)}(j'/i)$ implicitly depends on $W_d^{(2)}$ according to the definition (2.145) of $j' \equiv d_j'^{(2)}$, and similarly $\mathcal{S}_{bd}^{(2)}(i'/j)$ also implicitly depends on $W_c^{(1)}$. Let us recall here that $\sum_{i=1}^{n_1}$ runs over the spin-orbitals of $W_c^{(1)}$ and $\sum_{j=1}^{n_2}$ over the spin-orbitals of $W_d^{(2)}$.

(2.151) simplifies drastically when $a = c$ and $b = d$: indeed, we have

$$\mathcal{S}_{aa}^{(1)} \left(\frac{i'}{i} \right) = \left\langle W_a^{(1)} \,\middle|\, \left(W_a^{(1)} \,\middle|\, \frac{i'}{i} \right) \right\rangle = 0$$

since the spin-orbital $j' \equiv d_j'^{(2)}$ is then orthogonal to all the spin-orbitals of $W_a^{(1)}$ according to its very definition (2.145), and similarly $\mathcal{S}_{bb}^{(2)}(i'/j) = 0$. Hence

$$\langle W_a^{(1)} W_b^{(2)} | P_{(1)} | W_a^{(1)} W_b^{(2)} \rangle = \sum_{i=1}^{n_1} \sum_{j=1}^{n_2} (S_{ij})^2 \tag{2.153}$$

(here $\sum_{i=1}^{n_1}$ runs over the spin-orbitals of $W_a^{(1)}$ and $\sum_{j=1}^{n_2}$ over the spin-orbitals of $W_b^{(2)}$).

We now pursue the evaluation of the matrix element (2.150) by introducing appropriate charge distributions, as we did in the previous subsection for the matrix elements of V, i.e. by using the charge density operators $\boldsymbol{\rho}^{(s)}(\vec{r}^s)$:

$$\rho_{ac}^{(1)}\left(\frac{j'}{i}\right)(\vec{r}) = \langle W_a^{(1)}|\, \boldsymbol{\rho}^{(1)}(\vec{r}) \left|\left(W_c^{(1)} \left| \frac{j'}{i} \right)\right)\right\rangle \tag{2.154}$$

$$\rho_{bd}^{(2)}\left(\frac{i'}{j}\right)(\vec{r}) = \langle W_b^{(2)}|\, \boldsymbol{\rho}^{(2)}(\vec{r}) \left|\left\{ W_d^{(2)} \left| \frac{i'}{j} \right)\right)\right\rangle \tag{2.155}$$

these distributions appear as overlap intermolecular charge densities, by contrast with the intramolecular densities $f_{ac}^{(1)}$, $f_{bd}^{(2)}$ introduced previously. Then (2.150) appears as a sum of electrostatic interactions between these various formal charge densities:

$$\begin{aligned}\langle W_a^{(1)} W_b^{(2)}|\, VP_{(1)}\, |W_c^{(1)} W_d^{(2)}\rangle = \sum_{i=1}^{n_1} \sum_{j=1}^{n_2} \Bigg\{ \iint \frac{\rho_{ac}^{(1)}\left(\frac{j'}{i}\right)(\vec{r}^1)\rho_{bd}^{(2)}\left(\frac{i'}{j}\right)(\vec{r}^2)}{|\vec{r}^1 - \vec{r}^2|}\, \mathrm{d}\vec{r}^1\, \mathrm{d}\vec{r}^2 \\ + S_{ij}\left[\int V_{ac}^{(1)}(\vec{r}^2)\rho_{bd}^{(2)}\left(\frac{i'}{j}\right)(\vec{r}^2)\, \mathrm{d}\vec{r}^2 + \int V_{bd}^{(2)}(\vec{r}^1)\rho_{ac}^{(1)}\left(\frac{j'}{i}\right)(\vec{r}^1)\, \mathrm{d}\vec{r}^1\right] \\ + S_{ij}^2\langle W_a^{(1)} W_b^{(2)}|\, V\, |W_c^{(1)} W_d^{(2)}\rangle \Bigg\}\end{aligned} \tag{2.156}$$

where the last matrix element of V is given by (2.130), and $V_{ac}^{(1)}(\vec{r})$ (respectively $V_{bd}^{(2)}(\vec{r})$) denote the electrostatic potential created at the point $\vec{r}$ by the charge density $f_{ac}^{(1)}$ (respectively $f_{bd}^{(2)}$):

$$V_{ac}^{(1)}(\vec{r}) = \int \frac{f_{ac}^{(1)}(\vec{r})}{|\vec{r}^1 - \vec{r}|}\, \mathrm{d}\vec{r}^1 \qquad V_{bd}^{(2)}(\vec{r}) = \int \frac{f_{bd}^{(2)}(\vec{r})}{|\vec{r} - \vec{r}^2|}\, \mathrm{d}\vec{r}^2 \tag{2.157}$$

The formula (2.156) completes our evaluation of the matrix elements of $VP_{(1)}$ in terms of electrostatic interactions between formal charge densities. Note that the S^2-like behaviour clearly appears in (2.156), since each overlap distribution is of order S, as a consequence of the definitions (2.137) and (2.134, 2.135). Note also that the matrix elements of $VP_{(k)}$ (higher-order in exchange) could also be expressed by formulas analogous to (2.156) but involving overlap charge distributions resulting from k transfers instead of 1.

IV. THE SHORT-RANGE FIRST-ORDER ('EXCHANGE') TERM

A. The case of approximate eigenfunctions: the 'complementary exchange term'

According to the results of Section II.B.3 (see equations (2.95, 2.96, 2.97), the Heitler–London (first-order) interaction energy $\varepsilon_1^{\mathrm{HL}}$ may be explicitly split into the Rayleigh–Schrödinger first-order term $\varepsilon_1^{\mathrm{RS}}$ (electrostatic energy) and a so-called 'exchange' term $\varepsilon_1^{\mathrm{ex}}$, both of them being explicitly expressed in terms of matrix elements of the interaction operator V:

$$\varepsilon_1^{\mathrm{HL}} = \varepsilon_1^{\mathrm{RS}} + \varepsilon_1^{\mathrm{ex}} \qquad \varepsilon_1^{\mathrm{RS}} = \langle \phi_0 | V | \phi_0 \rangle$$
$$\varepsilon_1^{\mathrm{ex}} = \frac{-\langle \phi_0 | V \mathscr{A}'_{\mathrm{int}} | \phi_0 \rangle + \langle \phi_0 | V | \phi_0 \rangle \langle \phi_0 | \mathscr{A}'_{\mathrm{int}} | \phi_0 \rangle}{1 - \langle \phi_0 | \mathscr{A}'_{\mathrm{int}} | \phi_0 \rangle} \tag{2.158}$$

where $\phi_0 = W_0^{(1)} W_0^{(2)}$.

But these expressions were based upon the assumption that $W_0^{(1)}$ and $W_0^{(2)}$ are the exact (ground state) eigenfunctions of the subsystems. In practice, these eigenfunctions are almost never exactly known, and it is necessary to look more closely at the possible consequences of this situation. The interaction energy is indeed rather small with respect to the total energies of the molecules, so that the problem of calculating it with a small relative error may be a non-trivial task (for example remember the 'basis extension' effect and the slow convergence of the difference series for the He . . . He interaction energy, which were discussed at the end of Section III.A.1 above: these difficulties were precisely due to the non-exact character of the available ground-state function of the helium atom).

A first possibility is, of course, to use directly the formulas (2.158) with the approximate available functions $W_0^{(1)}$, $W_0^{(2)}$. A second possibility will be obtained by starting again from the formula (2.75) for the total Heitler–London energy $E_0 = \langle \psi_0 | H | \psi \rangle / \langle \psi_0 | \psi_0 \rangle$. It is, of course no more possible to isolate the exact unperturbed energy $E_0^0 = E_0^{(1)} + E_0^{(2)}$ since it is not known, but it is still possible to isolate in E_0 the *approximate* unperturbed energy

$$\bar{E}_0^0 = \bar{E}_0^{(1)} + \bar{E}_0^{(2)} \tag{2.159}$$

where we use the energy mean values $\bar{E}_0^{(s)}$ corresponding to the available approximate functions $W_0^{(s)}$:

$$\bar{E}_0^{(s)} = \langle W_0^{(s)} | H^{(s)} | W_0^{(s)} \rangle \tag{2.160}$$

We may indeed write according to (2.86):

$$\begin{aligned} \langle \psi_0 | H | \psi_0 \rangle &= \langle \phi_0 | H \mathscr{A}_{\mathrm{int}} | \phi_0 \rangle \\ &= \langle \phi_0 | H_0 \mathscr{A}_{\mathrm{int}} | \phi_0 \rangle + \langle \phi_0 | V \mathscr{A}_{\mathrm{int}} | \phi_0 \rangle \\ &= \langle \phi_0 | H_0 | \phi_0 \rangle - \langle \phi_0 | H_0 \mathscr{A}'_{\mathrm{int}} | \phi_0 \rangle + \langle \phi_0 | V \mathscr{A}_{\mathrm{int}} | \phi_0 \rangle \end{aligned} \tag{2.161}$$

and similarly

$$\begin{aligned} \langle \psi_0 | \psi_0 \rangle &= \langle \phi_0 | \mathscr{A}_{\mathrm{int}} | \phi_0 \rangle = \langle \phi_0 | \phi_0 \rangle - \langle \phi_0 | \mathscr{A}'_{\mathrm{int}} | \phi_0 \rangle \\ &= 1 - \langle \phi_0 | \mathscr{A}'_{\mathrm{int}} | \phi_0 \rangle \end{aligned} \tag{2.162}$$

Moreover

$$\begin{aligned} \langle \phi_0 | H_0 | \phi_0 \rangle &= \langle W_0^{(1)} W_0^{(2)} | H^{(1)} + H^{(2)} | W_0^{(1)} W_0^{(2)} \rangle \\ &= \langle W_0^{(1)} | H^{(1)} | W_0^{(1)} \rangle \langle W_0^{(2)} | W_0^{(2)} \rangle \\ &\quad + \langle W_0^{(1)} | W_0^{(1)} \rangle \langle W_0^{(2)} | H^{(2)} | W_0^{(2)} \\ &= \langle W_0^{(1)} | H^{(1)} | W_0^{(1)} \rangle + \langle W_0^{(2)} | H^{(2)} | W_0^{(2)} \rangle \\ &= \bar{E}_0^{(1)} + \bar{E}_0^{(2)} = \bar{E}_0^0 \end{aligned} \tag{2.163}$$

Hence finally:

$$E_0 = \frac{\langle\psi_0|\,H\,|\psi_0\rangle}{\langle\psi_0\,|\,\psi_0\rangle} = \frac{\bar{E}_0^0 - \langle\phi_0|\,H_0\mathscr{A}'_{\text{int}}\,|\phi_0\rangle}{1-\langle\phi_0|\,\mathscr{A}'_{\text{int}}\,|\phi_0\rangle} + \frac{\langle\phi_0|\,V\mathscr{A}_{\text{int}}\,|\phi_0\rangle}{\langle\phi_0|\,\mathscr{A}_{\text{int}}\,|\phi_0\rangle}$$

or, using $\bar{E}_0^0/1-\langle\phi_0|\,\mathscr{A}'_{\text{int}}\,|\phi_0\rangle = \bar{E}_0^0(1+\langle\phi_0|\,\mathscr{A}'_{\text{int}}\,|\phi_0\rangle/(1-\langle\phi_0|\,\mathscr{A}'_{\text{int}}\,|\phi_0\rangle))$:

$$E_0 - \bar{E}_0^0 = \frac{\langle\phi_0|\,(\bar{E}_0^0 - H_0)\mathscr{A}'_{\text{int}}\,|\phi_0\rangle}{\langle\phi_0|\,\mathscr{A}_{\text{int}}\,|\phi_0\rangle} + \frac{\langle\phi_0|\,V\mathscr{A}_{\text{int}}\,|\phi_0\rangle}{\langle\phi_0|\,\mathscr{A}_{\text{int}}\,|\phi_0\rangle} \qquad (2.164)$$

The second term of $(E_0-\bar{E}_0^0)$ is the expression (2.96) of $\varepsilon_1^{\text{HL}}$ for the case of exact eigenfunctions, while the first term is a correction which would actually vanish if ϕ_0 was the exact eigenfunction of H_0 (then indeed $\bar{E}_0^0 = E_0^0$ and $(H_0 - E_0^0)\,\phi_0 = 0$). Note that this correction also decreases exponentially as a function of R, owing to the presence of the interchange operator $\mathscr{A}'_{\text{int}}$ in the numerator. For definiteness, we shall use for it the name 'complementary exchange term' and denote it by $\varepsilon^{\text{c-ex}}$.

Obviously, it is desirable to get this complementary term as small as possible, since this smallness would strongly suggest that both approximate evaluations (2.158) and (2.164) are close to the exact Heitler–London interaction $\varepsilon_1^{\text{HL}}$.

As in the previous Section III.B, we shall now consider the main part (of order S^2) of $\varepsilon^{\text{c-ex}}$ obtained by retaining in $\mathscr{A}'_{\text{int}}$ the one-exchange term $P_{(1)}$:

$$\begin{aligned}\varepsilon^{\text{c-ex}} &= \frac{\langle\phi_0|\,(\bar{E}_0^0 - H_0)P_{(1)}\,|\phi_0\rangle + O(S^4)}{1+O(S^2)} \\ &= -\langle\phi_0|\,(H_0 - \bar{E}_0^0)P_{(1)}\,|\phi_0\rangle + O(S^4)\end{aligned} \qquad (2.165)$$

We shall evaluate this one-exchange approximation $\varepsilon^{\text{c-ex(1)}} = -\langle\phi_0|\,(H_0-E_0^0)P_{(1)}\,|\phi_0\rangle$ for the case where $W_0^{(1)}$ and $W_0^{(2)}$ are Slater determinants. For that purpose we use the formula (2.149) for $P_{(1)}[W_0^{(1)}W_0^{(2)}]$:

$$\begin{aligned}\varepsilon^{\text{c-ex(1)}} = -\sum_{i=1}^{n_1}\sum_{j=1}^{n_2}\Bigg\{&\left\langle W_0^{(1)}W_0^{(2)}\right|\,H_0-\bar{E}_0^0\left|\left(W_0^{(1)}\,\middle|\,\frac{j'}{i}\right)\left(W_0^{(2)}\,\middle|\,\frac{i'}{j}\right)\right\rangle \\ &+S_{ij}\Bigg[\left\langle W_0^{(1)}W_0^{(2)}\right|\,H_0-\bar{E}_0^0\left|W_0^{(1)}\left(W_0^{(2)}\,\middle|\,\frac{i'}{j}\right)\right\rangle \\ &+\left\langle W_0^{(1)}W_0^{(2)}\right|\,H_0-E_0^0\left|\left(W_0^{(1)}\,\middle|\,\frac{j'}{i}\right)W_0^{(2)}\right\rangle\Bigg] \\ &+S_{ij}^2\langle W_0^{(1)}W_0^{(2)}|\,H_0-\bar{E}_0^0\,|W_0^{(1)}W_0^{(2)}\rangle\Bigg\}\end{aligned} \qquad (2.166)$$

Now, in the last term

$$\langle W_0^{(1)}W_0^{(2)}|\,H_0-\bar{E}_0^0\,|W_0^{(1)}W_0^{(2)}\rangle \equiv \langle\phi_0|\,H_0-\bar{E}_0^0\,|\phi_0\rangle = 0$$

according to the very definition (2.159, 2.160) of $\bar{E}_0^0 = \langle\phi_0|\,H_0\,|\phi_0\rangle$. In the other matrix elements, we put $H_0 - \bar{E}_0^0 = (H^{(1)} - \bar{E}_0^{(1)}) + (H_2 - \bar{E}_0^{(2)})$, so that they reduce to products of matrix elements pertaining to the subsystem electron variables:

$$\langle W_0^{(1)} W_0^{(2)} | (H^{(1)} - \bar{E}_0^{(1)}) + (H^{(2)} - \bar{E}_0^{(2)}) \left| \left(W_0^{(1)} \left| \frac{j'}{i} \right) \left(W_0^{(2)} \left| \frac{i'}{j} \right) \right\rangle \right.\right.$$

$$= \langle W_0^{(1)} | H^{(1)} - \bar{E}_0^{(1)} \left| \left(W_0^{(1)} \left| \frac{j'}{i} \right) \right\rangle \left\langle W_0^{(2)} \right| \left(W_0^{(2)} \left| \frac{i'}{j} \right) \right\rangle \right.$$

$$+ \langle W_0^{(1)} | \left(W_0^{(1)} \left| \frac{i'}{j} \right) \langle W_0^{(2)} | H^{(2)} - E_0^{(2)} \left| \left(W_0^{(2)} \left| \frac{i'}{j} \right) \right\rangle \right.\right.$$

which vanishes since the scalar products are zero:

$$\left\langle W_0^{(1)} \middle| \left(W_0^{(1)} \left| \frac{j'}{i} \right) \right\rangle = 0 \qquad \left\langle W_0^{(2)} \middle| \left(W_0^{(2)} \left| \frac{i'}{j} \right) \right\rangle = 0 \right.\right.$$

j' (respectively i') is indeed orthogonal to all spin-orbitals of $W_0^{(1)}$ (respectively $W_0^{(2)}$) according to the definition (2.145). Finally we have:

$$\langle W_0^{(1)} W_0^{(2)} | (H^{(1)} - \bar{E}_0^{(1)}) + (H^{(2)} - \bar{E}_0^{(2)}) \left| W_0^{(1)} \left(W_0^{(2)} \left| \frac{i'}{j} \right) \right\rangle \right.$$

$$= \langle W_0^{(1)} | H^{(1)} - \bar{E}_0^{(1)} | W_0^{(1)} \rangle \left\langle W_0^{(2)} \middle| \left(W_0^{(2)} \left| \frac{i'}{j} \right) \right\rangle \right.$$

$$+ \langle W_0^{(1)} | W_0^{(1)} \rangle \langle W_0^{(2)} | H^{(2)} - \bar{E}_0^{(2)} \left| \left(W_0^{(2)} \left| \frac{i'}{j} \right) \right\rangle \right.$$

$$= \langle W_0^{(2)} | H^{(2)} \left| \left(W_0^{(2)} \left| \frac{i'}{j} \right) \right\rangle \right.$$

and similarly

$$\langle W_0^{(1)} W_0^{(2)} | H_0 - \bar{E}_0^0 \left| \left(W_0^{(1)} \left| \frac{j'}{i} \right) W_0^{(2)} \right\rangle = \langle W_0^{(1)} | H^{(1)} \left| \left(W_0^{(1)} \left| \frac{j'}{i} \right) \right\rangle \right.\right.$$

so that (2.166) reduces to

$$\varepsilon^{\text{c-ex}(1)} = -\sum_{i=1}^{n_1} \sum_{j=1}^{n_2} S_{ij} \left[\langle W_0^{(1)} | H^{(1)} \left| \left(W_0^{(1)} \left| \frac{j'}{i} \right) \right\rangle \right.\right.$$

$$\left. + \langle W_0^{(2)} | H^{(2)} \left| \left(W_0^{(2)} \left| \frac{i'}{j} \right) \right\rangle \right.\right] \tag{2.167}$$

We may deduce from (2.167) the following property:

The one-exchange approximation to the 'complementary exchange' term vanishes when the approximate functions used for representing the subsystems are the exact Hartree–Fock Slater determinants.

Indeed the function $(W_0^{(1)} \mid j'/i)$ is a mono-excited determinant with respect to $W_0^{(1)}$ (j' is orthogonal to the subspace of the occupied orbitals from which $W_0^{(1)}$ is built, and Brillouin's theorem (see, for example, Pilar (1968) Ch. 13, Section 13.5, p. 364) just asserts that the matrix element

$\langle W_0^{(1)} | H^{(1)} | W_* \rangle$ vanishes for any mono-excited determinant W_* when $W_0^{(1)}$ is the exact Hartree–Fock determinant. The same argument applies to the subsystem (2), so that all the matrix elements involved in (2.167) actually vanish if $W_0^{(1)}$ and $W_0^{(2)}$ are the respective exact Hartree–Fock determinants for the two subsystems.

Note that the proof does no more hold for the further terms of (2.165) involving two exchanges, since we would obtain determinants involving two transfers, i.e. two-excited with respect to the ground state determinants, and the corresponding matrix elements may be non-zero. For more than two transfers, they would again vanish, because the molecular hamiltonian $H^{(s)}$ is bielectronic and therefore has zero matrix elements between determinants which differ by more than two spin-orbitals.

This property explains the results of Murrell and Shaw (1967b, 1968b) for the He . . . He interaction: these authors computed ε_1^{RS}, $\varepsilon_1^{ex(1)}$ (the one-exchange approximation to ε_1^{ex}) and ε_1^{HL} for the two different functions W_0, the first one being not very accurate (a single Slater orbital with optimized exponent), and the second one being much closer to the exact Hartree–Fock solution (an orbital built as a linear combination of four Slater orbitals). The difference $\varepsilon_1^{HL} - (\varepsilon_1^{RS} + \varepsilon_1^{ex(1)}) = \varepsilon^{c\text{-}ex(1)} + O(S^4)$ is actually much smaller in the second case.

It also appears from these calculations that these various terms are strongly underestimated when using the less accurate function (ε_1^{HL} by a factor of about two, and $\varepsilon_1^{ex(1)}$ by a factor of about five in the equilibrium region). Thus reliable results can be computed only by using accurate enough functions for the subsystems, and under such conditions the complementary exchange term becomes almost negligible, so that both possibilities (2.158) and (2.164) for evaluating the exchange term become practically equivalent.

Dacre and McWeeny (1970) evaluated ε_1^{HL} as the finite sum of terms corresponding to increasing numbers of electron interchange, and Gerratt (1976) expressed this expansion in the framework of the 'spin-coupled' formalism (where the space and spin variables are explicitly separated). These developments hold for open-shell as well as closed-shell subsystems.

On another hand simplified methods for evaluating the short-range energy (ε_1^{HL} or ε_1^{ex}) in the framework of the Molecular Orbital theory (for closed-shell systems) have been proposed by Murrell and Varandas (1975), Ellis (1976).

An explicit formula for the Heitler–London energy ε_1^{HL} is given in Appendix C for the case where the subsystems eigenfunctions $W_0^{(s)}$ are approximated by Slater determinants. This formula involves the usual molecular integrals (kinetic, nuclear, electronic) between the (non-orthogonalized) orbitals of the subsystems and the elements of the inverse of the overlap matrix (built from the overlap integrals between these orbitals).

In the next section, we shall derive a compact formula (originally given by Williams, Schaad and Murrell, 1967) for $\varepsilon_1^{ex(1)}$ in terms of intermolecular overlap charge densities. This formula is well-suited for various theoretical analyses, e.g. the pair-wise additivity properties.

B. An expanded formula for the one-exchange term $\varepsilon_1^{ex(1)}$ (case of Slater determinants)

By introducing the expansion (2.138) $\mathscr{A}'_{int} = P_{(1)} - P_{(2)} + \ldots$ (see (2.380, 2.381) in Appendix A) in (2.158) and retaining the terms of order S^2, we obtain

$$\varepsilon_1^{ex} = \varepsilon_1^{ex(1)} + O(S^4) \tag{2.168}$$

$$\varepsilon_1^{ex(1)} = -\langle\phi_0|\, VP_{(1)}\, |\phi_0\rangle + \langle\phi_0|\, V\, |\phi_0\rangle\langle\phi_0|\, P_{(1)}\, |\phi_0\rangle \tag{2.169}$$

As in the previous section, we shall evaluate more explicitly $\varepsilon_1^{ex(1)}$ for the case where the functions $W_0^{(s)}$ (whose product is ϕ_0) are Slater determinants. The formula (2.156) gives us the expression for $\langle\phi_0|\, VP_{(1)}\, |\phi_0\rangle$ and (2.153) the expression for $\langle\phi_0|\, P_{(1)}\, |\phi_0\rangle$. Now, the last term in (2.156) (or, for that matter, (2.150)) is

$$\langle\phi_0|\, V\, |\phi_0\rangle\left(\sum_{i=1}^{n_1}\sum_{j=1}^{n_2} S_{ij}^2\right) \tag{2.170}$$

since

$$\langle W_0^{(1)} W_0^{(2)}|\, V\, |W_0^{(1)} W_0^{(2)}\rangle \equiv \langle\phi_0|\, V\, |\phi_0\rangle$$

may be factorized, and, on the other hand, (2.153) gives:

$$\langle\phi_0|\, P_{(1)}\, |\phi_0\rangle \equiv \langle W_0^{(1)} W_0^{(2)}|\, P_{(1)}\, |W_0^{(1)} W_0^{(2)}\rangle = \sum_{i=1}^{n_1}\sum_{j=1}^{n_2} S_{ij}^2 \tag{2.171}$$

so that the second term $\langle\phi_0|\, V\, |\phi_0\rangle\langle\phi_0|\, P_{(1)}\, |\phi_0\rangle$ of (2.169) exactly cancels the third contribution (2.170) to $\langle\phi_0|\, VP_{(1)}\, |\phi_0\rangle$. Thus, the right-hand side of (2.169) reduces to the first two terms of (2.156), that we write:

$$\varepsilon_1^{ex(1)} = -\sum_{i=1}^{n_1}\sum_{j=1}^{n_2}\left\{\iint \frac{\rho_{ij}^{(1)}(\vec{r}^1)\rho_{ij}^{(2)}(\vec{r}^2)}{|\vec{r}^1 - \vec{r}^2|}\, d\vec{r}\, d\vec{r}^2 + S_{ij}\left[\int V^{(1)}(\vec{r}^2)\rho_{ij}^{(2)}(\vec{r}^2)\, d\vec{r}^2 + \int V^{(2)}(\vec{r}^1)\rho_{ij}^{(1)}(\vec{r}^1)\, d\vec{r}^1\right]\right\} \tag{2.172}$$

where we use the simplified notations $\rho_{ij}^{(s)}$ for the overlap charge densities defined by (2.154, 2.155):

$$\rho_{ij}^{(1)}(\vec{r}) \equiv \rho_{00}^{(1)}\left(\frac{j'}{i}\right)(\vec{r}) = \langle W_0^{(1)}|\, \boldsymbol{\rho}^{(1)}(\vec{r})\, \left|\left(W_0^{(1)}\, \middle|\, \frac{j'}{i}\right)\right\rangle \tag{2.173}$$

$$\rho_{ij}^{(2)}(\vec{r}) \equiv \rho_{00}^{(2)}\left(\frac{i'}{j}\right)(\vec{r}) = \langle W_0^{(2)}|\, \boldsymbol{\rho}^{(2)}(\vec{r})\, \left|\left(W_0^{(2)}\, \middle|\, \frac{i'}{j}\right)\right\rangle \tag{2.174}$$

and (2.137) gives us the explicit formulas in terms of the orbitals, denoted $b_i^{(1)}$ for $W_0^{(1)}$ and $b_j^{(2)}$ for $W_0^{(2)}$:

$$\rho_{ij}^{(1)}(\vec{r}) = -b_i^{(1)}(\vec{r})b_j^{\prime(2)}(\vec{r})\delta_{ij}^{\text{spin}} \tag{2.175}$$

$$\rho_{ij}^{(2)}(\vec{r}) = -b_j^{(2)}(\vec{r})b_i^{\prime(1)}(\vec{r})\delta_{ij}^{\text{spin}} \tag{2.176}$$

As concerns $V^{(1)}(\vec{r}) \equiv V_{00}^{(1)}(\vec{r})$ and $V^{(2)}(\vec{r}) \equiv V_{00}^{(2)}(\vec{r})$, they also exactly correspond to the general definition (2.157), namely they denote the electrostatic potentials created at some point r by the ground-state charge distributions of the molecules (1) or (2):

$$V^{(1)}(\vec{r}) = \int \frac{f_{00}^{(1)}(\vec{r})}{|\vec{r}^1 - \vec{r}|} \mathrm{d}\vec{r}^1 = \sum_{\mu}^{(1)} \frac{Z_\mu}{|\vec{r}_\mu - \vec{r}|} - \sum_{i=1}^{n_1} \int \frac{|b_i(\vec{r}^1)|^2}{|\vec{r}^1 - \vec{r}|} \mathrm{d}\vec{r}^1 \tag{2.177}$$

$$V^{(2)}(\vec{r}) = \int \frac{f_{00}^{(2)}(\vec{r}^2)}{|\vec{r}^2 - \vec{r}|} \mathrm{d}\vec{r}^2 = \sum_{\nu}^{(2)} \frac{Z_\nu}{|\vec{r}_\nu - \vec{r}|} - \sum_{j=1}^{n_2} \int \frac{|b_j(\vec{r}^2)|^2}{|\vec{r}^2 - \vec{r}|} \mathrm{d}\vec{r}^2 \tag{2.178}$$

Some care must be exercised in the definition of the orthogonal components $b_i^{\prime(1)}$ and $b_j^{\prime(2)}$. The basic formula (2.145) indeed concerns the spin-orbitals $a_i^{(1)}$ and $a_j^{(2)}$. Let us consider for example $a_j^{\prime(2)}$:

$$a_j^{\prime(2)} = a_j^{(2)} - \sum_{k=1}^{n_1} S_{kj} a_k^{(1)} \tag{2.179}$$

where

$$S_{kj} = \langle a_k^{(1)} \mid a_j^{(2)} \rangle = \langle b_k^{(1)} \mid b_j^{(2)} \rangle \delta_{kj}^{\text{spin}} \tag{2.180}$$

It is therefore clear that all non-zero terms of the sum over k correspond to spin-orbitals $a_k^{(1)}$ having the same spin part σ_j as $a_j^{(2)}$, so that this spin function may be factorized:

$$a_j^{\prime(2)} = \left(b_j^{(2)} - \sum_{k=1}^{n_1} S_{kj} b_k^{(1)}\right)\sigma_j \tag{2.181}$$

which allows us to define an orbital $b_j^{\prime(2)}$ corresponding to the spin-orbital $a_j^{\prime(2)}$ written as $b_j^{\prime(2)}\sigma_j$:

$$b_j^{\prime(2)} = b_j^{(2)} - \sum_{k=1}^{n_1} S_{kj} b_k^{(1)} \tag{2.182}$$

where S_{kj}, given by (2.139), *involves the spin orthogonality factor and does not reduce to the pure orbital overlap* $\langle b_k^{(1)} \mid b_j^{(2)} \rangle$. A similar definition would hold for $a_i^{\prime(1)}$ and $b_i^{\prime(1)}$:

$$a_i^{\prime(1)} = a_i^{(1)} - \sum_{l=1}^{n_2} S_{il} a_l^{(2)} \tag{2.183}$$

$$b_i^{\prime(1)} = b_i^{(1)} - \sum_{l=1}^{n_2} S_{il} b_l^{(2)} \tag{2.184}$$

with $S_{il} = \langle a_i^{(1)} \mid a_l^{(2)} \rangle = \langle b_i^{(1)} \mid b_l^{(2)} \rangle \delta_{il}^{\text{spin}}$ according to (2.180).

The formula (2.172) was given without an explicit proof by Williams, Schaad and Murrell (1967) with slight differences of notation: these authors present their formula as a generalization of the formula proved by Murrell, Randic and Williams (1965, formula (37)) which was valid only for subsystems involving one spin-orbital or two spin-orbitals with different spins; they defined their overlap densities $\rho_{ij}^{(s)}$ and their potentials $V^{(s)}$ with signs opposite to those used here, and they also defined their $\rho_{ij}^{(s)}$'s in terms of the spin-orbitals (and not the orbitals, as is done here), so that all the integrals analogous to (2.172) and (2.177, 2.178) involve the space *and spin* variables (and not only the space variables as in the present formulas). This form using directly spin-orbitals and space–spin variables was also used in our previous works (Claverie, 1973, 1976).

Whether expressed in terms of orbitals as in (2.172) or in terms of spin-orbitals as in the works quoted, this formula for $\varepsilon_1^{\mathrm{ex}(1)}$ is especially appropriate for theoretical analysis: thus it has been used for investigating the pair-wise additivity of $\varepsilon_1^{\mathrm{ex}(1)}$ for more than two molecules (Williams *et al.*, 1967) or the possibility of evaluating the interaction $\varepsilon_1^{\mathrm{ex}(1)}$ between two molecules as a sum of pair-wise interactions between molecular subunits (atoms, bonds) (Claverie, 1973, 1976). These studies will be summarized in Section E below.

As a conclusion of this section and the previous one, we may notice that the repulsive character of the exchange terms (whether $\varepsilon_1^{\mathrm{HL}} - \varepsilon_1^{\mathrm{RS}}$ or $\varepsilon_1^{\mathrm{ex}}$) does not appear as clearly evident from their explicit expressions obtained up to now. This property may rather be derived from the Hellman–Feynman theorem (e.g. see Hirschfelder, Curtiss and Bird (1954) ch. 13) which gives the forces acting on the nuclei rather than the energy: Salem (1961) has indeed shown that the short-range 'exchange' repulsion appeared as the result of an (exponentially varying) decrease of the electronic charge density between the (closed-shell) subsystems, this decrease resulting in a lowered screening of the nuclear repulsion by the electronic density, and therefore in an increase of this repulsion (varying exponentially with distance except at very short distances).

C. An example of the search for simplified formulas: the helium–helium case

The helium atom is of course the simplest of the closed shell systems, but there are more general grounds for considering the interaction between two such atoms (or, more generally, between rare gas atoms): indeed, if it is assumed that an approximate evaluation of the 'exchange repulsion' between two molecules may be obtained as a sum of atom–atom terms, then the evaluation of this repulsion between rare gas atoms may give a useful guide for building such atom–atom formulas.

Many works have been devoted to the helium–helium interaction. References to the earlier works are given by Kestner and Sinanoglu (1966),

Gilbert and Wahl (1967); numerous *ab initio* studies were published recently: Marchand (1970), Bertoncini and Wahl (1970), Schaeffer III *et al.* (1970), McLaughlin and Schaeffer III (1971), Conway and Murrell (1972), Liu and McLean (1973), Daudey *et al.* (1974), Salahub (1974), Murrell and Varandas (1975), Wormer *et al.* (1975). Works more closely connected with our present purpose, namely getting simplified analytical formulas for the 'exchange' repulsion are those by Griffing and Wehner (1955) (see also Slater (1963), Ch. 6 and Appendix 10), Mason and Hirschfelder (1957), Murrell and Shaw (1967b, 1968b). The application of the 'electron gas model' to the evaluation of the short-range 'exchange' repulsion will be considered in the next section.

Each helium atom ($s=1, 2$) is described by a Slater determinant $(1/\sqrt{2})\,|b^{(s)}\alpha b^{(s)}\beta|$. We shall consider here two choices of the orbitals $b^{(s)}$, for which Murrell and Shaw (1967b, 1968b) computed $\varepsilon_1^{\text{RS}}$ (the electrostatic energy, which reduces here to penetration part $\varepsilon_1^{\text{pen}}$), $\varepsilon_1^{\text{ex(1)}}$ (the first-order exchange part limited to one exchange) and $\varepsilon_1^{\text{HL}}$ (the Heitler–London first-order interaction) (Murrel and Shaw denote these terms E_{coul}, E_{exch} and E respectively):

$$(1)\ b^{(s)} = 1s(1{\cdot}6875) \tag{2.185}$$

$$(2)\ b^{(s)} = 0{\cdot}884315\ 1s(1{\cdot}4) - 0{\cdot}051269\ 2s(1{\cdot}4) + 0{\cdot}175261\ 1s(3{\cdot}0) + 0{\cdot}010673\ 2s(3{\cdot}0) \tag{2.186}$$

$ns(\zeta)$ denotes the normalized Slater orbital $C_n r^{n-1} e^{-\zeta r}$; (2.185) is the SCF Slater orbital and (2.186) is the SCF 4-component orbital of Roothaan, Sachs, and Weiss (1960).

The explicit formulas for $\varepsilon_1^{\text{RS}} = \varepsilon_{\text{el}}$ and $\varepsilon_1^{\text{ex(1)}}$ were given by Murrell and Shaw (1967b) (equation 1 and 2 of their paper):

$$\varepsilon_{\text{el}} = \frac{4}{R} + 2(N_{11}^{(2)} + N_{22}^{(1)}) + 4M_{12}^{12} \tag{2.187}$$

$$\varepsilon_1^{\text{ex(1)}} = 2[S^2(3M_{12}^{12} + N_{11}^{(2)} + N_{22}^{(1)}) - S(M_{12}^{22} + M_{11}^{21} + N_{12}^{(1)} + N_{12}^{(2)}) - M_{11}^{22}] \tag{2.188}$$

where we used our notations introduced in Appendixes C and D equations (2.429) to (2.433), $S = S_{12}$; (2.187) is identical with (2.458). Note that, owing to the symmetry of the problem, we have $N_{11}^{(2)} = N_{22}^{(1)}$, $N_{12}^{(1)} = N_{12}^{(2)}$ and $M_{12}^{22} = M_{11}^{21}$. Note also that (2.188) is different from the expression (2.465) given in Appendix D for $\varepsilon_{1(S^2)}^{\text{HL-ex}}$, the exchange part (limited to order S^2) of $\varepsilon_1^{\text{HL}}$: the difference is the complementary exchange term $\varepsilon_1^{\text{c-ex}}$ (limited to order S^2). In agreement with the result proved in the previous Section B, this difference between $\varepsilon_1^{\text{HL}}$ and $(\varepsilon_1^{\text{RS}} + \varepsilon_1^{\text{ex}})$ (denoted E and $(E_{\text{coul}} + E_{\text{exch}})$ respectively by Murrell and Shaw (1967b, 1968b)) is actually much smaller for the four-component orbital than for the one-component orbital: the latter is indeed less close to the exact Hartree–Fock orbital than the former.

It is easily seen that $\varepsilon_1^{\text{ex-1}}$ given by (2.188) approximately varies like S^2/R: except at short distance, the nuclear attraction integrals $n_{11}^{(2)}=n_{22}^{(1)}$ and Coulomb integral M_{12}^{12} may be approximated by $1/R$; for *ns* Slater orbitals the hybrid nuclear integral $n_{12}^{(1)}$ and the hybrid two-electron integral M_{11}^{21} vary like S/R; finally the exchange integral M_{11}^{22} may be approximated by $S^2(C+\log\rho)/\rho$ (where $\rho=\zeta R$): see formula (2.484) in Appendix D. Upon inserting these approximate formulas in (2.188), all terms vary like S^2/R, except the $S^2\log\rho/\rho$ part of the exchange integral. It is therefore not astonishing that Murrell *et al.* (1965) and Murrell and Shaw (1967b) were able to represent $\varepsilon_1^{\text{ex}(1)}$ by the formula kS^2/R (k being a suitably fitted constant).

As explained at the end of Section III.A (formula (2.124)), we want to express the whole short-range part of the total interaction energy ($\varepsilon_1^{\text{pen}}+\varepsilon_1^{\text{ex}}$). Since the multipolar part of the electrostatic energy $\varepsilon_1^{\text{RS}}=\varepsilon_{\text{el}}$ vanishes in the present case, $\varepsilon_1^{\text{HL}}=\varepsilon_1^{\text{RS}}+\varepsilon_1^{\text{HL-ex}}$ reduces to this desired short-range part, and we shall now check various approximate formula for $\varepsilon_1^{\text{HL}}$ against the exact values of Murrell and Shaw (1967b, 1968b). In Appendix D, we derive simplified formulas by approximating the various integrals involved in the exact expression (limited to order S^2) (2.462); on another hand we fitted a simple exponential formula $B\mathrm{e}^{-bR}$ on the exact results $\varepsilon_1^{\text{HL}}$, simply by performing a linear least squares fitting of $\log\varepsilon_1^{\text{HL}}$ with respect to R:

$$\log\varepsilon_1^{\text{HL}}\simeq\log B-bR \tag{2.189}$$

This fitting was performed by using the eight values of R considered by Murrell and Shaw (1967b) (R from 4·0 to 7·5 with a step of 0·5 a.u.).

In most cases, we give the results of the approximate formula and the relative errors (percentage) with respect to the exact values.

From the qualitative point of view, it may be pointed out that the repulsive character of $\varepsilon_1^{\text{HL}}$ is due to the S-term of the formula (2.465), namely $S(-T_{12}+4n_{12}^{(1)}-2M_{11}^{12})$ which involves the kinetic, nuclear and two-electron 'hybrid' integrals: $-T_{12}$, $n_{12}^{(1)}$ and M_{11}^{12} are positive (at least at medium and long-range), and since $n_{12}^{(1)}\simeq M_{11}^{12}$ (both vary asymptotically like S/R) we have $4n_{12}^{(1)}-2M_{11}^{12}\simeq 2n_{12}^{(1)}$, hence the positiveness of the S-term. The other two terms (the S^2-term and the exchange integral contribution $-M_{11}^{22}/2$) are negative and may only reduce the positive 'hybrid' S-term.

More generally, the essential qualitative feature of the short-range term, namely its exponential-like behaviour as a function of the distance R, clearly appears as a consequence of the exponential behaviour of the overlap, hybrid and exchange integrals, a behaviour which may itself be attributed to the exponentially decreasing behaviour at large distance of the bound state functions (Ahlrichs, 1972b, 1973a,b); in the present orbital approximation of the exact state functions, we have exponentially decreasing orbitals and therefore exponentially small two-centre densities $b^{(1)}(\vec{r})b^{(2)}(\vec{r})$ which are involved in all above mentioned types of integrals and are responsible for their exponential decrease at large R.

1. The one-component orbital

The numerical results are given in Table 2.1. Besides the exact values of ε_1^{HL}, we give the values of $4S^2/R$, those of the formula (2.486, 2.487) of Appendix D):

$$\varepsilon_1^{HL} \simeq 4\frac{S^2}{R}[0{\cdot}3125\rho + 0{\cdot}2977843 - \log\rho)/2] \tag{2.190}$$

and those of the exponential fitting Be^{-bR} for which we calculated the parameter values (corresponding to atomic units for energy and distance):

$$B = 70{\cdot}561508 \qquad b = 2{\cdot}7520048 \tag{2.191}$$

We see that the representation of ε_1^{HL} by a formula of the type kS^2/R could not be accurate here, since the ratio $(4S^2/R)/\varepsilon_1^{HL}$ varies by a factor of 2 (from 0·671 to 0·343) from $R = 4{\cdot}0$ to $R = 7{\cdot}5$. This corresponds to the fact that the complementary exchange term is non-negligible for the one-component orbital, and, contrary to ε_1^{ex}, it cannot be represented accurately by an S^2/R type formula.

The approximate analytical formula (2.190) slightly underestimates the value of ε_1^{HL}; this corresponds essentially to the overestimation of the exchange integral of (2.465) by the approximation (2.487) (this overestimation clearly appears in the Table 2.6 in Appendix D). The accuracy increases with R and would be sufficient (about 5 per cent in the equilibrium region for practical purposes, were it not the fact that the one-component orbital does not allow an accurate evaluation of the true ε_1^{HL} (in the equilibrium region, this ε_1^{HL} (one-component) is about two times smaller than ε_1^{HL} (four-component), which seems physically correct when the total interaction energy is calculated by using it).

Finally, it is quite interesting that the simple exponential formula (2.189, 2.191) also gives an accuracy of about 5 per cent in the equilibrium region (note that a still better accuracy would be obtained by excluding from the least-squares fitting the two extreme values $R = 4$ and $R = 7{\cdot}5$).

2. The four-component orbital

The accurate values obtained by Murrell and Shaw (1967b, 1968b) show that the one-component orbital gives a repulsion energy ε_1^{HL} markedly smaller than that given by the four-component orbital (for which the obtained values lead to a total interaction energy in good agreement with experimental results). This difference is obviously due to the too high value of the unique Slater exponent of the one-component orbital: owing to the lack of flexibility of this function (ζ is the single parameter), the value of ζ which minimizes the energy of the isolated atom leads to an orbital too concentrated towards the nucleus and decreasing too fast at large distance. On the contrary, with a four-component orbital, the Slater components with large $\zeta(3{\cdot}0)$ allow for a good representation of the energy of the atom,

Table 2.1. The Heitler–London interaction energy for two helium atoms: case of the one-component Slater orbital. Comparison between the values of ε_1^{HL} (last column) and the approximate values given by $4S^2/R$, by the formula (2.190), and by the exponential fit (formulas (2.189, 2.191)). The values of R and of the energy are expressed in atomic units

R	$\rho = \zeta R$	$S \times 10^3$	$\frac{4S^2}{R} \times 10^5$	$\frac{4S^2/R}{\varepsilon_1^{HL}}$	$(2.190) \times 10^5$	$\frac{\Delta E}{E} \times 100$ [a] (%)	$B \exp(-bR)$ $\times 10^5$	$\frac{\Delta E}{E} \times 100$ [a] (%)	$\varepsilon_1^{HL} \times 10^5$ [b]
4·0	6·75	26·857051	72·130120	0·6710	94·021317	−12·538	116·90842	8·752	107·50
4·5	7·59375	14·007545	17·441007	0·5953	26·305858	−10·219	29·529463	0·783	29·30
5·0	8·4375	7·183882	4·128653	0·5327	7·098260	−8·409	7·458737	−3·758	7·75
5·5	9·28125	3·632575	0·959680	0·4808	1·857254	−6·951	1·883975	−5·612	1·996
6·0	10·125	1·814833	0·219575	0·4371	0·473282	−5·796	0·475866	−5·281	0·5024
6·5	10·96875	0·897319	0·495497	0·3999	0·117884	−4·856	0·120197	−2·988	0·1239
7·0	11·8125	0·439672	0·011046	0·3682	0·028784	−4·054	0·030360	1·201	0·0300
7·5	12·65625	0·213726	0·002436	0·3431	0·006906	−2·727	0·007669	8·008	0·0071

[a] ΔE = (approximate value) − (exact value). Approximate value = value in the previous column. E = exact value = ε_1^{HL} (last column).
[b] exact values taken from Murrell and Shaw (1967b).

without reducing the quality of the orbital at large distance, since this quality is insured by the components with smaller $\zeta(1{\cdot}4)$. The practical conclusion is that, in order to use *ab initio* computations for deriving formulae for the short-range repulsion, it is necessary to use atomic orbitals with sufficient quality, so as to insure a correct behaviour of the state function at large distance. Dacre and McWeeny (1970) arrive at the same conclusion in the discussion of their results for the neon–neon interaction.

Let us now consider the numerical results of Table 2.2. Besides the accurate values of ε_1^{HL}, we give the results for the following approximate formulas:

Table 2.2. The Heitler–London interaction energy for two helium atoms: case of the four-components Slater orbital. Comparison between the accurate values of ε_1^{HL} (last column) and the approximate values given by $4S^2/R$, by the formulas (2.496) and (2.193), by $1{\cdot}1\times(S_1)^2/R$, by $1{\cdot}1\times(2.193)_1$ (where the subscript 1 means that S has been replaced by S_1), and by the exponential fit (formulas (2.189, 2.194)). The values of R and of the energy are expressed in atomic units

R	$S\times10^3$	$[4S^2/R)\times10^5$	$\Delta E/E\times100$ [a](%)	$(2.496)\times10^5$	$\Delta E/E\times100$ [a](%)	$(2.193)\times10^5$	$\Delta E/E\times100$ [a](%)	$\varepsilon_1^{HL}\times10^5$ [b]
4·0	41·913407	175·67337	8·373	116·05853	−28·403	162·48194	0·236	162·1
4·5	24·104373	51·646295	7·529	34·375158	−28·430	48·125221	0·198	48·03
5·0	13·671262	14·952272	7·108	10·118769	−27·516	14·166277	1·478	13·96
5·5	7·661422	4·268900	2·544	2·957976	−28·946	4·141167	−0·524	4·163
6·0	4·248629	1·203390	−0·299	0·858302	−28·889	1·201623	−0·445	1·207
6·5	2·334283	0·335315	−1·552	0·247143	−27·439	0·346000	1·585	0·3406
7·0	1·271908	0·092443	−9·636	0·070610	−30·978	0·098854	−3·369	0·1023
7·5	0·687883	0·025236	−8·894	0·020016	−27·739	0·028023	1·166	0·0277

[a] ΔE = (approximate value) − (exact value)
Approximate value = value in the previous column
E = exact value = ε_1^{HL} (last column)
[b] exact values taken from Murrell and Shaw (1967b)

Table 2.2 (second part)

R	$S_1\times10^3$	$[1{\cdot}1\times4(S_1)^2/R]\times10^5$	$\Delta E/E\times100$ [a](%)	$1{\cdot}1\times[(2.193)$ with $(S_1)^2]\times10^5$	$\Delta E/E\times100$ [a](%)	$B\exp(-bR)\times10^5$	$\Delta E/E\times100$ [a](%)	$\varepsilon_1^{HL}\times10^5$ [b]
4·0	38·286065	161·24051	−0·530	149·132848	−7·999	165·80949	2·284	162·1
4·5	22·302129	48·633200	1·256	45·317548	−5·647	48·155511	0·261	48·03
5·0	12·781414	14·376079	2·980	13·620374	−2·433	13·985648	0·184	13·96
5·5	7·224238	4·175170	0·292	4·050240	−2·709	4·061806	−2·431	4·163
6·0	4·034709	1·193784	−1·011	1·192032	−1·240	1·179657	−2·655	1·207
6·5	2·229967	0·336617	−1·169	0·347343	1·980	0·342604	0·588	0·3406
7·0	1·221190	0·093739	−8·369	0·100240	−2·014	0·099501	−2·736	0·1023
7·5	0·663290	0·025811	−6·819	0·0286613	3·469	0·028898	4·324	0·0277

[a] ΔE = (approximate value) − (exact value)
Approximate value = value in the previous column
E = exact value = ε_1^{HL} (last column)
[b] exact value taken from Murrell and Shaw (1967b)

—The formula

$$\varepsilon_1^{HL} \simeq 4S^2/R \tag{2.192}$$

(according to the general proposal kS^2/R of Murrell *et al.* (1965) and also Murrell and Shaw (1967b). This type of formula is now more accurate than it was for the one-component orbital, as it could be expected since the complementary exchange term is now much smaller. It varies significantly too fast (relative error of about 8 per cent at $R=4$, and -9 per cent at $R=7{\cdot}5$) but takes correct values at the equilibrium distance.

—On the contrary the formula (2.496) derived in Appendix D under-estimates $\varepsilon_1^{\mathrm{HL}}$ rather strongly (by almost 30 per cent), but in a very uniform way, so that by introducing an empirical multiplicative factor it is possible to get a very good agreement with the exact values of $\varepsilon_1^{\mathrm{HL}}$; the value 1·4 for this factor was found satisfactory (it is fortuitous that this value coincides with the value of the lowest Slater exponent ζ_1 occurring in the four-component orbital):

$$\varepsilon_1^{\mathrm{HL}} \simeq 1{\cdot}4\frac{4S^2}{R}\left[0{\cdot}1276601R+0{\cdot}843156-\frac{\log R}{2}\right] \qquad (2.193)$$

(2.193) indeed gives a relative error lying between $-0{\cdot}5$ per cent and 1·6 per cent (except the value $-3{\cdot}37$ per cent at $R=7{\cdot}0$).

—In order to have a simpler formula, we considered the possibility of using, instead of the complete overlap integral S, the 'restricted' overlap integral $S_1=S[\zeta_1]$ obtained by retaining only the first two components of the orbital (2.185), namely those with the exponent $\zeta_1=1{\cdot}4$ (see formulas (2.490, 2.491) in Appendix D). S_1 is slightly smaller than S, so that, by using $1{\cdot}1S_1^2$ instead of S^2 in the previous formulas (2.192, 2.193) we obtain quite satisfactory results ($1{\cdot}1\times 4S_1^2/R$ even gives better results than $4S^2/R$).

—Finally we give the results corresponding to the least-squares fitted exponential formula Be^{-bR} for which we calculated the parameter values

$$B=32{\cdot}757942 \qquad b=2{\cdot}4728079 \qquad (2.194)$$

The accuracy of this very simple formula is rather remarkable, since the relative error remains smaller than 3 per cent in the range 4·0 to 7·0 of the values of R (the 4·32 per cent error at $R=7{\cdot}5$ is of no practical concern anyway). It must be emphasized that this accuracy is higher than the accuracy of the exponentially fitted formula for the case of the one-component orbital, despite the fact that the exact analytical expression of $\varepsilon_1^{\mathrm{HL}}$ is considerably more involved for the four-component orbital, than for the one-component orbital.

We may now draw some conclusions from this study of simplified formulas for the short-range energy in the helium–helium case.

(a) Searching simplified formulas by trying to approximate the explicit analytical formulas for the integrals between the atomic orbitals involved apparently becomes a prohibitive task, except for extremely simple cases (like the one-component orbital for the He atom). Thus, it is quite significant that, already when dealing with the four-component orbital, we were obliged to multiply our simplified formula by a suitable empirical constant (truly enough, the final formula thus obtained was the most accurate among all the approximate formulas considered here, but there is no guarantee that such an accuracy would be found by following a similar procedure in more complicated cases, namely heavier atoms).

(b) Since it appears anyway necessary to compute the exact values of $\varepsilon_1^{\mathrm{HL}}$ for some interatomic distances in order to fit and check the simplified formulas, a first systematic procedure would be the fitting of the single constant k' in the type of formula proposed by Murrell, Randic and Williams (1965) namely $k'(4S^2/R)$ for the helium–helium case and

$$\varepsilon_1^{\mathrm{HL}} \simeq \frac{k'}{R} \sum_i^{(1)} \sum_j^{(2)} n_i^{\mathrm{occ}} n_j^{\mathrm{occ}} (S_{ij})^2 \tag{2.195}$$

in the general case (n_i^{occ} and n_j^{occ} denote the occupation numbers of the orbitals i and j: for helium, $n_i^{\mathrm{occ}} = n_2^{\mathrm{occ}} = 2$, hence the factor 4 that we put in evidence before S^2). It is interesting that, in the He . . . He case, k' could be taken equal to 1·0 (for the 'good' four-component orbital). Murrell and Teixeira-Dias (1970a) investigated other formulas (using the penetration part of the coulomb integrals) for interactions between non-closed-shell systems, but these attempts are not relevant for our purpose, especially owing to the simplicity of the exponential-type (Born–Mayer) formula to be considered next.

(c) From the purely practical point of view, the possibility of using a simple exponential formula Be^{-bR} (fitted over a convenient range of values of R) must be strongly emphasized. It is, for sure, the simplest formula having the theoretically correct exponential behaviour (while being much faster to compute than an analytical formula like (2.195)), but it appears capable of a surprising accuracy despite its simplicity when compared with the complexity of an accurate calculation of $\varepsilon_1^{\mathrm{HL}}$. Actually the *ab initio* perturbation calculations by Conway and Murrell (1972, 1974), Murrell and Varandas (1975), and the *ab initio* SCF results of Gilbert and Wahl (1967), Matcha and Nesbet (1967) also show the feasibility of an exponential fitting for the short-range interactions between Helium, Neon and Argon; see also the discussion of an 'effective electron' model (Jansen, 1962, 1965) by Kolos and Les (1972) and the discussion of the electron-gas model by Cohen and Pack (1974). Murrell and Varandas (1975) also discuss explicitly the exponential-type formulas for the interactions between He, Ne and Ar. This possibility of finding simple formulas whose accuracy against the exact results is empirically ascertained, although it is not yet mathematically proved, has been stressed at the end of the Section III.A above. It is practically possible to compute exact interactions $\varepsilon_1^{\mathrm{HL}}$ between atoms and to check whether an exponential fitting is accurate enough in all cases. But the results thus obtained could be used for the interactions between arbitrary molecules only if the short-range interaction between such molecules may be expressed in terms of contributions from small subunits (atoms or bonds): the direct accurate computation of $\varepsilon_1^{\mathrm{HL}}$ between two molecules (in order to perform a subsequent fitting) would be too long and the procedure would almost completely lose its interest. From the practical point of view, it is indeed necessary to have a limited number of parameters, the fitting of which being made once for all, so that heavy computations (e.g. accurate atom–atom interactions) may be accepted for this step of the procedure; but afterwards no such heavy computation should be necessary for dealing with the interactions between arbitrary molecules. Actually, the calculation of the short-range interaction as a sum of atom–atom contribution is widely used in the so-called 'empirical' formulas, and some comments concerning the validity of this approximation will be presented in Section E below.

D. The 'electron gas' model

As we have seen in the previous section, the computation of accurate values of the short-range interaction energy, as least between atoms (by assuming rare-gas-type electronic configuration in order to get the short-range repulsion corresponding to closed-shell systems) appears as a practical prerequisite for the checking and fitting of simplified analytical formulas (moreover, it is necessary to use rather accurate atomic ground state functions in order to get meaningful results: in the case of He, the use of a single-exponent orbital was clearly insufficient). This is still a feasible task for atoms, although becoming more and more cumbersome when the number of electrons increases, but it looks rather prohibitive for molecules (except the simplest ones); such values of the short-range energy would, however, be quite interesting, at least for checking purposes (essentially concerning the use of local-type interactions between atoms or bonds). Thus, any method allowing an approximate but still reliable evaluation of this short-range interaction (between possibly large, or at least medium-size systems) would be welcome. The electron gas model may offer such a possibility. The essential idea of this model consists in evaluating the energy of a system in terms of its total electron density $\rho_{\text{ef}}(\vec{r})$, by using the expression of the energy as a functional of the density which is known for the *uniform* electron gas. *The basic assumption involved is therefore the use of this functional relation for a system with non-uniform density*. More specifically, the total energy per electron of a uniform electron gas with density $\rho = N/\Omega$ (N = total number of electrons, Ω = total volume) is expressed as the sum of a kinetic energy $E_k(\rho)$ a coulomb energy $E_c(\rho)$, an exchange energy $E_{ex}(\rho)$ (the sum of which being referred to as the Hartree–Fock energy $E_{HF} = E_k + E_c + E_{ex}$) and the correlation energy $E_{corr}(\rho)$ (for the study of the electron gas, see for example, Fetter and Walecka (1971) and the other works quoted in references 8, 9 of Gordon and Kim (1972)). For each of these contributions, the total energy is

$$E_{tot}(\rho) = NE(\rho) = \Omega\rho E(\rho) \tag{2.196}$$

hence the *energy per unit volume* $E^{d}(\rho)$ (the superscript d stands for density):

$$E^{d}(\rho) = E^{tot}(\rho)/\Omega = \rho E(\rho) \tag{2.197}$$

which will be used as a *volumic energy density* for the non-uniform case, i.e. the total energy will then be expressed by the integral:

$$\bar{E}_{\text{tot}} = \int E^{\text{d}}[\rho(\vec{r})]\,\text{d}\vec{r} = \int \rho(\vec{r})E[\rho(\vec{r})]\,\text{d}\vec{r} \tag{2.198}$$

where the constant density ρ of (2.197) has simply been replaced by the density function $\rho(\vec{r})$ pertaining to the non-uniform system.

Often, only a part of the total energy density has been used. Thus, in the so-called statistical theories of the atom, the Thomas–Fermi (TF) model retains only $E_k + E_c$, while the Thomas–Fermi–Dirac (TFD) model uses the complete Hartree–Fock energy $E_{\text{HF}} = E_k + E_c + E_{\text{ex}}$; then the theory consists in searching the density $\rho(\vec{r})$ which minimizes the approximate expression (2.198) of the total energy. Actually, the expression for the coulomb energy per electron may be taken as follows, with Z_μ denoting the nuclear charge and $\vec{r}_\mu$ the position of the nucleus μ:

$$E_c[\rho] = -\sum_\mu \frac{Z_\mu}{|\vec{r}-\vec{r}_\mu|} + \frac{1}{2}\int \frac{\rho(r')}{|\vec{r}-\vec{r}'|}\,\text{d}\vec{r}' \tag{2.199}$$

which corresponds through (2.198) to the coulomb energy of the non-uniform system including the spurious self-interaction term discussed below in formulas (2.204, 2.205):

$$E_c = -\sum_\mu Z_\mu \int \frac{\rho(\vec{r})}{|\vec{r}-\vec{r}_\mu|}\,\text{d}\vec{r} + \frac{1}{2}\int\int \frac{\rho(\vec{r})\rho(\vec{r})}{|\vec{r}-\vec{r}'|}\,\text{d}\vec{r}\,\text{d}\vec{r}' \tag{2.200}$$

For a system described by a Slater determinant, it is immediately seen that the double integral of (2.200) is nothing but the *unrestricted* sum

$$\sum_i \sum_j J_{ij}$$

of the coulomb integrals, i.e. it involves the spurious self-interaction term

$$\sum_i J_{ii}$$

besides the true coulomb energy

$$\sum\sum_{i \neq j} J_{ij}$$

(i and j refer here to spin-orbitals).

It is only for the other parts of the energy (per electron) that we must use the expressions corresponding to the uniform electron gas:

$$E_k[\rho] = C_k \rho^{2/3} \quad \text{with} \quad C_k = \tfrac{3}{10}(3\pi^2)^{2/3} \tag{2.201}$$

$$E_{\text{ex}}[\rho] = C_{\text{ex}} \rho^{1/3} \quad \text{with} \quad C_{\text{ex}} = -\tfrac{3}{4}\left(\frac{3}{\pi}\right)^{1/3} \tag{2.202}$$

Further information concerning the statistical theories of the atom may be found in Gombas (1949) and March (1957); see also the recent papers by Santos and Villagra (1972), Alonso and Fonseca (1974), Kim and Gordon (1974a), Junker and Menendez (1975), and references therein. Note that, in the TF and TFD models the correlation energy is not considered; actually, there is no exact analytical expression for it even for the uniform electron gas, but such expressions exist for the low and high density limits, and reasonable interpolation between both regions of density may be used (Gordon and Kim, 1972; equations 20 to 22).

Let us now consider the application of this model to the problem of the interatomic (or eventually intermolecular) interactions. A first possibility is to use from the beginning the electron gas model, i.e. to use the TF or TFD electron densities for the isolated atoms. Then the electron density of the diatomic supersystem is simply taken as the sum of the isolated atom densities:

$$\rho(\vec{r}) = \rho^{(1)}(\vec{r}) + \rho^{(2)}(\vec{r}) \tag{2.203}$$

(2.203) is the first approximation of the statistical perturbation theory for the density of the supersystem (Gombas, 1949).

Then the energy $\bar{E}$ of the supersystem, as well as those ($\bar{E}^{(1)}$, $\bar{E}^{(2)}$) of the isolated subsystems, are calculated from ρ, $\rho^{(1)}$ and $\rho^{(2)}$ according to (2.198), and the difference $\bar{E} - (\bar{E}^{(1)} + \bar{E}^{(2)})$ is explicitly calculated; of course, by using (2.200) for the coulomb part of the energy, we get the exact electrostatic energy corresponding to the unperturbed densities $\rho^{(1)}$ and $\rho^{(2)}$, i.e. the expression (2.130) with $f_{00}^{(s)}(\vec{r}) = \rho_{\text{nuc}}^{(s)}(\vec{r}) - \rho^{(s)}(\vec{r})$, $s = 1, 2$. For applications of this first possibility, see e.g. Santos and Villagra (1972) and the works quoted in reference 10 of Gordon and Kim (1972), especially Abrahamson (1969) (some mistakes in earlier work were mentioned by Günter (1964)). It may be mentioned that Abrahamson was able to express his results in the Born–Mayer (exponential) form $B\mathrm{e}^{-bR}$ with a satisfactory accuracy, in agreement with our statement at the end of the previous section concerning the usefulness of this representation.

The second possibility consists in using accurate electron densities for the isolated systems instead of the TF or TFD approximate densities; afterwards the procedure for obtaining the interaction energy remains the same. The atomic electron densities were obtained from accurate Hartree–Fock functions. The necessary integrations may be performed numerically (in order to avoid systematic errors, it is convenient to integrate numerically the difference $E^{\mathrm{d}}[\rho] - (E^{\mathrm{d}}[\rho^{(1)}] + E^{\mathrm{d}}[\rho^{(2)}])$ of the energy densities rather than performing separate numerical integrations for these three densities, as emphasized by Gordon and Kim (1972)). This second approach was followed by Gaydaenko and Nikulin (1970) and also Nikulin (1971) (who took into account $E_{\mathrm{c}}^{\mathrm{d}}, E_{\mathrm{k}}^{\mathrm{d}}, E_{\mathrm{ex}}^{\mathrm{d}}$), and by Gordon and Kim (1972) who also included $E_{\mathrm{corr}}^{\mathrm{d}}$ besides the three previous terms. They developed their model extensively: interactions between rare gases and monoatomic ions (Kim and

Gordon, 1974b), between these ions themselves (Kim and Gordon, 1974c), calculation of the polarization and dispersion energies according to a Drude (harmonic oscillator) model by using the short-range potential (given by the electron gas model) for evaluating the short-range corrections to these second-order energies (Kim and Gordon, 1974d). The Gordon–Kim method was also used by Gianturco and Dilonardo (1975) for computing interactions between rare gas atoms and ions.

It must now be emphasized that the zeroth-order approximation to the electron density defined by (2.203) implies the impossibility of representing the second-order long-range terms (polarization and dispersion). It is easily seen (e.g. by using the local charge density operators defined by the formulas (2.125, 2.126, 2.127) in Section III.B.1 above) that taking the simple product function $W_0^{(1)}W_0^{(2)}$ for the supersystem (i.e. the zeroth-order function of the standard RS treatment) actually leads to the density $\rho = \rho^{(1)} + \rho^{(2)}$ introduced by (2.203). Actually, apart from the coulomb term which is calculated exactly and gives the usual electrostatic energy, the other terms (E_{k}, E_{ex}, E_{corr}) lead to contributions which decrease exponentially at large distances: this is due to the fact that the integrand $E^{\mathrm{d}}[\rho] - (E^{\mathrm{d}}[\rho^{(1)}] + E^{\mathrm{d}}[\rho^{(2)}])$ becomes everwhere exponentially small when the distance is large between densities $\rho^{(1)}$ and $\rho^{(2)}$ which themselves decrease exponentially at large distances from their respective centres. Thus, the electron gas model of the interaction, based upon (2.203), must be considered essentially as a simplified procedure for getting the short-range part of the total interaction energy, and Kim and Gordon (1974d) actually supplemented their previous electron gas model results by an evaluation of the dispersion and polarization terms (calculated according to a Drude model, as mentioned above) in order to get reasonable values of the total interaction energy. In this respect, it must also be emphasized that Gaydaenko and Nikulin (1970) were able to represent their short-range results by Born–Mayer (exponential) formulas $B\mathrm{e}^{-bR}$; they generally give two sets of parameters corresponding to very short and short distances respectively, but for all atoms considered ($Z=2$ (Helium) to $Z=37$ (Krypton)) they deal only with distances smaller than the Van der Waals equilibrium distance.

But some modifications of the Gordon–Kim procedure must be performed in order to use it for the evaluation of the short-range part of the interaction, as shown by the works of Rae (1973), and also Cohen and Pack (1974). First, we should not include the correlation part of the energy (density $E_{\mathrm{corr}}^{\mathrm{d}}$) but only $E_{\mathrm{c}}^{\mathrm{d}}$, $E_{\mathrm{k}}^{\mathrm{d}}$, $E_{\mathrm{ex}}^{\mathrm{d}}$. Indeed E_{corr} appears in the work of Gordon and Kim (1972) as a substitute for the long-range attractive term ε_2^m, which cannot be obtained owing to the approximation (2.203) (this substitute cannot be perfect, of course, since it decreases exponentially and therefore gives much too small values at large distance). ε_2^m denotes the (suitably truncated) multipolar part of the second-order RS perturbation term $\varepsilon_2^{\mathrm{RS}}$ (see equation (2.123) in Section III.A.2). Actually, the use of $E_{\mathrm{c}}^{\mathrm{d}}$, $E_{\mathrm{k}}^{\mathrm{d}}$, $E_{\mathrm{ex}}^{\mathrm{d}}$ together with

(2.203) should give an approximation to our ε_1^{HL}. But when the corresponding interaction energy is computed, it still gives a curve with a small negative minimum instead of an everywhere repulsive (positive) one (Kim and Gordon, 1974d, Figures 4 and 5 and p. 11; Cohen and Pack, 1974, Figures 1 to 4, pp. 2733–2734); on the contrary, the Hartree–Fock energy (which, for rare gases, is rather close to the Heitler–London interaction energy ε_1^{HL} as pointed out in Section III.A.2 above, in the discussion following equation (2.123)) gives an everywhere positive, repulsive, potential curve (Cohen and Pack, 1974, Figures 1 to 4), and the perturbation calculations of ε_1^{HL} also exhibit this feature (Murrell and Shaw, 1967b, 1968b; Conway and Murrell, 1972, 1974; Murrell and Varandas, 1975). Cohen and Pack convincingly argue that this discrepancy between the original electron gas model and the Hartree–Fock (or Heitler–London) accurate values may be removed by the modification of the exchange energy proposed by Rae (1973) for systems containing a finite number of electrons. The introduction of this modification runs as follows. The sum of the Coulomb and Exchange energy (two-electron terms) for any system of electrons described by a Slater determinant is

$$E_{2-\mathrm{el}} = \sum_{i \neq j}\sum (J_{ij} - K_{ij}) \tag{2.204}$$

where $J_{ij} = \langle a_i(1)a_j(2)|\, 1/r_{12}\, |a_i(1)a_j(2)\rangle$ and $K_{ij} = \langle a_i(1)a_j(2)|\, 1/r_{12}\, |a_j(1)a_i(2)\rangle$ are the coulomb and exchange integrals corresponding to the pair of *spinorbitals* (a_i, a_j). Since $J_{ii} = K_{ii}$, (2.204) may as well be written by suppressing the restriction $i \neq j$:

$$E_{2-\mathrm{el}} = \sum_i \sum_j (J_{ij} - K_{ij}) \tag{2.205}$$

and this unrestricted summation makes easier the evaluation of the Coulomb part

$$\sum_i \sum_j J_{ij}$$

and of the Exchange part

$$\sum_i \sum_j K_{ij}$$

when the specific forms of J_{ij} and K_{ij} pertaining to the electron gas are introduced (the two alternative formulas (2.204) and (2.205) are also very well-known in quantum chemistry). In particular, the simple form (2.202) for the exchange energy density corresponds to the *unrestricted* sum

$$\sum_i \sum_j K_{ij}$$

On another hand the coulomb energy (2.200) corresponds to the unrestricted sum

$$\sum_i \sum_j J_{ij}$$

for a Slater determinant, which seems at first sight to be consistent with the above form of the exchange energy. But a problem arises when we consider the interaction energy: the self-coulomb term of the supersystem, namely

$$\sum_i J_{ii} = \sum_i^{(1)} J_{ii} + \sum_i^{(2)} J_{ii}$$

is just equal to the sum of the self-coulomb terms of the subsystems, and consequently the coulomb part of the interaction is obtained correctly by using whether the unrestricted or the restricted coulomb terms. Thus, in a completely consistent model, the same property should hold for the exchange part of the energy. But since the evaluation of the exchange part is not exact, while that of the coulomb part is exact, it is by no means evident that this invariance of the exchange part will hold, and we must then check which procedure (use of the unrestricted or of the restricted sum) gives the best results. Now, when the number N of electrons is very large, the self-exchange term

$$\sum_i K_{ii}\left(\equiv \sum_i J_{ii}\right)$$

is negligible with respect to the double sum (the ratio decreases like $1/N$), and there is no more problem. But Rae (1973, 1975) showed that this is no more the case when N is not large: he evaluated the exchange energy per electron E'_{ex} corresponding to the *restricted* sum

$$\sum_{i \neq j} \sum K_{ij}$$

as:

$$E'_{ex}[\rho] = \gamma(N) E_{ex}[\rho] \tag{2.206}$$

$$\gamma(N) = 1 - \tfrac{8}{3}\delta + 2\delta^2 - \tfrac{1}{3}\delta^4 \tag{2.207}$$

where $E_{ex}[\rho]$ is given by (2.202) (which corresponds to the unrestricted summation whatever N), and $\delta = \delta(N)$ is the (unique) positive root of (Rae (1975) uses $\beta = 2\delta$ instead of δ):

$$\delta^3 - \tfrac{9}{8}\delta^4 + \tfrac{1}{4}\delta^6 = \frac{1}{4N} \tag{2.208}$$

Thus $\delta \simeq (4N)^{-1/3}$ when N is very large, and $\gamma(N)$ actually converges to 1 when $N \to \infty$. But, as emphasized by Rae, the correction is important at small and moderate N: $E'_{ex}/E_{ex} = 0{\cdot}794843$ for $N = 500$, $0{\cdot}317452$ for $N = 10$ and $0{\cdot}132392$ for $N = 4$. For evaluating the exchange interaction energy

$\Delta E'^{ex}$, Rae (1973) used the factor $\gamma(N)$ (where $N=N_1+N_2$ is the total number of electrons of the super-system) for multiplying both energy densities $E_{ex}[\rho^{(1)}+\rho^{(2)}]$ and $(E_{ex}[\rho^{(1)}]+E_{ex}[\rho^{(2)}])$, thus obtaining:

$$\Delta E'^{ex}=\gamma(N)\Delta E^{ex}=\gamma(N)\int[E_{ex}[\rho^{(1)}+\rho^{(2)}]-(E_{ex}[\rho^{(1)}]+E_{ex}[\rho^{(2)}])]\,d\vec{r} \quad (2.209)$$

This procedure amounts to consider the variation of the corrected exchange energy of the super-system between its actual finite separation and the infinite separation, since in this limit the vanishing of the overlap between the densities $\rho^{(1)}$ and $\rho^{(2)}$ has the consequence that

$$E[\rho^{(1)}+\rho^{(2)}]=E[\rho^{(1)}]+E[\rho^{(2)}]$$

for any functional $E[\rho]$. It could be argued that it is not quite satisfactory to use the factor $\gamma(N)=\gamma(N_1+N_2)$ in the limit of infinite separation, and that the factors $\gamma(N_1)$ and $\gamma(N_2)$ should then be involved (Rae, 1975). This point will be briefly discussed below, but we shall first comment the results obtained by using the simple recipe (2.209).

The exchange interaction energy calculated with E'_{ex} is significantly different from that calculated with E_{ex} for systems with the size of the rare gas atoms, thus the invariance of the exchange interaction with respect to the inclusion of self-exchange does *not* hold for such systems. It is therefore necessary to check against accurate computations (of ε_1^{HL} or ε_{HF}) which of these procedures (inclusion or not of the self-exchange part) gives the best result in the framework of the electron-gas model. This has been done by Cohen and Pack (1974, Section II.B, Figures 1 to 4) and their conclusion is clear: the Rae modification (exclusion of the self-exchange part) gives a much better agreement with the accurate ε_{HF} in the equilibrium region: for the like pairs of He, Ne and Ar, the Rae modification always gives a positive energy, only slightly overestimated with respect to ε_{HF}, while the original Gordon–Kim procedure (without E_{corr}) gives a too low positive energy which finally becomes negative when the distances approaches the value of the minimum. It is encouraging that the accuracy of the Rae modification seems better for the heavier atoms (Ne, Ar) than for He, because it is just for these heavier atoms that approximate methods are needed for avoiding as much as possible the cumbersome *ab initio* computations.

Finally, Cohen and Pack (1974) checked that, by adding the long-range multipolar energy ε_2^m (which reduces to the dispersion term for rare gases) to the short-range energy ε_{GKR} (determined according to Gordon–Kim–Rae procedure just described), they actually obtained satisfactory potential curves (only slightly too shallow), while Rae (1973) obtained too deep minima (especially for the heavier atoms Ar and Kr) by adding to the previous energy $(\varepsilon_{GKR}+\varepsilon_2^m)$ the short-range correlation energy of the electron gas model originally introduced by Gordon and Kim (1972).

We shall therefore conclude that the Gordon–Kim–Rae (GKR) procedure may give satisfactory estimates of the (electrostatic+short-range)-energy $\varepsilon_1^m + \varepsilon_1^{pen} + \varepsilon_1^{ex}$ in formula (2.124), especially for not too small systems. More precisely, since the Coulomb energy is calculated exactly in the GKR procedure (electrostatic interaction between the nuclei and the electronic charge densities $\rho^{(1)}$ and $\rho^{(2)}$), the GKR procedure amounts to representing the 'exchange interaction' ε_1^{ex} by the interaction energy of the electron gas model obtained from the kinetic energy density $E_k^d[\rho]$ and the modified exchange energy density $E_{ex}'^d[\rho]$ defined by (2.197, 2.198, 2.201, 2.202, 2.206, 2.207, 2.208). It must be emphasized that the replacement of $E_{ex}[\rho]$ by $E'_{ex}[\rho]$ implies no significant increase of the computation time in the numerical integration (2.198), since the ratio E'_{ex}/E_{ex} does depend only on the number N of electrons and not on the density ρ: thus, the use of E'_{ex} instead of E_{ex} simply amounts to using, instead of C_{ex} given by (2.202), a constant $C'_{ex}(N) = C_{ex}(1 - 8\delta/3 + 2\delta^2 - \delta^4/3)$ depending on N.

We reconsider now, according to Rae (1975), the approximation involved in (2.209), namely the use of the same factor $\gamma(N)$ for both finite and infinite separations. A general formula involving three correction factors may be written:

$$\Delta E'^{ex} = \int [\gamma E_{ex}[\rho^{(1)} + \rho^{(2)}] - \gamma_1 E_{ex}[\rho^{(1)}] + \gamma_2 E_{ex}[\rho^{(2)}]]\, d\vec{r} \qquad (2.210)$$

The formula (2.209) corresponds to $\gamma = \gamma_1 = \gamma_2 = \gamma(N_1 + N_2)$, and a seemingly more natural choice could be $\gamma = \gamma(N_1 + N_2)$, $\gamma_1 = \gamma(N_1)$, $\gamma_2 = \gamma(N_2)$: but $\Delta E'^{ex}$ would then not go to zero at infinite separation, because in this limit $E_{ex}[\rho^{(1)} + \rho^{(2)}] = E_{ex}[\rho^{(1)}] + E_{ex}[\rho^{(2)}]$, so that (2.210) gives:

$$\lim_{R\to\infty} \Delta E'^{ex} = \int [(\gamma - \gamma_1) E_{ex}[\rho^{(1)}] + (\gamma - \gamma_2) E_{ex}[\rho^{(2)}]]\, d\vec{r} \qquad (2.211)$$

Thus, as emphasized by Rae (1975), this limit may be zero only if $\gamma \to \gamma_1$ in regions of space where $\rho^{(2)}/\rho^{(1)} \to 0$ and $\gamma \to \gamma_2$ in the regions where $\rho^{(1)}/\rho^{(2)} \to 0$. A possible recipe fulfilling this condition consists in choosing:

$$\gamma_1 = \gamma(N_1) \qquad \gamma_2 = \gamma(N_2) \qquad (2.212)$$

and in defining γ through a linear interpolation between γ_1 and γ_2, depending on the charge densities:

$$\gamma = (\gamma_1 \rho^{(1)} + \gamma_2 \rho^{(2)})/(\rho^{(1)} + \rho^{(2)}) \qquad (2.213)$$

it must be stressed that, according to the definition (2.212, 2.213), γ is now a function of the position $\vec{r}$ through the densities $\rho^{(1)}$ and $\rho^{(2)}$, and cannot be factorized outzide the integral in (2.210). Upon insertion of (2.212, 2.213), it is easily found that (2.210) becomes (cf. formula (16) of Rae,

1975):

$$\Delta E'^{\mathrm{ex}} = \int \Bigg[\gamma(N_1)\rho^{(1)}\left(\frac{E_{\mathrm{ex}}[\rho^{(1)}+\rho^{(2)}]}{\rho^{(1)}+\rho^{(2)}} - \frac{E_{\mathrm{ex}}[\rho^{(1)}]}{\rho^{(1)}}\right) + \gamma(N_2)\rho^{(2)}\left(\frac{E_{\mathrm{ex}}[\rho^{(1)}+\rho^{(2)}]}{\rho^{(1)}+\rho^{(2)}} - \frac{E_{\mathrm{ex}}[\rho^{(2)}]}{\rho^{(2)}}\right)\Bigg] \mathrm{d}\vec{r} \qquad (2.214)$$

For like pairs of rare gas atoms, results obtained from (2.209) may be found in Rae (1973), and results obtained from (2.214) in Rae (1975). The last procedure gives correct interaction energy curves, while the former gave too deep minima (especially for argon and krypton); but, as mentioned above, Rae includes the correlation part of the electron gas energy (in both his procedures), while Cohen and Pack (1974), by skipping this correlation term and simultaneously using the simpler evaluation (2.209) for the corrected exchange term, also obtained satisfactory results. Thus, with respect to this procedure of Cohen and Pack, the two supplementary features of Rae (1975), namely the use of (2.214) instead of (2.209) for the exchange term and the inclusion of the correlation term, would appear to cancel each other approximately, and the simpler procedure of Cohen and Pack (1974) would therefore appear rather interesting for actual practice.

It would be possible, for example, to perform, according to this procedure, a task analogous to that of Gaydaendo and Nikulin (1970), namely to compute systematically the short-range interactions between all usually encountered atoms, and to fit once for all the parameters (B, b) of Born–Mayer exponential formulas Be^{-bR} for representing these interactions, at least in the equilibrium region, which is the interesting region for us. The results of Gaydaenko and Nikulin cannot be used directly for two reasons: first, they were apparently obtained without the Rae modification (exclusion of the self-exchange part) which just becomes important (with respect to the total energy $\varepsilon_{\mathrm{GKR}}$) in the equilibrium region (see the Figures 1 to 4 of Cohen and Pack, 1974), and second, they considered only the short-range region and not the equilibrium region: 0·6 to 3·0 for He ($R_e \simeq 5{\cdot}6$ a.u.); 1·0 to 4·0 for Ne ($R_e \simeq 5{\cdot}8$); 1·0 to 5·0 for Ar ($R_e \simeq 7{\cdot}1$); 1·6 to 5·5 for Kr ($R_e \simeq 7{\cdot}4$) (for these approximate values of the experimental equilibrium distances R_e, see, for example, the curves reported by Rae (1973) and the references therein).

Since the analytical form of the GKR short-range interaction is simpler than the accurate $\varepsilon_1^{\mathrm{ex}}$ of (2.124), it also offers new possibilities for investigating approximation properties, the possibility of expressing the short-range interaction between two molecules as a sum of interactions between local subunits (atoms or bonds) and the influence of the atomic electron population upon such local-type short-range terms (see Section VI.A.4 below). Owing to the importance of such pair-wise additivity properties for the practical evaluation of the interaction energy between medium- or large-size molecules, they will be discussed in the next section.

E. Pair-wise additivity properties

We must distinguish two kinds of additivity properties: the pair-wise additivity of the interactions between several molecules, and the expression of the interaction between two molecules as a sum of interaction terms between local subunits (atoms or bonds).

1. Pair-wise additivity for several molecules

The pair-wise additivity between several molecules has been studied by Williams, Schaad, and Murrell (1967) by using the expressions of the type (2.172) (see Section IV.B) for the one-exchange energy $\varepsilon_1^{ex(1)}$ between three molecules on one hand, and the three pairs of molecules on the other hand (the expression (2.172) holds for molecules represented by Slater determinants, but the 'complementary' one-exchange energy (see Section IV.A above) may be shown to be exactly additive (Claverie, 1973, Ch. III, Section 3.C.1.b, pp. 207–209)). Then Williams *et al.* note that the difference, namely the non-additive part of the energy would vanish by assuming the Mulliken approximation, namely $b_i b_j = \frac{1}{2} S_{ij}(b_i^2 + b_j^2)$, for the product of orbitals involved in the overlap charge densities $\rho_{ij}^{(s)}$ defined by (2.175, 2.176). Even without resorting to the Mulliken approximation it is possible to argue that the non-additive part becomes small with respect to the additive part, at intermediate and large distance (Claverie, 1973, Ch. III, Section 3.C.1.a, pp. 206–207), and these arguments may be extended to the case of systems represented by an arbitrary function (instead of a single Slater determinant) by using the density matrix formalism (Claverie, 1973, Ch. III, Section 3.C.1.c, pp. 209–222). Actually, Williams *et al.* also computed the non-additive part for three helium atoms (for the equilateral triangle and linear configurations) and found it practically negligible already for moderate distances (the non-additive part is less than one per cent for $R > 3{\cdot}5$ a.u.) (there is apparently an error of sign in the Figure 2 of Williams *et al.*: the deviation from pair-wise additivity is negative for the equilateral triangle as written on p. 4921 and shown by the results of Bader *et al.* (1971)), so that the ordinate axis of Figure 2 corresponds to [100 (pair exchange − total exchange)/total exchange)], namely the opposite of the printed expression).

The SCF *ab initio* computations (using a gaussian basis) of Bader, Novaro and Beltran–Lopez (1971), Novaro and Beltran–Lopez (1972) confirm this result qualitatively for the He_2 system.

From the quantitative point of view, these authors find a larger non-additivity than Williams *et al.* Now, the computation by Novaro and Beltran–Lopez is rather accurate: they find pair interactions (their total E_2 divided by 3) very close to the values found by Gilbert and Wahl (1967) (for $R \leqslant 3$) and by McLaughlin and Schaeffer III (1971) (for $R \geqslant 3$); but Williams *et al.* give about $0{\cdot}027/3 = 0{\cdot}009$ (according to their Figure 3), and this small value is evidently due to the use of a single Slater orbital (with exponent 1.7) for representing each helium atom (indeed, according to our discussion in Section IV.C above, the use of such a non-accurate orbital results in underestimated values of the exchange interaction). The results of Williams *et al.* appear to be farther from the exact values than that of Bader *et al.* More accurate computations could clear up the point (e.g. the perturbation terms of Williams *et al.* should be computed by using the four-component orbital (2.186) instead of the one-component orbital).

Kolos and Les (1972) also found very small non-additive effects by using a modification of the 'effective electron' model introduced by Jansen (1962, 1965), Jansen and Lombardi (1965, 1967). They discuss the previous work of Margenau and Stamper (1966) (see also Margenau and Kestner (1969) Ch. 5) who computed three- and four-body forces between rare gas atoms (Ar, Kr, Xe) by using the original Jansen (1962, 1965) model: while Margenau and Stamper found a slow convergence of the multibody series (four-body energy as large as the three-body one, with the opposite sign, for a square configuration of four atoms), in disagreement with the accurate *ab initio* results for helium atoms (Novaro and Beltran–Lopez, 1972), Kolos and Les (1972) showed that this disagreement was due to the approximate character of the original Jansen model, and that with their suitably modified model (that they check against the accurate results for the helium case) the convergence of the multibody series is always good, even for rather short distances ($R = 3$ a.u.), and the three- and four-body (non-additive) terms become actually negligible at distances corresponding to the equilibrium region (between 5 and 7 a.u. for the rare gases). We may thus conclude that the pair-wise additivity (between several molecules) of the 'exchange' interaction holds with quite sufficient accuracy already in the equilibrium region (and of course at large distances); note that the penetration energy $\varepsilon_1^{\text{pen}}$ is exactly additive, since it is simply the short-range part of the electrostatic energy which is itself exactly additive. Recent papers by Les (1976) and Jeziorski, Bulski and Piela (1976) again consider the evaluation of the short-range repulsion for a system of several closed-shell atoms or molecules.

2. *Pair-wise additivity relative to molecular subunits*

Let us now consider the second problem, namely the expression of the short-range interaction energy between two molecules as a sum of local interaction terms. The theoretical analysis will make use of the fact that the ground state of many molecules may be quite well approximated by a single Slater determinant built from localized orbitals (orbitals localized on the atoms: core and lone pair orbitals, and orbitals localized on the bonds). It is indeed possible to localize almost completely the canonical orbitals (given by an SCF procedure) through an appropriate unitary transformation: Lennard-Jones and Pople (1950, 1951); Edmiston and Ruedenberg (1963a,b); Gilbert T. L. (1964); Ruedenberg (1965); Diner (1967), Diner *et al.* (1967, 1970); England and Ruedenberg (1971), England *et al.* (1971), England and

Gordon (1971, 1972), Amos *et al.* (1975). Moreover, the localized orbitals are transferable to some extent between different molecules, and their use gives a theoretical basis for understanding the transferability of various properties: see the review paper by O'Leary *et al.* (1975) and the book edited by Chalvet *et al.* (1975) (especially the papers therein by Millié *et al.*, p. 59 and by Daudey *et al.*, p. 155). The degree of localizability, as measured by the ratio of the sum of the squares of the main coefficients to the sum of the squares of all coefficients, has been found very high for orbitals obtained by the CNDO Method (Diner *et al.*, 1970). 99 per cent for σ orbitals, 98 per cent for π orbitals in linear polyenes, 95 per cent for σ lone pair orbitals; this ratio decreases (80 per cent to 90 per cent) for π lone pair orbitals and π orbitals in cyclic conjugate (aromatic) molecules, in which case there may be several equivalent localized structures: thus, for benzene, the π orbitals may be localized (each on two atoms) to a degree of 80 per cent (further details concerning the localization of orbitals are given by Diner (1967) and Diner *et al.* (1967)).

The use of localized orbitals gives a logical basis for the analysis of the intermolecular interaction energy in terms of local contributions. Before dealing with the first-order exchange *energy* itself, we may already mention that Salem (1961) evaluated the corresponding *forces* (by calculating the variations of charge density due to the interaction and applying the Hellman–Feynman theorem) and actually obtained the result that the localized orbitals displayed additive contributions to these forces (this conclusion remains valid for the interaction between conjugated systems like butadiene and ethylene (Claverie, 1973, Ch. III, Section 3.C.2)).

Now the expression (2.172) derived in Section B above for the one-exchange term $\varepsilon_1^{\text{ex}(1)}$ is appropriate for an analysis in terms of local contributions. This analysis has been already presented (Claverie, 1973, Ch. III Section 3.C.2; 1976, Section 4) and will not be repeated here in full detail. We only recall the principle of the argument. The overlap charge distributions $\rho_{ij}^{(s)}$ defined by (2.175, 2.176) in Section A may be decomposed into 'local' and 'non-local' parts denoted $\rho_{ij}^{\prime(s)}$ and $\rho_{ij}^{\prime\prime(s)}$ respectively:

$$\rho_{ij}^{(1)} = \rho_{ij}^{\prime(1)} + \rho_{ij}^{\prime\prime(1)} \tag{2.215}$$

with

$$\rho_{ij}^{\prime(1)} = [b_i^{(1)} b_j^{(2)} - S_{ij} (b_i^{(1)})^2] \delta_{ij}^{\text{spin}} \tag{2.216}$$

$$\rho_{ij}^{\prime\prime(1)} = \left[- \sum_{k \neq i}^{(1)} S_{kj} b_i^{(1)} b_k^{(1)} \right] \delta_{ij}^{\text{spin}} \tag{2.217}$$

and similarly for $\rho_{ij}^{(2)}$. (In the above references Claverie (1973, 1976) these charge densities are expressed in terms of the *spin-orbitals* a_i instead of the corresponding orbitals b_i).

The densities $\rho_{ij}^{\prime(s)}$ depend only on the pair of spin-orbitals i and j, while the densities $\rho_{ij}^{\prime\prime(s)}$ involve all spin-orbitals of molecules (s) respectively.

Since we use localized orbitals, the densities $\rho_{ij}'^{(s)}$ depends only on the two regions (bonds, or eventually atoms) corresponding to the orbitals $b_i^{(1)}$ and $b_j^{(2)}$ in molecules (1) and (2) respectively. This property does not hold for the remaining densities $\rho_{ij}''^{(s)}$. But these must have smaller values than the densities ρ' in the important cases, because they involve, besides the overlap integrals S_{kj}, the products $b_i b_k$ of *different localized orbitals*: such products decrease rather fast when the regions i and k are not close, and even for neighbour regions, these densities $b_i b_k$ are of dipolar nature, since their total charge $\int b_i(\vec{r}) b_k(\vec{r})\, d\vec{r} = 0$ according to the orthogonality of b_i and b_k. Now, the essential part of the short-range repulsion comes from those regions i and j which are the closest for a given configuration of the two molecules, and for these regions ρ_{ij}'' will contribute significantly less than ρ_{ij}', not only because of the low values and dipolar character of the densities $b_i b_k$, but also because the integrals S_{kj} will be smaller than S_{ij} since the distances between k and j are larger than between i and j. For the same reason, it may be expected that, in the interaction $[\rho_{ij}^{(1)} \mid V^{(2)}]$ involved in (2.172), the local part $V_j^{(2)}$ of the potential $V^{(2)}$ created by the region j will play the dominant role (and similarly $V_i^{(1)}$ would be dominant in $[\rho_{ij}^{(2)} \mid V^{(1)}]$). Thus, we may finally expect that the dominant part of (2.172) will be given by the terms

$$-\{\rho_{ij}'^{(1)} \rho_{ij}'^{(2)} + S_{ij}([V_i^{(1)} \mid \rho_{ij}'^{(2)}] + [V_j^{(2)} \mid \rho_{ij}'^{(1)}])\} \tag{2.218}$$

corresponding to the closest region i of (1) and j of (2), and these terms (2.218) depend only on the regions i and j since they involve only $\rho_{ij}'^{(1)}$, $\rho_{ij}'^{(2)}$, $V_i^{(1)}$ and $V_j^{(2)}$. If we denote e_{ij} this dominant local type interaction between the regions i and j, *which is valid essentially when these regions are the closest*, it is clear that the formula

$$\varepsilon_1^{\text{ex}(1)} \simeq \sum_i^{(1)} \sum_j^{(2)} e_{ij} \tag{2.219}$$

will be a convenient approximation, because for every possible configuration of the two molecules, this formula will correctly represent the interaction between the closest regions of the molecules, and it does not matter that the *negligible* contributions coming from the other *non-closest* pairs of regions are not accurately represented by the local-type formula e_{ij}.

A similar (and simpler) analysis may be proposed in the framework of the electron-gas model described in the previous Section D. Indeed the electronic density corresponding to a Slater determinant is (see (2.134) and (2.135) in Section III.B.1):

$$\sum_i b_i(\vec{r})|^2 = \sum_I n_I^{\text{occ}} |b_I(\vec{r})|^2$$

(where i refers to the spin-orbitals while I refers to the orbitals). Since these distributions are localized when we use localized orbitals, the total density in the neighbourhood of a given region u is essentially given by one (or a small number) of these localized distributions: let us denote $\rho_u^{(1)}(\vec{r})$ these region densities for molecule (1) and $\rho_v^{(2)}(\vec{r})$ for molecule (2). Then, when the molecules are in a relative configuration for which the regions at closest distance are respectively u and v, the short-range interaction will be essentially given, according to (2.198) (see Section D above), by:

$$e_{uv} = \int (\rho_u^{(1)} + \rho_v^{(2)}) E[\rho_u^{(1)} + \rho_v^{(2)}] \, d\vec{r} \\ - \left(\int \rho_u^{(1)} E[\rho_u^{(1)}] \, d\vec{r} + \int \rho_v^{(2)} E[\rho_v^{(2)}] \, d\vec{r} \right) \tag{2.220}$$

and the other regions u', v' of the molecules, lying farther apart, will give significantly smaller terms $e_{u'v'}$: thus, by the same argument as used above, the sum

$$\sum_u^{(1)} \sum_v^{(2)} e_{uv}$$

may be expected to give a reasonable estimate of the short-range interaction for all possible configurations; this is quite clear when a single region–region interaction is predominant, but even if two neighbour regions u, u' for example are simultaneously the closest to the region v of the other molecule, the sum $e_{uv} + e_{u'v}$ may still give an evaluation of the exact interaction which is sufficient for practical use (it must be kept in mind that the short-range interaction terms vary exponentially with distance, so that a relatively small variation of intermolecular distance suffices to compensate even for a factor of two).

It may be concluded from this discussion based upon the use of localized orbitals that the first-order 'exchange' (short range) energy $\varepsilon_1^{ex(1)}$ between two molecules should appear, with a reasonable accuracy, as a sum of independent interaction terms between localized regions (bonds, and eventually lone pairs). Since these terms involve S^2 (square of the overlap between such localized orbitals), it will be sufficient in practice, to take into account the valence orbitals (since the inner shell orbitals, having larger Slater exponents, give much smaller overlap integral). Similarly, in the framework of the electron-gas model, the inner shell orbitals will bring negligible contributions to the electron density in the intermolecular region, where the difference of the energy densities

$$E^d[\rho^{(1)} + \rho^{(2)}] - (E^d[\rho^{(1)}] + E^d[\rho^{(2)}])$$

takes its largest values.

It would be desirable to go one step further and to express $\varepsilon_1^{ex(1)}$ as a sum of atom–atom short-range contributions: this is a very common practice, owing to its computational simplicity (for example, see Kitaigorodskii (1961a,b; 1965; 1966), Kitaigorodskii and Mirskaya (1961, 1964a,b); Kitaigorodskii *et al.* (1969), Kitaigorodskii and Ahmed (1972); Liquori (1966, 1969), Giglio (1969, 1970 and references therein), Coiro *et al.*

(1972); Ramachandran and Sasisekharan (1968); Poltev and Sukhorukov (1968b, and references therein); Scheraga (1968), Momany *et al.* (1974a,b), Shipman *et al.* (1975, 1976), Burgess *et al.* (1975, 1976); Williams (1972b); Hagler, Huler and Lifson (1974); Popov (1975, and references therein). Now Salem (1961) already noticed that the bond–bond interactions cannot be expressed rigorously as a sum of atom–atom interactions, owing to the occurrence of the square of the overlap integrals between the bond orbitals: denoting two such bond orbitals $\phi = c_1\chi_1 + c_2\chi_2$ and $\phi' = c_1'\chi_1' + c_2'\chi_2'$ we have $S_{\phi\phi'} = c_1c_1'\sigma_{11'} + c_1c_2'\sigma_{12'} + c_1'c_2\sigma_{1'2} + c_2c_2'\sigma_{22'}$ (where $\sigma_{11'} = \langle\chi_1 \mid \chi_1'\rangle$, and so on), and this is a sum of atom–atom terms, but $(S_{\phi\phi'})^2$ is not owing to the cross terms. Taking the example of two hydrogen molecules (Salem, 1961): $c_1 = c_2 = c_1' = c_2' = c$, $\chi_1 = \chi_2 = \chi_1' = \chi_2' = 1s$ orbital, we consider the following three configurations:

(1) linear configuration, with the closest atoms at distance R: $S_{\phi\phi'} \simeq c^2\sigma(R)$, $(S_{\phi\phi'})^2 = [c^2\sigma(R)]^2$.

(2) parallel configuration (the two molecules at distance R): $S_{\phi\phi'} \simeq 2c^2\sigma(R)$, $(S_{\phi\phi'})^2 = 4[c^2\sigma(R)]^2$

(3) crossed configuration (all four distances $R_{11'} = R_{12'} = R_{21'} = R_{22'} = R$): $S_{\phi\phi'} = 4c\sigma(R)$, $(S_{\phi\phi'})^2 = 16[c^2\sigma(R)]^2$.

On the other hand, if we use a sum of atom–atom terms $f(R)$ (varying exponentially with distance) and keeping also in all cases the dominant terms only, we shall get in these three cases respectively: (1) $f(R)$ (2) $2f(R)$ (3) $4f(R)$. It is therefore not possible to reproduce $(S_{\phi\phi'})^2$ in the three cases; the best choice is certainly to reproduce the intermediate case (2), by choosing $f(R) = 2(c^2\sigma(R)^2$; then, in case (1), $f(R)$ will give twice $(S_{\phi\phi'})^2$, and in case (3), $4f(R)$ will give half $(S_{\phi\phi'})^2$. The atom–atom approximation could at first sight seem too bad, but the above discrepancy could become smaller for molecules involving more than two atoms, and, as already noticed above, the exponential variation may compensate a factor of 2 through a rather small variation of intermolecular distance. In this connection, it may be mentioned that, in order to improve the atom–atom type formulas, Williams (1965) proposed to use, for a bonded hydrogen atom, a centre of force slightly displaced from the position of the proton towards the other atom of the bond. Thus the usefulness of sum of atom–atom terms as a first approximation may be understood. If a higher accuracy was desired, bond–bond interactions terms could be tried: thus, as suggested by the example above, instead of using the formula

$$\sum_{i=1}^{2} \sum_{j'=1'}^{2'} f(R_{ij'})$$

we could use the relationship $c^2\sigma(R) = (f(R)/2)^{1/2}$ for deriving the bond–bond interaction

$$(S_{\phi\phi'})^2 = c^2\left[\sum_{i=1}^{2} \sum_{j'=1'}^{2'} \sigma(R_{ij'})\right]^2 = \left[\sum_{i=1}^{2} \sum_{j'=1'}^{2'} (f(R_{ij'})/2)^{1/2}\right]^2$$

Analogous formulas could be built for molecules involving several bonds. The use of such improved representations of the short-range interactions could prove itself more especially useful for the case of hydrogen bonds where the equilibrium distance is usually short and the configuration of the closest bonds in the interacting molecules is nearer to the linear configuration than to the parallel one.

As a matter of conclusion, it must be mentioned that some criticisms are made from time to time to the practice of using sums of atom–atom terms for evaluating the interaction energy between molecules (for example, see a recent paper by Harris *et al.* (1975), and the paper by Mies (1965) quoted therein). Actually, the situation is rather complex, and there is no reason why the accuracy of this approximation should be exactly the same for the various contributions to the interaction energy (the short-range first-order 'exchange' repulsion and the long-range terms: electrostatic, polarization, dispersion). Accordingly, the discussion of these criticisms will be deferred to Section VI.A.6 below, after the simplified formulas for the long-range terms have also been considered. We shall content ourselves here with the following comment concerning the short-range term only: it is probably true that a sum of atom–atom terms does not give a highly accurate evaluation of the intermolecular short-range interaction, but it is nevertheless not devoid of any foundation, because it appears as an approximation of theoretically more founded formulas, namely those involving a sum of interactions between appropriate subunits of the molecules (bonds, atoms bearing lone pairs).

V. THE LONG-RANGE TERMS (ELECTROSTATIC, POLARIZATION, DISPERSION)

We shall now deal with the long-range contributions, namely the multipole parts ε_1^m and ε_2^m (see formula (2.124) in Section III.A.1) of the first and second-order terms $\varepsilon_1^{\mathrm{RS}}$ and $\varepsilon_2^{\mathrm{RS}}$ of the long-range Rayleigh–Schrödinger perturbation treatment (see formula (2.10) in Section II.A.1). Since their theoretical treatment and the search for simplified formulas have been much more investigated than it was done for the short-range 'exchange' term, we shall here deal with these terms more briefly than we did for the short-range term.

A. A basic concept: the multi-centered multiple expansion

The main problem with the perturbation expansion (2.10) is to find a simplified expression for the matrix elements $\langle W_0^{(1)} W_0^{(2)} | V | W_a^{(1)} W_b^{(2)} \rangle$ ($a, b = 0$ or not). Following the pioneering works of Eisenschitz and London (1930) and London (1930a,b), most authors treated these matrix elements by replacing the potential by its multipole expansion, often limited to its first term (dipole–dipole): see Margenau (1939a); Kihara (1958) (dispersion and third-order term); Hirschfelder, Curtiss and Bird (1954, part III); Fontana (1961); Dalgarno and Davison (1966); Dalgarno (1967) (dispersion); Buckingham (1967) (electrostatic and polarization terms); Hirschfelder and Meath (1967); Kihara (1967, part II) (electrostatic term); Margenau and Kestner (1969, Ch. 2); Isnard *et al.* (1976), and references therein. Now these works dealt with the interactions between atoms or very small molecules, at distances large with respect to the molecular sizes themselves, while our own purpose is to evaluate the interaction energy between molecules of arbitrary size and up to medium-range (equilibrium region). With respect to such a purpose, the above-mentioned procedure (replacing V by its multipole expansion around a *single centre* for each molecule, or equivalently, multipole expansion of each molecular charge density around a *single molecular centre*) has two essential drawbacks:

(a) from the practical point of view, it becomes more and more difficult to use multipoles of higher and higher order, as it would in principle be necessary, since the intermolecular distance is no more large with respect to molecular sizes.

(b) more fundamentally, a multipole expansion represents exactly the potential of a charge distribution only outside a sphere centred at the origin (chosen for the expansion) and containing entirely the charge distribution. This is a serious defect when the molecular shape is quite different from a sphere, because there are regions outside the molecule which are closer to the origin than the end-points of the molecule, and the (one-centre) multipole expansion cannot represent accurately the potential in such regions. Moreover, since the molecular charge densities extend to infinity (with an exponential decrease at large distance: see Ahlrichs (1972b, 1973a, 1973b) and O'Connor (1973)), the use of the multipole expansion of V leads to a *divergent* series in powers of $1/R$ (Dalgarno and Lewis, 1956; Fukui and Yamabe, 1968; Kreek and Meath, 1969; Singh *et al.*, Kreek *et al.*, 1970; Pan and Meath, 1971; Riera and Meath, 1972; Young, 1975); it is of course *asymptotic*, i.e. when truncated at some convenient order, the series gives an approximation which becomes better and better when R increases (see also the recent studies by Deal and Young, 1976 and Ahlrichs, 1976). But when R on the contrary decreases and becomes of the order of magnitude of the molecular sizes, the minimal error is still too large and adding higher order multipolar terms would only make the situation worse.

Thus, we are led to the conclusion that a direct multipole expansion of the potential V with a single centre in each molecule loses its practical usefulness for usual-size molecules in the equilibrium region (see Pollak and Rein (1967), Pack *et al.* (1972)).

On the contrary, the formulas for the matrix elements of V derived in Section III.B.1 following the Longuet-Higgins (1956) treatment introduce no *a priori* simplification of V. The simplifications will concern only the molecular charge densities involved, and a much greater flexibility of the approximation will be preserved. Thus, in the MO-LCAO framework, the electronic parts (2.135, 2.137) of the charge densities will finally reduce (upon replacing the Molecular Orbitals b_i by Linear Combinations of Atomic Orbitals) to a sum of monoatomic and diatomic elementary distributions (and among these the dominant ones will be associated with bonded pairs of atoms). Then, each of these *small-size* atom- or bond-distributions may be expressed with sufficient accuracy by a multipole expansion around its *own appropriate centre*, so that the total molecular charge distribution will appear as a sum of *several* multipole expansion pertaining to *different* centres (hence the name *multi-centred*), and such a representation is, by its very definition, perfectly adapted to the molecular shape, whatever its complexity. By the way, it may be recalled that London (1942) already proposed to represent the charge densities (transition densities in his case, since he dealt with the dispersion term) by suitable sets of point charges (so-called 'monopoles'). But really important developments concerning the use of multi-centred multipole expansions occurred only rather recently (Rein *et al.*, 1972; Rabinowitz and Rein, 1972; Rein, 1973; Claverie, 1973, 1976; Hall, 1973; Tait and Hall, 1973; Shipman, 1975; Amos and Yoffe, 1975c; Amos and Crispin, 1976b,c): see the discussion at the end of the next Section B.

B. The electrostatic term

The analysis of the electrostatic interaction, namely the first-order term $\varepsilon_1^{RS} = \langle W_0^{(1)} W_0^{(2)} | V | W_0^{(1)} W_0^{(2)} \rangle$ of the expansion (2.10) (Section II.A.1) is a straightforward application of the expressions derived in Section III.B.1 for the matrix elements of V. We indeed have, according to (2.130, 2.131, 2.132):

$$\varepsilon_1^{RS} = \langle W^{(1)} W^{(2)} | V | W^{(1)} W^{(2)} \rangle = \iint \frac{f_{00}^{(1)}(\vec{r}^1) f_{00}^{(2)}(\vec{r}^2)}{|\vec{r}^1 - \vec{r}^2|} \, d\vec{r}^1 \, d\vec{r}^2 \qquad (2.221)$$

where $f_{00}^{(s)}(\vec{r}) = \langle W_0^{(s)} | \boldsymbol{\rho}^{(s)}(\vec{r}) | W_0^{(s)} \rangle$ is the (mean) charge distribution of the molecule (s). If $W_0^{(s)}$ is a Slater determinant:

$$W_0^{(s)} = \frac{1}{\sqrt{n!}} \det |a_1^{(s)}(1) \ldots a_n^{(s)}(n)| \qquad (2.222)$$

where the $a_i^{(s)} = b_i^{(s)} \sigma_i$ are the spin-orbitals, $f_{00}^{(s)}$ is given by the formula (2.134):

$$f_{00}^{(s)}(\vec{r}) = \sum_{\mu}^{(s)} Z_\mu^{(s)} \, \delta(\vec{r} - \vec{r}_\mu^{(s)}) - \sum_{i=1}^{n} |b_i^{(s)}(\vec{r})|^2 \qquad (2.223)$$

As explained in Section A, we do not introduce the multipole expansion of $f_{00}^{(s)}(\vec{r})$ around some molecular centre chosen in the molecule (s). We assume that the molecular orbitals b_i are expressed in terms of atomic orbitals $\{\chi_\alpha\}$, as is almost always the case in practice (note that we do not need here to use localized molecular orbitals). Then, inserting

$$b_i(\vec{r}) = \sum_{\alpha} c_{i\alpha}\chi_\alpha(\vec{r}) \tag{2.224}$$

into (2.223), we shall obtain $f_{00}(\vec{r})$ as a sum of nuclear point charges and electronic distributions among which the dominant ones will be strongly localized (for the sake of simplicity, we omit from now on the superscript (s) labelling the molecule: it is not necessary, since we analyse the charge distribution of each molecule quite independently from the other):

$$f_{00}(\vec{r}) = \sum_{\mu} Z_\mu \delta(\vec{r} - \vec{r}_\mu) - \sum_{\alpha}\sum_{\beta} p_{\alpha\beta}\chi_\alpha^*(\vec{r})\chi_\beta(\vec{r}) \tag{2.225}$$

where the $p_{\alpha\beta}$'s are the elements of the charges and bond orders matrix:

$$p_{\alpha\beta} = \sum_{i(\text{spin-orbitals})} c_{i\alpha}^* c_{i\beta} \tag{2.226}$$

(if we use orbitals b_I with occupation numbers n_I^{occ}, this formula of course becomes

$$p_{\alpha\beta} = \sum_{I(\text{orbitals})} n_I^{\text{occ}} c_{I\alpha}^* c_{I\beta})$$

The first term of (2.225)

$$\sum_{\mu} Z_\mu \delta(\vec{r} - \vec{r}_\mu)$$

corresponds to the set of the nuclear charges Z_μ located at the points $\vec{r}_\mu$. The elementary charge distributions $\chi_\alpha^*(\vec{r})\chi_\beta(\vec{r})$ appearing in (2.225) have the announced property: owing to the strongly localized character of the (atomic) orbitals $\chi_\alpha(\vec{r})$, these distributions are non-negligible only if χ_α and χ_β belong to the same atom or to neighbouring atoms. Then, the important elementary distributions, having a small size (size of an atom or of a bond) will be quite accurately approximated by a very limited multipole expansion (the first or the first two non-zero multipoles) for the purpose of calculating interactions at distance of the order of a few ångströms. Thus, instead of having a multipole expansion of $f_{00}(\vec{r})$ around a single centre, we shall represent $f_{00}(\vec{r})$ as a sum of *several* multipole expansions (centred at the atoms, and eventually also at appropriate points of the bonds: see below) each one of these expansions containing very few terms (charge, dipole, quadrupole at most). As a consequence, the electrostatic interaction between two molecules will appear as a sum of interactions between these *local* multipoles.

The elementary charge distributions $\chi_\alpha^*\chi_\beta$ may be classified into three types (Claverie, 1973, Ch. III, Section 4.B): the one-centre 'diagonal' ($|\chi_\alpha|^2$)

and 'non-diagonal' ($\chi_\alpha^*\chi_{\alpha'}$) distributions, and the two-centre $\chi_\alpha^*\chi_\beta$ distributions. The two types of one-centre distributions have only a finite number of non-zero multipoles (with respect to their centre) since the atomic orbitals have spherical harmonics as angular parts: indeed, let us write explicitly $\chi_\alpha(\vec{r})=f_1(\vec{r})Y_{l_1}^{m_1}(\theta,\phi)$ and $\chi_{\alpha'}(r)=f_2(r)Y_{l_2}^{m_2}(\theta,\phi)$ (e.g. see Messiah (1959) Vol. I, Appendix B, Section IV-10); the components Q_n^m of the multipole moment of order n of a distribution $\rho(r,\theta,\phi)$ are given by (see Hirschfelder, Curtiss and Bird (1954), Ch. 12, p. 840, equations (12.1)–(12.18b)):

$$Q_n^m=\iiint \rho(r,\theta,\phi)r^n Y_n^m(\theta,\phi)r^2\sin\theta\,\mathrm{d}r\,\mathrm{d}\theta\,\mathrm{d}\phi \tag{2.227}$$

with $\rho(r,\theta,\phi)=\chi_\alpha^*(\vec{r})\chi_{\alpha'}(\vec{r})$, we may separate in (2.227) the angular part:

$$\iint (Y_{l_1}^{m_1}(\theta,\phi))^* Y_{l_2}^{m_2}(\theta,\phi)Y_n^m(\theta,\phi)\sin\theta\,\mathrm{d}\theta\,\mathrm{d}\phi \tag{2.228}$$

which may be different from 0 only if

$$\begin{gathered}|l_1-l_2|\leqslant n\leqslant l_1+l_2\\ -m_1+m_2+m=0\end{gathered} \tag{2.229}$$

(see Messiah (1960) Vol. II, Appendix D, Section I-3; the condition (2.229) is the well-known addition theorem for the angular momenta). Thus, *only the multipole moments with order n between $|l_1-l_2|$ and l_1+l_2 may be non-zero.* We shall now examine the three kinds of distributions separately:

1. The 'diagonal' distributions $|\chi_\alpha|^2$

Their main multipole is the charge:

$$S_{\alpha\alpha}=\int|\chi_\alpha(\vec{r})|^2\,\mathrm{d}\vec{r}=1$$

In practice, it will be very rare to consider higher-order multipole moments (for an s orbital, all these further moments are zero; for a p orbital, only the quadrupole is non-zero beyond the charge). Quite generally, for an orbital with azimuthal quantum number l, only multipoles between orders 0 and $2l$ may be non-zero, according to the general formula (2.229).

In the case where several charge distributions centred on the same atom are considered (as occurs in the complete molecular charge distribution), some cancellation occurs, according to Unsöld's theorem (see Slater (1960) Ch. 7, Section 7-3, p. 182): this theorem indeed states that the sum of the square of the moduli of all the angular parts (spherical harmonics) of the atomic orbitals, which correspond to a same azimuthal quantum number l, is a constant:

$$\sum_{m=-l}^{l}|Y_l^m(\theta,\phi)|^2=\frac{2l+1}{4\pi} \tag{2.230}$$

and therefore

$$\sum_{m=-l}^{l} |f(r)Y_l^m(\theta, \phi)|^2 = \frac{2l+1}{4\pi}|f(r)|^2 \tag{2.237}$$

Let us recall that the functions are normalized as follows:

$$\int |f(r)|^2 r^2 \, dr = 1,$$

$$\int_0^{2\pi} d\phi \int_0^{\pi} \sin\theta \, d\theta Y_l^m(\theta, \phi)^2 = 1)$$

Now, the density $r^2|f(r)|^2$ is spherically symmetric and therefore all its multipole moments are zero beyond the first one, namely the total charge equal to 1. For example, let us consider the three $2p$ orbitals: $2p_x = Cxe^{-\zeta r}$, $2p_y = Cye^{-\zeta r}$, $2p_z = Czxe^{-\zeta r}$ (C = normalization factor): we have $(2p_x)^2 + (2p_y)^2 + (2p_z)^2 = C^2(x^2+y^2+z^2)e^{-2\zeta r} = C^2r^2e^{-2\zeta r} = (Cre^{-\zeta r})^2$, which is strictly the charge density corresponding to a $2s$ orbital (but with total charge 3). This result extends to the case of hybrid orbitals, since the charge density

$$\sum_{\alpha} |\chi_\alpha|^2$$

is invariant with respect to a unitary transformation of the orbitals: putting

$$h_i = \sum_{\alpha} c_{i\alpha}\chi_\alpha$$

(where the matrix $c_{i\alpha}$ is unitary), we indeed obtain:

$$\sum_i |h_i|^2 = \sum_i \left(\sum_\alpha \sum_\beta c_{i\alpha}^* c_{i\beta} \chi_\alpha^* \chi_\beta\right) = \sum_\alpha \sum_\beta \chi_\alpha^* \chi_\beta \left(\sum_i c_{i\alpha}^* c_{i\beta}\right)$$

$$= \sum_\alpha |\chi_\alpha|^2$$

Thus, for any (unitary) hybridization of s, p_x, p_y, p_z, we have

$$\sum_{i=1}^{4} |h_i|^2 = s^2 + 3[s(\xi_p)]^2$$

As emphasized by Mantione (1969a, 1971), this implies that Hanna (1968) was not correct when he considered only the quadrupole moments of the $2p_z$ orbitals of the carbon atoms in benzene: if the Slater exponents are assumed the same for all three $2p$ orbitals (x, y, z), the quadrupole moment of the $2p_z$ orbital will be exactly cancelled by the moments of the orbitals in the plane (hybridized or not), provided that all orbitals are assumed equally populated. The total quadrupole moment may be different from zero only if the cancellation is made imperfect, either because the orbitals are unequally populated, or because the Slater exponents of the $2p$ orbitals are assumed different (this possibility has been considered by Mantione (1969a) for the case of benzene).

2. The 'non-diagonal' one-centre distributions $\chi_\alpha^*\chi_{\alpha'}$

The orbitals χ_α and $\chi_{\alpha'}$ are different but belong to the same atom; they may be supposed orthogonal. Then the total charge of such a distribution is

$$S_{\alpha\alpha'} = \int \chi_\alpha^*(\vec{r})\chi_{\alpha'}(\vec{r})\,\mathrm{d}\vec{r} = 0$$

In practice, the distribution will therefore be approximated by its dipole moment. Explicit formulas for dipole moment integrals between Slater orbitals are given by Hamilton (1957). Anyway, only a finite number of multipole moments may be non-zero. If a basis of s and p orbitals is used, this dipole will be non-zero for the distributions (sp); if a basis of hybrid orbitals was used, dipole moments corresponding to (sp) distributions would already appear in the 'diagonal' distributions (there is indeed no intrinsic character in the separation of the one-centre distributions into the 'diagonal' and 'non-diagonal' ones: a change of the atomic orbital basis (e.g. by hybridization) modifies both parts). Note that, as for the quadrupole moment, the total dipole moment due to all atomic distributions may be non-zero only if the spherical symmetry is broken, either because some Slater exponents are different, or because the orbital populations are not equal (this occurs especially for atoms bearing lone pairs).

In the CNDO–SCF method (e.g. see Pople and Beveridge (1970) or Murrell and Harget (1972)), the atomic dipole moments are well-known under the name of 'hybridization moments'. The total charge distribution of the molecule may be approximated by the set of *atomic net charges* (nuclear charge + total electronic charge given by the 'diagonal' distributions $|\chi_\alpha|^2$ belonging to the atom) *and* of the *atomic dipoles*: the total dipole moment is actually computed by including the contribution of these atomic dipoles, and it would be *inconsistent* to compute electrostatic interactions between molecules studied by this method by using the net charges only (the interactions between net charges and atomic dipoles and between the atomic dipoles themselves must also be included). Of course, it is possible to replace each atomic dipole by suitable charges, for example, by putting such charges on the atom and on the atoms bonded to it, and by choosing them in such a way as to reproduce the atomic dipole (see Appendix F); the practical advantage is that the molecular charge distribution is then represented by a set of point charges only (without dipoles anymore), but these charges are 'effective' ones, and differ from the original CNDO net charges, through the fictitious charges which mimic the atomic dipoles. In Hückel-type methods, on the contrary, it may be possible to consider the net charges only.

If the inter-atomic overlap integrals are no more neglected (for example, in an *ab initio* method), then the third type of distribution also occurs:

3. The 'non-diagonal' two-centre distributions $\chi_\alpha^* \chi_\beta$

The charge of such a distribution is the overlap integral $S_{\alpha\beta} = \int \chi_\alpha^*(\vec{r})\chi_\beta(\vec{r})\, d\vec{r}$. In contrast with the two previous cases, the multipole expansion now contains an infinite number of non-zero terms. Here, too, it will be quite exceptional to consider multipoles beyond the charge and the dipole (as above; Hamilton, 1957). There is now some arbitrariness in the approximate representation of such a distribution; four possibilities are listed below:

(a) use of the middle M of the segment $\alpha\beta$ as the centre for the multipole expansion: charge $S_{\alpha\beta}$ at M and dipole of the distribution *calculated with respect to* M (the first non-zero multipole moment does not depend on the choice of the origin, but the following do depend on it).

(b) if $S_{\alpha\beta} \neq 0$, this charge is put at the point M' of the segment $\alpha\beta$ such that the longitudinal component of the dipole moment (calculated with M' as the origin) is zero. Then we have at M' the charge $S_{\alpha\beta}$ and eventually a dipole orthogonal to the segment $\alpha\beta$.

(c) the charge $S_{\alpha\beta}$ is divided in two equal charges $S_{\alpha\beta}/2$, which are placed at the atoms bearing the orbitals α and β (Mulliken's (1949, 1955) approximation). The purpose of this procedure is to avoid the introduction of new multipole centres besides the atoms, thus avoiding to increase the number of distances between such centres belonging to different molecules (distances which must be computed for getting the electrostatic interaction energy between the molecules). But the dipole moment of the distribution $\chi_\alpha^* \chi_\beta$ will *not* be reproduced, unless it is zero with respect to the middle M of the segment $\alpha\beta$. If this peculiar case is not realized, the non-zero dipole moment should be put at the point M; it is known as the 'homopolar' dipole (Coulson and Rogers, 1961).

(d) the previous discussion immediately suggests to divide the charge $S_{\alpha\beta}$ in two unequal parts q_α and q_β (placed at atoms bearing α and β) such that the longitudinal component $\mu_{\alpha\beta}^L$ of the dipole moment is reproduced. If $\mu_{\alpha\beta}^L$ denotes the algebraic component on the axis $\overline{\alpha\beta}$ (and consequently $\mu_{\beta\alpha}^L = -\mu_{\alpha\beta}^L$), it is easily found that:

$$q_\alpha = \tfrac{1}{2}S_{\alpha\beta} - \mu_{\alpha\beta}^L / R_{\alpha\beta} \qquad q_\beta = \tfrac{1}{2}S_{\alpha\beta} - \mu_{\beta\alpha}^L / R_{\alpha\beta} \tag{2.232}$$

where $R_{\alpha\beta}$ is the distance between the atoms bearing the orbitals χ_α and χ_β. This procedure is known as the Löwdin approximation (Daudel *et al.* (1952), Löwdin (1953, 1956); Howard (1963)). In principle, the transversal component $\mu_{\alpha\beta}^T$ would remain (at the middle of $\alpha\beta$, say), but in practice the $\mu_{\alpha\beta}^T$'s corresponding to the different atomic orbitals belonging to the same pair of atoms have a tendency to cancel each other, so that it is a reasonable approximation to neglect them.

As a conclusion of this discussion, it must therefore be emphasized that using *only the Mulliken atomic net charges* (given, for example, by an *ab initio* computer program) for the calculation of an electrostatic interaction would be *incorrect for two reasons*: (1) the atomic dipoles must be taken into account (as we emphasized previously for the case of CNDO methods) and (2) the homopolar dipoles $\vec{\mu}_{\alpha\beta}$ must be taken into account, or else atomic net charges obtained according to Löwdin's procedure should be used. An *ab initio* computer program which gives only Mulliken net charges may therefore be quite misleading as concerns a reasonable representation of the molecular charge distribution in terms of charges and dipoles.

The Löwdin approximation may be written for the charge distribution $\chi_\alpha\chi_\beta$ itself (instead of being applied only to its charge $S_{\alpha\beta}$):

$$\chi_\alpha\chi_\beta \simeq S_{\alpha\beta}(a\,|\chi_\alpha|^2 + b\,|\chi_\beta|^2) \tag{2.233}$$

where $a = q_\alpha/S_{\alpha\beta}$ and $b = q_\beta/S_{\alpha\beta}$, and this opens the way to a more general representation of $\chi_\alpha\chi_\beta$ suggested by Ruedenberg (1951): the orbital χ_β is replaced by its projection on the subspace spanned by all available atomic orbitals of the atom μ which bears χ_α :

$$\chi_\beta \simeq \sum_{\alpha'}^{(\mu)} S_{\alpha'\beta}\chi_{\alpha'}$$

hence

$$\chi_\alpha\chi_\beta \simeq \sum_{\alpha'}^{(\mu)} S_{\alpha'\beta}\chi_\alpha\chi_{\alpha'}$$

similarly, we can project χ_α on the subspace of the orbitals of the atom ν which bears χ_β, thus getting

$$\chi_\alpha\chi_\beta \simeq \sum_{\beta'}^{(\nu)} S_{\alpha\beta'}\chi_{\beta'}\chi_\beta$$

and we can finally use a 'mixture' of these two representations with weights a and b $(a+b=1)$:

$$\chi_\alpha\chi_\beta = a\sum_{\alpha'}^{(\mu)} S_{\alpha'\beta}\chi_\alpha\chi_{\alpha'} + b\sum_{\beta'}^{(\nu)} S_{\alpha\beta'}\chi_{\beta'}\chi_\beta \tag{2.234}$$

Then, upon replacing the one-centre distributions $\chi_\alpha\chi_{\alpha'}$ and $\chi_\beta\chi_{\beta'}$ by their multipole expansions, we would finally obtain an approximate representation of the distribution $\chi_\alpha\chi_\beta$ by a set of multipoles located at the atoms bearing χ_α and χ_β.

A generalized decomposition of the type (2.234) has been proposed by Jug (1973, 1975) for the splitting of the overlap charge $S_{\alpha\beta}$ into atomic charges

$$\left(S_{\alpha\beta} = \sum_{\alpha'} f_{\alpha'}S_{\alpha\alpha'} + \sum_{\beta'} f_{\beta'}S_{\beta\beta'}\right)$$

this work is therefore concerned with the definition of atomic charges, rather than with a complete description of the molecular charge distribution (for example, the atomic 'hybridization' dipoles, dealt with under heading (2) above, are not considered). The same purpose, namely a suitable definition of atomic charges is also pursued in a series of paper by Politzer and Harris (1970), Politzer and Stout (1971), Politzer (1971), Politzer and Mulliken (1971) Politzer and Reggio (1972), Politzer and Politzer (1973): these authors define each atomic charge by integrating the electronic density over an appropriate region 'belonging' to the atom. Another method for defining atomic charges was also proposed recently by Fliszar *et al.* (1974a,b).

By contrast with these methods for defining atomic charges, it must be emphasized that our own purpose is the representation of the *molecular charge distribution* by a set of multipoles, preferably located at the atoms, and involving eventually dipoles, quadrupoles besides charges. The criterion for the quality of such a representation is of course the reproduction of the electrostatic field created by the exact molecular charge distribution at intermediate and large distance. When the charge distributions have small size (atom or bond), the lowest multipoles actually give a sufficient accuracy (e.g. see the calculation for $H_2 \ldots H_2$ by Bender *et al.* (1972)). In this connection, it may be mentioned that Cohan (1969) proposed to introduce the multipole expansion for each one of the charge distributions $|b_i(r)|^2$ associated with the *molecular orbitals* $b_i(r)$ (see equation (2.134) in Section III.B.2.a); but, if *delocalized* molecular orbitals are used, this procedure is not yet satisfactory, because the corresponding charge distributions then extend over the whole molecule, exactly as the total charge distribution does, and the criticism made in the previous Section A also holds for each of these orbital charge distribution: they are too extended and cannot be represented properly by single-centre multipole expansion up to sufficiently small distances (no matter how a centre is chosen for each of these orbital distributions). The second really essential step is to introduce elementary charge distributions with the smallest possible size (and not too numerous, of course!). This could be done by resorting to *localized* molecular orbitals in Cohan's procedure, as done by Amos and Crispin (1976b,c), but it is just as simple to introduce the expansion in terms of atomic orbitals (when working in the LCAO scheme) as done in the present section. This procedure was developed by Rein *et al.* (1972), Rabinowitz and Rein (1972), Rein (1973) and by the author (Claverie, 1973, 1976).

In the special case where only spherical gaussians are used as basis functions, the procedure simplifies considerably, because the product of two spherical gaussians is itself a spherical gaussian (whether the centres of these gaussians coincide or not), so that all multipole expansions of the elementary charge distribution reduce to their first term, namely the charge: thus, the electronic charge distribution may be represented by a set of point charges, but this simplicity is gained at the expense of a large increase of the number of these point charges, which would be much more numerous than the number of the atoms in the molecule (Hall, 1973; Tait and Hall, 1973). Thus, simplified representations, involving a number of point charges equal to the number of atoms have been proposed afterwards by Shipman (1975) and also Amos and Yoffe (1975c).

Of course, once the electronic density is approximated by a set of multipoles, the nuclear charges are added to the electronic charges located on the atoms thus generating the so-called atomic *net* charges, and the calculation of the electrostatic interaction between two (or more) molecules is reduced to a straightforward computation of the interaction between the charges, dipoles, quadrupoles associated with every molecule (e.g. see Rein, 1973, formulas 43 to 51).

Moreover, when two molecules lie sufficiently far apart, the refined multi-centre representation of the molecular charge distribution is no more necessary, and a sufficiently accurate evaluation of the electrostatic interaction energy may be obtained by resorting to the usual one-centre multipole expansion (moreover, since the intermolecular distance is large, it will be actually sufficient to use the first non-zero multipole only).

C. The representation of transition densities by bond dipoles and the decomposition of the molecular polarizability into bond polarizabilities

The expressions (2.10) for the polarization and dispersion energies may be transformed in such a way as to involve the polarizability (static or dynamic) as is well known in the long-range domain (e.g. see Buckingham (1967), Dalgarno and Davison (1966), Dalgarno (1967), Margenau and Kestner (1969) Ch. 2, pp. 51–53). We shall extend this type of treatment to the region of intermediate distances, more rigorously than was done by Claverie and Rein (1969). For that purpose it is first necessary to decompose the total molecular polarizability as a sum of local contributions (for the sake of simplicity, we shall consider the case of the static polarizability, but the principle of the analysis is valid for the dynamical polarizability as well).

A not too strong static electric field $\vec{\mathscr{E}}$ creates in a molecule an induced dipole $\vec{\mu}_{\text{ind}}$ which depends linearly on $\vec{\mathscr{E}}$ (linear response approximation):

$$\vec{\mu}_{\text{ind}} = \bar{\bar{A}}\vec{\mathscr{E}} \qquad (2.235)$$

where $\bar{\bar{A}}$ is called the (static) polarizability tensor (the dynamical tensor $\bar{\bar{A}}(\omega)$ plays a similar role when the field is a sinusoidal function of time $\vec{\mathscr{E}} = \vec{\mathscr{E}}_0 \cos \omega t$). Standard Rayleigh–Schrödinger perturbation theory gives for the tensor $\bar{\bar{A}}$ the expression (see e.g. Kauzmann (1957) Ch. 16, p. 646, formula (A-19a) with $\nu = 0$):

$$\bar{\bar{A}} = 2\,\text{Re} \sum_I{}' \frac{\vec{\mu}_{0I} \otimes \vec{\mu}_{I0}}{E_I - E_0} \qquad (2.236)$$

where $|I\rangle$ denotes the eigenfunctions of the molecule, E_I their eigenvalues, and $\vec{\mu}_{0I}$ the 'transition dipole moment'

$$\vec{\mu}_{0I} = \langle 0|\, \vec{\mu}\, |I\rangle \tag{2.237}$$

where $\vec{\mu}$ denotes the dipole moment operator:

$$\vec{\mu} = \sum_{\nu} Z_\nu \vec{r}_\nu - \sum_{i} \vec{r}_i \tag{2.238}$$

and $\sum_I'$ means that the ground-state $|0\rangle$ is excluded from the summation. The symbol $\otimes$ denotes the dyadic or tensor product (for example, see Morse and Feshbach (1953) Vol. I, Ch. 1, p. 54): if $\vec{U}$ and $\vec{V}$ are arbitrary vectors, $\vec{U} \otimes \vec{V}$ is the tensor defined by its action upon any vector $\vec{X}: (\vec{U} \otimes \vec{V})\vec{X} = \vec{U}(\vec{V} \cdot \vec{X})$; in an orthonormal basis, the matrix elements of $\vec{U} \otimes \vec{V}$ are $(\vec{U} \otimes \vec{V})_{ij} = U_i V_j^*$ (see also Appendix E).

Now, we shall represent the excited states $|I\rangle$ in the framework of the so-called 'excitonic' treatment. (Denis *et al.*, 1973, 1975): using *localized* orbitals (denoted by $u, v, w \ldots$), the ground-state $|0\rangle$ will be represented by a Slater determinant W_0, and we shall introduce the '*localized*' monoexcited determinants $W_{u-w'}$ where an occupied orbital of W_0 has been replaced by an 'excited' localized orbital w' (the prime will always indicate a virtual orbital); if it is necessary to indicate that w' belongs to the same region that some occupied orbital v, we shall use the notation $w'(v)$. These 'localized' determinants $W_{u-w'}$ will be denoted $|\alpha\rangle$ for brevity. Then the true monoexcited states $|I\rangle$ will be obtained, in a first approximation, by performing the Configuration Interaction of these basic states $|\alpha\rangle$:

$$|I\rangle = \sum_{\alpha} c_{I\alpha}\, |\alpha\rangle \tag{2.239}$$

where the matrix $c_{I\alpha}$ is unitary since both sets $\{|I\rangle\}$ and $\{|\alpha\rangle\}$ are orthonormal. States $|\alpha\rangle$ with the same mean energy (e.g. corresponding to analogous localized excitations on identical chemical bonds) will give a cluster of true eigenstates $|I\rangle$ under the effect of the Configuration Interaction.

Now, using (2.239), formula (2.236) becomes

$$\bar{\bar{A}} = 2\,\mathrm{Re} \sum_{\alpha}' \sum_{\beta}' \vec{\mu}_{0\alpha} \otimes \vec{\mu}_{\beta 0} \left(\sum_{I} \frac{c_{I\alpha} c_{I\beta}^*}{E_I - E_0} \right) \tag{2.240}$$

We shall simplify the bracketed factor

$$s_{\alpha\beta} = \sum_{I} c_{I\alpha} c_{I\beta}^* / (E_I - E_0) \tag{2.241}$$

by using the above-mentioned fact that the eigenstates $|I\rangle$ may be classified into clusters corresponding to states $|\alpha\rangle$ with (almost) equal energies. We may distinguish two cases:

(1) $|\alpha\rangle$ and $|\beta\rangle$ have markedly different energies and therefore correspond to *different* clusters of eigenstates $|I\rangle$: then, there is *no* state $|I\rangle$ such that $c_{I\alpha}$ and $c_{I\beta}$ are *both* important, so that $s_{\alpha\beta}$ may be expected to be small.

(2) $|\alpha\rangle$ and $|\beta\rangle$ have (almost) equal energies and correspond to a single cluster of states $|I\rangle$; then $c_{I\alpha}$ and $c_{I\beta}$ will both be important for the states $|I\rangle$ of this cluster, and we shall therefore use a kind of 'closure approximation' by replacing the energies E_I by the value E_α common to the 'localized' excited states $|\alpha\rangle$ generating the cluster: (2.241) then gives:

$$s_{\alpha\beta} \simeq \frac{1}{E_\alpha - E_0} \sum_I c_{I\alpha} c_{I\beta}$$

i.e.

$$s_{\alpha\beta} \simeq \delta_{\alpha\beta}/(E_\alpha - E_0) \tag{2.242}$$

and this formula appears appropriate also for the first case.

Then (2.40) becomes

$$\bar{\bar{A}} = 2\,\mathrm{Re} \sum_\alpha{}' \frac{\vec{\mu}_{0\alpha} \otimes \vec{\mu}_{\alpha 0}}{E_\alpha - E_0} \tag{2.243}$$

$$= 2\,\mathrm{Re} \sum_u \sum_{w'} \frac{\mu_{u-w'} \otimes (\vec{\mu}_{u-w'})^*}{\Delta E_{u-w'}} \tag{2.244}$$

$$= \sum_u \bar{\bar{A}}_u \tag{2.245}$$

with

$$\bar{\bar{A}}_u = 2\,\mathrm{Re} \sum_{w'} \vec{\mu}_{u-w'} \otimes (\vec{\mu}_{u-w'})^* / \Delta E_{u-w'} \tag{2.246}$$

where

$$\vec{\mu}_{u-w'} = \langle W_0 | \vec{\mu} \, | W_{u-w'} \rangle \quad \text{and} \quad \Delta E_{u-w'} = E_{u-w'} - E_0 \tag{2.247}$$

$\bar{\bar{A}}_u$ clearly appears as a polarizability associated with the localized orbital u. If there are several orbitals in the same region (σ and π orbitals for example) the corresponding tensors $\bar{\bar{A}}_u$ may be added in order to define a polarizability for this region (in practice, the regions are the chemical bonds). We have therefore found a theoretical qualitative justification for the possibility of expressing the experimental polarizabilities of molecules as a sum of suitable bond polarizabilities (Le Fèvre, 1965; see also the theoretical arguments presented by Amos and Musher (1968) and also Amos and Crispin (1975, 1976a,b,c)). From the practical point of view, an essential property of the bond polarizabilities is their *transferability* from a molecule to another. Atomic polarizabilities may also be used, but they are much less transferable: thus, Caillet and Claverie (1974, 1975) used them for computational simplicity, they were calculated for each molecule by splitting bond polarizabilities into atomic parts. It must be emphasized that the definition (2.246) of $\bar{\bar{A}}_u$ involves a summation over *all* virtual localized orbitals w' and not only over those of the region of u: in this context, the transferability of bond polarizabilities values would imply that the contribution of the virtual orbitals w' belonging to other regions (which cannot be neglected) is practically independent of the detailed nature of the bonds surrounding the bond *u* (this type of property has actually been found by Zwanzig (1963) for an idealized model, namely a chain of coupled harmonic oscillators). This type of argument may help in understanding the relationships between the experimental and theoretical aspects of molecular polarizability. These relationships indeed seemed obscure during a long time:

On one hand, from the experimental point of view, the possibility of expressing the molecular polarizability as a sum of transferable bond polarizabilities seems well established for a wide variety of molecules (see for example the early work of Vickery and Denbigh (1949) concerning the alcanes); going one step further, tables of transferable *principal* polarizabilities (and not only the mean polarizability) were compiled (Denbigh (1940), Le Fèvre and Le Fèvre (1955), Le Fèvre (1965); see also Hirschfelder, Curtiss and Bird (1954) Ch. 13, Section 2-c). For most bonds, only the longitudinal (α^L or $\alpha^{\|}$) and transversal (α^T or $\alpha^{\perp}$) polarizabilities are distinguished; for some cases (aromatic C—C bonds for example), the three principal polarizabilities are distinguished (see Le Fèvre (1965) Table 22). According to some experimental results on alcanes the additivity of bond polarizability tensors would be less accurate than that of the mean polarizabilities (Powers *et al.* (1961)), which is not surprising since the former property is more refined than the latter.

On the other hand, the theoretical computations of the molecular polarizability based upon MO–SCF perturbation theory (as examples of recent methods, see Amos and Yoffe (1975d), Nir *et al.* (1973)) long remained rather disconnected from this feature of the experimental findings (such a situation is altogether not exceptional and holds as well for other decomposition of molecular properties into local (bond) properties: see the book edited by Chalvet *et al.* (1975) and especially the paper by Daudey *et al.* (1975) therein (p. 155) concerning the partition of the molecular energy into local contributions). For example, Murrell and Teixeira-Dias (1970b) computed the polarizabilities of methane, ethane and propane by using an SCF–CNDO method, and their results do not agree with the assumption of additive transferable bond tensors; but these computed values are quite different from the experimental values of Powers *et al.* (1961), which are much closer to the relation (between the anisotropies) which would hold in the case of strict additivity. Thus, the non-additivity may seemingly be overestimated when approximate calculations of the molecular polarizability are used, and the corresponding methods may therefore be quite inappropriate for the analysis of the approximate additivity supported by experimental evidence. Pitzer (1959) even considered as fortuitous this empirical additivity, arguing that the local electric field $\vec{\mathscr{E}}_{\text{loc}}$ experienced by a given bond is significantly different from the external field $\vec{\mathscr{E}}$ owing to the fields created on that bond by the induced moments existing on the other bonds.

This idea is sound, and was expressed very early, as emphasized by Applequist *et al.* (1972) (see references 7, 8 therein) who used 'basic' atomic polarizabilities and took into account the mutual polarization superimposed to the direct effect of the external field in order to get the total effective polarizability, which is therefore different from the sum of the basic atomic polarizabilities (further applications are given by Applequist and Carl (1973). Similar procedures have been applied for bond polarizabilities by Rowell and Stein (1967), Mortensen (1968), and for a chain of harmonic oscillators (Drude model) by Zwanzig (1963). Now, in all these cases, the total polarizability appears as a sum of local (atomic or bond) 'effective' contributions, the difference with respect to the 'basic' polarizability attributed to the isolated subunit corresponding to the mutual polarization of the subunits: as pointed out above, our formula (2.246) for every bond polarizability tensor exhibits a similar feature, since it involves a summation over *all* virtual localized orbitals and not only those belonging to the bond under consideration. The essential question is therefore the extent of *transferability* from one molecule to another of the '*effective*' (and not only the 'basic') polarizabilities. Now, it is perfectly possible that the bonds surrounding the given bond u contribute (through their virtual localized orbitals) to the mean bond polarizability α_u (or the tensor $\bar{\bar{A}}_u$) by an amount which, although non-negligible, *varies relatively little from one molecule to the other.* Zwanzig (1963) actually found such a result for his Drude model: the effective polarizability associated with each oscillator is significantly different from that assumed for the isolated oscillator, but the difference is essentially brought about by the neighbour oscillators, so that a 'saturation effect' appears: the effective polarizability of each oscillator reaches already for moderate lengths a value which increases only slightly when the length is further increased. Similarly, the bonds surrounding a given bond are not widely different from one molecule to another (there are few possible sets of valency angles between one bond and the neighbour ones), and this approximate invariance of the immediate surrounding, associated with the saturation effect (small contributions from distant parts of the molecule) offers a theoretical basis for the understanding of the approximate transferability of *effective* bond polarizabilities (the effective atomic polarizabilities which may always be defined for any given molecule, seem less transferable, and this is in agreement with the fact that our theoretical analysis was made in terms of localized *bond* (and eventually lone pair) orbitals: it is not possible to localize all valency orbitals of the ground state SCF function on the atoms instead of the bonds.

It must also be emphasized that the almost exclusive use of delocalized orbitals has contributed in the past to make the situation somewhat obscure. Thus Salem (1962a) and Sternlicht (1964) were led to consider that the non-saturated hydrocarbons like polyenes are a class quite different from the saturated hydrocarbons owing to the occurrence in the former compounds of delocalized orbitals. Sternlicht speaks of 'σ-electrons', 'localized' on atoms or bonds, and of 'π-electrons', 'mobile and distributed on many atoms': such qualifications are clearly abusive concerning *electrons* and could only be applied (except 'mobile') to *orbitals.*

The abusive character of the confusion between electrons and orbitals was repeatedly emphasized by various authors (e.g. see Daudel, 1966). Nevertheless it cannot be denied that this persistent confusion reflects in some way a legitimate care of the quantum chemist, namely knowing 'what the electrons really do' in a molecule. But this innocent-looking question cannot be really answered in the framework of present day quantum mechanics (so that a highly orthodox answer would dismiss the question itself as nonsensical). Whatever the future issue, this question has much more far-reaching implications than it could be suspected at first sight (Claverie and Diner, 1976; Diner and Claverie, 1976). In actual fact, the π-orbitals of the polyenes may be localized very strongly, almost as well as the σ-orbitals of the saturated alcanes (Diner *et al.*, 1970) and all the developments of the present section concerning the decomposition of the molecular polarizability could be applied as well to these conjugated molecules (the situation remains more controversial for aromatic molecules, where the localizability of the π-orbitals is less strong). These considerations lead to question the validity of the results of Davies (1952) concerning the polarizability of long polyenes; using Hückel functions for the π-orbitals (under the assumption of equal bond lengths). Davies found a total polarizability proportional to L^3 (where L denotes the length of the polyene), while the saturation effect would always predict a proportionality to L (for large L). Actually, closer inquiry reveals (Malrieu, private communication) that Davies' result is an artefact of the Hückel method for equal bond length, namely the energy of the lowest transition goes to zero like $1/L$ while the corresponding transition moment increases to infinity like L; on the contrary, the experimental results suggest that both quantities have finite limits when L increases, and this result may be found already in the framework of the Hückel method provided that *alternating bond lengths* are used (Kutzelnigg, 1966); with such finite values, the total polarizability recovers its normal behaviour (proportionality to L for large L). Consequently, Davies's results for the polarizability of polyenes (as well as Coulson and Davies' (1952) results for the dispersion energy between polyenes) may be dismissed as artefacts due to the use of equal bond lengths in the framework of the Hückel method.

Fortunately, in the recent years, the theoretical understanding of the use of bond polarizabilities has made significant progress: already Amos and Musher (1968) justified this procedure for saturated hydrocarbons, by using a self-consistent perturbation method without resorting to localized orbitals; later on, the use of such orbitals allowed to get stronger results, both in the framework of usual perturbation theory as done in the present section (the essential content of which was first presented by Claverie (1973), Ch. III, Section 4.C), and in the framework of self-consistent perturbation procedures as done by Amos and Crispin (1975, 1976a).

D. The polarization term

1. *General treatment*

The results of the previous section may be easily applied to the polarization parts (see formula (2.10)) of the second-order long-range energy $\varepsilon_2^{\mathrm{RS}}$. Let us consider the first term (the second term could be treated in exactly the same way); in order to agree with the notations of the previous Section C, we note $W_I^{(1)}$ instead of $W_a^{(1)}$ the excited eigenfunctions of the molecule (1) and $\Delta E_I^{(1)}$ instead of $[E_a^{(1)} - E_0^{(1)}]$:

$$E_{\mathrm{pol}}^{(1)} = -\sum_I{}'^{(1)} \frac{|\langle W_0^{(1)} W_0^{(2)} |\, V\, | W_I^{(1)} W_0^{(2)} \rangle|^2}{\Delta E_I^{(1)}} \tag{2.248}$$

Inserting (2.239): $W_I^{(1)} = \sum_\alpha c_{I\alpha} W_\alpha^{(1)}$, we get

$$\langle W_0^{(1)} W_0^{(2)} |\, V\, | W_I^{(1)} W_0^{(2)} \rangle = \sum_\alpha c_{I\alpha} V_\alpha^{(1)} \tag{2.249}$$

with

$$V_\alpha^{(1)} = \langle W_0^{(1)} W_0^{(2)} |\, V\, | W_\alpha^{(1)} W_0^{(2)} \rangle \tag{2.250}$$

$$= \iint \frac{f_{0\alpha}^{(1)}(\vec{r}^{\,1}) f_{00}^{(2)}(\vec{r}^{\,2})}{|\vec{r}^{\,1} - \vec{r}^{\,2}|} \, \mathrm{d}\vec{r}^{\,1} \, \mathrm{d}\vec{r}^{\,2} \tag{2.251}$$

according to (2.130). Then, using (2.249), we transform (2.248):

$$E_{\mathrm{pol}}^{(1)} = -\sum_\alpha{}' \sum_\beta{}' V_\alpha^{(1)} V_\beta^{(1)*} \left(\sum_I{}' \frac{c_{i\alpha} c_{I\beta}^*}{\Delta E_I^{(1)}} \right) \tag{2.252}$$

where the bracketed expression is nothing but the $s_{\alpha\beta}$ defined above by (2.241). Using its approximation (2.242), we get:

$$E_{\mathrm{pol}}^{(1)} \simeq -\sum_\alpha{}' \frac{|V_\alpha^{(1)}|^2}{\Delta E_\alpha^{(1)}} = -\sum_u^{(1)} \sum_{w'}^{(1)} \frac{|V_{u-w'}^{(1)}|^2}{\Delta E_{u-w'}^{(1)}} \tag{2.253}$$

where u and w' have the same meaning as in the previous section (u: occupied localized orbital, w': virtual localized orbital).

Now, we shall use the localized character of these orbitals in order to get a simple expression of the electrostatic energy

$$V_{u-w'}^{(1)} = \iint \frac{f_{u-w'}^{(1)}(\vec{r}^{\,1}) f_{00}^{(2)}(\vec{r}^{\,2})}{|\vec{r}^{\,1} - \vec{r}^{\,2}|} \, \mathrm{d}\vec{r}^{\,1} \, \mathrm{d}\vec{r}^{\,2} \tag{2.254}$$

$$\simeq -\vec{\mathscr{E}}_u^{(2)} \cdot \vec{\mu}_{u-w'}^{(1)} \tag{2.255}$$

where $\vec{\mu}_{u-w'}^{(1)}$ denotes the dipole moment of the transition distribution

$$f_{u-w'}^{(1)}(\vec{r}) = -u(\vec{r}) w'(\vec{r}) \delta_{uw'}^{\sigma}$$

(see (2.137); for simplicity we suppose that real orbitals are used) and $\vec{\mathscr{E}}_u^{(2)}$ denotes the electric field created by the molecule (2) at some point of the region u where the transition dipoles $\vec{\mu}_{u-w'}^{(1)}$ (u fixed, w' variable) will be placed (for a bond, the middle may be chosen, but this is not compulsory, and other choices could be tried for heteronuclear bonds). Of course, the approximation of (2.254) by (2.255) becomes less and less exact when w' is localized at larger and larger distances from u, but this less good approximation will concern an increasingly smaller interaction, since the values of $f^{(1)}_{u-w'}(\vec{r})$ decrease exponentially when the distance between the *localized* orbitals u and w' increases: consequently (2.255) may be expected to approximate accurately enough the important terms among all the $V^{(1)}_{u-w'}$. Then

$$(V^{(1)}_{u-w'})^2 = (\vec{\mathscr{E}}^{(2)}_u \cdot \vec{\mu}^{(1)}_{u-w'})^2 = \vec{\mathscr{E}}^{(2)}_u \cdot (\vec{\mu}^{(1)}_{u-w'} \otimes \vec{\mu}^{(1)}_{u-w'}) \cdot \vec{\mathscr{E}}^{(2)}_u \tag{2.256}$$

according to the definition of the dyadic product (see previous Section C and Appendix E, equation (2.503). Then (2.253) becomes:

$$E^{(1)}_{\text{pol}} = -\sum_u^{(1)} \vec{\mathscr{E}}^{(2)}_u \cdot \left(\sum_{w'}^{(1)} \frac{\vec{\mu}^{(1)}_{u-w'} \otimes \vec{\mu}^{(1)}_{u-w'}}{\Delta E^{(1)}_{u-w'}} \right) \cdot \vec{\mathscr{E}}^{(2)}_u \tag{2.257}$$

$$= -\tfrac{1}{2} \sum_u^{(1)} \vec{\mathscr{E}}^{(2)}_u \cdot \bar{\bar{A}}^{(1)}_u \cdot \vec{\mathscr{E}}^{(2)}_u \tag{2.258}$$

where $\bar{\bar{A}}^{(1)}_u$ actually appears as the local (bond) polarizability tensor according to (2.246). We recognize here the classical expression of a polarization energy for the molecule (1) (hence the name given to the original expression (2.248) in (2.10)), and this energy actually appears, within the approximations made, as the sum of polarization energies calculated independently for each region (bond) of the molecule corresponding to localized orbitals. In practice, we use bond polarizabilities given by systematics deduced from experimental molecular polarizabilities (Le Fèvre, 1965).

The electric fields $\vec{\mathscr{E}}^{(2)}_u$ created by the molecule (2) are evaluated from the same representation of the electrostatic charge distribution which is used for calculating electrostatic interaction: appropriate simplified representations in terms of atomic charges, dipoles . . . have been discussed above (Section B).

The special form taken by $\vec{\mathscr{E}} \cdot \bar{\bar{A}} \cdot \vec{\mathscr{E}}$ when $\bar{\bar{A}}$ is axially symmetric (with respect to the bond axis) is easily found (Rein, Claverie and Pollak (1968), Appendixes A-I and A-IV; Claverie (1973), Ch. III, Section 4.E.2):

$$\vec{\mathscr{E}} \cdot \bar{\bar{A}} \cdot \vec{\mathscr{E}} = \alpha^T (\vec{\mathscr{E}})^2 + (\alpha^L - \alpha^T)(\vec{\mathscr{E}} \cdot \vec{e}^L) \tag{2.259}$$

where α^L and α^T are the longitudinal and transverse polarizabilities respectively, and $\vec{e}^L$ is a unit vector collinear to the longitudinal axis.

2. The simplified expression at large distance

When the intermolecular distance is large (with respect to molecular sizes), all the fields $\vec{\mathscr{E}}^{(2)}_u$ become almost equal to a field $\vec{\mathscr{E}}^{(2)}$ created by the molecule (2) at some point (chosen as 'centre of force') in molecule (1). Then (2.258) becomes:

$$E^{(1)}_{\text{pol}}(\text{long-range}) = -\tfrac{1}{2} \vec{\mathscr{E}}^{(2)} \cdot \left(\sum_u^{(1)} \bar{\bar{A}}^{(1)}_u \right) \cdot \vec{\mathscr{E}}^{(2)} \tag{2.260}$$

$$= -\tfrac{1}{2} \vec{\mathscr{E}}^{(2)} \cdot \bar{\bar{A}}^{(1)} \cdot \vec{\mathscr{E}}^{(2)} \tag{2.261}$$

where $\bar{\bar{A}}^{(1)}$ denotes the total polarizability tensor of molecule (1). It must be emphasized that (2.261) is exact independently of the validity of the approximations which allowed us to obtain the decomposition (2.258): (2.261) may be derived directly from (2.248) and (2.236) in a rigorous way (Buckingham (1967); Claverie (1973), Ch. III, Section 4.E.3). Let us indeed consider again (2.248); we know, according to formula (2.130), that the matrix element in the numerator $\langle W_0^{(1)} W_0^{(2)} | V | W_I^{(1)} W_0^{(2)} \rangle$ is equal to the electrostatic interaction between the transition density $f_{0I}^{(1)}$ of molecule (1) and the ground state density $f_{00}^{(2)}$ of molecule (2). At large intermolecular distance, this interaction reduces to the energy $-\vec{\mu}_{0I}^{(1)} \cdot \vec{\mathscr{E}}^{(2)}$ of the transition dipole $\vec{\mu}_{0I}^{(1)}$ (namely the dipole moment of $f_{0I}^{(1)}$) in the electric field $\vec{\mathscr{E}}^{(2)}$ created by the molecule (2) at the 'centre of force' chosen in molecule (1) (the exact position of this centre is not important since the distance is large with respect to the molecular sizes). Then (2.248) becomes:

$$E_{\text{pol}}^{(1)} = -{\sum_I}'^{(1)} \frac{(-\vec{\mu}_{0I}^{(1)} \cdot \vec{\mathscr{E}}^{(2)})^2}{\Delta E_I^{(1)}} = -{\sum_I}'^{(1)} \frac{\vec{\mathscr{E}}^{(2)} \cdot (\vec{\mu}_{0I}^{(1)} \otimes \vec{\mu}_{I0}^{(1)}) \cdot \vec{\mathscr{E}}^{(2)}}{\Delta E_I^{(1)}}$$

$$= -\vec{\mathscr{E}}^{(2)} \cdot \left({\sum_I}'^{(1)} \frac{\vec{\mu}_{0I}^{(1)} \otimes \vec{\mu}_{I0}^{(1)}}{\Delta E_I^{(1)}} \right) \cdot \vec{\mathscr{E}}^{(2)} = -\tfrac{1}{2} \vec{\mathscr{E}}^{(2)} \cdot \bar{\bar{A}}^{(1)} \cdot \vec{\mathscr{E}}^{(2)} \qquad (2.262)$$

where we used the expression (2.236) of the polarizability tensor $\bar{\bar{A}}^{(1)}$.

Buckingham (1967) studied the long-range polarization energy for induced multipoles of arbitrary order, and also considered the non-linear effects arising from higher-order perturbation terms (hyperpolarizabilities).

3. *Pair-wise non-additivity*

If we have more than two molecules, it is easily seen that the polarization energy of everyone keeps the same form as previously:

$$E_{\text{pol}}^{(s)} = -\tfrac{1}{2} {\sum_u}^{(s)} \vec{\mathscr{E}}_u^{[s]} \cdot \bar{\bar{A}}_u^{(s)} \cdot \vec{\mathscr{E}}_u^{[s]} \qquad (2.263)$$

but now $\vec{\mathscr{E}}_u^{[s]}$ is the electric field created at the region (bond), u by *all* molecules different from (s), namely

$$\vec{\mathscr{E}}_u^{[s]} = \sum_{p \neq s} \vec{\mathscr{E}}_u^{(p)}$$

Thus the fields $\vec{\mathscr{E}}_u^{(p)}$ created at the region u by the molecules (p) combine additively, but the polarization energies are *not* additive since they are *quadratic* functions of the fields:

$$\vec{\mathscr{E}}_u^{[s]} \cdot \bar{\bar{A}}_u \cdot \vec{\mathscr{E}}_u^{[s]} = \sum_{p \neq s} \vec{\mathscr{E}}_u^{(p)} \cdot \bar{\bar{A}}_u \cdot \vec{\mathscr{E}}_u^{(p)} + \sum_{p \neq s} \sum_{q \neq s,p} \vec{\mathscr{E}}_u^{(p)} \cdot \bar{\bar{A}}_u \cdot \vec{\mathscr{E}}_u^{(q)} \qquad (2.264)$$

and the non-diagonal terms ($p \neq q$) represent the difference with respect to exact additivity. This non-additivity may be important in relative value (e.g.

see Pullman *et al.* (1967): for 3 molecules, the non-additivity may reach 100 per cent), but since the polarization term is rather small with respect to the electrostatic and dispersion terms (which are pair-wise additive), the pair-wise additivity of the total interaction may still hold approximately (see Sections E.4 and F).

4. Use of atomic polarizabilities

While the evaluation of the electrostatic energy as an interaction between atomic charges and dipoles requires the computation of all intermolecular atom–atom distances, the evaluation of the polarization energy as a sum of bond polarization terms requires the calculation of all intermolecular atom–bond distances (in our applications, we almost always used the middle of the bond as the 'centre of force' where the induced dipole moment was assumed to be localized; the possibility of considering a different centre of force for non-symmetric bonds is considered by Caillet and Claverie (1975, footnote on p. 450), and also in Section E.5 below). Now, it would be computationally advantageous to avoid the calculation of these atom–bond distances besides the atom–atom ones, and this may be realized if we use the atoms as polarizable subunits instead of the bonds. Since the atomic contributions to the polarizabilities empirically appear much less transferable than the bond contributions (such a situation is altogether in agreement with our analysis of Section C which lead us to polarizability contributions associated with the bonds), we propose to continue to use the bond polarizabilities as basic data, and to define from them atomic polarizabilities by sharing each bond polarizability between the two atoms defining the bonds, and then adding for every atom all polarizability increments brought by the various bonds which start from the atom considered. For example, Caillet and Claverie (1975) use the following procedure: 'effective' numbers of electrons $n_{u,\lambda}$ pertaining to an atom λ involved in a bond u are defined as $n_{u,\lambda} = (N_u/2 + N_\lambda^c/\nu_\lambda)$, where: N_u is the 'number of bonding electrons' (e.g. 2 for a C–C bond, 4 for a C═C bond, 3 for a C$\overset{\cdot\cdot\cdot}{-}$C aromatic bond; more complicated π systems are considered by Caillet and Claverie, 1975); N_λ^c is the number of valence electrons of the atom λ which are *not* involved in some bond (lone pair(s) electrons) and ν_λ denotes the number of bonds starting from the atom λ (i.e. the number of atoms bonded to λ). Then the bond polarizability (tensor $\bar{\bar{A}}_u$ or mean polarizability α_u) is shared between the atoms λ and κ (joined by the bond u) according to the weights $n_{u,\lambda}$ and $n_{u,\kappa}$:

$$\bar{\bar{A}}_{u,\lambda} = \frac{n_{u,\lambda}}{n_{u,\lambda} + n_{u,\kappa}} \bar{\bar{A}}_u \qquad \bar{\bar{A}}_{u,k} = \frac{n_{u,\kappa}}{n_{u,\lambda} + n_{u,\kappa}} \bar{\bar{A}}_u$$

Then the total atomic polarizability of every atom is obtained by summing the increments thus brought about by the bonds starting from this atom (we denote this $u \in \lambda$):

$$\bar{\bar{A}}_\lambda = \sum_{u \in \lambda} \bar{\bar{A}}_{u,\lambda}$$

In actual applications, of course, we applied this procedure to the mean polarizability rather than to the anisotropic polarizability tensor, in order to have as simple computations as possible.

E. The dispersion term

This term is the last one in formula (2.10), and we rewrite it as

$$E_{\text{disp}} = -\sum_I{}^{\prime(1)} \sum_J{}^{\prime(2)} \frac{|\langle W_0^{(1)} W_0^{(2)}| V | W_I^{(1)} W_J^{(2)}\rangle|^2}{\Delta E_I^{(1)} + \Delta E_J^{(2)}} \tag{2.265}$$

Thus, only transition densities are involved in the expression of the matrix elements according to (2.130):

$$\langle W_0^{(1)} W_0^{(2)}| V | W_I^{(1)} W_J^{(2)}\rangle = \iint \frac{f_{0I}^{(1)}(\vec{r}^1) f_{0J}^{(2)}(\vec{r}^2)}{|\vec{r}^1 - \vec{r}^2|} \, \mathrm{d}\vec{r}^1 \, \mathrm{d}\vec{r}^2 \tag{2.266}$$

Thus, contrary to the electrostatic and polarization terms, the dispersion term has a 'universal' character, i.e. it remains non-negligible whether the multipole moments of the ground state charge distribution are zero or not. Actually, the dispersion term may be considered as the result of the correlation between the instantaneous dipole moments of the two systems, but this appears clearly only when the theory starts from the concept of dynamic polarizability (see Ninham *et al.* (1970), Mitchell *et al.* (1972)) rather than starting from the RS perturbation term as we do here.

It is much more difficult to obtain simple accurate expressions for the dispersion term than it was for the electrostatic and polarization ones, and this may be the reason why the works dealing with the dispersion term are much more numerous.

We shall follow here the same line of thought as we did for the polarization term, namely expressing the dispersion term as a sum of contributions from interactions between localized regions of the two molecules. We shall make use for that purpose of the dynamic polarizability, which was extensively used by numerous authors for studying the dispersion interaction between atoms or very small molecules. Finally, some other approaches will be reviewed, by considering them from our present point of view, namely the elaboration of simplified formulas suitable for molecules with arbitrary sizes and shapes.

1. *General expression*

We transform the expression (2.265) as we did in the previous section for the polarization term, namely by introducing the locally excited determinants $W_\alpha^{(1)}$, $W_\beta^{(2)}$:

$$W_I^{(1)} = \sum{}^{\prime(1)} c_{I\alpha}^{(1)} W_\alpha^{(1)} \qquad W_J^{(2)} = \sum{}^{\prime(2)} c_{J\beta}^{(2)} W_\beta^{(2)} \tag{2.267}$$

hence:

$$\langle W_0^{(1)} W_0^{(2)}| V | W_I^{(1)} W_J^{(2)}\rangle = \sum_\alpha{}^{\prime(1)} \sum_\beta{}^{\prime(2)} c_{I\alpha}^{(1)} c_{J\beta}^{(2)} V_{\alpha\beta} \tag{2.268}$$

where, according to (2.130):

$$V_{\alpha\beta}=\langle W_0^{(1)}W_0^{(2)}|V|W_\alpha^{(1)}W_\beta^{(2)}\rangle=\iint\frac{f_{0\alpha}^{(1)}(\vec{r}^1)f_{0\beta}^{(2)}(\vec{r}^2)}{|\vec{r}^1-\vec{r}^2|}\,d\vec{r}^1\,d\vec{r}^2 \quad (2.269)$$

Then, inserting (2.268) in (2.265), we get:

$$E_{\text{disp}}=-\sum_\alpha{}'^{(1)}\sum_{\alpha'}{}'^{(1)}\sum_\beta{}'^{(2)}\sum_{\beta'}{}'^{(2)}V_{\alpha\beta}V_{\alpha'\beta'}\times\left[\sum_I{}^{(1)}\sum_J{}^{(2)}\frac{c_{I\alpha}^{(1)}c_{I\alpha'}^{(1)}c_{J\beta}^{(2)}c_{J\beta'}^{(2)}}{\Delta E_I^{(1)}+\Delta E_J^{(2)}}\right] \quad (2.270)$$

(for simplicity, we write this derivation in the case of real orbitals, so that all coefficients c's and matrix elements V's are real).

The essential difficulty with the expression (2.270) is that the bracketed factor is not explicitly separable into a product of quantities depending on one molecule only, a condition which is necessary if we want to express in a simple way the dispersion energy as a function of properties of the isolated molecule (on the contrary, in formula (2.252) for $E_{\text{pol}}^{(1)}$, the bracketed factor depended only on a single molecule). In order to realize such a separation, we shall mention two methods.

2. *The 'mean excitation energy' approximation*

This approximate procedure consists in replacing the various denominators $\Delta E_I^{(1)}+\Delta E_J^{(2)}$ by some appropriate mean value, which is conveniently written as $\bar{U}^{(1)}+\bar{U}^{(2)}$ (this is the so-called 'closure approximation') (Unsöld, 1927) used very early by London (1930b, 1937) in his work on dispersion energy (see also Buckingham (1967)). Then $1/(\bar{U}^{(1)}+\bar{U}^{(2)})$ may be factorized in (2.270), and the numerator of the bracketed expression actually separates into a product of two sums, each of them depending on a single molecule, and (2.270) becomes (see Claverie and Rein (1969), Claverie (1973) Ch. III, Section 4.F.2):

$$E_{\text{disp}}\simeq-\frac{1}{\bar{U}^{(1)}+\bar{U}^{(2)}}\sum_\alpha{}'^{(1)}\sum_{\alpha'}{}'^{(1)}\sum_\beta{}'^{(2)}\sum_{\beta'}{}'^{(2)}V_{\alpha\beta}V_{\alpha'\beta'}\times\left(\sum_I{}^{(1)}c_{I\alpha}^{(1)}c_{I\alpha'}^{(1)}\right)\left(\sum_J{}^{(2)}c_{J\beta}^{(2)}c_{J\beta'}^{(2)}\right) \quad (2.270)$$

now $\sum_I^{(1)}c_{I\alpha}^{(1)}c_{I\alpha'}^{(1)}=\delta_{\alpha\alpha'}$ and $\sum_J^{(2)}c_{J\beta}^{(2)}c_{J\beta'}^{(2)}=\delta_{\beta\beta'}$, hence (2.271) becomes:

$$E_{\text{disp}}\simeq-\frac{1}{U^{(1)}+U^{(2)}}\sum_\alpha{}'^{(1)}\sum_\beta{}'^{(2)}(V_{\alpha\beta})^2 \quad (2.272)$$

Then, as we did for the polarization term, we take into account the localized character of the excitations (and therefore of the corresponding transition densities) $\alpha(u-w')$ and $\beta(v-x')$ in their respective molecules, and we reduce $V_{\alpha\beta}$ to the interaction between the transition dipoles $\vec{\mu}_{u-w'}^{(1)}$ and

$\vec{\mu}^{(2)}_{v-x'}$, namely

$$V_{\alpha\beta} = V_{u-w';v-x'} \simeq -\vec{\mu}^{(1)}_{u-w'} \frac{\bar{\bar{T}}_{uv}}{(R_{uv})^3} \vec{\mu}^{(2)}_{v-x'} \tag{2.273}$$

where

$$\bar{\bar{T}}_{uv} = 3\vec{r}_{uv} \otimes \vec{r}_{uv} - \bar{\bar{1}} \quad \text{with} \quad \vec{r}_{uv} = \frac{\vec{R}_{uv}}{R_{uv}} \tag{2.274}$$

$\vec{R}_{uv} = \vec{R}^{(2)}_v - \vec{R}^{(1)}_u$ denotes the vector joining the points $\vec{R}^{(1)}_u$ and $\vec{R}^{(2)}_v$ where the transition dipoles $\vec{\mu}^{(1)}_{u-w'}$ and $\vec{\mu}^{(2)}_{v-x'}$ are located respectively in the regions u and v. Let us recall that this tensor $\bar{\bar{T}}$ expresses the electric field $\vec{\mathscr{E}}$ created by a dipole $\vec{\mu}$ at a distance $\vec{R}$ (from the dipole) by the formula:

$$\vec{\mathscr{E}} = (\bar{\bar{T}}\vec{\mu})/R^3 \tag{2.275}$$

Then (2.272) becomes

$$E_{\text{disp}} \simeq -\frac{1}{\bar{U}^{(1)} + \bar{U}^{(2)}} \sum_u^{(1)} \sum_v^{(2)} \frac{1}{(R_{uv})^6} \sum_{w'}^{(1)} \sum_{x'}^{(2)} (\vec{\mu}^{(1)}_{u-w'} \cdot \bar{\bar{T}}_{uv} \cdot \vec{\mu}^{(2)}_{v-x'})^2 \tag{2.276}$$

or, using the identity (2.514) of Appendix E:

$$E_{\text{disp}} \simeq -\frac{1}{\bar{U}^{(1)} + \bar{U}^{(2)}} \sum_u^{(1)} \sum_v^{(2)} \frac{1}{(R_{uv})^6} \text{Tr} \left[\bar{\bar{T}}_{uv} \left(\sum_{w'}^{(1)} \vec{\mu}^{(1)}_{u-w'} \otimes \mu^{(1)}_{u-w'} \right) \times \bar{\bar{T}}_{uv} \left(\sum_{x'}^{(2)} \vec{\mu}^{(2)}_{v-x'} \otimes \vec{\mu}^{(2)}_{v-x'} \right) \right] \tag{2.277}$$

where Tr denotes the Trace (sum of the diagonal elements). On the other hand, the same closure approximation, when applied to the expressions (2.244, 2.246) for the polarizability, gives (let us recall that all matrix elements have been chosen real):

$$\bar{\bar{A}}^{(1)}_u \simeq \frac{2}{\bar{U}^{(1)}} \sum_{w'}^{(1)} (\vec{\mu}^{(1)}_{u-w'} \otimes \vec{\mu}^{(1)}_{u-w'}) \qquad \bar{\bar{A}}^{(2)}_v \simeq \frac{2}{\bar{U}^{(2)}} \sum_{x'}^{(2)} (\vec{\mu}^{(2)}_{v-x'} \otimes \vec{\mu}^{(2)}_{v-x'}) \tag{2.278}$$

Using the formulas (2.278), we could transform (2.277) in two different ways: either we replace $(\sum^{(1)}_{w'})$ by $[\bar{U}^{(1)}/2]\bar{\bar{A}}^{(1)}_u$, and similarly for $(\sum^{(2)}_{x'})$, or conversely we replace $\bar{U}^{(1)}$ in terms of $\bar{\bar{A}}^{(1)}_u$ and $(\sum^{(1)}_{w'})$ (similarly for $\bar{U}^{(2)}$). The second procedure was proposed by Salem (1960); it is easily applied when the systems (1) and (2) are highly symmetric, because the tensors involved in (2.278) are then isotropic, so that the formula (2.278) reduce to scalar relations; Salem (1960) has indicated a generalization for the non-isotropic case. This second procedure is actually equivalent to assuming a peculiar approximation of the dynamic polarizability, as we shall see below (Section 3), so that we develop here the first procedure only. Then, (2.277) becomes

$$E_{\text{disp}} \simeq -\frac{1}{4}\frac{\bar{U}^{(1)}\bar{U}^{(2)}}{\bar{U}^{(1)}+\bar{U}^{(2)}} \sum_u^{(1)} \sum_v^{(2)} \frac{1}{(R_{uv})^6} \operatorname{Tr}[\bar{\bar{T}}_{uv}\bar{\bar{A}}_u^{(1)}\bar{\bar{T}}_{uv}\bar{\bar{A}}_v^{(2)}] \tag{2.279}$$

In actual practice, the regions u and v refer to bonds, since the bond polarizabilities are rather transferable and tables of values deduced from experimental molecular polarizabilities are available (Le Fèvre, 1965). But for computational reasons, the use of atom–atom $1/(R_{\mu\nu})^6$ terms may be desirable. In the present case, it would be sufficient to replace the sum $\sum_u^{(1)}\sum_v^{(2)}$ over bonds by a sum $\sum_\mu^{(1)}\sum_\nu^{(2)}$ over atoms and to use for $\bar{\bar{A}}^{(1)}, \bar{\bar{A}}^{(2)}$ atomic polarizabilities (which may be built by splitting bond polarizabilities as indicated above (Section D.4)).

Let us consider the explicit form of (2.279) when $\bar{\bar{A}}_u^{(1)}$ and $\bar{\bar{A}}_v^{(2)}$ are axially symmetric (Rein *et al.* (1968), Appendix A-III):

$$\bar{\bar{A}} = \alpha_L(\vec{e}_L\otimes\vec{e}_L)+\alpha_T(\bar{\bar{1}}-\vec{e}_L\otimes\vec{e}_L) = \alpha_T\bar{\bar{1}}+\delta(\vec{e}_L\otimes\vec{e}_L) \tag{2.280}$$

where α_L and α_T are the longitudinal and transverse polarizabilities $\delta = \alpha_L - \alpha_T$ and $\vec{e}_L$ is the unit vector defining the symmetry axis.

For the sake of simplicity, we omit the subscripts u, v in the subsequent evaluation of the trace involved in (2.279):

$$\begin{aligned}\operatorname{Tr}(\bar{\bar{T}}\bar{\bar{A}}^{(1)}\bar{\bar{T}}\bar{\bar{A}}^{(2)}) &= \operatorname{Tr}[\bar{\bar{T}}(\alpha_T^{(1)}\bar{\bar{1}}+\delta^{(1)}\vec{e}_L^{(1)}\otimes\vec{e}_L^{(1)}) \\ &\quad\times \bar{\bar{T}}(\alpha_T^{(2)}\bar{\bar{1}}+\delta^{(2)}\vec{e}_L^{(2)}\otimes\vec{e}_L^{(2)})] \\ &= \alpha_T^{(1)}\alpha_T^{(2)}\operatorname{Tr}[(\bar{\bar{T}})^2]+\delta^{(1)}\alpha_T^{(2)}\operatorname{Tr}[\bar{\bar{T}}(\vec{e}_L^{(1)}\otimes\vec{e}_L^{(1)})\bar{\bar{T}}] \\ &\quad+\alpha_T^{(1)}\delta^{(2)}\operatorname{Tr}[(\bar{\bar{T}})^{(2)}(\vec{e}_L^{(2)}\otimes\vec{e}_L^{(2)})] \\ &\quad+\delta^{(1)}\delta^{(2)}\operatorname{Tr}[\bar{\bar{T}}(\vec{e}_L^{(1)}\otimes\vec{e}_L^{(1)})\bar{\bar{T}}(\vec{e}_L^{(2)}\otimes\vec{e}_L^{(2)})]\end{aligned} \tag{2.281}$$

We introduce now:

$$\operatorname{Tr}[(\bar{\bar{T}})^2] = 6 \tag{2.282}$$

$$\operatorname{Tr}[\bar{\bar{T}}(\vec{e}_L^{(1)}\otimes\vec{e}_L^{(1)})\bar{\bar{T}}] = (\bar{\bar{T}}\,.\,\vec{e}_L^{(1)})^2 = \vec{e}_L^{(1)}\,.\,(\bar{\bar{T}})^2\,.\,\vec{e}_L^{(1)} \tag{2.283}$$

$$\operatorname{Tr}[(\bar{\bar{T}})^2(\vec{e}_L^{(2)}\otimes\vec{e}_L^{(2)})] = (\bar{\bar{T}}\,.\,\vec{e}_L^{(2)})^2 = \vec{e}_L^{(2)}\,.\,(\bar{\bar{T}})^2\,.\,\vec{e}_L^{(2)} \tag{2.284}$$

see formulas (2.506, 2.507) in Appendix E; since the Trace does not depend on the order of the matrices in the product, we have $\operatorname{Tr}[(\bar{\bar{T}})^2(\vec{e}\otimes\vec{e})] = \operatorname{Tr}[\bar{\bar{T}}(\vec{e}\otimes\vec{e})\bar{\bar{T}}]$. As concerns (2.282), we have:

$$\bar{\bar{T}}^2 = 9(\vec{r}\otimes\vec{r})^2-6(\vec{r}\otimes\vec{r})+\bar{\bar{1}} = 3\vec{r}\otimes\vec{r}+\bar{\bar{1}} \tag{2.285}$$

since $(\vec{r}\otimes\vec{r})^2 = \vec{r}\otimes\vec{r}$ (Appendix E); therefore $\operatorname{Tr}(\bar{\bar{T}})^2 = 3\operatorname{Tr}(\vec{r}\otimes\vec{r})+\operatorname{Tr}\bar{\bar{1}} = 3+3 = 6$, since (see Appendix E):

$$\operatorname{Tr}(\vec{r}\otimes\vec{r}) = 1 \quad\text{and}\quad \operatorname{Tr}\bar{\bar{1}} = 3 \tag{2.286}$$

We also have (formula (2.514) in Appendix E):

$$\mathrm{Tr}\,[\bar{\bar{T}}(\vec{e}_{\mathrm{L}}^{(1)}\otimes\vec{e}_{\mathrm{L}}^{(1)})\bar{\bar{T}}(\vec{e}_{\mathrm{L}}^{(2)}\otimes\vec{e}_{\mathrm{L}}^{(2)})]=(\vec{e}_{\mathrm{L}}^{(1)}\,.\,\bar{\bar{T}}\,.\,\vec{e}_{\mathrm{L}}^{(2)})^{2} \tag{2.287}$$

Now, we evaluate explicitly the scalar products involved in (2.283, 2.284, 2.286) by introducing the expressions (2.274) and (2.285) of $\bar{\bar{T}}$ and $(\bar{\bar{T}})^2$ respectively:

$$\vec{e}_{\mathrm{L}}^{(1)}\,.\,\bar{\bar{T}}\,.\,\vec{e}_{\mathrm{L}}^{(2)}=3(\vec{r}\,.\,\vec{e}_{\mathrm{L}}^{(1)})(\vec{r}\,.\,\vec{e}_{\mathrm{L}}^{(2)})-(\vec{e}_{\mathrm{L}}^{(1)}\,.\,\vec{e}_{\mathrm{L}}^{(2)}) \tag{2.288}$$

$$\vec{e}_{\mathrm{L}}\,.\,(\bar{\bar{T}})^{2}\,.\,\vec{e}_{\mathrm{L}}=3(\vec{r}\,.\,\vec{e}_{\mathrm{L}})^{2}+1 \tag{2.289}$$

Then, inserting in (2.281) the values of the Traces given by (2.282), 2.283, 2.284, 2.288, 2.289) we obtain the desired explicit formula

$$\begin{aligned}\mathrm{Tr}\,[\bar{\bar{T}}\bar{\bar{A}}^{(1)}\bar{\bar{T}}\bar{\bar{A}}^{(2)}]&=6\alpha_{\mathrm{T}}^{(1)}\alpha_{\mathrm{T}}^{(2)}\\&\quad+\delta^{(1)}\alpha_{\mathrm{T}}^{(2)}[3(\vec{r}\,.\,\vec{e}_{\mathrm{L}}^{(1)})^{2}+1]+\alpha_{\mathrm{T}}^{(1)}\delta^{(2)}[3(\vec{r}\,.\,\vec{e}_{\mathrm{L}}^{(2)})^{2}+1]\\&\quad+\delta^{(1)}\delta^{(2)}[3(\vec{r}\,.\,\vec{e}_{\mathrm{L}}^{(1)})(\vec{r}\,.\,\vec{e}_{\mathrm{L}}^{(2)})-(\vec{e}_{\mathrm{L}}^{(1)}\,.\,\vec{e}_{\mathrm{L}}^{(2)})]^{2}\end{aligned} \tag{2.290}$$

Of course, the formula (2.290) may be rewritten in terms of the mean polarizability

$$\bar{\alpha}=(\alpha_{\mathrm{L}}+2\alpha_{\mathrm{T}})/3=\alpha_{\mathrm{T}}+\delta/3 \tag{2.291}$$

instead of the transverse polarizability α_{T}. Then the formula (2.290) may be inserted in (2.279) for evaluating the traces $\mathrm{Tr}\,[\bar{\bar{T}}_{uv}\bar{\bar{A}}_{u}^{(1)}\bar{\bar{T}}_{uv}\bar{\bar{A}}_{v}^{(2)}]$.

If the polarizabilities are now supposed to be isotropic:

$$\bar{\bar{A}}_{u}^{(1)}=\alpha_{u}^{(1)}\bar{\bar{1}}\qquad\bar{\bar{A}}_{v}^{(2)}=\alpha_{v}^{(2)}\bar{\bar{1}} \tag{2.292}$$

we have the still simpler formula:

$$\mathrm{Tr}\,[\bar{\bar{T}}_{uv}\bar{\bar{A}}_{u}^{(1)}\bar{\bar{T}}_{uv}\bar{\bar{A}}_{v}^{(2)}]=\alpha_{u}^{(1)}\alpha_{v}^{(2)}\,\mathrm{Tr}\,(\bar{\bar{T}})^{2}=6\alpha_{u}^{(1)}\alpha_{v}^{(2)} \tag{2.293}$$

according to (2.282), and therefore (2.279) becomes

$$E_{\mathrm{disp}}\simeq-\frac{3}{2}\frac{\bar{U}^{(1)}\bar{U}^{(2)}}{\bar{U}^{(1)}+\bar{U}^{(2)}}\sum_{u}^{(1)}\sum_{v}^{(2)}\frac{\alpha_{u}^{(1)}\alpha_{v}^{(2)}}{(R_{uv})^{6}} \tag{2.294}$$

At large intermolecular distance R, all distances $\vec{R}_{uv}$ may be replaced by $\vec{R}$ and all tensors $\bar{\bar{T}}_{uv}$ may be replaced by the unique tensor $\bar{\bar{T}}=3(\vec{r}\otimes\vec{r})-1$, where $\vec{r}=\vec{R}/R$. Then (2.279) becomes (Claverie and Rein, 1969):

$$\begin{aligned}E_{\mathrm{disp}}&\simeq-\frac{1}{4}\frac{\bar{U}^{(1)}\bar{U}^{(2)}}{\bar{U}^{(1)}+\bar{U}^{(2)}}\frac{1}{R^{6}}\mathrm{Tr}\left[\bar{\bar{T}}\Big(\sum_{u}\bar{\bar{A}}_{u}^{(1)}\Big)\bar{\bar{T}}\Big(\sum_{v}\bar{\bar{A}}_{v}^{(2)}\Big)\right]\\&=-\frac{1}{4}\frac{\bar{U}^{(1)}\bar{U}^{(2)}}{\bar{U}^{(1)}+\bar{U}^{(2)}}\frac{1}{R^{6}}\mathrm{Tr}\,[\bar{\bar{T}}\bar{\bar{A}}^{(1)}\bar{\bar{T}}\bar{\bar{A}}^{(2)}]\end{aligned} \tag{2.295}$$

If the polarizabilities are supposed isotropic, we get from (2.294) (with $R_{uv}=R$) or from (2.295) (with $\bar{\bar{A}}^{(1)}=\alpha^{(1)}\bar{\bar{1}}$ and $\bar{\bar{A}}^{(2)}=\alpha^{(2)}\bar{\bar{1}}$):

$$E_{\mathrm{disp}}\simeq-\frac{3}{2}\frac{\bar{U}^{(1)}\bar{U}^{(2)}}{\bar{U}^{(1)}+\bar{U}^{(2)}}\frac{\alpha^{(1)}\alpha^{(2)}}{R^{6}} \tag{2.296}$$

This is the well-known formula of London (1930b, 1937). The essential drawback of the 'mean excitation energy' procedure is of course the uncertainty in the choice of this 'mean excitation energy'. The rule proposed by London (1930b, 1937), namely to take for $\bar{U}^{(s)}$ the ionization potential of (s), is satisfactory for the atoms H and He, but not for the heavier rare gas atoms, for which a value about twice the ionization potential would be more satisfactory (Dalgarno and Davison, 1966; Huron and Claverie, 1969; Pitzer, 1959). The values given by Pitzer (1959) for the dispersion constant C_6 (corresponding to $E_{\text{disp}} = -C_6/R^6$) are deduced from experimental virial coefficients; they appear slightly too large with respect to the theoretical accurate C_6 constants given by Dalgarno and Davison (1966), probably because Pitzer takes into account implicitly the contributions of further attractive terms ($-C_8/R^8$, $-C_{10}/R^{10}$) to the virial coefficient. Moreover, in the case of a molecule, the use of a single mean value $\bar{U}^{(s)}$ neglects possible differences between the bonds. The second method may avoid these defects.

3. The dynamic polarizability method

This method consists in expressing *rigorously* the dispersion energy in terms of the dynamic polarizabilities of the two molecules. It has been widely applied for atoms (and very small molecules): see Dalgarno and Kingston (1961), Kingston (1964), Barker and Leonard (1964), Hirschfelder *et al.* (1964, Section X.C), Karplus and Kolker (1964), Dalgarno and Davison (1966), Dalgarno (1967), Dalgarno *et al.* (1967), Epstein (1968b), Gordon (1968b), McQuarrie *et al.* (1969), Tang and Karplus (1968), Tang (1968, 1969, 1970, 1971), Langhoff and Karplus (1970a,b), Langhoff *et al.* (1971), Starkschall and Gordon (1971, 1972a). Accurate methods based on the direct use of the basic formula (2.265) have also been proposed: Bell (1965, 1966), Alexander (1970a,b), Kramer (1970); Fukui and Yamabe (1968), Starkschall and Gordon (1972b)).

Kutzelnigg (1968, 1969) developed for the calculation of the coefficients $C_6, C_8, C_{10} \ldots$ (of $-R^{-6}, -R^{-8}, -R^{-10}, \ldots$ respectively) a method based upon the determination of the asymptotic natural states of the supersystem, through the minimization of the Hylleraas functional associated with the second-order perturbation energy. The method was applied to hydrogen and alkali atoms, by using for the latter monoelectronic pseudo-potentials (Kutzelnigg, 1969). This line of work was pursued by Maeder (1972).

When the dynamic polarizabilities are used, two approaches are actually possible:

(a) to express directly the dispersion as an interaction between the fluctuating instantaneous dipoles (and higher-order multipoles if desired), a procedure which obviously introduces the dynamic polarizabilities on physical grounds (Mavroyannis and Stephen, 1962; Dzyaloshinskii *et al.*, 1961; McLachlan, 1963a,b,c; 1965; see also the related approaches of Jehle

(1965), Linder (1965, 1967), Linder and Rabenold (1972), Ninham *et al.* (1970), Mitchell *et al.* (1972), and also Langbein (1973)).

(b) to start from the formula (2.265) given by the RS perturbation expansion, and to transform it mathematically so that the theoretical expressions for the dynamic polarizabilities appear explicitly. This way was actually followed in the works quoted first (at the beginning of the present Section 3), and we shall also use it in the present treatment (taken from Claverie (1973) Ch. III, Section 4.F) whose aim is to develop an approximate generalization (of these previous treatments) for arbitrary molecules, by introducing bond dynamic polarizabilities.

Let us first recall the definition of the dynamic polarizability (e.g. see Kauzmann (1957), Ch. 16, Section A.c pp. 645–646): when a sinusoidal electric field $\vec{\mathscr{E}} = \vec{\mathscr{E}}_0 \cos \omega t$ is applied to a molecule, an induced dipole moment $\vec{\mu} = \vec{\mu}_0 \cos \omega t$ appears, with the same angular frequency ω, and $\vec{\mu}$ depends linearly on $\vec{\mathscr{E}}$ when $\vec{\mathscr{E}}$ is not too strong (linear response approximation; for an introduction to the general theory of linear response, see for example Landau and Lifshitz (1958, Ch. XII, Sections 122 to 124)):

$$\vec{\mu} = \bar{\bar{A}}(\hbar\omega)\vec{\mathscr{E}} \quad \text{or} \quad \vec{\mu}_0 = \bar{\bar{A}}(\hbar\omega)\vec{\mathscr{E}}_0 \tag{2.297}$$

where $\bar{\bar{A}}(\hbar\omega)$ is the *dynamic polarizability tensor* at angular frequency ω, and its theoretical expression is:

$$\bar{\bar{A}}(\hbar\omega) = 2\sum_I{}' \frac{(\vec{\mu}_{0I} \otimes \vec{\mu}_{I0})\Delta E_I}{(\Delta E_I)^2 - (\hbar\omega)^2} \tag{2.298}$$

where the notation is the same as in formulas (2.236, 2.237) of Section C: $\Delta E_I = E_I - E_0$ and the eigenstates $|I\rangle$ are chosen real, so that the transition dipole moments $\vec{\mu}_{0I} = \vec{\mu}_{I0}$ are also real. Note that, for $\omega = 0$, we actually recover the expression (2.236) of the static polarizability tensor.

We now start again from (2.270). We consider the mathematical identity (which may be easily obtained either by direct integration along the real axis or by contour integration in the complex plane):

$$\frac{1}{a+b} = \frac{2}{\pi}\int_0^{+\infty} \frac{ab}{(a^2+\xi^2)(b^2+\xi^2)}\,\mathrm{d}\xi \qquad (a, b > 0) \tag{2.299}$$

and we apply it to the factor $1/(\Delta E_I^{(1)} + \Delta E_J^{(2)})$ in (2.270). Then the separation of the bracketed factor becomes possible, since the integrand in (2.299) is the *product* of two terms $a/(a^2+\xi^2)$ and $b/(b^2+\xi^2)$, each of them depending on a single molecule:

$$E_{\text{disp}} = -\sum_\alpha{}' \sum_{\alpha'}{}' \sum_\beta{}' \sum_{\beta'}{}' V_{\alpha\beta} V_{\alpha'\beta'} F(\alpha, \alpha'; \beta, \beta') \tag{2.300}$$

with

$$F(\alpha, \alpha'; \beta, \beta') = \frac{2}{\pi}\int_0^{+\infty} \mathrm{d}\xi\, s_{\alpha\alpha'}^{(1)}(\xi)\, s_{\beta\beta'}^{(2)}(\xi) \tag{2.301}$$

where

$$s^{(1)}_{\alpha\alpha'}(\xi)=\sum_I{}'^{(1)}\frac{c^{(1)}_{I\alpha}c^{(1)}_{I\alpha'}\Delta E^{(1)}_I}{(\Delta E^{(1)}_I)^2+\xi^2}$$
$$s_{\beta\beta'}(\xi)=\sum_J{}'^{(2)}\frac{c^{(2)}_{J\beta}c^{(2)}_{J\beta'}\Delta E^{(2)}_J}{(\Delta E^{(2)}_J)^2+\xi^2} \quad (2.302)$$

Now, these quantities $s_{\alpha\alpha'}(\xi)$ are analogous to the $s_{\alpha\alpha'}$ encountered in Section C (equation (2.241)) for the analysis of the static polarizability into local (bond) contributions (we see in fact that $s_{\alpha\alpha'}(0)=s_{\alpha\alpha'}$): they actually play the same role in the analysis of the 'polarizability at imaginary frequency' $\bar{\bar{A}}(i\xi)$, which is obtained by analytic continuation of the dynamic polarizability $\bar{\bar{A}}(\hbar\omega)$ (thus, in (2.298), we need only to replace $-(\hbar\omega)^2$ by $+\xi^2$).

Then, the same argument (based on the consideration of localized orbitals) that was used in Section C for getting the approximate expression (2.242) of $s_{\alpha\beta}$ may be applied here to $s_{\alpha\alpha'}(\xi)$, thus giving:

$$s_{\alpha\alpha'}(\xi)\simeq\frac{\Delta E^{(1)}_\alpha}{(\Delta E^{(1)}_\alpha)^2+\xi^2}\,\delta_{\alpha\alpha'} \qquad s^{(2)}_{\beta\beta'}(\xi)\simeq\frac{\Delta E^{(2)}_\beta}{(\Delta E^{(2)}_\beta)^2+\xi^2}\,\delta_{\beta\beta'} \quad (2.303)$$

Inserting (2.303) in (2.300, 2.301), we get:

$$E_{\text{disp}}\simeq-\sum_\alpha{}'^{(1)}\sum_\beta{}'^{(2)}(V_{\alpha\beta})^2\left(\frac{2}{\pi}\int_0^{+\infty}\mathrm{d}\xi\frac{\Delta E^{(1)}_\alpha}{(\Delta E^{(1)}_\alpha)^2+\xi^2}\frac{\Delta E^{(2)}_\beta}{(\Delta E^{(2)}_\beta)^2+\xi^2}\right) \quad (2.304)$$

Then, denoting explicitly $W^{(1)}_\alpha=W^{(1)}_{u-w'}$ (localized monoexcitation $u-w'$) and $W^{(2)}_\beta=W^{(2)}_{v-x'}$, the matrix element $V_{\alpha\beta}$ is

$$V_{\alpha\beta}=V_{(u-w')(v-x')}=\iint\frac{f^{(1)}_{u-w'}(\vec{r}^{\,1})f^{(2)}_{v-x'}(\vec{r}^{\,2})}{|\vec{r}^{\,1}-\vec{r}^{\,2}|}\,\mathrm{d}\vec{r}^{\,1}\,\mathrm{d}\vec{r}^{\,2} \quad (2.305)$$

and, for the same reason as previously (discussion in Section D between equations (2.254), (2.255), and (2.256)), this electrostatic interaction may be reduced to the interaction between the dipoles of the transition distributions $f^{(1)}_{u-w'}(\vec{r})$, $f^{(2)}_{v-x'}(\vec{r})$, namely:

$$V_{(u-w')(v-x')}\simeq-\vec{\mu}^{(1)}_{u-w'}\cdot\frac{\bar{\bar{T}}_{uv}}{(R_{uv})^3}\cdot\vec{\mu}^{(2)}_{v-x'} \quad (2.306)$$

and:

$$[V_{(u-w')(v-x')}]^2\simeq\frac{1}{(R_{uv})^6}\,\mathrm{Tr}\,[\bar{\bar{T}}_{uv}(\vec{\mu}^{(1)}_{u-w'}\otimes\vec{\mu}^{(1)}_{u-w'})\bar{\bar{T}}_{uv}(\vec{\mu}^{(2)}_{v-x'}\otimes\vec{\mu}^{(2)}_{v-x'})] \quad (2.307)$$

according to the tensor identity (2.514) of Appendix E:

$$(\vec{X}\,.\,\bar{\bar{B}}\,.\,\vec{Y})(\vec{Y}\,.\,\bar{\bar{B}}\,.\,\vec{X}) = \mathrm{Tr}\,\bar{\bar{B}}(\vec{Y}\otimes\vec{Y})\bar{\bar{B}}(\vec{X}\otimes\vec{X}) \tag{2.308}$$

Then, replacing in (2.304) $\sum_\alpha'^{(1)}$ by $\sum_u^{(1)}\sum_{w'}^{(1)}$ and $\sum_\beta'^{(2)}$ by $\sum_v^{(2)}\sum_{x'}^{(2)}$, and inserting (2.307), we easily get:

$$E_{\text{disp}} \simeq -\frac{1}{2\pi}\sum_u^{(1)}\sum_v^{(2)}\frac{1}{(R_{uv})^6}\int_0^{+\infty} \mathrm{d}\xi\,\mathrm{Tr}\,[\bar{\bar{T}}_{uv}\bar{\bar{A}}_u^{(1)}(i\xi)\bar{\bar{T}}_{uv}\bar{\bar{A}}_v^{(2)}(i\xi)] \tag{2.309}$$

where

$$\bar{\bar{A}}_u^{(1)}(i\xi) = 2\sum_{w'}^{(1)}\frac{(\vec{\mu}_{u-w'}^{(1)}\otimes\vec{\mu}_{u-w'}^{(1)})\Delta E_{u-w'}^{(1)}}{(\Delta E_{u-w'}^{(1)})^2+\xi^2} \tag{2.310}$$

$$\bar{\bar{A}}_v^{(2)}(i\xi) = 2\sum_{x'}^{(2)}\frac{(\vec{\mu}_{v-x'}^{(2)}\otimes\vec{\mu}_{v-x'}^{(2)})\Delta E_{v-x'}^{(2)}}{(\Delta E_{v-x'}^{(2)})^2+\xi^2} \tag{2.311}$$

appear as the analytic continuation at imaginary frequency of *bond dynamic polarizabilities* $\bar{\bar{A}}_u^{(1)}(\hbar\omega)$, $\bar{\bar{A}}_v^{(2)}(\hbar\omega)$ (recently, Amos and Crispin (1976a) also presented a theoretical justification for the expression of the total dynamic polarizability as a sum of bond contributions: they use localized orbitals and then apply an SCF-perturbation treatment). Thus we have succeeded to express the total dispersion interaction between the two molecules as a sum of independent dispersion-like interactions between the regions (bonds) corresponding to the localized orbitals.

We shall now consider first the special case of the isotropic polarizabilities, and introduce a suitable approximation for them, which makes possible the derivation of an explicit analytical formula. Afterwards, we shall deal with the general case (non-isotropic polarizabilities) by introducing similar approximations (for the principal polarizabilities).

If the bond dynamic polarizabilities are assumed isotropic:

$$\bar{\bar{A}}_u^{(1)}(i\xi) = \alpha_u^{(1)}(i\xi)\bar{\bar{1}} \qquad \bar{\bar{A}}_v^{(2)}(i\xi) = \alpha_v^{(2)}(i\xi)\bar{\bar{1}} \tag{2.312}$$

(2.309) becomes, according to (2.293):

$$E_{\text{disp}} \simeq -\frac{3}{\pi}\sum_u^{(1)}\sum_v^{(2)}\frac{1}{(R_{uv})^6}\int_0^{+\infty}\alpha_u^{(1)}(i\xi)\alpha_v^{(2)}(i\xi)\,\mathrm{d}\xi \tag{2.313}$$

According to the exact formulas (2.310, 2.311), it appears convenient to approximate the dynamic polarizabilities by a rational function $P_M(\hbar\omega)/Q_N(\hbar\omega)$ (where P_M and Q_N are polynomials of degree M and N respectively). Dalgarno and Davison (1966) have used finite sums

$$\sum_{m=1}^{p} g_m/(U_m^2+\xi^2)$$

of terms analogous to those of (2.298, 2.310, 2.311), fitted on experimental data for the index of refraction. The so-called Padé approximants have also been used extensively (Tang and Karplus, 1968; Tang, 1970; Langhoff and Karplus, 1970a,b,c; Langhoff *et al.*, 1971). The Padé approximant $\mathcal{P}(N, M) = P_M/Q_N$ is defined by the requirement that the $N+M+1$ first terms of the power series expansion of $\mathcal{P}(N, M)$ are identical with those of the exact function to be approximated. Besides the original thesis of Padé (1892), the reader may refer to the review paper by Baker (1965), the book edited by Baker and Gammel (1970), and the papers by Goscinski and Brändas (1971), and Longman (1971) (who gives a computational procedure). Here, we shall content ourselves with the simplest approximation (see Mavroyannis and Stephen (1962) equation (19)):

$$\alpha(\hbar\omega) \simeq \frac{a}{1-(b\hbar\omega)^2} \quad \text{or} \quad \alpha(i\xi) \simeq \frac{a}{1+(b\xi)^2} \tag{2.314}$$

Two ways of choosing the coefficients a and b will be mentioned (Claverie (1973), Ch. III, Section 4.F.3) below, and in both cases it is required that $\alpha(0) = \bar{\alpha}$ (the mean polarizability) is correctly reproduced, which means that

$$a = \bar{\alpha} \quad \text{hence} \quad \alpha(i\xi) \simeq \frac{\bar{\alpha}}{1+(b\xi)^2} \tag{2.315}$$

Tang (1969), Kramer and Herschbach (1970) studied thoroughly the approximate representations of the type (2.313) for $\alpha(i\xi)$. They examined (as we do here) several possible choices of the parameter b, and payed special attention to the bounding properties of this approximate representation.

Inserting the approximation (2.315) for $\alpha_u^{(1)}$ and $\alpha_v^{(2)}$ in (2.313), we evaluate readily the integral by using (2.299), thus getting:

$$E_{\text{disp}} \simeq -\frac{3}{2} \sum_u^{(1)} \sum_v^{(2)} \frac{1}{(R_{uv})^6} \frac{\alpha_u^{(1)} \alpha_v^{(2)}}{b_u^{(1)} + b_v^{(2)}} \tag{2.316}$$

This result is very similar to the formula (2.294) obtained by the 'mean energy' approximation, since $\bar{U}^{(1)}\bar{U}^{(2)}/(\bar{U}^{(1)} + \bar{U}^{(2)}) = 1/(1/\bar{U}^{(1)} + 1/\bar{U}^{(2)})$; thus $b_u^{(1)}$ and $b_v^{(2)}$ replace respectively $1/\bar{U}^{(1)}$ and $1/\bar{U}^{(2)}$. But, as announced previously, we now have *different* constants $b_u^{(1)}$ and $b_v^{(2)}$ corresponding to every bond (instead of a single constant $1/\bar{U}^{(s)}$ for all bonds of the molecule (s)).

We now examine two possible recipes for choosing these coefficients $b = 1/\bar{U}$ (since $b = 1/\bar{U}$ has the dimension of the inverse of an energy, it could be named 'inverse effective energy').

a. Mavroyannis–Stephen (1962) *approximation.* It is required that the approximation (2.314) gives the correct values in the limits $\xi = 0$ and $\xi \to +\infty$. These two conditions lead respectively to:

$$a = \alpha(0) = \bar{\alpha} \tag{2.317}$$

$$a/b^2 = n \quad \text{hence} \quad b = \sqrt{\bar{\alpha}/n} \tag{2.318}$$

where n denotes the number of electrons of the system. (2.318) is obtained as follows: we have $\bar{\alpha} = (A_{11} + A_{22} + A_{33})/3$; then (2.298) gives:

$$\bar{\alpha}(i\xi) = \frac{2}{3} \sum_I{}' \frac{\Delta E_I |\mu_{0I}|^2}{(\Delta E_I)^2 + \xi^2} = \sum_I{}' \frac{f_I}{(\Delta E_I)^2 + \xi^2} \tag{2.319}$$

where

$$f_I = \tfrac{2}{3} |\vec{\mu}_{0I}|^2 \Delta E_I \tag{2.320}$$

is the so-called oscillator strength (in atomic units) of the transition $(0I)$, and we have

$$\sum_I{}' f_I = n \tag{2.321}$$

according to the Thomas–Reiche–Kuhn sum rule (Kauzmann, 1957, Ch. 16; Dalgarno and Lynn, 1957; Slater, 1960, p. 157 and Appendix 11; Dalgarno, 1962; Hirschfelder *et al.*, 1964; Dalgarno and Davison, 1966, Section III.B; Gordon, 1968a). If we apply the formula (2.316) with the choice (2.318) of the b's to the dispersion interaction between atoms, we get

$$E_{\text{disp}} \simeq -\frac{3}{2} \frac{1}{R^6} \frac{\alpha^{(1)} \alpha^{(2)}}{(\alpha^{(1)}/n^{(1)})^{1/2} + (\alpha^{(2)}/n^{(2)})^{1/2}} \tag{2.322}$$

which is formally identical with the Slater–Kirkwood (1931) formula obtained by a variational procedure, but in their case $n^{(1)}$ and $n^{(2)}$ denoted the numbers of electrons pertaining to the valence shell only (instead of the total numbers of electrons). Now, Langhoff and Karplus (1970b, Section II.C) have shown that the choice (2.317, 2.318) corresponds to using for $\alpha(i\xi)$ the Padé approximant $\mathscr{P}(1, 0)$, and this approximant is an upper bound, so that the resulting formula (2.322) is an upper bound to the exact dispersion term

$$-\tfrac{3}{2}(1/R^6) \int_0^{+\infty} \alpha^{(1)}(i\xi) \alpha^{(2)}(i\xi) \, d\xi$$

indeed, $\alpha(i\xi)$ is positive for every ξ according to the definition (2.319, 2.320), so that the replacement of $\alpha^{(1)}(i\xi)$ and $\alpha^{(2)}(i\xi)$ by upper bounds generates an upper bound of the integral of their product. Actually, the use of the total number of electrons for $n^{(1)}$ and $n^{(2)}$ would give grossly overestimated values of the dispersion term $-C_6/R^6$ between rare gas atoms, for which accurate enough values of the coefficient C_6 are available (e.g. see Tang (1970) and also Starkschall and Gordon (1971)). The use of the Slater–Kirkwood recipe ($n^{(l)}$ = electron population of the valence shell of the atom l) would give a correct order of magnitude. More precisely, Kim and Gordon (1974d, Table II) find that this recipe ($n = 2$ for He, $n = 8$ for Ne, Ar, Kr, Xe) still slightly overestimates the *theoretical* C_6 for light atoms (He, Ne). Nevertheless Pitzer (1959, Table II) and Salem (1960, Table 2) found that a slightly different recipe ($n=2$ for He, $n=6$ for Ne, Ar, Kr, Xe) underestimated their C_6 values deduced from *experimental* potentials, and this underestimation persists for Ar, Kr and Xe when using the previous recipe; the reason, of course, is that the experimental values are deduced from Lennard-Jones potentials fitted essentially for the region of the Van der Waals minimum, where the dispersion terms beyond R^{-6} (namely R^{-8}, R^{-10}) are no more negligible (see the formula (2.121)) for He . . . He at the beginning of Section III; see also Dalgarno and Davison (1966, Section VI.B); Dalgarno (1967); Fukui and Yamabe (1968); McQuarrie *et al.* (1969); Starkschall and Gordon (1972a,b)). As a consequence, the '*effective*' C_6 coefficient of the Lennard-Jones potential, which must simulate the total dispersion interaction (and not only its dipole–dipole R^{-6} part), must be larger than the *theoretical* C_6 coefficient. Now, from the practical point of view, if we use only an R^{-6} term, it is better to use an 'effective' coefficient in order to have an evaluation of the complete dispersion term. Thus, the Slater–Kirkwood recipe (as used by Kim and Gordon (1974d), namely taking $n=$ number of electrons in the valence shell) appears convenient at least up to the neon atom (for heavier atoms, suitably increased 'effective' numbers should be used: e.g., twice the values of N_{eff} given by Kim and Gordon (1974d)).

We may now propose a similar recipe for the bond parameters $b_u^{(1)}$, $b_v^{(2)}$ involved in the formula (2.316): we use (2.318) with $n_u^{(1)}$ (respectively $n_v^{(2)}$) equal to the number of 'valence electrons' pertaining to the bond u (respectively v); if there are lone pair(s) on some atom, we may share the corresponding number $N_{\text{l.p.}}$ of electrons equally between the ν bonds which start from this atom (thus each of these bonds receives a supplementary effective number $N_{\text{l.p.}}/\nu$). As pointed out previously for rare gas atoms, this recipe would hold up to the neon atom (first and second-row atoms), but for heavier atoms it would be appropriate to add supplementary effective numbers corresponding to the inner shells. Thus, by using the accurate results available for rare gas atoms as a basis for reasonable extrapolation and by checking against experimental results, it should be a feasible task to elaborate a set of 'effective' numbers of electrons for all atoms and bonds, which would allow an evaluation of the dispersion energy between molecules much more satisfactory than the evaluation based upon the use of a single 'mean excitation energy' for each molecule.

b. Salem–Tang–Karplus approximation. Tang and Karplus (1968) remark that $1/(\Delta E+\hbar\omega)+1/(\Delta E-\hbar\omega)=2\Delta E/[(\Delta E)^2-(\hbar\omega)^2]$, so that the definition (2.298) of $\bar{\bar{A}}(\hbar\omega)$ may be rewritten:

$$\bar{\bar{A}}(\hbar\omega)=\bar{\bar{A}}^{+}(\hbar\omega)+\bar{\bar{A}}^{+}(-\hbar\omega) \quad \text{with} \quad \bar{\bar{A}}^{+}(\hbar\omega)=\sum_I{}' \frac{\vec{\mu}_{0I}\otimes\vec{\mu}_{I0}}{\Delta E_I+\hbar\omega} \tag{2.323}$$

hence, for the mean polarizability

$$\alpha(\hbar\omega)=\alpha^+(\hbar\omega)+\alpha^+(-\hbar\omega) \quad \text{with} \quad \alpha^+(\hbar\omega)=\frac{1}{3}\sum_I{}' \frac{|\vec{\mu}_{0I}|^2}{\Delta E_I+\hbar\omega} \tag{2.324}$$

Then Tang and Karplus represent $\alpha^+(\hbar\omega)$ by Padé approximants, and the simplest one is

$$\alpha^+(\hbar\omega)\simeq\frac{a'}{1+b'\hbar\omega} \tag{2.325}$$

where we may determine a' and b' by requiring that $\alpha^+(\hbar\omega)$ is correctly reproduced for $\omega=0$ and $\omega\to\infty$. Now, $a'=\alpha^+(0)$ and, according to (2.324), $\bar{\alpha}=\alpha(0)=2\alpha^+(0)$, hence:

$$a'=\tfrac{1}{2}\bar{\alpha} \tag{2.326}$$

On the other hand, as $\omega\to+\infty$ (2.324) gives (the subscript $j=1, 2, 3$ will denote the coordinate axes x, y, z):

$$\begin{aligned}\hbar\omega\alpha^+(\hbar\omega)\to\tfrac{1}{3}\sum_I{}' |\vec{\mu}_{0I}|^2 &= \tfrac{1}{3}\sum_{j=1}^{3}\sum_I{}' \langle 0|\boldsymbol{\mu}_j|I\rangle\langle I|\boldsymbol{\mu}_j|0\rangle \\ &= \tfrac{1}{3}\sum_{j=1}^{3} [\langle 0|\boldsymbol{\mu}_j^2|0\rangle-\langle 0|\boldsymbol{\mu}_j|0\rangle^2]\end{aligned}$$

according to the closure relation

$$\sum_I{}' |I\rangle\langle I| = \sum_I |I\rangle\langle I| - |0\rangle\langle 0| = 1-|0\rangle\langle 0|$$

Now, denoting for brevity $\langle 0|\,\mathbf{A}\,|0\rangle=\langle\mathbf{A}\rangle$, we have

$$\langle 0|\,\boldsymbol{\mu}_j^2\,|0\rangle-\langle 0|\,\boldsymbol{\mu}_j\,|0\rangle^2=\langle\boldsymbol{\mu}_j^2\rangle-\langle\boldsymbol{\mu}_j\rangle^2=\langle(\boldsymbol{\mu}_j-\langle\boldsymbol{\mu}_j\rangle)^2\rangle=(\Delta\boldsymbol{\mu}_j)^2$$

(we use the symbol Δ for standard deviation), thus finally

$$\hbar\omega\alpha^+(\hbar\omega)\to\tfrac{1}{3}\sum_{j=1}^{3}(\Delta\boldsymbol{\mu}_j)^2=\tfrac{1}{3}(\Delta\vec{\boldsymbol{\mu}})^2$$

Getting the same behaviour with (2.325) therefore requires:

$$\frac{a'}{b'}=\tfrac{1}{3}(\Delta\vec{\mu})^2 \quad \text{or} \quad b'=\frac{3}{2}\frac{\bar{\alpha}}{(\Delta\vec{\mu})^2} \tag{2.327}$$

Now, (2.323) gives for $\alpha(\hbar\omega)$ the approximate formula:

$$\alpha(\hbar\omega)\simeq\frac{2a'}{1-(b'\hbar\omega)^2} \quad \text{and} \quad \alpha(i\xi)\simeq\frac{2a'}{1+(b'\xi)^2} \tag{2.328}$$

which is of the type (2.314); actually, the numerator $2a'=\bar{\alpha}$ is the same as in the Mavroyannis–Stephen choice, the only difference being the coefficient b'

given by (2.327) instead of b given by (2.318). Then (2.316) gives, for the case of single-centre systems (atoms or diatomic molecules)

$$E_{\text{disp}} \simeq -\frac{1}{R^6} \frac{\bar{\alpha}^{(1)}\bar{\alpha}^{(2)}}{\bar{\alpha}^{(1)}/(\Delta\vec{\boldsymbol{\mu}}^{(1)})^2 + \bar{\alpha}^{(2)}/(\Delta\vec{\boldsymbol{\mu}}^{(2)})^2} \tag{2.329}$$

which is identical with the formula (12) given by Salem (1960) (indeed, for an atom or any system with zero permanent dipole $\langle\vec{\boldsymbol{\mu}}\rangle = 0$, we have $(\Delta\vec{\boldsymbol{\mu}})^2 = \langle(\vec{\boldsymbol{\mu}} - \langle\vec{\boldsymbol{\mu}}\rangle)^2\rangle = \langle\vec{\boldsymbol{\mu}}^2\rangle$): hence the name given to the present choice. It must be emphasized that Salem (1960) obtained this formula in a quite different way, that we called the 'second procedure' in the framework of the mean energy method (Section (2) above), namely to eliminate $\bar{U}^{(1)}$ and $\bar{U}^{(2)}$ between the equations (2.277) and (2.278): indeed, for a single-centre system, (2.278) gives

$$\bar{\alpha} \simeq \frac{2}{3\bar{U}} \sum_I{}' |\vec{\mu}_{0I}|^2 = \tfrac{2}{3}(\Delta\vec{\boldsymbol{\mu}})^2/\bar{U}$$

hence

$$\bar{U} = \tfrac{2}{3}(\Delta\vec{\boldsymbol{\mu}})^2/\bar{\alpha} = 1/b',$$

so that the formula (2.294) becomes identical with (2.329) for this choice of $\bar{U}$. It is interesting that *this Salem's* (1960) *procedure turns out to be equivalent to the simplest rational approximation of* $\alpha^+(\hbar\omega)$ *in the framework of the Tang–Karplus* (1968) *method for representing the dynamic polarizability.*

Kramer and Herschbach (1970) also pointed out that the approximation $\alpha^a(i\xi)$ of $\alpha(i\xi)$ defined by (2.326, 2.327, 2.328) leads to Salem's formula, and that this approximate $\alpha^a(i\xi)$ could be defined directly (i.e. without using the splitting (2.324) $\alpha(\hbar\omega) = \alpha^+(\hbar\omega) + \alpha^+(-\hbar\omega)$) in the following way: assuming an approximate α^a of the type (2.315), the coefficient b is fixed by the requirement that:

$$\int_0^{+\infty} \alpha^a(i\xi)\,\mathrm{d}\xi = \int_0^{+\infty} \alpha(i\xi)\,\mathrm{d}\xi \tag{2.330}$$

indeed, with α^a given by (2.315), the left-hand side is equal to $\frac{1}{2}\pi(\bar{\alpha}/b)$; on the other hand, according to (2.319, 2.320), the right-hand side is equal to

$$\sum_I{}' \frac{\pi}{2} f_I/\Delta E_I = \frac{\pi}{2} S(-1)$$

according to the definition (2.331) below; hence $\bar{\alpha}/b = S(-1)$ or $b = \bar{\alpha}/S(-1) = b'$ according to (2.335) below.

Following Bell's (1965, 1966) procedure, Alexander (1970a) evaluated the coefficients C_6, C_8, C_{10}, ... (of the multipole expansion of the dispersion term) directly in terms of the lowest transition energy $\Delta E_1 = E_1 - E_0$ and of the *oscillator strength sums* (of each subsystem):

$$S(k) = \sum_I{}' f_I(\Delta E_I)^k \tag{2.331}$$

where the oscillator strength f_I is defined by (2.319, 2.320). Moreover, Alexander was able to derive upper and lower for these coefficients and for the 'dipole–dipole' coefficient C_6, the first term of his formula (21) gives the following upper bound:

$$C_6 \leqslant \tfrac{3}{8}[S^{(1)}(-2)S^{(2)}(-1) + S^{(1)}(-1)S^{(2)}(-2)] \tag{2.332}$$

which reduces, for two identical systems, to:

$$C_6 \leqslant \tfrac{3}{4}S(-2)S(-1) \tag{2.333}$$

Now $S(-2)$ and $S(-1)$ are given by well-known sum rules (for example, see Dalgarno and Davison, 1966, Section III.B; Dalgarno and Lynn, 1957):

$$S(-1) = \tfrac{2}{3}\langle(\vec{\boldsymbol{\mu}} - \langle\boldsymbol{\mu}\rangle)^2\rangle = \tfrac{2}{3}(\Delta\vec{\boldsymbol{\mu}})^2 \qquad S(-2) = \bar{\alpha} \tag{2.334}$$

then the coefficient b' defined by (2.326, 2.327) may be written

$$b' = \bar{\alpha}/S(-1) = S(-2)/S(-1) \tag{2.335}$$

Now, the upper bounds (2.332, 2.333) may be written:

$$C_6 \leqslant \tfrac{3}{8}[\bar{\alpha}^{(1)}S^{(2)}(-1) + \bar{\alpha}^{(2)}S^{(1)}(-1)] = \tfrac{1}{4}[\bar{\alpha}^{(1)}(\Delta\vec{\boldsymbol{\mu}}^{(2)})^2 + \bar{\alpha}^{(2)}(\Delta\vec{\boldsymbol{\mu}}^{(1)})^2] \tag{2.336}$$

and, for identical systems:

$$C_6 \leqslant \tfrac{3}{4}\bar{\alpha}S(-1) = \tfrac{1}{2}\bar{\alpha}(\Delta\vec{\boldsymbol{\mu}})^2 \tag{2.337}$$

Now, according to (2.329), *the coefficient C_6 (of $-1/R^6$) in Salem's formula for two identical systems reduces exactly to the upper bound* (2.337) *as pointed out by Alexander* (1970a, Section III, text between formulas (24) and (25)). Kramer and Herschbach (1970) also recognized that the right hand side of (2.333) is an upper bound to C_6 for *identical* systems: indeed, they introduce, through their equation (21),

$$\omega_1 = S(-1)/S(-2) = S(-1)/\alpha = 1/b'$$

and their formula (22) then gives:

$$C_6 < \tfrac{3}{4}\alpha^2\omega_1 = \tfrac{3}{4}\alpha^2(S(-1)/\alpha) = \tfrac{3}{4}\alpha S(-1)$$

(as mentioned previously, Kramer and Herschbach's analysis is based upon the approximation of the dynamic polarizability by a formula of the type (2.315).) *This is no more true when the systems are different*, and it is easily seen that Salem's C_6 coefficient is then actually *smaller* than Alexander's upper bound (2.336):

$$C_6(\text{Salem}) = \bar{\alpha}^{(1)}\bar{\alpha}^{(2)}/[\bar{\alpha}^{(1)}/(\Delta\vec{\boldsymbol{\mu}}^{(1)})^2 + \bar{\alpha}^{(2)}/(\Delta\vec{\boldsymbol{\mu}}^{(2)})^2] = 1/(B+C)$$

where we put

$$B = 1/[\bar{\alpha}^{(2)}(\Delta\vec{\boldsymbol{\mu}}^{(1)})^2] \quad \text{and} \quad C = 1/[\bar{\alpha}^{(1)}(\Delta\vec{\boldsymbol{\mu}}^{(2)})^2]$$

while the upper bound $C_6^{\text{up}}(A)$ (2.332) then writes $C_6^{\text{up}}(A) = (1/B + 1/C)/4 = (B+C)/4BC$; hence

$$\begin{aligned} C_6^{\text{up}}(A) - C_6(\text{Salem}) &= [1/(B+C)][(B+C)^2/4BC - 1] \\ &= [1/(B+C)](B-C)^2/4BC \geqslant 0 \end{aligned}$$

It may be mentioned that Musher (1963) obtained for C_6 Alexander's expression (2.332) (this is easily seen by specializing Musher's formula (17) for the case of spherical systems:

$$\alpha_{xx}(s) = \alpha_{yy}(s) = \alpha_{zz}(s) = \alpha^{(s)} = S^{(s)}(-2),$$
$$\langle xx\rangle(s) = \langle yy\rangle(s) = \langle zz\rangle(s) = \tfrac{1}{3}(\Delta\vec{\boldsymbol{\mu}}^{(s)})^2 = \tfrac{1}{2}S^{(s)}(-1)$$

Musher evaluated directly the basic perturbation expression (2.265) by using the method proposed by Dalgarno and Lewis (1955) and Schwartz (1959) in an approximate way, so as to make this method tractable. Musher noticed that he recovered Salem's (1960) formula for two like systems, but he could not recognize that his expression gave an upper bound to the exact C_6 in all cases, and was therefore less interesting than Salem's one for non-identical systems, as it appears from the numerical results presented below.

The coefficient C_6 (Salem) and the upper bound $C_6^{\text{up}}(A)$ may indeed be readily calculated for all pairs of rare gas atoms (and also the hydrogen atom) by using the values of $S(-1)$ and $S(-2)$ taken from Kramer (1970) which are displayed in Table 2.3. The values obtained for C_6 (Salem) and $C_6^{\text{up}}(A)$ are displayed in Table 2.4, together with the values recommended by Dalgarno (1967) and upper and lower bounds proposed by Gordon (1968b), Tang (1970) and by Starkschall and Gordon (1971); it must be emphasized that, in contrast with Alexander's bounds, these latter bounds were derived by using the oscillator strength sums $S(2k)$ with *even* indexes only. We draw two main conclusions from these results:

(1) Alexander's upper bound (2.332) is a very good one. Contrary to that of Mavroyannis–Stephen (namely, according to (2.317, 2.318, 2.322):

$$C_6(\text{MS}) = \frac{3}{2}\frac{\bar{\alpha}^{(1)}\bar{\alpha}^{(2)}}{(b^{(1)} + b^{(2)})}, \quad \text{with} \quad b^{(s)} = \sqrt{\frac{\bar{\alpha}^{(s)}}{n^{(s)}}}$$

Table 2.3. The oscillator strength sums $S(-1)$ and $S(-2)$ for the hydrogen and rare gas atoms (in atomic units). These values were borrowed from Table I of Kramer (1970) (the value of $\bar{\alpha}$ for the Helium atom is corrected (1·385 instead of 1·375) since there is probably an error in Kramer's table, as shown by comparison with other tables of the values of $S(-2)$: Tang (1970) Table II, Starkschall and Gordon (1971) Table III)

	H	He	Ne	Ar	Kr	Xe
$\frac{2}{3}(\Delta\vec{\boldsymbol{\mu}})^2 = S(-1)$	2·0	1·5	3·125	7·595	9·89	13·17
$\bar{\alpha} = S(-2)$	4·5	1·385	2·668	11·09	16·77	27·09

Table 2.4. The coefficient C_6 of the dispersion term $(-C_6/R^6)$ for the hydrogen and rare gas atoms (all values in atomic units)

				Upper and lower bounds			
			Dalgarno recommended	Gordon, Tang[d]		Starkschall and Gordon[e]	
	C_6 Salem[a]	C_6^{up} (A)[b]	value[c]	lower	upper	lower	upper
H H	6·75	6·75	6·50	6·46	6·54		
He	2·952	3·57	2·83	2·79	2·85		
Ne	5·802	7·274	5·6	5·53	5·83		
Ar	20·176	21·134	20	19·6	20·8		
Kr	28·689	29·267	29	28·0	30·3		
Xe	42·456	42·542	42	41·1	45·7		
He He	1·558	1·558	1·47	1·43	1·48	1·4614	1·4622
Ne	3·119	3·123	3·0	2·88	3·22	2·82	3·24
Ar	9·666	10·183	9·6	9·29	10·51	9·42	9·78
Kr	13·303	14·570	13	12·9	15·4	13·18	13·56
Xe	18·884	22·078	19	18·3	22·9	19·19	20·93
Ne Ne	6·253	6·253	6·3	5·82	7·28	5·47	6·49
Ar	19·180	20·595	20	18·6	22·9	18·1	21·3
Kr	26·325	29·547	27	25·7	33·7	25·3	29·3
Xe	37·247	44·923	38	36·3	50·4	37·0	45·6
Ar Ar	63·171	63·171	65	62·6	75·2	62·7	66·9
Kr	88·398	88·893	91	88	110	88·5	93·5
Xe	128·13	131·92	130	127	165	130	144
Kr Kr	124·39	124·39	130	124	162	125	131
Xe	181·59	183·29	190	180	244	185	203
Xe Xe	267·58	267·58	270	263	366	275	315

[a] Coefficient of $-1/R^6$ in formula (2.329) (Salem, 1960)
[b] Formula (2.336) (Alexander, 1970a)
[c] Dalgarno (1967) Table V
[d] Gordon (1968b) Table III, Tang (1970) Table I
[e] Starkschall and Gordon (1971) Table I, column 'polarizability data'

$n^{(s)}$ being the total number of electrons of the subsystem s), Alexander's bound does not become too large for heavier atoms: for example, the values of C_6^{up} (MS) for like pairs of atoms (calculated with the values $\bar{a}$ of Table 2.3) are (in atomic units): 7·15 (H . . . H), 1·728 (He . . . He), 10·336 (Ne . . . Ne), 117·154 (Ar . . . Ar), 309·04 (Kr . . . Kr), 777·09 (Xe . . . Xe). It is still more remarkable that C_6^{up} (A), despite its simplicity (it requires the knowledge of only two oscillator strength sums $S(-1)$ and $S-2$)) appears to be, in a number of cases, a better upper bound than those derived (in a more cumbersome way) by Gordon (1968b), Tang (1970), Starkschall and Gordon (1971). In two cases (Ar . . . Xe and Xe . . . Xe), Alexander's 'upper bound' is even slightly smaller than Starkschall and Gordon's 'lower bound': this contradiction simply means, of course, that one of these 'bounds' or both, is (are) not accurate, owing to some lack of accuracy in the data (e.g. the oscillator strength sums) used for evaluating these bounds.

(2) Salem's formula also gives remarkably good results for C_6. The discrepancy with respect to Dalgarno's (1967) recommended values remains smaller than 5 per cent (except for He ... He, where it amounts to 6 per cent). Moreover, these values C_6 (Salem) always lie between the upper and lower bounds of Tang (1970) and Gordon (1968b) (and in most cases they also lie between the upper and lower bounds of Starkschall and Gordon (1971), which seem more refined, but are perhaps not rigorous bounds for the heavier atoms (see above). It is quite remarkable that such satisfactory results can be achieved by using only two properties of every atom, namely the polarizability $\bar{\alpha}$ and the dispersion of the dipole moment

$$(\Delta\vec{\boldsymbol{\mu}})^2 = \langle(\vec{\boldsymbol{\mu}} - \langle\vec{\boldsymbol{\mu}}\rangle)^2\rangle$$
$$= \langle(\vec{\boldsymbol{\mu}})^2\rangle$$

for atoms since then $\langle\vec{\boldsymbol{\mu}}\rangle = 0$. But since the further dispersion terms $-C_8/R^8$, $-C_{10}/R^{10}$ are not negligible in the region of the minimum (as recalled previously in connection with the Mavroyannis–Stephen choice), it would be imperative to complete Salem's R^{-6} term by some evaluation of these further terms, for example, by using the Starkschall and Gordon (1972b) procedure.

c. Use of bond–bond interaction terms for large molecules. It could be possible to evaluate the sums $\sum_{w'} |\vec{\mu}_{u-w'}|^2$ (or even $\sum_{w'} \mu_{i,u-w'}\mu_{j,u-w'}$) which correspond to the various bonds u and thus to obtain bond parameters b'_u which could be used for the interaction between arbitrary molecules (formula (2.316); different parameters could even be introduced, as Salem (1960) did, corresponding to the different principal values of every bond polarizability tensor (see also Claverie (1973) Ch. III, Section 4.F.3.b.β pp. 278–280). In view of the success of this 'Salem–Tang–Karplus' choice of the dynamic polarizability in the case of atom–atom interactions, the evaluation of such bond parameters appears as a promising line of research.

In Table 2.5, we give the three different coefficients $C(L, L)$, $C(L, T)$ and $C(T, T)$ for the anisotropic interaction between two hydrogen molecules ($C(X, Y)$, where $X, Y = L$ or T is defined by:

$$C(X, Y) = \frac{1}{2\pi}\int_0^{+\infty} \alpha_X(i\xi)\alpha_Y(i\xi)\, d\xi$$

see the definition by Victor and Dalgarno (1970), and compare the formulas (2.309, 2.341, 2.342, 2.351). The two choices for the b_L and b_T coefficients are presented, together with the tentatively accurate values of Victor and Dalgarno (1970) and Langhoff *et al.* (1971). The Mavroyannis–Stephen choice gives larger values, as it could be expected from the results for rare gas atoms, but there is no really significant overestimation for such small numbers of electrons, and both choices clearly give satisfactory results when compared with the more refined values.

Table 2.5. Comparison between several estimates of the coefficients of the dispersion energy for the $H_2 \ldots H_2$ interaction (all values are expressed in atomic units)

	α_L[a]	b_L[a]	α_T[a]	b_T[a]	$C(L, L)$	$C(L, T)$	$C(T, T)$
Choice (1) (Mavroyannis–Stephen) (with $n = 2$)	$\alpha_L = 6{\cdot}4$	$(\alpha_L/n)^{1/2} = 1{\cdot}79$	$\alpha_T = 4{\cdot}7$	$(\alpha_T/n)^{1/2} = 1{\cdot}533$	2·88	2·26	1·8
Choice (2) (Salem)	$\alpha_L = 6{\cdot}4$	$\alpha_L/(2\langle\boldsymbol{\mu}_L^2\rangle) = 2{\cdot}05$	$\alpha_T = 4{\cdot}7$	$\alpha_T/(2\langle\boldsymbol{\mu}_T^2\rangle) = 1{\cdot}546$	2·50	2·09	1·78
Langhoff, Gordon and Karplus[b]					2·732	2·063	1·564
Victor and Dalgarno[c]					2·984	2·227	1·670

a_X and b_X ($X = L$ or T) correspond to a representation of the dynamic polarizability $\alpha_X(i\xi) = a_X/[1+(b_X\xi)^2]$ of the type (2.314). Then the coefficient $C(X, Y)$ defined in the text is given by: $C(X, Y) = (1/4)a_Xa_Y/(b_X + b_Y)$ (see formula (9.10) in reference (b) and formula (1.2) in reference (c))

[a] numerical values of α_L, α_T, $\boldsymbol{\mu}_L^2$, $\boldsymbol{\mu}_T^2$ from Salem (1960)

[b] R. W. Langhoff, R. G. Gordon and M. Karplus, *J. Chem. Phys.*, **55,** 2126 (1971), Table III

[c] G. A. Victor and A. Dalgarno, *J. Chem. Phys.*, **53,** 1316 (1970)

Let us now consider more explicitly the general case of anisotropic polarizabilities. We start again from (2.309) and introduce the explicit representation of every bond polarizability tensor $\bar{\bar{A}}^{(s)}$ ($s = 1$ or 2; we drop the subscripts u and v for simplicity) in terms of its eigenvalues $\alpha_\kappa^{(s)}$ (principal polarizabilities) and its unit eigenvectors $\vec{e}_\kappa^{(s)}$ (principal directions) ($\kappa = 1, 2, 3$):

$$\bar{\bar{A}}^{(s)} = \sum_{\kappa=1}^{3} \alpha_\kappa^{(s)}(\vec{e}_\kappa^{(s)} \otimes \vec{e}_\kappa^{(s)}) \tag{2.338}$$

(see (2.501) in Appendix E).

We shall now assume that the eigenvectors $\vec{e}_\kappa$ do *not* depend on the frequency ω (otherwise, the use of (2.338) would have no special interest and we would be obliged to use directly (2.309) with all matrix elements of $\bar{\bar{A}}_u^{(1)}$ and $\bar{\bar{A}}_v^{(2)}$ depending on ξ). This assumption is reasonable, insofar as the symmetry properties of the bond are often sufficient to determine the eigenvectors common to all elementary dyadic products $\vec{\mu}_{u-w'} \otimes \vec{\mu}_{u-w'}$ (let us recall that the eigenvectors of a dyadic product $\vec{\mu} \otimes \vec{\mu}$ are $\vec{\mu}$ itself and all vectors orthogonal to $\vec{\mu}$): thus, for a diatomic molecules, all transitions moments $\vec{\mu}_{0I}$ are either colinear or orthogonal to the bond axis; now, if all dyadic products $\vec{\mu}_{u-w'} \otimes \vec{\mu}_{u-w'}$ have a common set of eigenvectors, $\bar{\bar{A}}_u(\hbar\omega)$ clearly has this set of eigenvectors whatever ω.

Then, the traces $\mathrm{Tr}\,[\bar{\bar{T}}\bar{\bar{A}}^{(1)}\bar{\bar{T}}\bar{\bar{A}}^{(2)}]$ in (2.309) become:

$$\begin{aligned}\mathrm{Tr}\,[\bar{\bar{T}}\bar{\bar{A}}^{(1)}(i\xi)\bar{\bar{T}}\bar{\bar{A}}^{(2)}(i\xi)] = {} & \\ \sum_{\kappa=1}^{3}\sum_{\lambda=1}^{3} & \alpha_\kappa^{(1)}(i\xi)\alpha_\lambda^{(2)}(i\xi)\,\mathrm{Tr}\,[\bar{\bar{T}}(\vec{e}_\kappa^{(1)} \otimes \vec{e}_\kappa^{(1)})\bar{\bar{T}}(\vec{e}_\lambda^{(2)} \otimes \vec{e}_\lambda^{(2)})] \\ & = \sum_{\kappa=1}^{3}\sum_{\lambda=1}^{3} \alpha_\kappa^{(1)}(i\xi)\alpha_\lambda^{(2)}(i\xi)(\vec{e}_\kappa^{(1)} \,.\, \bar{\bar{T}} \,.\, \vec{e}_\lambda^{(2)})^2\end{aligned} \tag{2.339}$$

where the trace involving the dyadic products was expressed according to (2.514) of Appendix E. Now, upon inserting (2.339), (2.309) becomes:

$$E_{\text{disp}} \simeq -\frac{1}{2\pi} \sum_u^{(1)} \sum_v^{(2)} \frac{1}{(R_{uv})^6} \sum_{\kappa=1}^{3} \sum_{\lambda=1}^{3} (\vec{e}_{u,\kappa}^{(1)} \cdot \bar{\bar{T}}_{uv} \cdot \vec{e}_{v,\lambda}^{(2)})^2 \times \int_0^{+\infty} \alpha_{u,\kappa}^{(1)}(i\xi)\alpha_{v,\lambda}^{(2)}(i\xi)\,\mathrm{d}\xi \tag{2.340}$$

Introducing now for every principal dynamic polarizability an approximation of the type (2.315):

$$\alpha_{u,\kappa}^{(1)}(i\xi) = \frac{\alpha_{u,\kappa}^{(1)}}{1+(b_{u,\kappa}^{(1)}\xi)^2} \qquad \alpha_{v,\lambda}^{(2)}(i\xi) = \frac{\alpha_{v,\lambda}^{(2)}}{1+(b_{v,\lambda}^{(2)}\xi)^2} \tag{2.341}$$

where $\alpha_{u,\kappa}^{(1)}$ and $\alpha_{v,\lambda}^{(2)}$ denote the principal static polarizabilities of the bonds u and v, we immediately get from (2.299):

$$\int_0^{+\infty} \alpha_{u,\kappa}^{(1)}(i\xi)\alpha_{v,\lambda}^{(2)}(i\xi)\,\mathrm{d}\xi = \frac{\pi}{2}\frac{\alpha_{u,\kappa}^{(1)}\alpha_{v,\lambda}^{(2)}}{b_{u,\kappa}^{(1)}+b_{v,\lambda}^{(2)}} \tag{2.342}$$

Hence, upon inserting (2.342) in (2.340):

$$E_{\text{disp}} \simeq -\frac{1}{4} \sum_u^{(1)} \sum_v^{(2)} \frac{1}{(R_{uv})^6} \sum_{\kappa=1}^{3} \sum_{\lambda=1}^{3} (\vec{e}_{u,\kappa}^{(1)} \cdot \bar{\bar{T}} \cdot \vec{e}_{v,\lambda}^{(2)})^2 \frac{\alpha_{u,\kappa}^{(1)}\alpha_{v,\lambda}^{(2)}}{b_{u,\kappa}^{(1)}+b_{v,\lambda}^{(2)}} \tag{2.343}$$

where, according to the definition (2.274) of $\bar{\bar{T}}$, we have

$$\vec{e}_{u,\kappa}^{(1)} \cdot \bar{\bar{T}} \cdot \vec{e}_{v,\lambda}^{(2)} = 3(\vec{e}_{u,\kappa}^{(1)} \cdot \vec{r})(\vec{e}_{v,\lambda}^{(2)} \cdot \vec{r}) - (\vec{e}_{u,\kappa}^{(1)} \cdot \vec{e}_{v,\lambda}^{(2)}) \tag{2.344}$$

We shall finally, as we did in the case of the 'mean excitation energy' method (end of Section 2 above), give the special formula for the practically important case of axially symmetric bond polarizabilities. We could use (2.343) by inserting $\alpha_{u,1}^{(1)} = \alpha_{v,L}^{(1)}$, $\alpha_{u,2}^{(1)} = \alpha_{u,3}^{(1)} = \alpha_{u,T}^{(1)}$ and similarly for the bond v in molecule (2), but it is simpler to start again from (2.309) and to use for the various traces $\mathrm{Tr}\,[\bar{\bar{T}}\bar{\bar{A}}_u^{(1)}\bar{\bar{T}}\bar{\bar{A}}_v^{(2)}]$ the formula (2.290) previously derived for the case of axially symmetric tensors $\bar{\bar{A}}_u^{(1)}$ and $\bar{\bar{A}}_v^{(2)}$ (in the next formulas we drop the subscripts u and v for simplicity):

$$\begin{aligned} \mathrm{Tr}\,[\bar{\bar{T}}A^{(1)}(i\xi)\bar{\bar{T}}\bar{\bar{A}}^{(2)}(i\xi)] &= 6\alpha_{\mathrm{T}}^{(1)}(i\xi)\alpha_{\mathrm{T}}^{(2)}(i\xi) \\ &+ \delta^{(1)}(i\xi)\alpha_{\mathrm{T}}^{(2)}(i\xi)[3(\vec{r}\cdot\vec{e}_{\mathrm{L}}^{(1)})^2+1] + \alpha_{\mathrm{T}}^{(1)}(i\xi)\delta^{(2)}(i\xi)[3(\vec{r}\cdot\vec{e}_{\mathrm{L}}^{(2)})^2+1] \\ &+ \delta^{(1)}(i\xi)\delta^{(2)}(i\xi)[3(\vec{r}\cdot\vec{e}_{\mathrm{L}}^{(1)})(\vec{r}\cdot\vec{e}_{\mathrm{L}}^{(2)}) - (\vec{e}_{\mathrm{L}}^{(1)}\cdot\vec{e}_{\mathrm{L}}^{(2)})]^2 \end{aligned} \tag{2.345}$$

Now, the integration over ξ of (2.345) concerns only the principal polarizabilities: $\alpha_{\mathrm{L}}(i\xi)$, $\alpha_{\mathrm{T}}(i\xi)$ and their difference $\delta = \alpha_{\mathrm{L}} - \alpha_{\mathrm{T}}$. Therefore, using the approximate representation (2.315) for $\alpha_{\mathrm{L}}(i\xi)$ and $\alpha_{\mathrm{T}}(i\xi)$ (see

(2.341)), we have the four integrals:

$$\int_0^{+\infty} \alpha_{\mathrm{T}}^{(1)}(i\xi)\alpha_{\mathrm{T}}^{(2)}(i\xi)\,\mathrm{d}\xi = \frac{\alpha_{\mathrm{T}}^{(1)}\alpha_{\mathrm{T}}^{(2)}}{b_{\mathrm{T}}^{(1)}+b_{\mathrm{T}}^{(2)}} \tag{2.346}$$

$$\int_0^{+\infty} [\alpha_{\mathrm{L}}^{(1)}(i\xi)-\alpha_{\mathrm{T}}^{(1)}(i\xi)]\alpha_{\mathrm{T}}^{(2)}(i\xi)\,\mathrm{d}\xi = \frac{\alpha_{\mathrm{L}}^{(1)}\alpha_{\mathrm{T}}^{(2)}}{b_{\mathrm{L}}^{(1)}+b_{\mathrm{T}}^{(2)}} - \frac{\alpha_{\mathrm{T}}^{(1)}\alpha_{\mathrm{T}}^{(2)}}{b_{\mathrm{T}}^{(1)}+b_{\mathrm{T}}^{(2)}} \tag{2.347}$$

$$\int_0^{+\infty} \alpha_{\mathrm{T}}^{(1)}(i\xi)[\alpha_{\mathrm{L}}^{(2)}(i\xi)-\alpha_{\mathrm{T}}^{(2)}(i\xi)]\,\mathrm{d}\xi = \frac{\alpha_{\mathrm{T}}^{(1)}\alpha_{\mathrm{L}}^{(2)}}{b_{\mathrm{T}}^{(1)}+b_{\mathrm{L}}^{(2)}} - \frac{\alpha_{\mathrm{T}}^{(1)}\alpha_{\mathrm{T}}^{(2)}}{b_{\mathrm{T}}^{(1)}+b_{\mathrm{T}}^{(2)}} \tag{2.348}$$

$$\begin{aligned}\int_0^{+\infty} [\alpha_{\mathrm{L}}^{(1)}(i\xi)-\alpha_{\mathrm{T}}^{(1)}(i\xi)][\alpha_{\mathrm{L}}^{(2)}(i\xi)-\alpha_{\mathrm{T}}^{(2)}(i\xi)]\,\mathrm{d}\xi &= \frac{\alpha_{\mathrm{L}}^{(1)}\alpha_{\mathrm{L}}^{(2)}}{b_{\mathrm{L}}^{(1)}+b_{\mathrm{L}}^{(2)}} - \frac{\alpha_{\mathrm{L}}^{(1)}\alpha_{\mathrm{T}}^{(2)}}{b_{\mathrm{L}}^{(1)}+b_{\mathrm{T}}^{(2)}} \\ &\quad - \frac{\alpha_{\mathrm{T}}^{(1)}\alpha_{\mathrm{L}}^{(2)}}{b_{\mathrm{T}}^{(1)}+b_{\mathrm{L}}^{(2)}} + \frac{\alpha_{\mathrm{T}}^{(1)}\alpha_{\mathrm{T}}^{(2)}}{b_{\mathrm{T}}^{(1)}+b_{\mathrm{T}}^{(2)}}\end{aligned} \tag{2.349}$$

We introduce for brevity the notation:

$$F_{\mathrm{X,Y}} = \frac{\alpha_{\mathrm{X}}^{(1)}\alpha_{\mathrm{Y}}^{(2)}}{b_{\mathrm{X}}^{(1)}+b_{\mathrm{Y}}^{(2)}} \tag{2.350}$$

where X, Y stand for one of the symbols L or T. Using the integrals (2.346, 2.347, 2.348, 2.349), the integral of (2.345) writes:

$$\begin{aligned}\int_0^{+\infty} \mathrm{Tr}\,[\bar{\bar{T}}\bar{\bar{A}}^{(1)}(i\xi)\bar{\bar{T}}\bar{\bar{A}}^{(2)}(i\xi)]\,\mathrm{d}\xi &= 6F_{\mathrm{T,T}} \\ &+[3(\vec{r}\,.\,\vec{e}_{\mathrm{L}}^{(1)})^2+1](F_{\mathrm{L,T}}-F_{\mathrm{T,T}})+[3(\vec{r}\,.\,\vec{e}^{(2)})^2+1](F_{\mathrm{T,L}}-F_{\mathrm{T,T}}) \\ &+[3(\vec{r}\,.\,\vec{e}_{\mathrm{L}}^{(1)})(\vec{r}\,.\,\vec{e}_{\mathrm{L}}^{(2)})-(\vec{e}_{\mathrm{L}}^{(1)}\,.\,\vec{e}_{\mathrm{L}}^{(2)})]^2(F_{\mathrm{L,L}}-F_{\mathrm{L,T}}-F_{\mathrm{T,L}}+F_{\mathrm{T,T}})\end{aligned} \tag{2.351}$$

Now, the special formula (for axially symmetric bond polarizability) is obtained by inserting in (2.309) the explicit formula (2.351) for every integral

$$\int_0^{+\infty} \mathrm{Tr}\,[\bar{\bar{T}}_{uv}\bar{\bar{A}}_u^{(1)}(i\xi)\bar{\bar{T}}_{uv}\bar{\bar{A}}_v^{(2)}(i\xi)]\,\mathrm{d}\xi$$

corresponding to every pair of bonds (u in molecule 1, v in molecule 2).

When the intermolecular distance R becomes large with respect to the molecular sizes, all bond–bond distances R_{uv} may be replaced by R, and all tensors $\bar{\bar{T}}_{uv}$ may be replaced by the unique tensor $\bar{\bar{T}}=3(\vec{r}\otimes\vec{r})-1$, where $\vec{r}=\vec{R}/R$. But the formulas (2.309) and (2.343) do not simplify to the same extent as (2.279) (which reduced to (2.295)), because the existence of *different* denominators $1/(b_u^{(1)}+b_v^{(2)})$, corresponding to the different bonds, precludes any simple factorization of

$$\sum_u \bar{\bar{A}}_u^{(1)} = \bar{\bar{A}}^{(1)} \quad \text{and} \quad \sum_v \bar{\bar{A}}_v^{(2)} = \bar{\bar{A}}^{(2)}$$

in the explicit formulas of the type (2.376) or (2.343); this factorization would occur in the basic formula (2.309), thus giving for large R:

$$E_{\text{disp}} \simeq -\frac{1}{2\pi}\frac{1}{R^6}\int_0^{+\infty} \text{Tr}\,[\bar{\bar{T}}\bar{A}^{(1)}(i\xi)\,\bar{\bar{T}}\bar{A}^{(2)}(i\xi)]\,\mathrm{d}\xi \tag{2.352}$$

but this result is mainly of formal interest. A substantial simplification of (2.343) occurs only if the average $\bar{E}_{\text{disp}}$ of E_{disp} over all orientations (assumed equally probable) is considered; the result is (Claverie, 1973, Ch. III, Section 4.F.6.b):

$$\bar{E}_{\text{disp}} \simeq -\frac{1}{6}\frac{1}{R^6}\left[\sum_u^{(1)}\sum_v^{(2)}\sum_{\kappa=1}^{3}\sum_{\lambda=1}^{3}\frac{\alpha_{u,\kappa}^{(1)}\alpha_{v,\lambda}^{(2)}}{b_{u,\kappa}^{(1)}+b_{v,\lambda}^{(2)}}\right] \tag{2.353}$$

where the factorized bracketed expression may be computed once for all for a given pair of molecules.

At this point, it is appropriate to mention the so-called 'retardation effect' (Casimir and Polder, 1948): when the intermolecular distance is large enough, it becomes necessary to take into account the finite velocity for the propagation of the electromagnetic field (which is responsible for the interaction) between the two molecules: the correlation between the fluctuating instantaneous dipoles in the two subsystems is thereby reduced, and the dispersion interaction decreases like $-1/R^7$ instead of $-1/R^6$ as $R \to \infty$. But this effect becomes non-negligible only for distances of the order of 100 Å, (Margenau and Kestner, 1969, Ch. 6, p. 209) so that it may be neglected when dealing with intermolecular interactions in the region of the equilibrium distance. We therefore refer the interested reader to specialized works dealing with this subject: Dzyaloshinskii *et al.* (1961), Mavroyannis and Stephen (1962), McLachlan (1963a,b), Power (1967), Margenau and Kestner (1969) Ch. 6, Feinberg and Sucher (1970), Ninham *et al.* (1970), Mitchell *et al.* (1972), Michels and Suttorp (1972, 1973), Suttorp (1973), Sun Chia Chung and Su Chan (1976). It must be mentioned that, rather recently, it has been possible to derive the dispersion interaction (including the retardation effect) in the framework of a purely classical theory, namely 'Stochastic (or random) electrodynamics' (Boyer, 1973, 1975). This theory is discussed, together with the quantum and semi-classical theories, in the review paper by Milonni (1976).

4. *Pair-wise additivity properties*

As for the case of the short-range interaction (Section IV.E) we must distinguish two kinds of such properties: the pair-wise additivity of the dispersion interaction between several molecules, and the expression of the dispersion interaction between two molecules as a sum of interaction terms between local subunits.

a. Pair-wise additivity between several molecules. The issue is rather clear here. The problem was dealt with by Margenau (1939a), who showed that, when the matrix elements $\langle W_0^{(1)} W_0^{(2)} | V | W_I^{(1)} W_J^{(2)} \rangle$ (involved in the basic perturbation expression (2.265) are replaced by their multipole expansion, the various multipole–multipole terms thus generated (dipole–dipole R^{-6} dipole–quadrupole R^{-8}, dipole–octopole and quadrupole–quadrupole R^{-10}) are all rigorously pair-wise additive (see also Kihara, 1958). The question was considered again by Wojtala (1964), Margenau and Stamper (1966), Margenau and Kestner (1969, Ch. 5). Wojtala (1964) emphasized that the pair-wise additivity was no more rigorous if the exact matrix elements were kept instead of resorting to their multipole expansion. Thus, this non-additivity may be ascribed to the penetration part of the integrals (let us recall that this part decreases exponentially at large distance). Since we may assume that a limited multipole expansion of the dispersion term insures a sufficient accuracy in the region of interest for us (equilibrium region or larger distances), the dispersion term will be practically pair-wise additive under such conditions.

b. Additivity of interaction terms between local subunits in two molecules. The issue is now more controversial. The general formulas (2.309, 2.313, 2.316, 2.340, 2.343, 2.351) appear as sums of interaction terms between local regions of the molecules corresponding to localized bonds, but the introduction of local dynamic polarizabilities relies upon the approximate procedure described in Section C. As pointed out there, the success of empirical systematics of bond polarizabilities supports the practical validity of this procedure. Zwanzig (1963) and Applequist *et al.* (1972) rightfully stress the difference between the polarizability of an isolated subunit and the effective polarizability of the subunit embedded in a molecule (the latter only would correspond to the empirical polarizability of the subunit), and Zwanzig, using the Drude oscillator model found that the dispersion energy calculated as a sum of local interactions using these 'effective' polarizabilities yielded results in satisfactory agreement with the exact (second-order) dispersion energy, thus supporting the calculations of Salem (1962a) concerning saturated chain molecules (other examples of the use of the Drude model are provided by Bade (1957), and also Kim and Gordon (1974d)). Longuet-Higgins and Salem (1961) previously studied the 'local additivity' (by applying the closure approximation to the basic expression (2.265) and without introducing local polarizabilities), and they concluded that local additivity holds provided that the electron correlation dies rapidly enough with distance within each molecule, and this is realized not only for saturated molecules but also for conjugated molecules with *alternating* bond lengths. Longuet-Higgins and Salem also find that this is no more true for conjugated molecules with *equal* bond lengths, thus agreeing with Coulson and Davies (1952) results, but as we pointed out above (end of Section C), this case is just the non-physical one. Sternlicht (1964) essentially confirms this view. The recent work by Amos and Crispin (1976a) also justifies the use of dynamic bond polarizabilities. The question of what polarizabilities of the subunits must be used is not yet completely clear, however. For the Drude model (two parallel chains of oscillators), Zwanzig (1963) found that, at *large* distances, the effective polarizabilities gave correct results, but Rabenold and Linder (1975), while agreeing with Zwanzig's result for large distances, found that for smaller distances (a few ångströms) the *unperturbed* or *isolated* subunit polarizabilities (which are smaller than the effective ones) gave satisfactory results (this appears essentially due to a cancellation between the three- and four-body effects). Such effects may depend on the relative orientation of the two assemblies of oscillators (simulating the real molecules) as well as on the geometry assumed for every assembly. For example, Yorke and De Rocco (1970) obtained for highly symmetrical assemblies of oscillators a result opposite to that of Zwanzig (1963) for parallel linear chains, namely, at large distances, the use of 'unperturbed' subunit polarizabilities still gave correct values of the dispersion energy (Yorke and De Rocco speak of a vanishing 'non-additivity' effect between the various subunits of every assembly).

On another hand, it is not clear to what extent the results concerning the Drude model may be transposed to the real bonds encountered in molecules. Actually, as emphasized above, the validity of the formula (2.309) (involving a sum of interactions between local subunits) depends only on the validity of the procedure defined in Section C, which expresses the excited eigenfunctions as a combination of excitations between localized orbitals. From the practical point of view, the use of a sum of local terms of the type (2.309, 2.316, 2.343, 2.351) with experimental bond polarizabilities and appropriate bond parameters $b = 1/\bar{U}$ indeed gives reasonable results (see Part VI below).

c. Expression of the dispersion interaction as a sum of atom–atom terms. The evaluation of the electrostatic energy as a sum of interactions between charges, dipoles, $\cdots$ located at the atoms (see Section B above) requires the computation of all distances between the atoms of the different molecules. A sum of dispersion terms between bonds (as in formula (2.313) requires the supplementary evaluation of all bond–bond distances (for example, between the middles of the bonds). As in the case of the polarization energy (see Section D.4 above), it would be advantageous, from the computational point of view, to dispose of a formula for the dispersion requiring only the interatomic distances (almost all the so-called 'empirical' formulas actually belong to this type). Two procedures may be proposed for deriving an atom–atom type formula from a bond–bond type formula like (2.313) or (2.316):

(1) we may consider a formula, analogous to (2.316), involving atom–atom terms:

$$E_{\text{disp}} \simeq -\frac{3}{2} \sum_{\mu}^{(1)} \sum_{\nu}^{(2)} \frac{1}{(R_{\mu\nu})^6} \frac{\alpha_\mu^{(1)} \alpha_\nu^{(2)}}{b_\mu^{(1)} + b_\nu^{(2)}} \tag{2.354}$$

where $\alpha_\mu^{(1)}, \alpha_\nu^{(2)}$ are atomic mean polarizabilities and $b_\mu^{(1)}, b_\nu^{(2)}$ are atomic 'inverse effective energies' (since the replacement of a sum of bond–bond terms by atom–atom terms is necessarily an approximate procedure, it does not seem necessary to introduce the refinement of non-isotropic atomic polarizabilities, hence the use of the formula (2.316) assuming isotropic polarizabilities). As mentioned previously in Section D.4, atomic polarizabilities α_μ may be obtained by sharing each bond mean polarizability $\bar{\alpha}_u$ between the two atoms $\lambda(u)$, $\mu(u)$ (connected by this bond) according to the relative weights $n_{u,\lambda}$ and $n_{u,\mu}$ which denote effective electron population (of the valence shell) contributed to the atoms λ and μ by the bond u (Caillet and Claverie, 1975, Section 2; also Section D.4 above). Then we only have to determine suitable atomic b_μ parameters. We could introduce some mean value of the various parameters b_u corresponding to the bonds starting from the atom. Another possibility is to use the formula (2.318):

$$b_\mu = \sqrt{\bar{\alpha}_\mu / n_\mu^{\text{eff}}} \tag{2.355}$$

where $\bar{\alpha}_\mu$ is the atomic mean polarizability (obtained as just mentioned above) and n_μ^{eff} will be some atomic 'effective number of electrons', e.g. the sum

$$n_\mu = \sum_{u(\mu)} n_{u,\mu}$$

of the numbers $n_{u,\mu}$ previously introduced (the notation $\sum_{u(\mu)}$ means that the sum runs over all bonds u starting from the atom μ). This recipe should give reasonable values at least up to the second-row atoms; for heavier atoms, as pointed out above in Section 3 (discussion of the Mavroyannis–Stephen choice), the 'effective numbers' of electrons should be somewhat different from the number of electrons in the valence shells. A possibility would be to choose n_u^{eff} for bonds in order to reproduce, according to (2.318), good values of the parameter b, such as those given by the Salem–Tang–Karplus choice: thus by using b'_u given by (2.327):

$$n_u^{\text{eff}} = \bar{\alpha}_u / b_u^2 = \bar{\alpha}(4(\Delta\vec{\mu}_u)^2)/(9\bar{\alpha}_u^2) = (4/9)(\Delta\vec{\mu}_u)^2/\bar{\alpha}_u$$

but care must be exercised according to whether the values of n_u^{eff} are intended to reproduce only the R^{-6} part of the dispersion term, or to reproduce also approximately the non-negligible R^{-8}, R^{-10} parts in the equilibrium region.

(2) The bond coefficient of $-1/(R_{uv})^6$ is, according to (2.316)

$$C^{uv} = \frac{3}{2} \frac{\alpha_u^{(1)} \alpha_v^{(2)}}{b_u^{(1)} + b_v^{(2)}} \tag{2.356}$$

We may share it directly between the atoms $\lambda^{(1)}, \mu^{(1)}$ and $\nu^{(2)}, \rho^{(2)}$ bonded by u and v respectively, and then add for each pair of atoms $\mu^{(1)}, \nu^{(2)}$ the various contributions $C_{\mu,\nu}^{u,v}$ thus obtained:

$$C_{\mu\nu} = \sum_{u(\mu)} \sum_{v(\nu)} C_{\mu,\nu}^{u,v} \tag{2.357}$$

These quantities $C_{\mu\nu}$ could then be used as coefficients of $-1/(R_{\mu\nu})^6$ in an atom–atom type formula:

$$E_{\text{disp}} \simeq -\sum_{\mu}^{(1)} \sum_{\nu}^{(2)} C_{\mu\nu}/(R_{\mu\nu})^6 \tag{2.358}$$

The sharing of $C^{u,v}$ into atomic contributions $C_{\mu,\nu}^{u,v}$ could be performed according to the relative weights $w_{\mu,\nu}^{u,v} = n_{u,\mu} n_{v,\nu}$ built from the effective numbers $n_{u,\mu}$ already introduced (for sharing the bond polarizabilities into atomic contributions). Of course, the initial C^{uv} coefficients may be effective ones (i.e. increased in such a way as to take into account the R^{-8}, R^{-10} terms), or else we could introduce analogous sharing procedures for C_8^{uv}, C_{10}^{uv} coefficients if these were explicitly used.

5. Choice of the 'centre of force'

Up to now, we have not discussed in detail the problem of the choice of the 'centre of force' for a given bond, namely the point at which the various transition dipoles $\vec{\mu}_{u-w'}$ are assumed to be located, so that this point is just used for defining the inter-bond distances $\vec{R}_{uv}$ appearing in the calculation of the dispersion term. Since the transition dipoles are used as an approximate representation of the corresponding transition charge densities, it would be obviously possible to choose the origin of the multipole expansion for every charge density in such a way as to make the representation by the dipole as accurate as possible (e.g. by minimizing the quadrupole part of the potential; if the charge density is axially symmetric, the quadrupole potential may even be reduced to zero by an appropriate choice of the origin on the axis). For a homonuclear diatomic molecule (or, approximately, for a homonuclear bond) the appropriate origin is, for obvious symmetry reasons, the middle of the bond (and this holds for all transition densities). But this is no more exact for heteronuclear bonds, and the cross-terms in $|V_{\alpha\beta}|^2$ obtained from a dipole–dipole term in one $V_{\alpha\beta}$ and a dipole–quadrupole term in the other $V_{\alpha\beta}$ may give rise to an R^{-7} term: Babloyantz (1959), Herman (1963, 1966), Margenau and Kestner (1969, Ch. 2, Section 2.6), Robert *et al.* (1969), Girardet *et al.* (1970) considered the interaction between a diatomic molecule and an atom; Buckingham (1965, 1967, 1968) derived general formulas, Bonamy and Robert (1970) studied the interaction $NH_3 \cdots He$, Girardet and Robert (1971) considered the dispersion interaction between two linear molecules, and Isnard *et al.* (1976) considered linear and tetrahedral molecules. These authors generally choose as the origin the centre of mass of the diatomic molecule and then evaluate the non-zero coefficient of the R^{-7} term. For our own purpose, we rather deal with bonds in molecules, and it appears more appropriate to choose the origin in the bond in such a way as to reduce as much as possible the coefficients of the R^{-7} term. It may be deduced from the works of Margenau and Kestner (1969, Ch. 2, Section 2.6) and Girardet *et al.* (1970) that a convenient choice is the centre of gravity of the electronic charge density corresponding to the valence orbitals (in the case of diatomic molecules); some analogous choice could be worked out for bonds in a molecule.

Finally, it must be emphasized that the replacement of a single centre of force in the bond by two centres of force located at the atoms (as occurs when the sum of bond–bond terms is replaced by a sum of atom–atom terms) is equivalent to introducing a whole series of supplementary terms $1/R^7, 1/R^8, \cdots$ corresponding to the initial single bond centre: thus, a sharing of the bond–bond coefficients into suitably unequal atom–atom coefficients would make possible to reproduce the $1/R^7$ part of the interaction expressed in terms of a single centre of force for every bond.

F. The perturbation terms beyond second-order

While the first- and second-order terms have been analyzed first for two molecules, and only then for several molecules, the higher-order terms have almost always been considered from the beginning for more than two molecules. The reason is that, at least for isotropic systems (e.g. rare gas atoms) the two-body terms are much smaller than the three-body (non-additive) term already at third-order, so that the attention was focused essentially on the many-body terms beyond second-order. As it was done for the second-order term, the higher-order terms may be decomposed into various contributions involving the permanent (ground state) multipole moments, or independent of them; of course, the complexity of this decomposition increases rapidly with the order of perturbation: the first thorough study of the third-order term was made by Stogryn (1971) (where references to previous works may be found), and he distinguishes two 'polarization-like' terms (involving permanent multipoles) and three 'dispersion-like' terms (one of them involving a permanent multipole). We shall consider separately the case of two molecules and the case of more numerous molecules.

1. Two molecules

The thorough study by Stogryn (1971) provides us with a numerical evaluation of the third-order terms (together with the first- and second-order) for the general case of asymmetric linear molecules (see the last Section 3.2 of his paper). The third-order energy then involve R^{-9} terms, and, from the numerical values given for the colinear configuration ($\varepsilon^{(1)} = -1{\cdot}033R^{-3}$, $\varepsilon^{(2)} = -245{\cdot}9R^{-6}$, $\varepsilon^{(3)} = -1137R^{-9}$, with R and $\varepsilon^{(n)}$ in atomic units), it appears that $\varepsilon^{(3)}$ may actually be neglected in practice already at intermediate distances if we content ourselves with a few per cent accuracy (for $R=3\,\text{Å}=5{\cdot}669$ a.u., $\varepsilon^{(3)}/\varepsilon^{(2)}=0{\cdot}0254$; for $R=4\,\text{Å}=7{\cdot}559$ a.u., $\varepsilon^{(3)}/\varepsilon^{(2)}=0{\cdot}0107$). This conclusion holds *a fortiori* for two spherically symmetric systems, in which case the first non-zero term of $\varepsilon^{(3)}$ involves R^{-11}, and no more R^{-9} (see the discussion under heading (4) in Section III.A above). Malrieu (1971a,b) used diagrammatical techniques for summing the most important contributions, and his results also indicate a fast convergency of the perturbation series: thus, for two isotropic systems, and by reducing all matrix elements $\langle W_a^{(1)} W_b^{(2)}|\, V\,|W_c^{(1)} W_d^{(2)}\rangle$ to their dipole part, his partial summation amounts to multiply the second-order (dispersion) term by a factor of about $1/(1-(\alpha^{(1)}\alpha^{(2)}/R^6))$, and it is easily seen from the polarizability values for the rare gas atoms (Table 2.3) that such a factor is already smaller than one per cent in the equilibrium region. These various results therefore strengthen the statements made in Section III.A concerning the practical possibility of neglecting the terms beyond second-order when evaluating the interaction between only two molecules.

2. *Several molecules*

The terms beyond second-order now play a more important role, owing to the many-body non-additive terms. Thus, at the third-order, even for isotropic systems, we have an R^{-9} three-body term, the so-called 'dipole–dipole–dipole' term (it involves a transition dipole on every system), which is given by (Axilrod and Teller, 1943; Muto, 1943):

$$E_{(3,3,3)} = C_{(3,3,3)} \frac{3\cos\gamma_1 \cos\gamma_2 \cos\gamma_3 + 1}{(R_{12})^3 (R_{23})^3 (R_{31})^3} \tag{2.359}$$

where $\gamma_1=(\vec{R}_{12}, \vec{R}_{13})$ denotes the angle between $\vec{R}_{12}$ and $\vec{R}_{13}$, and similarly $\gamma_2=(\vec{R}_{23}, \vec{R}_{21})$, $\gamma_3=(\vec{R}_{31}, \vec{R}_{32})$. A number of works were devoted to this 'triple–dipole' term: Axilrod (1951a,b), Midzuno and Kihara (1956), Kihara (1958), Kestner and Sinanoglu (1963), McLachlan (1963b), Dalgarno and Davison (1966) (this review paper displays numerical values of $C_{(3,3,3)}$ for rare gases), Margenau and Stamper (1966), Tang (1969), Margenau and Kestner (1969, Ch. 5). The $C_{(3,3,3)}$ coefficient may be expressed exactly in terms of the dynamic polarizabilities at imaginary frequencies, as it was the case for the C_6 coefficient in the dispersion term (see Section E.3 above) (McLachlan, 1963b; Linder and Hoernschemeyer, 1964):

$$C_{(3,3,3)} = \frac{3}{\pi} \int_0^{+\infty} \alpha^{(1)}(i\xi)\alpha^{(2)}(i\xi)\alpha^{(3)}(i\xi)\, d\xi \tag{2.360}$$

This formula was used by Tang (1969), who introduced simplified expressions of the type (2.315) for $\alpha^{(s)}(i\xi)$, and was therefore able to obtain a simple explicit formula for $C_{(3,3,3)}$ in terms of the static polarizabilities $\alpha^{(s)}$ and the 'inverse effective energies' $b^{(s)}$. Kramer and Herschbach (1970) also derived 'combination rules' from (2.360), while Langhoff and Karplus (1970b) used it for getting accurate values of $C_{(3,3,3)}$ for rare gas atoms and some diatomic molecules. According to (2.359), $E_{(3,3,3)}$ may be positive or negative for three molecules, but for molecules in a condensed phase (liquid, crystal) the total energy $E_{(3,3,3)}$ is positive (i.e. destabilizing) (Axilrod, 1951b; Kestner and Sinanoglu, 1963). As concerns its absolute value, the last two authors evaluate it to about 5 per cent to 10 per cent (according to the molecular species considered) of the total dispersion energy.

For non-isotropic molecules, the formula (2.359) must be modified, it will involve the various components of the non-isotropic polarizability tensors (Stogryn, 1970). Moreover, if the molecules have a non-zero permanent dipole (or higher-order multipole) the third-order energy involves new terms which contain non-additive parts (Stogryn, 1971). For molecules in a condensed phase, it could be guessed that the different orientations taken by the molecules are likely to make these new terms rather small, but, to the author's knowledge, a detailed study has not been performed for ascertaining this conjecture.

As concerns the higher-order perturbation terms, it may be mentioned that they were studied by Bade and Kirkwood (1957) for a linear lattice of isotropic harmonic oscillators (Drude model); then 3-, 4-, $\cdots$ N-body terms appear at the successive perturbation order. The nth order energy $\varepsilon^{(n)}$ ($n \geqslant 2$) turns out to be proportional to $(\alpha/R^3)^n$, where α is the polarizability of each oscillator and R the nearest-neighbour distance. Since α/R^3 is significantly smaller than 1 already for intermediate distance, the correction with respect to $\varepsilon^{(2)}$ brought about by the higher-order terms remains quite small (less than 4 per cent for all cases considered by Bade and Kirkwood), but it must be mentioned that the situation would be less favourable for a two- or three-dimensional lattice (because the ratios of the numbers of triplets, quadruplets, $\cdots$ n-uplets of molecules to the number of pairs are then higher than for a linear lattice).

Many-body interactions are also implicitly taken into account through the use of a macorscopic dielectric constant in the general theory of van der Waals interactions between macroscopic bodies, to be considered in the next section.

G. The general theory of Van der Waals forces for small or large bodies embedded in some surrounding medium ('Lifshitz theory')

The dynamic polarizability approach for the evaluation of the (second-order) dispersion term may be developed independently of the Rayleigh–Schrödinger perturbation method and generalized to a considerable extent. Thus, Casimir and Polder (1948) first applied the method of quantum electrodynamics to the evaluation of retarded dispersion forces. Later, Dzyaloshinskii, Lifshitz and Pitaevskii (1961) developed in an important review paper a so-called 'general theory of Van der Waals forces', including the interaction between macroscopic bodies (solids or liquids) in the presence of a surrounding medium and at finite temperature. The forces are calculated in terms of the fluctuating electromagnetic field associated with the various bodies, which are considered as *continuous media, characterized by a complex, frequency-dependent, dielectric constant.* The treatment may be specialized to microscopic systems (molecules), in which case the dynamic polarizability replaces the dielectric constant. Dzyaloshinskii *et al.* (1961) use the formalism of quantum electrodynamics, which makes their paper rather difficult to read. Several authors followed this line of work: Mavroyannis and Stephen (1962), and also McLachlan (1963a,b,c; 1965). Related approaches based on simpler methods were proposed by Jehle (1965, 1969), Linder (1965, 1967), Linder and Rabenold (1972); Ninham *et al.* (1970), Richmond and Ninham (1971a,b), Mitchell *et al.* (1971, 1972), Mitchell and Ninham (1972), Mahanty and Ninham (1973) and also Langbein (1973). Numerous applications were made, especially oriented towards complex systems (large molecules, multilayered systems) as encountered in biology: see the review papers by Parsegian (1973) and Israelachvili (1974), where appropriate references may be found.

In connection with the subject of the previous Section F, it may be pointed out that this kind of method implicitly involves many-body effects, owing to the use of macroscopic dielectric constants, whose values result from the behaviour of many *interacting* molecules in an external electromagnetic field.

But such a global approach must be used with appropriate caution. Thus, when dealing with the interaction between macroscopic liquid bodies (at finite temperature), it may be argued that the global approach gives the *total* interaction energy (actually the free energy, see, for example, Jehle (1965), Linder (1967), and also Linder and Rabenold (1972)) and not the dispersion contribution alone: indeed, the dipole (or permanent multipole) moment and the polarizability of the molecules both contribute to the macroscopic dielectric constant, so that the use of this constant involves the electrostatic and polarization contributions as well as the dispersion one. But the contributions thus obtained automatically are *statistical means with respect to the molecular motions*, since the dielectric constant just involves such a statistical averaging. Thus, if we now consider the interaction between molecular systems, a few Angströms apart, with more or less fixed orientations, the situation is quite different, because the electrostatic and polarization contributions now correspond to a *non-fluctuating part of the electromagnetic field*, so that the dynamic polarizability alone may supply the dispersion contribution only, the electrostatic and polarization contributions deserve a special evaluation for their own (taking into account the dielectric medium if there is any). The situation just described obviously holds as soon as ordered structures are considered, since statistical averaging over the molecular orientations is no more justified for the molecules which constitute these structures. This point was not stressed, e.g. by Ninham and Parsegian (1970) in their early applications of the 'general theory of Van der Waals forces' to multilayer systems (alternating slabs of lipids and water), but was later properly mentioned by Parsegian (1973) and Israelachvili (1974, pp. 346–347, 368, 383). It must also be kept in mind that the Lifshitz theory is in principle applicable only when the distances between the interacting bodies are large with respect to the interatomic distances in each body (for example, see Israelachvili, 1974, p. 378). We may therefore conclude that the connection between the 'microscopic' and 'macroscopic' approaches is far from complete, and both of them will probably have to be used, in a complementary way, for dealing with complex systems involving intermolecular distances which are not large.

VI. SOME EXAMPLES OF APPLICATIONS

As it was emphasized in the Introduction (Section I) of the Chapter, the intermolecular 'weak' interactions (which exist between closed-shell systems) exhibit the characters of *non-saturation, low directivity, low specificity,* in contrast with the 'strong' (or 'chemical', or 'valency') interactions, which appear between non-closed shell systems, and which exhibit the opposite characters. Owing to these properties of the weak interactions, it cannot be expected, in general, to find highly privileged binary interactions when dealing with an assembly of numerous molecules. Thus, in principle, either a complete statistical treatment of the assembly of molecules should be performed (case of a pure liquid, for example), or at least, if two (or a small number of) molecules are singled out, the effect of the surrounding molecules over the interaction between the privileged ones should be taken into account. Both cases point out the importance of statistical consideration when dealing with weak interactions (this is most necessary when the systems under study are at finite temperature). Ideally speaking, the interaction potentials studied in the three previous sections should be used as input data in appropriate statistical treatments. But the elaboration of such treatments for systems involving molecules with rather arbitrary size and shape is not an easy task.

As concerns the pure liquids, for example, we mentioned in the Introduction (Section I) that a really practical theory for molecular liquids had been proposed only quite recently: the Reference Interaction Site Model (RISM) (Andersen and Chandler, 1972; Lowden and Chandler, 1973, 1974; H. C. Andersen, 1975) which assumes as the intermolecular potential a sum of atom–atom terms, and may therefore be applied to arbitrary molecules but much work remains to be done before such a method is conveniently developed, checked and eventually becomes a routine procedure; on the contrary, the potentials based upon an expansion in terms of spherical harmonics around a single centre in each molecule (for example, see Rowlinson, 1969, Ch. 8) had a rather restricted range of practical validity (only not too strong departures from a spherical shape of the molecule could be treated in practice). A further advantage of the RISM method is that it could conceivably be extended to flexible molecules, by allowing internal degrees of freedom (e.g. rotations around bonds) subjected to suitable intramolecular potentials.

For more complex systems (which are, of course, encountered in most applications), the available (or practically usable) statistical treatments will almost inevitably be less well founded (this does not preclude a satisfactory accuracy from the practical point of view!). It is therefore quite understandable that, for such complicated systems, techniques of a somewhat 'global' or 'macroscopic' character have been proposed, such as those mentioned in Section V.G above: reaction-field techniques (Linder, 1967: Linder and Rabenold, 1972), use of macroscopic (frequency-dependent) dielectric constants (Dzyaloshinskii *et al.*, 1961; Parsegian, 1973; Israelachvili, 1974).

Since the purpose of this chapter is the study of the basic formulas expressing the interaction energy (between two molecules essentially), and not the statistical treatments based on such formulas, it will be necessary to select as examples of application essentially such cases where the statistical element is not important: interactions between a very small number of molecules, or interactions concerning molecules with some prescribed geometry, as occurs in a crystal. Thus, in a first section, we shall give an outline of the practical formulas ('semi-empirical', 'empirical') used for the applications, and, in the next sections, we shall briefly review some applications concerning the following topics: interactions between two molecules, interactions between a molecule and a solid surface, interactions between a solute molecule and the surrounding solvent, interactions in crystals, interactions involving macromolecules. The use of simplified (atom–atom) formulas of the intermolecular type for dealing with intramolecular conformational problems (especially for large molecules) will be mentioned, but the applications specifically oriented towards biological problems will not be considered here: other chapters of this book are devoted to this kind of applications.

All examples reported in this section are given for illustrative purpose only: no attempt has been made to make an exhaustive review of all applications of the semi-empirical or simplified formulas to the evaluation of intermolecular interactions.

A. Outline of simplified formulas

All applications reviewed in the present Part VI are based upon the use of formulas which evaluate the intermolecular interaction energy by using local-type terms, i.e. atom–atom terms essentially (in our own laboratory, atom–bond terms for the polarization energy and bond–bond terms for the dispersion energy have also been used). As an exception, we mention that Banerjee and Salem (1966) and Banerjee (1967) evaluated theoretically the benzene–benzene interactions (in the benzene crystal) without introducing such a decomposition: the dispersion energy was evaluated by considering each benzene molecule as a non-isotropic (axially symmetric) polarizable entity (with the centre of the molecule as the unique centre of force), while the short-range repulsion ('exchange') energy was evaluated from the molecular orbitals by using a Murrell-type formula kS^2/R (Section IV.C, formula (2.195)).

The semi-empirical computation schemes used by various authors at the present time are rather similar, and the differences often concern only the numerical values of parameters rather than the analytical form of the formulas. During a long time, it was a common practice among the users of simplified (so-called empirical) formulas to evaluate only the dispersion and repulsion energies by resorting to interatomic Lennard-Jones or Buckingham (6-exp) potentials (for example, see Kitaigorodskii, 1961a,b; Williams, 1966, 1967; Giglio and Liquori, 1967); such procedures were especially applied for evaluating 'non-bonded' interactions in problems of conformational analysis (Kitaigorodskii and Dashevskii, 1967; Dashevskii and Kitaigorodskii, 1967; Ramachandran and Sasisekharan, 1968; see also the References 8 to 19 in Poltev and Sukhorukov, 1968b). The neglect of the electrostatic and polarization contributions was more or less justified for molecules with small net charges like hydrocarbons, but the necessity of including them, in order to have a procedure with general validity, was progressively recognized (Kitaigorodskii and Mirskaya, 1964b, 1965; Kitaigorodskii, 1965, 1966; Scheraga, 1968; Popov *et al.*, 1968; Poltev and Sukhorukov, 1967, 1968; Coiro *et al.*, 1972; Morris, 1973, 1974; Hagler *et al.*, 1974; Momany *et al.*, 1974a (see also References 7–13 in his paper), Shipman *et al.*, 1975), Evans and Watts (1975, 1976). It is possible that some thoughtless use of atom–atom potentials, reduced to the dispersion–repulsion terms only, was the origin of some criticisms made to the use of atom-potentials in general; we shall therefore discuss some such criticisms in the Section A.6 below.

Following the recent account by Caillet and Claverie (1975) we now give a summary of practical formulas for the four main contributions (electrostatic, polarization, dispersion, short-range repulsion), together with some indications concerning the representation of the hydrogen bond, and we indicate appropriate modifications to be brought about at very short distances (in order to avoid spurious energy decreases when some interatomic distance goes to zero).

1. *Electrostatic energy*

According to the analysis of Section V.B, the molecular charge distribution is approximated by a set of multipoles located at several centres in the molecule; in practice, charges q_ν located at the atoms ν (so-called 'monopoles') have been almost exclusively used. Nevertheless Dreyfus (1970) also included dipoles and quadrupoles located at the atoms and the middles of the segments joining the atoms for a study of the hydrogen bond, and Rein and co-workers (see Rein, 1973, and references therein) have extensively studied the use of local multipole moments (essentially dipole and quadrupole) and the speed of the convergence of such multi-centred multipole expansions. It must be emphasized that the local charges ('monopoles') may be replaced by '*effective*' charges in such a way as to mimic the electrostatic potential due to the *theoretical* local charges, dipoles, quadrupoles ... (this explains why the exact Coulomb potential may be represented adequately by suitable point charges even at medium range, as pointed out by Tait and Hall (1973)).

Using (atomic) charges only (practical formulas involving dipoles and quadrupoles may be found, for example, in Rein (1973), we have:

$$E_{\mathrm{el}} = (\mathrm{Cste}) \sum_{\lambda}^{(1)} \sum_{\nu}^{(2)} \frac{q_\lambda^{(1)} q_\nu^{(2)}}{R_{\lambda\nu}} \tag{2.361}$$

where the (Cste) takes care of the units: thus, if the charges q_λ are expressed in electronic charge units, and the distances $R_{\lambda\nu}$ in ångström units, the corresponding energy unit for q^2/R would be $23{\cdot}064 \times 10^{-19}$ joule = $5{\cdot}5111 \times 10^{-22}$ kcal, and if we multiply this unit by the Avogadro number = $6{\cdot}023 \times 10^{23}$ in order to get kcal/mol, we have finally (Cste) = 331·934 (kcal/mol).

As concerns the atomic charges q_λ, some authors tried to get them from experimental dipole moments (for example, see Popov *et al.* (1968) and the EPEN procedure of Shipman *et al.* (1975) described below), but most frequently they are deduced from some quantum-mechanical calculation. (Information concerning the semi-empirical methods (which are almost indispensable for large molecules) may be found, for example, in the books by Pilar (1968), Murrell and Harget (1972), Pople and Beveridge (1970); see Sections V.A,B). Hückel method for the π charges (using the parameters of Berthod and Pullman (1964, 1965, 1966) for hetero-atoms) and Del Re's (1958) method for the σ charges are the simplest and were used successfully in our laboratory (Pullman *et al.*, 1966a,b, 1967; Claverie *et al.*, 1966; Claverie, 1968a,b; Pullman, A., and Pullman, B., 1968; Caillet and Claverie, 1974, 1975) and by others (Poltev and Sukhorukov, 1967, 1968). More recently, the Extended Hückel method was used (Huron and Claverie, 1971), and also the CNDO/2 method (Momany *et al.*, 1974a; Caillet and Claverie, 1975; Caillet *et al.*, 1976). Both these semi-empirical schemes were examined by Rabinowitz and Rein (1972), Swissler and Rein (1972). But, as emphasized at the end of Section V.B, these last two methods introduce *atomic dipoles* (the so-called 'hybridization moments') besides the atomic net charges for representing the molecular charge distribution: as a consequence, if it is desired to still use formula (2.361) with charges only, it is necessary to decompose the atomic dipoles into suitable atomic charges (see Appendix F), which are added to the atomic net charges in order to get 'effective' total charges (moreover, if *ab initio* calculations were used, it would also be necessary to take into account not only the charges but also the dipoles of the two-atoms charge distributions, and therefore to use Löwdin's atomic net charges rather than Mulliken's ones: see Section V.B).

Experimental values of dipole moments may be found in the book by McClellan (1963); experimental values of multipole moments for small molecules have been compiled by Stogryn and Stogryn (1966), together with some theoretical estimates; see also Buckingham *et al.* (1968).

For the quadrupolar benzene molecule, Evans and Watts (1975, 1976) use the following peculiar electrostatic representation: instead of charges located at the atoms, they introduce six 'centres of force' and put a quadrupole (equal to one-sixth of the total molecular quadrupole) at each of these centres.

Shipman, Burgess and Scheraga (1975) proposed an 'Empirical Potential using Electrons and Nuclei' (EPEN) where the charge distribution of a molecule is represented by a set of *integer* charges: the nuclear positive ones (for the heavy atoms, the nuclear charge minus 2 is used, in order to represent the (1*s*) inner shell electron pair) and electronic negative ones (corresponding to the valence electrons) located at appropriate points on the bonds, these points being chosen so as to reproduce experimental dipole moments for small molecules (then, these small molecules are used for defining molecular fragments, which may be assembled for getting the charges representing large molecules).

2. *Polarization* (*or induction*) *energy*

In Section V.D, the polarization energy of some molecule (say *s*) was obtained as a sum of polarization energies for the various bonds: (formulas (2.258) and (2.263):

$$E_{\mathrm{pol(s)}} = (\mathrm{Cste})(-\tfrac{1}{2}) \sum_{u}^{(s)} \vec{\mathscr{E}}_u^{[s]} \,.\, \bar{\bar{A}}_u^{(s)} \,.\, \vec{\mathscr{E}}_u^{[s]} \tag{2.362}$$

where $\bar{\bar{A}}_u$ is the polarizability tensor of the bond u and $\vec{\mathscr{E}}_u^{[s]}$ the electric field created at this bond by all surrounding molecules (as concerns the choice of the point on the bond where the electric field is to be evaluated, see Caillet and Claverie (1975, footnote on p. 450) and the Section V.D.5 in the present work). If the molecular charge distributions are represented by atomic charges as previously, we immediately have

$$\vec{\mathscr{E}}_u^{[s]} = \sum_{t \neq s} \sum_{\lambda}^{(t)} q_\lambda^{(t)} \frac{\vec{R}_{\lambda u}}{(R_{\lambda u})^3} \tag{2.363}$$

where $\vec{R}_{\lambda u}$ is the vector joining the atom λ in molecule (t) to the 'centre of polarizable charge' on the bond u of molecule (s).

Values of bond polarizabilities were deduced from the experimental molecular polarizabilities as by Le Fèvre (1965) (for most bonds we have axial symmetry, i.e. α_{T} (Transversal) = α_{V} (Vertical); if this is not the case, it is possible to replace α_{T} and α_{V} by their mean value in order to still use the special formula (2.259) for the axially symmetric case). The values for some bonds not listed in Le Fèvre's Table 22 may be evaluated from the polarizabilities of appropriate small molecules: Huron and Claverie (1969) thus obtained for O–H (from CH_3OH) $\alpha_L = 0{\cdot}61$ Å^3, $\alpha_T = 0{\cdot}76$ Å^3 and for N—O of the nitro-group $—NO_2$ they used $\alpha_L = 2{\cdot}48$ Å, $\alpha_T = 0{\cdot}62$ Å^3. For C═N we may use $\alpha_L = 2{\cdot}43$ Å^3, $\alpha_T = 1{\cdot}00$ Å^3 (values obtained by interpolating the available values for C═C and C═O).

The factor (Cste) in (2.362) takes care of the units; if the polarizabilities (who have the dimension of a volume) are expressed by using the unit volume (Å^3) corresponding to the unit length used for the distances R, this factor is the same as for the electrostatic energy.

Bond polarization energies were extensively used in our laboratory (Claverie and Rein, 1969; Rein *et al.*, 1968; Claverie, 1968b, 1970, 1971b; Huron and Claverie, 1969, 1971; Mantione, 1968, 1969, 1971; Mantione and Daudey, 1970). But, as explained in Section V.D.4, the use of atomic polarization energies is computationally more interesting, since it saves the calculation of the atom–bond distances $R_{\lambda u}$. Suitable values of the atomic polarizabilities may be obtained by sharing the bond polarizabilities into atomic contributions (see Section V.D.4 and the more complete treatment in Caillet and Claverie (1975)). The formulas differ from (2.362, 2.363) only by the replacement of the bond labels u by the atom labels λ.

Although each local electric field (2.363) is a sum of atomic contribution the resulting local polarization energy is *not* simply a sum of atom–atom terms, because it is a *quadratic* functional of the electric field. Thus, if we consider, for definiteness, atom terms $\vec{\mathscr{E}}_\lambda \,.\, \bar{\bar{A}}_\lambda^{(s)} . \vec{\mathscr{E}}_\lambda$ and if we insert the explicit expression

$$\vec{\mathscr{E}}_\lambda = \sum_{\mu \notin (s)} q_\mu \vec{R}_{\mu\lambda}/(R_{\mu\lambda})^3$$

analogous to (2.363), we get

$$-\frac{1}{2}\sum_{\lambda}^{(s)} \vec{\mathscr{E}}_\lambda \,.\, \bar{\bar{A}}^{(s)} \,.\, \vec{\mathscr{E}}_\lambda = -\frac{1}{2}\sum_{\lambda}^{(s)} \sum_{\mu \notin (s)} \sum_{\nu \notin (s)} q_\mu q_\nu \frac{\vec{R}_{\mu\lambda}}{(R_{\mu\lambda})^3} \,.\, \bar{\bar{A}}_\lambda \,.\, \frac{\vec{R}_\nu\lambda}{(R_\nu\lambda)^3} \tag{2.364}$$

which is a sum of three-atom terms. This higher complexity of the polarization energy is probably one of the reasons why it has been considered less frequently than the other contributions: Poltev and Sukhorukov (1967, 1968) and also Evans and Watts (1975, 1976) take it into account, but Momany *et al.* (1974), Hagler *et al.* (1974) and Shipman *et al.* (1975) do not. Fortunately, the polarization contribution often turns out to be significantly smaller than the other contributions in actual practice.

3. Dispersion and short-range repulsion energies

We consider here these two terms together because many semi-empirical formulas of the Lennard-Jones or Buckingham type actually proceed in this way. The reason for this situation is of course that, for systems with zero net charges, these two terms are the only non-vanishing ones. This holds for rare gas atoms, which were just among the first systems for which intermolecular potentials were investigated somewhat quantitatively; such semi-empirical investigations are altogether continuing: for example, see Robinson *et al.* (1971).

The dispersion energy is always represented by a $-A/R^6$ term, in agreement with the theoretical analysis. But it must be kept in mind that such an analysis also gives further terms $-C_8/R^8$, $-C_{10}/R^{10}, \ldots$, which are not negligible in the equilibrium region: for He...He, at the equilibrium distance $R = 5{\cdot}75$ a.u., these two terms $(-C_8/R^8, -C_{10}/R^{10})$ amount to $\frac{1}{3}$ of the main term $-C_6/R^6$ (see equation (2.121) borrowed from Murrell and Shaw (1967b, 1968b)). Consequently, either such supplementary terms should be explicitly added, on the supposition that appropriate 'centres of forces' have been chosen, so that the $1/R^7$ 'eccentricity effect' is kept negligible (see Section V.D.5) (this would make it necessary to evaluate the coefficients C_8 and C_{10}: for example, see Starkschall and Gordon (1972b), Cohen and Pack (1974)), or an 'effective' coefficient $A > C_6$ must be used in the $-1/R^6$ term kept alone. Although the latter procedure is obviously less satisfactory from the theoretical point of view, it was adopted, in view of its greater simplicity, in all semi-empirical formulas quoted below. It appears nevertheless possible, in the future, to use procedures such as that of Starkschall and Gordon (1972b) for deducing reasonable values of C_8 and C_{10} from that of C_6 for bond–bond interactions (and hence eventually for atom–atom ones): hence, the use of theoretically founded values of C_6, C_8, C_{10} could become practically possible.

The theoretical developments of the Section V.E suggested the use of a sum of bond–bond dispersion terms, but, as it was the case for the polarization energy, it is also practically interesting to use a sum of atom–atom terms (see Section V.D.4.c), and most semi-empirical formulas actually use such terms.

From this point of view, a similar situation holds for the short-range repulsion energy: the theoretical analysis of Section IV.E.2 would rather suggest to use a sum of contributions involving the localized orbitals of the two molecules (now, such orbitals correspond essentially to bonds, and eventually to atoms in the case of lone pair orbitals). Future developments could introduce formulas of this type, but up to now semi-empirical formulas have used sums of atom–atom terms.

As concerns the analytical form of these terms, essentially two types have been tried: the inverse power $1/R^m$ (with $m > 6$, the Lennard-Jones (1924) choice $m = 12$ being the most frequent) and the exponential e^{-bR} (R. A. Buckingham, 1938). As we saw in Part IV, only the latter form is theoretically founded, up to rather short interatomic distances (at extremely short distances ($R < 0{\cdot}5$ a.u., say) all quantum-mechanical integrals (involving the electrons) come close to a finite limit (reached for $R = 0$), so that the internuclear repulsion $Z_\mu Z_\nu / R_{\mu\nu}$ may become dominant), and the inverse power law must be considered only as a purely *empirical* way of representing a repulsion which increases steeply as R decreases. By contrast, we would rather refer to the exponential formula Be^{-bR} as a *semi-empirical* one, in order to stress that the analytical form is theoretically founded, while the parameter B and b must be fitted in some suitable way (but not necessarily with respect to experimental values of the interaction potential: accurately computed values may be used as well, as we illustrated in Section IV.C for the He . . . He interaction). Besides its computational simplicity, the $1/R^m$ formula (with $m > 6$) automatically insures that the sum of the dispersion and repulsion energy goes to $+\infty$ as $R \to 0$, as it must; on the contrary, with the exponential formula for the repulsion, the $-1/R^6$ dispersion term would give rise to a spurious maximum at some small value of R, and for still smaller values the total energy would spuriously decrease to $-\infty$: some modifications of the formula at such small distance are therefore needed for eliminating these unwanted features.

Let us now review some examples. Kitaigorodskii (1961a) introduced a Buckingham 6-exp formula (whose parameters were later corrected by Kitaigorodskii and Mirskaya, 1964a) for hydrocarbon molecules, and extensions to several heteroatoms were proposed later: Kitaigorodskii and Dashevskii (1968), Kitaigorodskii *et al.* (1969), Ahmed and Kitaigorodskii (1972); see also the reviews by Kitaigorodskii (1970), Kitaigorodskii and Mirskaya (1970), and the book by Kitaigorodskii (1973).

In our own laboratory, we used a sum of bond–bond dispersion-like terms according to the 'mean excitation energy method' (formulas (2.279, 2.290, 2.294) in Section V.E.2) (Claverie and Rein, 1969; Rein *et al.*, 1968; Claverie, 1968, 1970, 1971b); Huron and Claverie (1969) replaced the too small mean energies used before (namely the ionization potentials) by more correct ones, according to the arguments of Pitzer (1959), and thus obtained a satisfactory agreement with Kitaigorodskii's atom–atom type formula; other applications were made by Huron and Claverie (1971), Mantione (1968, 1969, 1971) and also Mantione and Daudey (1970). More recently, a formula based upon the dynamic polarizability approach (see (2.316) or (2.309, 2.351) in Section V.E.3) was developed by using the 'Mavroyannis–Stephen' choice with the number of valence electrons taken as the 'effective number' (formulas (2.317, 2.318)).

On another hand, we also used (in order to have an evaluation the short-range repulsion term) a Kitaigorodskii formula (see the recent account by Caillet and Claverie, 1975):

$$E_{\text{disp}} + E_{\text{rep}} = \sum_{\lambda}^{(1)} \sum_{\nu}^{(2)} E(\lambda, \nu) \tag{2.365}$$

$$E(\lambda, \nu) = k_\lambda k_\nu \left[-\frac{A}{z^6} + Be^{-\gamma z} \right] \tag{2.366}$$

$$z = R_{\lambda\nu}/R^0_{\lambda\nu} \qquad R^0_{\lambda\nu} = \sqrt{(2R^w_\lambda)(2R^w_\nu)} \tag{2.367}$$

where R_λ^w and R_ν^w are the Van der Waals radii of atoms λ and ν respectively (we used essentially the radii proposed by Bondi (1964)). Some grounds for defining $R_{\lambda\nu}^0$ by the geometric mean rule in (2.367) instead of the arithmetic mean rule $((2R_\lambda^w+2R_\nu^w)/2=R_\lambda^w+R_\nu^w)$ are given by Huron and Claverie (1969), Good and Hope (1970). The parameters A, B and γ (the previous publications (Kitaigorodskii, 1961a; Caillet and Claverie (1975) used the symbol α; we prefer to use here the symbol γ for this dimensionless parameter in order to avoid any confusion with the polarizability) do not depend on the atomic species: this necessary dependence is brought about by $R_{\lambda\nu}^0$ and by the factors k_λ, k_ν which allow the energy minimum to have different values according to the atomic species involved (Kitaigorodskii *et al.*, 1969; Huron and Claverie, 1974b). The values of these various parameters and of the Van der Waals radii used in recent applications are given by Caillet and Claverie (1974, 1975). The values of the dispersion energy given by the dispersion part $(-A/z^6)$ of the formula (2.365, 2.366) are in satisfactory agreement with those obtained from the bond–bond formula previously mentioned (i.e. the one based upon the dynamic polarizability approach). As a further refinement, we took into account the effect of the electronic population of each atom on the short-range repulsion term (the theory suggests the existence of such an effect, see for example Murrell, Kettle and Tedder (1965) pp. 339–340); if q_λ is the net charge and N_λ^{val} the number of valence electrons of the neutral atom, the actual electronic population is $N_\lambda^{\text{val}}-q_\lambda$; for every atom–atom repulsive term $E_{\text{rep}}(\lambda, \nu)$ we introduced the product of the ratios $(N_\lambda^{\text{val}}-q_\lambda)/N_\lambda^{\text{val}}$ and $(N_\nu^{\text{val}}-q_\nu)/N_\nu^{\text{val}}$, thus obtaining the modified short-range energy:

$$E_{\text{rep}}^{\text{modified}}(\lambda, \nu)=(1-q_\lambda/N_\lambda^{\text{val}})(1-q_\nu/N_\nu^{\text{val}})E_{\text{rep}}(\lambda, \nu) \qquad (2.368)$$

where $E_{\text{rep}}(\lambda, \nu)=k_\lambda k_\nu B \exp(-\gamma R_{\lambda\nu}/R_{\lambda\nu}^0)$ according to (2.366, 2.367).

This correction is rather small in most cases. It becomes noticeable mainly for hydrogen atoms bearing a non-negligible (positive) charge (for example, those bonded to oxygen and nitrogen atoms) because $N_H^{\text{val}}=1$.

The Buckingham (6-exp) potential (for atom–atom terms) was also used by Hill (1948), Williams (1965, 1966, 1967, 1972a,b), Rae and Mason (1968), Popov *et al.* (1968) (further references may be found in Popov (1975)), Poltev and Sukhorukov (1967, 1968a,b), Snook and Spurling (1970), Morris (1973, 1974), Evans and Watts (1975, 1976) (Evans and Watts represent the benzene molecule by six 'centres of force' lying at an adjustable distance from the centre of the molecule; the dispersion energy is evaluated according to the 'mean excitation energy' approximation: see Section V.E.2 and formulas (2.279, 2.290)). A variant of this potential combining the exponential and inverse-power law for the repulsion term (namely using $\exp(-\gamma R)/R^D$ instead of $\exp(-\gamma R)$) was used by Mason and Rice (1954), Bartell (1960), Scott and Scheraga (1966) and also Parsonage and Pemberton (1967); Coen *et al.* (1967) discussed this general kind of formulas, and some numerical comparisons are given by Di Nola and Giglio

(1970) (for further applications, see also Giglio and Liquori (1967), Giglio (1969, 1970) and also Coiro *et al.* (1972, 1973)). Apparently, the 'generalized' repulsion potential $\exp(-\gamma R)/R^D$ involves no decisive advantage, and many recent works continue to use the simple exponential formula. Venkatachalam and Ramachandran (1967) compared several semi-empirical potential functions.

The Lennard-Jones (6–12, i.e. $m=12$) potential was used by Momany *et al.* (1974a,b, 1975) and by Hagler *et al.* (1974); the last authors also fitted a Lennard-Jones (6–9) potential, and found it possible to reproduce the crystal experimental data (used for the fitting) practically as well as with the (6–12) potential. Morris (1973) actually emphasized that the structure of a crystal may be rather insensitive to the parameters used for the dispersion–repulsion potential; this observation explains why the principle of close-packing extensively used by Kitaigorodskii (1961c, 1965) was able to account for the structures of a large number of organic crystals.

In their potential EPEN (see the end of Section 1 above), Shipman *et al.* (1975) use an exponential formula $B\exp(-bR)$ for the repulsion and an R^{-6} formula for the dispersion, but the former term (exponential repulsion) is evaluated only between the points corresponding to the electronic charges, while the latter term (dispersion) is evaluated only between the centres of the heavy atoms.

We finally mention the convex-core potential proposed by Kihara (1953, 1955, 1967, 1970 and references therein). This potential function for the *whole* interaction between two like molecules is *not* expressed as a sum of atom–atom (or bond–bond) terms, but by a single term similar to the Lennard-Jones formula:

$$U(\rho)=U_0\left[\left(\frac{\rho_0}{\rho}\right)^{12}-2\left(\frac{\rho_0}{\rho}\right)^{6}\right] \tag{2.369}$$

where the distance variable ρ is not the distance between molecular centres but the minimum distance between suitable molecular 'cores' (e.g. for a diatomic molecule, this 'core' will be the rod joining the nuclei), and ρ_0 is some suitable parameter. This potential is interesting if the molecular core may be chosen convex, because several mathematical relations hold specifically for convex surfaces, and make possible an analytical evaluation of various integrals, among which those involved in the expression of the second virial coefficient (see the above mentioned works by Kihara (1967, 1970). The essential drawback of this approach (interesting by its simplicity) is, of course, the fact that the energy minimum $-U_0$ (reached for $\rho=\rho_0$) does not depend on the relative orientation of the molecules, a property which cannot hold exactly: for example, for two rod-like molecules the interaction energy is larger for parallel molecules than for colinear ones, even when the distance of the rods (molecular cores) is the same. This lack of flexibility of the convex–core potential therefore makes it unable to account for the tendency of rod-like molecule to associate parallel to each other due to the dispersion interaction. The convex-core potential thus appears more suited for applications involving some integration over the whole range of relative configurations, by introducing suitably fitted parameters U_0 and ρ_0: this is actually the kind of applications performed by Kihara (second virial coefficient for gases) and coworkers (solubility of gases, see Section D below).

4. Representation of the hydrogen bond

In order that the semi-empirical formulas are relevant for wide variety of problems, it is of course highly desirable that they are able to reproduce the occurrence of the 'hydrogen bond' interaction. During a long time, it was a common practice among the users of empirical formulas to introduce a special potential for representing the hydrogen bond: for example, see Scheraga (1968), Popov *et al.* (1968) (use of a Morse potential), McGuire *et al.* (1972) and Momany *et al.* (1974a, 1975) (use of a special Lennard-Jones (10–12) potential). A drawback of this procedure is that the atoms considered as able to participate in some hydrogen bond must be chosen *a priori.* More recently, however, the quantum-mechanical studies supported more and more conclusively the statement that the weak hydrogen 'bond' was actually an interaction of intermolecular type (and not some kind of chemical bonding), the specific character of which being simply a rather short equilibrium distance, a charge transfer contribution more important as usual (because of these shorter distances), and also a stronger directionality than usual (Dreyfus, 1970; Dreyfus and Pullman, 1970; Kollman and Allen, 1970, 1972 and also Daudey, 1974). It was therefore tempting to apply directly the usual semi-empirical formulas (including the electrostatic and polarization terms, which are especially important in this case), eventually with some supplementary term or some other modification taking into account a non-negligible charge transfer contribution at short distance.

With our formulas previously described (2.366, 2.367, 2.368), the positive net charge of the hydrogen atom results in a reduction of the repulsive term (formula (2.368)), but this did not appear sufficient, and we introduced a supplementary reduction of this term (through reducing B and increasing γ) below some prescribed interatomic distance (Caillet and Claverie, 1975, Section 2.d). This reduction of the repulsive term may be considered as a subterfuge for the introduction of an attractive charge-transfer term (which would be exponentially varying, too). It must be emphasized that, in order to get a correct (not too strong) interaction energy at the correct equilibrium distance, it turned out to be necessary to reduce also the parameter A of the dispersion term; but this is in agreement with the theoretical analysis (Murrell and Teixeira-Dias, 1971), according to which the multipole part $(-1/R^6, \ldots)$ of the dispersion energy overestimates it at short distances owing to the neglect of the penetration parts of the intermolecular integrals (the reduced electronic population of the Hydrogen atom with respect to a neutral one also certainly contributes to reduce the dispersion energy involving this atom). In actual calculations (Caillet and Claverie, 1974, 1975; Caillet *et al.*, 1976) the modified formula is applied for *any* pair of atoms involving one hydrogen and one 'heavy' atom, i.e. no *a priori* information (concerning the occurrence of hydrogen bond between special pairs of atoms) has to be introduced.

Hagler, Huler and Lifson (1974) and also Hagler and Lifson (1974) avoided the use of a special supplementary potential for the hydrogen bond; they even renounced to use some modification for the hydrogen–heavy atom interaction terms and fitted directly the parameters of their potentials (involving the electrostatic, dispersion and short-range repulsion contributions) for the various atoms (H, C, N, O) so as to reproduce the crystallographic data of a number of amide molecules.

5. *Modification of the* (6-exp) *potential at very short distances*

As mentioned above, a (6-exp) potential $U(R) = -A/R^6 + B\exp(-bR)$ cannot be correct as $R \to 0$ because $-A/R^6 \to -\infty$ while $B\exp(-bR) \to B$ so that $U(R)$ behaves in a wrong way for very small R: instead of increasing monotonously to infinity as R decreases to zero, $U(R)$ reaches high positive maximum (for $R = R_{\max}$, say) and then decreases to $-\infty$ (for example, see Egelstaff, 1967, Ch. 3, p. 28; Hirschfelder, Curtiss and Bird, 1954, Ch. 3). This incorrect behaviour must be eliminated if the applications of the interatomic potential cover the whole range of values of the distance, as it occurs for integrations involving the potential (statistical mechanical problems) or for minimization problems (otherwise, minimization procedures could end in spurious minima corresponding the coincidence of two atoms of the interacting molecules). A conceptually very simple procedure (for example, see Egelstaff, 1967, Ch. 3) is to replace $U(R)$ by a hard-core potential with radius $R_{\max}$ (i.e. to define $U(R) = +\infty$ for $R < R_{\max}$). The drawback is the high singularity (in the mathematical sense) thus introduced. For our own applications, we preferred to use a smoother modification, insuring the continuity of $U(R)$ and its first derivative (see Caillet *et al.*, 1976): a critical value z_c was chosen as the inflexion point of the function $-A/z^6 + B\exp(-\gamma z)$ situated between the maximum and the minimum; then, for $z < z_c$, the function $1/z^6$ was replaced by a quadratic function $(a_d z^2 + b_d)$ and the function $\exp(-\gamma z)$ by a function $(a_r/z + b_r)$, the parameters $a_d = -3/z_c^8$, $b_d = 4/z_c^6$ and $a_r = z_c^2\exp(-\gamma z_c)$, $b_r = (1-\gamma z_c)\exp(-\gamma z_c)$ being chosen so as to insure the continuity of the function and of its first derivative; finally, a similar change is applied to the polarization term: considering the electric field $(\vec{R}/R)/R^2$, we write $1/R^2 = 1/(R^0 z)^2$ (where $\vec{R} = \vec{R}_{\lambda\nu}$ and $R^0 = R^0_{\lambda\nu}$), and we replace, for $z < z_c$, $1/z^2$ by $(a_p z^2 + b_p)$ with the same requirement (continuity of the function and its first derivative), hence $a_p = -1/z_c^4$, $b_p = 2/z_c^2$.

6. Some comments about the legitimacy of the simplified 'atom–atom' formulas

From time to time, some criticisms are made to the use of simplified 'atom–atom' type formulas for the evaluation of intermolecular interactions, and the discussion of such criticisms will be a useful conclusion to the present Section A. Thus, Harris *et al.* (1975) report experimental findings concerning the structure of the complex Ar . . . OCS (they find a T-shaped structure with Ar on the side of the linear molecule, in contrast with the linear structure of the complexes Ar . . . ClF and Kr . . . ClF, with Cl in the centre) and state that the representation of the intermolecular potential as a sum of pair-wise additive atom–atom terms is devoid of 'serious theoretical justification' and that, in the cases under consideration, such a representation would be 'totally incorrect' since it predicts 'the closest packed equilibrium structure' (this would agree with Ar . . . OCS, but not with Ar . . .ClF or Kr . . . ClF); they also quote Mies (1965) who found that, for the He . . . H_2 system, this type of potential failed 'to reproduce the correct forces or transition probabilities for vibrational energy relaxation'. These criticisms suggest us the following comments:

(a) Harris *et al.* (1975) refer to the pair-wise additive interatomic potentials (as those of Lennard-Jones or Buckingham). But when the electrostatic and polarization terms are taken into account (as they must) besides the dipersion and repulsion ones, the situation is modified, because, as stated above in Section 2, the polarization energy, even when expressed in the framework of an approximation based upon interatomic distances, is *not* pair-wise additive. Thus, if we consider the examples of Harris *et al.*, the electric field (and therefore the polarization energy of the Argon atom) created by OCS on one hand and ClF on the other hand are different, and this difference could contribute to explain the difference of the shapes of the complexes. An actual calculation would be needed to clear up the point, but it is true anyway that no close-packing principle holds for the polarization energy, still less for the electrostatic energy (when it does not vanish) since the interatomic electrostatic contributions may be positive or negative according to the signs of the charges involved.

(b) For the He . . . H_2 interaction, only the dispersion and repulsion terms of the simplified formulas are non-vanishing, and the criticism of Mies (1965) would seem to apply. But it must be emphasized that we are confronted here with a very refined problem (vibrational energy relaxation) concerning systems of very small size: now, the simplified semi-empirical formulas are obviously not designed in order to get an extreme accuracy, but on the contrary in order to give the essential features of the intermolecular energy function for medium-size (or larger) molecules (these essential features being the positions corresponding to energy minima, with reasonable evaluations of these, and also reasonable results when some integration is performed for statistical-mechanical purposes). With respect to such limited goals, there is no evidence from the various applications made so far by using semi-empirical formulas (see the brief review below), that these would be seriously incorrect, provided, of course, that abusive simplifications are not used: thus, except for molecules with very small net charges, it is rather dangerous to neglect the electrostatic contribution and, to a lesser extent, the polarization one (thus, the 'hydrogen bonds' may be obtained by using a suitable set of simplified semi-empirical formulas, but they could not be by using a dispersion–repulsion formula only). We have emphasized in Sections IV.E and V.E that the theoretical analysis suggested a sum of bond–bond contributions (rather than atom–atom) for the short range repulsion and dispersion, and this would agree with Mies' (1965) criticism; but this situation would not prevent the atom–atom type formulas to display a convenient approximation for not too refined purposes. It must also be mentioned that some refinements (not considered up to now) of the atom–atom terms are conceivable: thus, non-isotropic atomic polarizability tensors (obtained by sharing the bond polarizability tensors) could be introduced instead of the isotropic (mean) atomic polarizabilities used so far for the sake of simplicity.

We should therefore conclude that such criticisms like those of Harris *et al.* (1975) may usefully remind us that any simplified, semi-empirical formula may eventually fail for some refined problem, but these warnings do not invalidate in any way the relevance of such formulas for treating a variety of problems involving medium-sized (or large) molecules, for which a knowledge of the 'gross' features of the intermolecular potential functions is sufficient.

B. Interactions between two molecules

It must be recalled that we consider here only examples of applications of simple, explicit analytical formulas. Direct quantum-mechanical calculations, whether *ab initio* or more approximate, are outside our present scope. Examples of *ab initio* calculations have been given in the Introduction (Part I), and applications to solute–solvent interactions are reviewed by A. and B. Pullman (1975). As concerns the approximate calculations, we quote some applications to the hydrogen bond (Van Duijneveldt and Murrell, 1967; Van Duijneveldt, 1968, 1969) and to the interaction between non-polar groups (hydrogen atoms) (Murrell and Shaw, 1968a; Lekkerkerker and Laidlaw, 1970b; Laidlaw *et al.*, 1971; Magnasco and Musso, 1971a,b). Applications to biological molecules (especial the purine and pyrimidine bases) are treated in the chapter by Rein in this book.

Following the remarks of Dewar and Thompson (1966) and Malrieu and Claverie (1968) concerning the role of the usual Van der Waals interactions in the so-called charge-transfer complexes (see Section III.A.3 above), Mantione (1968a, 1969b) evaluated these interactions (by using simplified formulas of the type described in the previous section) for the complexes between tetracyanoethylene (TCNE) and a number of aromatic hydrocarbons and heterocyclic compounds. The results were in satisfactory agreement with the experimental thermodynamic parameters (free energy F and internal energy U) corresponding to the formation of the complex. Strictly speaking, the result of the theoretical evaluation gives U for the isolated complex (in the gas phase), and Mantione (1969b) could compare these theoretical energies with their experimentally available counterparts for complexes between TCNE and four methylated benzene compounds. Most frequently, the experimental measurements concern complexes in solution, and the comparison with the theoretical U becomes qualitative only: Mantione (1968a) thus obtained a linear relationship between the theoretical interaction energy and the experimental enthalpy H measured in solution (the correlation is, of course, less good when the experimental free energy F is used instead of H). In two other papers, Mantione (1968b, 1969a) studied the contribution of the polarization to the dipole moment of several 'charge-transfer' complexes. Mantione (1968a) also studied the interaction between riboflavine and several purine molecules in order to investigate the optimum relative configuration of the riboflavine and purine moieties in the FAD (flavine-adenine-dinucleotide) and its analogues.

A number of calculations of the interaction energy between the purine and pyrimidine bases were performed by several authors, owing to their biological importance: for example, see Pullman *et al.* (1967) (study of planar triplets of bases), Claverie (1968b) (study of the helical structure of DNA), Pollak and Rein (1968) (study of the energies of Crick's wobble hypothesis), Ninio (1971) and the Appendix by Claverie (1971b) (interactions between stacked triplet of bases and their bearing upon the understanding of the genetic code), the thesis by Caillet (1972), the review papers by A. and B. Pullman (1968) and by Rein (1973), and finally the chapter by Rein in this book.

For small rigid molecules, semi-empirical intermolecular potentials have been used (and fitted) for reproducing the second virial coefficient: for example, see Hirschfelder, Curtiss and Bird (1954, Ch. 3) and, for the special case of water, Eisenberg and Kauzmann (1969, Ch. 2).

Evans and Watts (1975) evaluated the second virial coefficient for benzene in terms of their potential involving six 'centres of force' (different from the atoms) for each molecule. Their purpose was in fact to fit the adjustable parameters of their potentials. A further application to the crystal of benzene (Evans and Watts, 1976) involving three-body terms will be reported in Section E below.

Lamotte (1973) and Lamotte *et al.* (1974) studied the interaction between naphthalene molecules and solvent alkane molecules and found preferred orientations, as suggested by experimental findings about spectral solvent shifts (Lamotte, 1973; Lamotte and Joussot-Dubien, 1973) and in agreement with the model proposed by Groszek (1970) for the adsorption of hydrocarbons on graphite.

As a final example, Morris (1974) calculated the interaction energy between two anthracene molecules, in order to elucidate the structure of the anthracene dimer, and stressed the necessity of including the electrostatic energy for a proper treatment: the dispersion energy alone would favour a strictly parallel configuration ('sandwich dimer') while the electrostatic energy would favour a configuration where the molecular planes are perpendicular (the large symmetry axes remaining parallel); the experimental configuration actually is intermediate between these two extremes. Nevertheless, a possible role of the solvent should not be overlooked, and Surel-Pée (1974) made a first attempt to include the interaction with solvent molecules in a study concerning the anthracene dimer (in alkane solvent): according to the calculations for the dimer *in vacuo*, the sandwich configuration remains the most stable (because the gain of electrostatic energy is smaller than the loss of dispersion energy when the configuration departs from the sandwich one); but in a model calculation involving three pentane molecules besides the two anthracene ones, a second energy minimum (besides the 'sandwich' one) was found, corresponding to an angle of about 60° between the small axes of the anthracene molecules (with one of the pentane molecules inserted in this dihedral angle between the planes of the anthracene molecules). Thus, the calculations of Surel-Pée would suggest that the 'open' configuration of the dimer is due to the solvent effect rather than to some intrinsic stability for the isolated dimer. Whatever the issue, it is very likely that the solvent or environment effect cannot be safely neglected when dealing with such loose complexes (striking environment effects will be reported in Section E below devoted to crystals). Thus, except when dealing with complexes in the gas phase (on which the experimental data are rather scanty), it may be a controversial matter to relate the calculations concerning two molecules with the actual physical situation for which experimental data are available.

C. Interactions between one molecule and a surface

The study of surface state has become very recently a subject of rapidly increasing interest, and researchers in that domain try to establish a connection between the widely different methods which were developed in the past in solid-state physics on one hand and in quantum chemistry on the other hand. Here we do not deal with such questions and we simply recall some calculations of the interaction energy between the numerous molecules of a surface (solid or liquid) and one molecule outside.

Poshkus (1965) calculated the interaction energy between a graphite surface and rare gas atoms or some non-polar molecules (H_2, N_2, C_6H_6), and evaluated from them various thermodynamic parameters pertaining to the adsorption process (a comparison with the interactions in the gas phase was also performed). At the same discussion of the Faraday Society (concerning Intermolecular Forces), Casanova *et al.* (1965) reported experimental results concerning the adsorption of argon on graphitized carbon black, Everett (1965) discussed theoretical and experimental results concerning the interaction between adsorbed molecules, and Kiselev (1965) presented a classification of the various kinds of molecules and adsorptive surfaces.

Kouznetsov and Guiochon (1969) calculated the adsorption potential on graphitized carbon black of large molecules (terpenic compounds and aromatic hydrocarbons). Rothstein (1970) treated the thermodynamics of a gas–solid complex by using an approximate cell-model (for the adsorbed gas atoms) with a potential calculated as a sum of atom–atom terms (the actual application concerned the adsorption of xenon and myoglobin). Pisani *et al.* (1974), Battezzati *et al.* (1975) study the equilibrium configuration and surface motion of hydrocarbon molecules (methane, ethane, propane, ethylene, benzene, cyclohexene and cyclohexane) adsorbed on graphite; they use for the calculation of the interaction energy between the molecule and graphite an interesting method proposed by Steele (1973): instead of summing directly over all carbon atoms of a graphite reticular plane the atom–atom terms pertaining to every atom of the molecule, the coefficients of the Fourier series expansion of this total atom–graphite interaction potential are calculated first, and the potential is then computed for the necessary position by using this Fourier expansion. This procedure is made possible by the two-dimensional periodic character of the potential (with respect to the displacements of the atom parallel to the graphite planar surface). In practice, a few terms of the Fourier expansion display a sufficient accuracy, and this procedure offers two advantages: first, the computing time for the evaluation of the energy is strongly reduced, and second the explicit analytical form of the potential given by its Fourier expansion makes easy the analytical calculation of first and second derivatives, which are quite interesting in order to use highly efficient minimization procedures, and are altogether necessary for a normal mode analysis of the motion of the adsorbed molecules. It must be mentioned that Battezzati *et al.* (1975) use only a Lennard–Jones potential for their atom–atom term; some supplementary cautions should be taken when using an electrostatic term, because of the slower decrease of this term with increasing distance. Finally, it may be noted that the equilibrium configuration found for propane by Battezzati *et al.* (1975, Table 2.b) agrees with the optimum configurations found for naphthalene–alkane complexes by Lamotte *et al.* (1974) and with the model proposed earlier by Groszek (1970) for the adsorption of hydrocarbons on graphite (see previous Section B).

D. Solute–solvent interactions

As explained at the beginning of the present Section VI, we shall not deal here with the most elaborate statistical-mechanical theories of liquid mixtures. Indeed, up to recent years, these theories dealt essentially with simple liquids, whose molecules (or atoms, in the case of rare gases or metals) interacted through central forces, with a *single* centre per molecule (for example, an account of this stage of the theory may be found in Rowlinson,

1969, Ch. 9 or Hill, 1960, Ch. 18 to 20); see also Gurikov (1968a,b). It may be especially mentioned that the success of the so-called perturbation theory for the pure simple liquids motivated its application to liquid mixtures (Barker and Henderson, 1968; Leonard *et al.*, 1970; Grundke *et al.*, 1973; Henderson, 1974). But the application to mixtures of non-simple liquids would be possible only if good 'reference states' could be calculated for the pure non-simple liquids, and methods aiming to that purpose are not yet fully developed (see the RISM method, which was briefly discussed in the Introduction of the Chapter (Part I) and at the beginning of the present Part VI).

Consequently, despite the fact that the refined statistical-mechanical treatments for the liquid state (pure or mixture) look presently much more promising than they did, say, ten years ago, much work remains to be done before they can be applied in a routine fashion to problems involving medium-size molecules.

We shall therefore mention here only some practical applications of intermolecular interactions calculations involving more or less drastic simplifications as concerns the statistical-mechanical treatment. We shall consider only the case of *dilute solutions* (each solute molecule is surrounded by solvent molecules only). Two types of calculations may be distinguished: those in which a certain number of solvent molecules surrounding the solute one (for example, the first solvation shell) are explicitly considered ('discrete model'), and those in which a more drastic simplification is performed, namely the solvent discrete molecules are replaced by a continuum medium endowed with suitable properties (hence the name 'continuum model'). Actually, it may be easier to incorporate convenient statistical features in a continuum model than in a discrete one: thus, we shall see that a continuum model for the electrostatic energy directly gives the (Helmholtz) free energy rather than the energy only (then, energy and entropy may be easily obtained).

By contrast, a discrete model appears naturally suited for giving energy (or enthalpy) only: unless some kind of statistical treatment is added, entropy (and the thermodynamic functions involving it) will not be obtained. For the so-called 'regular solutions' (for example, see Hildebrand and Scott (1962), Hildebrand *et al.* (1970)), the evaluation of the energy would indeed suffice, since the entropy changes are evaluated independently of the intermolecular potentials, under the assumption of a purely random distribution of the various molecules. But such an assumption is not likely to be satisfactory for a solvent like water with strongly anisotropic interactions. The entropy changes associated with the modification of the statistical behaviour of water molecules around a solvent are well-known ('iceberg effect' accompanying the solvation of non-polar gases in water (Frank and Evans, 1945)), and certainly play a role in the so-called 'hydrophobic bonding' (Kauzmann, 1959; Némethy and Scheraga, 1962b); see further references in Section 3 below). These points must be kept in mind when the results of discrete model calculations (without statistical treatment) are applied to the study of solvent effects.

1. Discrete-model calculations

The principle of these studies is to put some solvent molecules around the solute molecule (usually not more than a first solvation shell, since the computational effort increases considerably with the number of molecules) and to evaluate the interaction energy between the solute and the solvent molecules, and also between the solvent molecules themselves (the optimal configurations are indeed determined by the total energy, and the statistical behaviour as well). In an early work (Huron and Claverie, 1969), the solvent molecules were put around the solute as reasonably as possible and then the solute–solvent interaction energy was calculated. But such a procedure involved a too great uncertainty, and a computer program performing the minimization of the *total* energy (solute–solvent + solvent–solvent) with respect to the configurations (position and orientation) of all solvent molecule was put into use (Daudey, unpublished results). It turns out that the total energy exhibits a number of different local minima (which may be obtained by varying the initial configurations of the solvent molecules), although *some* especially favourable configurations may appear almost unchanged in the various solvation shells corresponding to these different minima of the total energy (such favourable configurations define 'solvation sites' pertaining to the solute molecule (hydration site when the solvent is water); for direct quantum mechanical studies of such interactions, see A. Pullman (1975) and the review by A. and B. Pullman (1975)). The method was used for explaining the solvent effect on the absorption and fluorescence spectra of the solute molecule (Mantione and Daudey, 1970; Mantione, 1971). It was also used by Huron and Claverie (1971) for evaluating the solvation energy in nitrobenzene of a series of gases (saturated and non-saturated hydrocarbons, CO_2, SH_2, SO_2). The solvation energy is calculated as the difference between two large numbers: the solute–solvent interaction energy, and the cavitation energy (associated with the creation in the solvent of a cavity able to accommodate the solute molecule). This situation is unfavourable as concerns numerical accuracy, and it turned out necessary, in that study, to use for every solute a Boltzmann average of the different solute–solvent energy values corresponding to the local minima (obtained from various initial configurations by the minimization procedure, as mentioned above). Such a procedure obviously appears as a crude substitute for the calculation of the exact statistical mean energy.

As another example, the discrete model is used presently in our laboratory (together with the continuum model, see the end of the next section) for investigating the effect of the solvent (water) on the relative stability of two conformations (gauche and trans) of the acetylcholine molecule. At this point, it may be mentioned that the interaction energies involved in the discrete model may be obtained from direct quantum-mechanical calculations (of the 'supermolecules' and the isolated molecules) instead of using the kind of simplified formulae considered in the present work (see the review by A. and B. Pullman, 1975). This method of calculation is much more sophisticated, and therefore also much more time-consuming; besides its own interest, it gives some check for calculations based upon simplified formulas.

We shall finally mention that Dashevskii and Sarkisov (1974) considered the possibility of performing the statistical-mechanical evaluation of the relevant thermodynamic functions in terms of simplified intermolecular potentials by using the Monte-Carlo method; in their paper, they are more especially interested in the solvation and hydrophobic interaction of non-polar molecules in water (actually they consider a hard sphere and a methane molecule as examples of such solutes). The Monte-Carlo method is, of course, rather time-consuming, but it may give quite reliable results for all thermodynamic function (energy, free energy, entropy) and such results obtained for simple systems (as those considered by Dashevskii and Sarkisov) are extremely interesting.

2. Continuum-model calculations

The importance of the computational effort required by discrete model calculations if numerous solvent molecules are considered (and especially if some statistical treatment, requiring a large number of configurations, is desired) constitutes a strong incentive to develop 'continuum models' where the complexity of the statistical-mechanical behaviour of the solvent molecules is replaced by the suitable properties of a continuum medium (polarizability per unit volume, dielectric constant, thermal expansion coefficient . . .) The continuum models are not restricted to purely energetic quantities: thus, in the case of the electrostatic interaction, these models directly give the free energy rather than the internal energy itself (see below).

The conceptual basis of the continuum models is clear: at large distances from the solvent molecule, the discrete summations of the intermolecular interactions over the solvent molecules may be rightfully replaced by appropriate continuous integrals. The essential difficulty lies in the use of such an integral also at intermediate distances from the solute molecule, where it may be feared that such a procedure is unable to reflect correctly the complexity of the behaviour of the discrete solvent molecules. A natural improvement would consist in treating the closest solvent molecules (for example, the first solvation shell) as discrete molecules, the rest of the solvent being considered as a continuum. Some steps towards such a 'discrete-continuum model' will be mentioned below.

As concerns the continuum models themselves, two kinds of problems may be distinguished: to define suitable integrals corresponding to the various terms of the intermolecular potential (electrostatic, polarization, dispersion, short-range repulsion), and to use a suitable cavity, adapted to the shape of the solute molecule (the domain outside this cavity is assumed to be filled with the continuum medium, and the integrals therefore extend over it). In early works using a continuum model, only some part of the interaction potential was considered, and very simple shapes of the cavity (spheres or ellipsoids) were used, for mathematical convenience. A first series of works dealt with the electrostatic interaction only: a set of charges (or eventually higher-order multipoles) inside the cavity, representing the charge distribution of the solute (see Sections V.B and VI.A.1), interacts with the solvent outside the cavity, considered as a dielectric continuous medium (characterized by a dielectric constant). Thus, the problem is to solve the Poisson equation (which reduces to the Laplace equation outside the cavity since there are net charges only inside) with the appropriate boundary conditions at the surface of the cavity. After the early work of Born (1920) (a charge at the centre of a sphere), Kirkwood (1934) considered a set of charges in a sphere, and later in an ellipsoid (Kirkwood and Westheimer, 1938; Westheimer and Kirkwood, 1938; see also the collection of reprints of Kirkwood, 1968); later on, the same problem was further investigated by Hill (1944), Abbott and Bolton (1952), Wada (1954), Buckley and Maryott (1954) and also Buckley (1958) who gave various applications. Very recently, Buff *et al.* (1975) considered the interaction between *two* spheres (containing point charges) embedded in a dielectric medium.

Besides the electrostatic term, the polarization term was also included by Onsager (1936) in his well-known paper about the theory of the macroscopic dielectric constant for polar substances (the system considered was simply a polarizable dipole at the centre of a sphere), and the case of a general (polarizable) charge distribution in a sphere was treated by Bonnor (1951).

In all these treatments, the connection with thermodynamic quantities is made by considering a fictitious charging process, i.e. the charges of the central molecule (and also, eventually, those of the solvent, which contribute to its dielectric constant) are multiplied by a 'charging parameter', which is varied continuously from 0 to 1 in order to 'turn on' the interaction. As a consequence of this procedure, the (Helmholtz) free energy is obtained (as we announced above), and the energy and entropy may be easily deduced from it. The use of such fictitious charging processes is a powerful theoretical tool; it must, however be pointed out that the use of this tool may involve subtle difficulties, as exemplified by the discussion of Kirkwood (1934, footnote on p. 356), concerning two different charging processes, the Debye–Hückel one and the Güntelberg one. In the former, all charges (or, more generally, all 'constants of forces') of the solute and solvent molecules are varied simultaneously, and the *total free energy* F the system is obtained; in the latter, only the charges of the solute are varied, and, in principle, the *chemical potential* ($\partial F/\partial N$) should be obtained (Fowler, 1936, Sections 8.81 and 13.6 to 13.7; Eyring *et al.*, 1964, Ch. 14). Some aspects of these questions in the solute–solvent problem are considered by Huron and Claverie (1972, Appendix C; 1974b, Appendix B) and by Dashevskii and Sarkisov (1974), but these questions cannot yet be considered as definitely settled in the case of approximate treatments. The 'charging parameter' method works quite successfully in the Debye–Hückel theory of ionic solutions, which essentially amounts to evaluating the interaction between a charge placed at the centre of a sphere (the ion) with the surrounding medium characterized not only by a dielectric constant (which represents the solvent) but also a non-zero ionic strength (which represents the other ions); thus, the Poisson–Boltzmann equation is used instead of the Poisson equation (which corresponds to the case of zero ionic strength): about the Debye–Hückel theory, for example, see Fowler (1936, Section 8.81 pp. 269–274 and Sections 13.6, 13.61, 13.62, 13.7, pp. 541–552), Eyring *et al.* (1964, Ch. 14, p. 401) and Part II of Parsegian's (1973) review; about the 'potential of mean force' which is an important concept in these theories, see Fowler (1936, Section 8.8, pp. 267–269) and Hill (1956, Section 31, pp. 193–195 and Section 37, pp. 237–239). Actually, Kirkwood (1934) considered the interaction of an arbitrary charge distribution inside a sphere with a solvent having a non-zero ionic strength besides its dielectric constant. An example of application involving a cylindrical cavity embedded in a solvent with non-zero ionic strength will be given in Section F below: this model was introduced for studying the problem of the intercalation of amino-acridines between the base pairs of DNA. It is an interesting feature of the continuum model for the electrostatic interactions that it may, in principle, be extended to solutions with non-zero ionic strength (in practice, of course, the difficulty of solving the Poisson–Boltzmann equation instead of the Poisson one should be carefully examined, but if a numerical method is used for solving the former equation in a way similar the one mentioned below for solving the latter, it is unlikely that considerable differences of computational effort do occur).

As concerns now the dispersion term of the intermolecular potential, it was considered by other workers for non-polar molecules. The status of this term appears somewhat simpler, since it is always negative (while the sign of the electrostatic term is orientation-dependent). This feature suggests a very simple continuous-like calculation: Every infinite sum $-\sum_j C_{ij}/(R_{ij})^6$, where i refers to some atom of the solute and j to the atoms of the solvent, is replaced by an integral

$$-c_i \int_{\Omega_{\mathrm{E}}} \frac{1}{|\vec{r}-\vec{R}_i|^6}\,\mathrm{d}v,$$

where c_i is a suitable constant (which accounts for the volume density of the various kinds of atoms of the solvent molecules), and Ω_{E} denotes the domain of space outside the cavity (E stands for external). Such a procedure was indeed used by Bakshiev *et al.* (1968) in order to investigate the effect of non-polar solvents on the optical spectrum of the solute; the cavity was assumed spherical. On the other hand Linder (1960, 1961, 1962) developed reaction-field methods for evaluating this solute-continuous medium dispersion energy (these methods resemble more closely to those previously described for the electrostatic energy: in this case, the reaction-field indeed naturally appears as the electric field created in the solute cavity by the surrounding dielectric, as a consequence of its polarization by the charges of the solute). Here, too, the solute cavity was assumed spherical for practical purposes.

Somewhat later, Linder (1967) presented a unified theoretical framework (always based upon reaction-field techniques) where the three long-range contributions (electrostatic, polarization, dispersion) were treated. But no detailed practical application was reported, and, particularly, none of the technical problems associated with the possible use of non-spherical solute cavities was considered.

By contrast, Huron and Claverie (1972, 1974a,b) focused their attention upon the use of *realistic cavities with arbitrary shape;* moreover, all four contributions to the simplified intermolecular potential (the three long-range ones and the short-range repulsion) were taken into account. The cavity associated with the solute is defined by the Van der Waals surface, namely the union of Van der Waals spheres centred at every atom of the solute molecule (more precisely, the Van der Waals radii used in the intermolecular potential (see Section A.3 above) are multiplied by a suitable constant, so that the volume of the molecular cavity agrees with the molar volume of the pure solute at the same temperature). The dispersion and repulsion energies are evaluated by integrating the corresponding potentials $-1/|\vec{r}-\vec{R}_i|^6$ and $\exp(-b|\vec{r}-\vec{R}_i|)$ over the external volume Ω_E, for every atom i of the solute (Huron and Claverie, 1972). The volume integrals are transformed into surface integrals (which are themselves calculated numerically) and suitable multiplicative constants are introduced, so that the continuum method reproduces the results of discrete summations for the energy between a 'solute' sphere and 'solvent' spheres packed around it (the radius of the 'solute' sphere being eventually different from the radius of the 'solvent' spheres). This already allows the treatment of non-polar molecules, and satisfactory values of the vaporization energies were thus obtained for a number of saturated hydrocarbons. Then, a method for evaluating the electrostatic free energy was developed according to the principles outlined above (solution of the Poisson equation with suitable boundary conditions on the surface Σ of the cavity). Owing to the complicated shape of this surface (made from pieces of spherical surfaces), an approximate solution must be searched by a numerical method (the electrostatic potential is expanded inside Σ and outside Σ by using two different suitable bases, limited to a finite number of functions) (Huron and Claverie, 1974a). When calculating all interaction terms (except the polarization of the solute, which is markedly smaller than the total interaction energy, and which could be calculated easily if necessary), satisfactory values were obtained for the vaporization energies of a number of various polar molecules (Huron and Claverie, 1974b). When the 'solute' molecule is actually different from the 'solvent' ones, it is necessary to evaluate also the cavitation energy (corresponding to the creation in the pure solvent of the cavity accommodating the solute molecule). The process of introducing a solute molecule into a solvent may be conveniently divided in two steps (e.g. see Uhlig, 1937; Eley, 1938; Pierotti, 1963): (1) creation of a suitable cavity in the solvent; (2) introduction into this cavity of the solute molecule which then interacts with the solvent. It must be emphasized that this second step involves not only the 'switching-on' of the solute–solvent interaction, but also the change of the energy of the solvent itself, due to its reorganization in the neighbourhood of the solute molecule. The cavitation energy can be evaluated, for example, by using a suitable value for the surface tension: we are indeed dealing with cavities of molecular size, so that the macroscopic surface tension cannot be used directly, and curvature corrections must certainly be introduced (Choi *et al.*, 1970). For example, this 'surface tension approach' to the cavitation problem was followed by Sinanoglu and Abdulnur (1964, 1965), Claverie *et al.* (1966), Sinanoglu (1967a,b; 1968), and also Halicioglu and Sinanoglu (1969). Another approach was proposed by Pierotti (1963, 1965) in his studies of the solubility of non-polar gases (in non-associated solvents and in water): namely, he used the theoretical evaluation of the free enthalpy of cavitation in a fluid of hard spheres made by Reiss *et al.* (1959, 1960), Helfand *et al.* (1960), Reiss and Mayer (1961) and Mayer (1963). Of course, all these simplified evaluation of the cavitation thermodynamic functions must be used with adequate care, as pointed out by Dashevskii and Sarkisov (1974) who studied through 'computer experiment' (Monte-Carlo method) the solvation of a hard sphere or a methane molecule in water.

At the present time, the continuum model just described is being applied in our laboratory, to the problem of the conformation of acetylcholine in water, together with discrete model (as mentioned above). This will allow a thorough comparison of both methods on a non-trivial case. In a previous attempt concerning this system Beveridge *et al.* (1974) indeed used a too crude model (a dipole at the centre of a sphere) whose results cannot be considered as reliable for such a complicated molecule, and a quantum-mechanical *ab initio* calculation of the interaction energy of acetylcholine with five strongly interacting water molecules actually leads to rather different results (Pullman *et al.*, 1975).

This is a clear-cut example where sufficiently refined continuum and discrete models must be used in order to settle the problem in a mutually consistent way.

3. *Other miscellaneous treatments*

As mentioned previously, it is tempting to develop a 'discrete-continuum' model: a finite number of solvent molecules are treated explicitly as discrete molecules, and the surrounding solvent is represented by a continuous medium. Some steps have been made in this direction by Beveridge and Schnuelle (1974, 1975) and also Schnuelle and Beveridge (1975).

The convex-core potential of Kihara (1953, 1955, 1967, 1970) (see the end of Section A.3) was applied to the study of the solubility of some gases: Ne, Ar, Kr, Xe, N_2, CO_2, ethane, cyclopropane in cyclohexane (Kihara and Jhon, 1970); Ne, Ar, Kr, Xe, N_2, CO_2, BF_3 methane, acetylene in benzene (Kihara and Yamazaki, 1971); Ne, Ar, Kr, Xe, N_2, O_2 methane in water (Jhon *et al.*, 1972) (this paper uses the significant structure theory (Eyring *et al.*, 1964, Ch. 12; Eyring and Jhon, 1969); according to the review by Rowlinson (1971) of Eyring and Jhon's (1969) monograph, there is some controversy concerning the theoretical status of the significant structure theory; anyway, even if the theory did have some semi-empirical character (as compared, for example, with the approaches based upon distribution functions, see Fitts, 1972, Section 6.2.3.), this would not prevent it from being successful for specific applications, as reported by Eyring and Jhon (1969)). The simplicity of the convex-core potential (2.369) allows an analytical evaluation of the dispersion and repulsion energies between the solute molecule and the solvent ones. The price to pay for this simplicity is the lack of refinement of the potential and the impossibility to apply this procedure to non-convex molecules (a case which occurs inevitably when dealing with long and/or ramified molecules).

Another line of work dealt with the statistical-mechanical aspects of solute–solvent problems, with less emphasis upon specific intermolecular potential functions: for example, see the theory of the solubility of non-polar gases in non-polar solvents and in water by Pierotti (1963, 1965) and the theories of hydrophobic bonding (Kauzmann, 1959; Némethy and Scheraga, 1962a,b; Salem, 1964; Némethy, 1967; Ben Naim, 1971a,b, 1972a,b; Wilhelm and Battino, 1972; Hermann, 1971, 1972, 1975a,b). The book by Lewin (1974) considers essentially entropic (rather than energetic) effects. Rather empirical procedures have also been proposed for dealing with the solute–solvent problems (for example, see Weintraub and Hopfinger, 1974), but their degree of accuracy is uncertain at the present time.

A problem related to the one considered in this section is the study of the *effective interaction* between two solute molecules embedded in a solvent. Some steps in this direction (which involve the consideration of many-body effects beyond the second-order perturbation terms, see Section V.F) were made by Kestner and Sinanoglu (1963, 1966), Sinanoglu and Abdulnur (1964, 1965), Sinanoglu (1967a,b, 1968).

Finally, it may be mentioned that another series of works, based upon the macroscopic theory of Van der Waals forces (Dzyaloshinskii *et al.* (1961); see the beginning of Section V.E.3 and the Section V.6), studied the interaction between (quasi-macroscopic) layers of molecules separated by some other medium (for example, lipid layers separated by an aqueous medium, as it occurs when studying membranes): for example, see

Derjaguin *et al.* (1965), Scheludko *et al.* (1965), and the review papers by Parsegian (1973) and Israelachvili (1974). But it must be kept in mind that this kind of approach concerns (at least up to the present time) the interaction between sets of numerous molecules separated by *large* distances (much greater than 5 Å, say): as a consequence, such methods could not be applied (in their present stage) to the problem dealt with in the present section, namely the interaction between one solute molecule and the surrounding solvent. Some connection should be easier to establish with the 'microscopic' approach to the problem of the effective interaction between two molecules embedded in a solvent, which was referred to in the previous paragraph.

E. Interactions in crystals

The evaluation of interactions between molecules in crystals has certainly been, up to now, the main field of application of the simplified formulas for the interaction energy. Actually, most of the authors quoted in Section A.3 used simplified or 'semi-empirical' formulas for applications to crystals. Indeed, as long as a complete thermodynamical description of the crystal is not wanted, calculations of the lattice energy, which involve no statistical-mechanical treatment, are sufficient; anyway, such lattice energy calculations would be the basis of statistical mechanical treatments giving a more elaborate description of the crystal at finite temperature; but they still involve the summation of a large number of interactions, hence a strong motivation for using the simplest possible formulas.

Owing to the equivalence of all molecules in a crystal (boundary effects are neglected), the total energy of a lattice may be simply connected with the interaction between any single molecule and the surrounding ones. The precise relationship of course depends upon the pair-wise property of the intermolecular potential. Thus, with a potential of the kind discussed in Section A, the total interaction energy of the lattice is:

$$E=\sum_{p<q}\sum u_{pq}^{(2)}+\sum_{p} u_p^{\mathrm{pol}} \tag{2.370}$$

where $u_{pq}^{(2)}$ denotes the sum of the pair-wise additive terms (electrostatic, dispersion, short-range repulsion) between the molecules p and q, while u_p^{pol} denotes the polarization energy of the molecule p in the *total* electric field created by all surrounding molecules. Now

$$\sum_{p<q}\sum u_{pq}^{(2)}=\tfrac{1}{2}\sum_{p\neq q}\sum u_{pq}^{(2)} \tag{2.371}$$

and, owing to the equivalence of all molecules,

$$\sum_{q}^{[p]} u_{pq}^{(2)}$$

(where $[p]$ means that $q \neq p$ in the summation) is independent of p, and the same property holds for u_p^{pol}. Hence, denoting

$$U^{(2)} = \sum_q^{[p]} u_{pq}^{(2)} \quad \text{and} \quad U^{\text{pol}} = u_p^{\text{pol}} \tag{2.372}$$

and denoting N the total number of molecules, we get:

$$E = \frac{N}{2} U^{(2)} + NU^{\text{pol}} \tag{2.373}$$

and the energy per molecule

$$E/N = \tfrac{1}{2} U^{(2)} + U^{\text{pol}} \tag{2.374}$$

Many-body contributions to the energy could be handled in a similar way. Let us consider, for example, a three-body $u_{pqr}^{(3)}$. The sum over the different triplets of molecules becomes successively:

$$\sum\sum\sum_{p<q<r} u_{pqr}^{(3)} = \frac{1}{3!} \sum\sum\sum_{(p \neq q \neq r)} u_{pqr}^{(3)}$$

where $(p \neq q \neq r)$ means that all three indexes are different

$$= \frac{1}{3!} \sum_p \left[\sum^{[p]} \sum_{q \neq r}^{[p]} u_{pqr}^{(3)} \right] = \frac{1}{3!} \sum_p \left[2! \sum^{[p]} \sum_{q<r}^{[p]} u_{pqr}^{(3)} \right]$$

$$= \tfrac{1}{3} \sum_p \left[\sum^{[p]} \sum_{q<r}^{[p]} u_{pqr}^{(3)} \right] = \tfrac{1}{3} NU^{(3)} \tag{2.375}$$

where $U^{(3)}$ denotes the bracketed sum, which is the same whatever the molecule p, and the contribution to E/N will be $U^{(3)}/3$.

Thus, the contribution of an n-body term to the lattice energy generally involves an $(n-1)$uple summation. In principle, the polarization energy in its fully developed form would involve two-body and three-body terms (see Section V.D.3, equation (2.264)); but, fortunately, this three-body character only appears through the *quadratic* dependence of the polarization energy on the electric field, which is itself a two-body additive (vector) quantity. Thus, the total electric field at a given molecule (or, more precisely, the total electric fields created on the various atoms or bonds of this molecule) may still be evaluated by a simple summation, like the two-body contribution $U^{(2)}$, and the polarization energy is evaluated only after the end of the summation processes (see formulas (2.263) in Section V.D.3 and (2.364) in Section VI.A.2).

But the consideration of a genuine three-body term (terms beyond the second-order of perturbation, see Section V.F) would require a double summation, which means a significant increase of computation time. It is therefore essential to assess the relative magnitude of such contributions with respect to those involving a simple summation only (namely $U^{(2)}$ and

U^{pol}). Available studies deal with atoms or non-polar molecules in a one-dimensional lattice (Bade and Kirkwood, 1957) and in three-dimensional liquids and crystals (Kestner and Sinanoglu, 1963) indicate that the n-body contributions ($n > 2$), although small, are not completely negligible: Kestner and Sinanoglu (1963) evaluate them from 5 per cent to 10 per cent of the dispersion contribution, with the opposite sign (positive, i.e. destabilizing). This order of magnitude is confirmed by the explicit computation of the three-body interaction for the benzene crystal (Evans and Watts, 1976, Table 2). Similarly, for the case of the rate gases, Dymond, Rigby and Smith (1965) state that it is not possible to reconcile exactly the experimental data for the gaseous state and the crystalline state if the terms beyond the two-body ones are ignored in the latter case. It is, of course, always possible to fit purely two-body potentials on crystal data, but the potentials thus obtained should be considered as 'effective' two-body (see Sinanoglu, 1967, 1968) rather than the exact two-body part of the true potential. All applications quoted below proceed this way, except the calculations of Evans and Watts (1976) concerning the benzene crystal, and our own works (Caillet and Claverie (1974, 1975), Caillet *et al.* (1976)), where we consider our two-body potential as the true one, and we simply evaluated the higher-order terms to 7 per cent of the dispersion contribution (with the opposite sign): this is the value given by Sinanoglu and Kestner for methane. But if this value holds for the 'non-polar' part of the interaction, other terms involving the permanent multipoles would occur when considering 'polar' molecules (Stogryn, 1971); by 'polar' molecules, we mean not only dipolar molecules, but also molecules with a non-negligible higher-order multipole moment, i.e. molecules with non-negligible net charges. Fortunately, the molecules of a crystal lattice necessarily exhibit different orientations with respect to any given 'central' molecule, so that the electric fields (and hence the energy terms depending on them) tend to cancel each other (while the dispersion terms always add up). Thus, already the usual electrostatic energy contributes significantly less than the dispersion to the total lattice energy (Kitaigorodskii and Mirskaya, 1964a,b, 1965; Zorkii *et al.*, 1970), the polarization energy is still smaller (Caillet and Claverie (1975) Section 3.D, results for nitrobenzene: electrostatic energy: −1·95 kcal/mol; polarization: −0·42; dispersion: −20·4; repulsion: +6·66 higher-order terms (−7 per cent of dispersion): 1·43; total: −14·68), and it may therefore be hoped that the higher-order terms connected with the non-zero charges will be actually negligible for most molecules. The results of Caillet and Claverie (1975) for carbon dioxide would suggest that it is an exception (electrostatic: −2·51; polarization: −0·42; dispersion: −8·78; repulsion: 5·70), but it is actually a small molecule (hence giving a moderate dispersion energy) with very large net charges, as manifested by its high quadrupole moment and the electrostatic energy is more important than usual with respect to the total energy (see also Kitaigorodskii *et al.*, 1969). But this situation does not hold for benzene, and the result of Sen and Basu (1968) who found for the benzene crystal an electrostatic contribution as large as the experimental sublimation energy (about 10 kcal/mol) means that these authors used an overestimated value of the quadrupole moment.

As a general rule, the application of the lattice energy calculations involves the search of the minimum (or rather minima) of this energy as a function of the various degrees of freedom (lattice parameters, orientations of the molecules in the unit cell, and eventually conformational parameters of the molecules themselves). The minima thus found are then used for interpreting (or eventually predicting) the experimental crystal structure. Most frequently, the lattice symmetry and parameters are supposed to be known, and the theoretical study deals with the other degrees of freedom (for example, see Giglio and Liquori, 1967; Giglio, 1969).

Many authors performed calculations involving the dispersion and short-range repulsion terms only, by limiting themselves to weakly polar molecules. Thus, crystals of hydrocarbons were extensively studied: Williams, 1968), Rae and Mason (1968), Kitaigorodskii (1961a,b, 1965), Kitaigorodskii and Mirskaya (1961, 1964a), Kitaigorodskii and Ahmed (1972), Liquori and co-workers (for example, see references in Di Nola and Giglio, 1970). Crystals involving other atoms besides hydrogen and carbons (e.g. halogenated hydrocarbons, orthorhombic sulphur trioxide, cubic α-nitrogen, orthorhombic cyanogen) were also studied with the same kind of formulas (see Giglio, 1970; Di Nola and Giglio, 1970; Coiro *et al.*, 1972, 1973; and references in these papers; Ahmed and Kitaigorodskii, 1972).

In order to deal with definitely 'polar' molecules, the electrostatic term was introduced (but not the polarization term) by Kitaigorodskii and Mirskaya (1964b) (energy of a dipole lattice), Kitaigorodskii *et al.* (1969) (crystal of carbon dioxide), Kitaigorodskii (1973, Ch. II). A special potential for the hydrogen bond was also added by Momany *et al.* (1974a,b, 1975), who dealt with crystals of hydrocarbons, carboxylic acids, amines, amides and amino-acids (these papers provide an extensive bibliography of works concerning the applications of simplified intermolecular potentials to crystal problems). But Hagler *et al.* (1974), Hagler and Lifson (1974) were able to derive a 'consistent force field' for amide crystals without introducing a special potential for the hydrogen bond: thus, they used only the electrostatic, dispersion and repulsion terms, with all parameters (including the atomic charges) obtained from a least squares fitting with respect to experimental data (molecular dipole moment and heat of sublimation of several amide crystals). But it must be pointed out that this procedure gave very different dispersion–repulsion parameters for the hydrogen bonded to carbon and that bonded to nitrogen: in the latter case, these parameters were found practically zero, and this seemingly astonishing result could simply be interpreted as a subterfuge for the modification of the interatomic potential at short distances, as is discussed in Section A.4 above (simultaneous lowering of dispersion and repulsion at short distance with respect to the values which would be obtained from the unmodified medium-range formula).

Evans and Watts (1976) evaluated the lattice energy of the benzene crystal by using their six-centre model of the benzene molecule. They include all relevant two-body terms (electrostatic, polarization, dispersion, short-range repulsion) and also calculate explicitly the three-body non-additive interaction. Their results are in good agreement with experimental data, and with our own calculations (Caillet and Claverie, 1975).

In another series of works, the variation of the internal degrees of freedom of the molecules (conformational parameters: bond lengths, bond angles, torsional angles) was also included, by adding suitable potential terms corresponding to the intramolecular energy. Thus, Warshel and Karplus (1972) proposed a method for getting this intramolecular energy for conjugated molecules (as well as for saturated ones) (QCFF/PI: Quantum Mechanical Consistent Force Field for π systems; see further developments in Warshel, 1973); for the 'saturated part' of the molecule (σ-system), more usual atom–atom (intramolecular) terms were used (Warshel and Karplus, 1972). The simultaneous use of inter- and intramolecular potential makes it possible to study the optimal molecular configuration in the crystal, the coupling between intramolecular and lattice vibrations (analysis of the normal modes) (Huler and Warshel, 1974, 1975). The application of intermolecular potentials to the study of lattice dynamics is also dealt with in the book edited by Califano (1975) (see the papers by Walmsley, Dows, Bonadeo and d'Alessio, Pawley, Schnepp).

Burgess *et al.* (1975), Shipman *et al.* (1976) also applied directly their EPEN procedure (Shipman *et al.* (1975), see Sections A.1 and A.3 above), which involves the electrostatic, dispersion and repulsion terms, to the study of crystal packing and lattice energies for ammonia and several alkanes, amines and alcohols (other applications to conformational problems will be reported in the next Section F).

In our laboratory, some problems involving crystals were treated by using our previously described formulas (see Section A and the beginning of the present Section E): the pair-wise additive terms (electrostatic, polarization, short-range repulsion), the polarization term (calculated by using the total electric field created by all surrounding molecules), the short-range modification of the dispersion and repulsion terms for the interaction between a hydrogen atom and a heavy-atom (Section A.4), the very-short-range modification of the repulsion term (in order to insure that the interatomic potentials remains repulsive as the interatomic distance decreases to zero), and finally the evaluation of the higher-order (many-body) terms (roughly estimated to −7 per cent of the dispersion energy). This procedure was applied for checking purposes to the crystals of methane, carbon dioxide, benzene and nitrobenzene (calculation of the lattice energy and verification of the existence of a minimum of the theoretical energy corresponding closely enough to the experimental geometry of the crystal) (Caillet and Claverie, 1975).

Rather than to study the crystal for itself, the purpose of these calculations was to study the changes of the configuration of some subunit (binary complex or flexible molecule) when isolated on one hand, and when embedded in the crystal (made from the same subunits) on the other hand. Taking into account the possibility of such changes may avoid a number of false problems.

Thus, by making an extensive compilation of stacking patterns observed in the crystals of nucleic bases or nucleosides, Bugg *et al.* (1971) found a partial (rather than complete) stacking in almost all cases, and were at pains for explaining this fact in terms of binary interactions alone. Later on, Motherwell and Isaacs (1972) performed calculations of the interaction energy for *columns* of stacked molecules (but not for three-dimensional real lattices) and were led to predict complete stacking for the bases and uncomplete stacking for the nucleosides (which appeared as a consequence of the steric hindrance due to the ribose moieties). But this result was still at variance with experiment, since Bugg *et al.* (1971) listed several cases of partial stacking for crystals of purine bases without sugar (9-methyladenine, guanine, purine in the N_9H tautomeric form). Thus, we performed calculations at the same time for binary complexes, columns and three-dimensional lattices of 9-methyladenine (Caillet and Claverie, 1974): the configuration of two stacked adenine in a column and in the binary complex for the minimum of energy was practically the same, in agreement with Motherwell and Isaacs' results (anyway, this configuration does not correspond to exact stacking, owing to the electrostatic repulsion which would exist between like atoms: a suitable lateral displacement allows to change this repulsion into an attraction, without reducing significantly the dispersion attractive term). But for the real three-dimensional crystal, the minimum of the total energy was actually found in agreement with experiment, namely with the molecules slightly displaced and tilted from an exactly parallel configuration. Thus, it is simply the requirement of the energy minimization for the *three-dimensional lattice* which explains the departure from a binary minimum configuration; and this departure appears quite natural if we remark that, for non-spherical molecules, it is impossible, on purely geometrical grounds, to put the molecules in a three-dimensional lattice in such a way that any pair of neighbour molecules realizes an optimum configuration for itself (thus the planes of two adenine molecules cannot be parallel for *all* pairs of neighbour molecules): thus, some compromise must obviously be found in the three-dimensional lattice (while this is not yet the case for a column of molecules, where nothing hinders each pair of neighbours to take the optimum binary configuration). The nitrobenzene crystal offers a still more striking example than adenine does (Caillet and Claverie, 1975). Apparently, it is only for binary complexes with a very specific geometry (like the hydrogen-bonded ones) that negligible changes of this geometry occur in the crystal.

The same conceptual framework applies as well to the problem of the conformational changes (with respect to the isolated molecule) due to the crystalline environment: there is obviously nothing astonishing if the requirement of minimizing the *total* energy of the crystal (conformational intramolecular energy plus the intermolecular energy of the lattice) eventually results in some 'compromise' between these two parts of the energy, and therefore in a conformation of the molecules in the crystal somewhat different from the optimum configuration for an isolated molecule. We actually found such a situation in a very recent study concerning the adrenaline molecule (Caillet *et al.*, 1976). Quantum-mechanical calculations (whether by *ab initio* or PCILO method) of the energy of the isolated molecule for adrenaline and other phenethylamines, and also for indolalkylamines, predict for the most stable conformer a perpendicular conformation ($\tau_1 = \pm 90°$, $\tau_2 = \pm 60°$: 'perpendicular-gauche'; $\tau_1 = \pm 90°$, $\tau_2 = 180°$: 'perpendicular-trans') (see Pullman *et al.*, 1972); Pullman, Berthod and Courrière, 1974; Pullman, Courrière and Berthod, 1974; Port and Pullman, 1974). Now, in crystals containing these molecules (they cocrystallize frequently with some other molecule or ion), the most stable conformer predicted for the isolated molecule is often, but not always, found (Carlström *et al.*, 1973; B. Pullman, 1977): sometimes, the 'planar-trans' conformation ($\tau_1 = 0°$, $\tau_2 = 180°$) is found, for example, in the mixed crystal (−)adrenaline–hydrogen(+)tartrate (Carlström, 1973), although the more normal' perpendicular-trans conformation is found in the crystal of (−)-adrenaline (A. M. Andersen, 1975). Other examples are quoted by Caillet *et al.* (1976). Now, the planar-trans conformation does not even correspond to a local minimum of the conformational energy map for the isolated molecules. All these facts suggest that the planar-trans conformation results from a compromise between the intra-molecular energies and the lattice energy in order to get the minimum total energy, and the calculations for the crystal of (−)adrenaline–hydrogen(+)tartrate actually confirm this guess: the planar-trans and perpendicular-trans conformers have energies larger than that of the perpendicular-gauche conformer by 4 and 3 kcal/mol respectively; but, with respect to the lattice energy calculated for the experimental mixed crystal involving the planar-trans conformer, the best lattice energies calculated for the hypothetical mixed crystals involving the perpendicular-trans and perpendicular-gauche conformers were found *higher* by 9·4 and 15·9 kcal/mol respectively: thus, as concerns the total energy, the experimental crystal actually lies below the two others by $9{\cdot}4 - 3{\cdot}0 = 6{\cdot}4$ and $15{\cdot}9 - 4{\cdot}0 = 11{\cdot}9$ kcal/mol respectively (Caillet *et al.* (1976). Thus, we have here a striking example of the possible effects of a crystalline environment upon some properties of embedded molecules.

F. Miscellaneous applications to conformational and macromolecular problems

We would mention here only very briefly the use of simplified formulas (similar to those used for *inter*molecular interactions) for studying the molecular conformations (which involves the variation of the *intra*molecular energy). Contrary to the situation for intermolecular problems (at large and medium distances), the legitimacy of using such formulas for intramolecular problems is not yet convincingly proved. It is not possible to apply exactly the same kind of approximations used for intermolecular problems, because the distances between the closest non-bonded atoms in a molecule are significantly shorter than usual intermolecular equilibrium distances. This does not mean that suitable simplified formulas do not exist for intramolecular energy variations accompanying conformational changes: thus, for treating the distances (shorter than usual) encountered in hydrogen-bonded complexes, we found it possible to use formulas suitably modified at such short distances. Clearly, from the qualitative point of view, electrostatic, polarization, dispersion (i.e. correlation) and short-range repulsion effects exist between the so-called non-bonded atoms of a molecule; but the charge-transfer may be important, too, and the disentangling of some terms may become quite difficult at short distance. Thus, theoretical derivations of simplified formulas do not exist at the present time, and existing attempts using localized orbitals (which seem to be appropriate for this purpose) stress the non-trivial nature of this problem (Daudey *et al.*, 1975; Rojas, 1976).

Despite this theoretical gap, simplified formulas more or less similar to those used in the intermolecular case have been applied rather widely to conformational energy calculations. In some early conformational studies of macromolecules, extremely simple hard-core potentials (accounting for steric hindrances only) were used (for example, see Ramachandran and Sasisekharan, 1968). But later on most authors introduced 'non-bonded interactions' of intermolecular type (that is, a sum over all pairs of non-bonded atoms of dispersion, repulsion and eventually electrostatic terms), supplemented by potentials involving bonded atoms and therefore reflecting the bond pattern of the molecule (potentials assocated with the variation of bond lengths, of valence angles, and especially with the rotation around single bonds, or torsion): for example, see Kitaigorodskii and Dashevskii (1967), the review by Scheraga (1968), Warshel and Karplus (1972) (who also develop a special simplified procedure (QCFF/PI) for dealing with π orbital systems, in order to treat all possible conformational changes in conjugated molecules), Huler and Warshel (1974), Popov *et al.* (1968), Popov (1975), Momany *et al.* (1974a,b, 1975).

Owing to the theoretical uncertainties which beset these simplified formulas for conformational energy changes, it is desirable to have direct quantum-mechanical evaluations of the intra-molecular energy, at least for checking purposes.

The development, in our laboratory, of the PCILO (Perturbative Configuration Interaction using Localized Orbitals) method in the semi-empirical framework of the CNDO hypotheses made such calculations feasible enough, by reducing considerably the computation time with respect to other methods (see for example the review by B. and A. Pullman (1974) concerning such calculations for amino-acid residues of proteins). For example such calculations made it possible to correct some earlier predictions (made by using simplified formulas) which missed the possibility of intramolecular hydrogen bonds, because these simplified formulas were actually uncomplete (no electrostatic term) and were therefore unable to reproduce the hydrogen-bonded interactions. It may be hoped that, through this checking with respect to direct quantum-mechanical computations, it will be possible to derive simplified formulas having sufficient reliability.

Up to now, simplified formulas have been applied to medium-size organic molecules (see Kitaigorodskii, 1973, Ch. VII): thus Dashevskii and Kitaigorodskii (1967) and also Kitaigorodskii and Dashevskii (1968) investigated the conformations of several polycyclic aromatic hydrocarbons, Burgess *et al.* (1975) consider some alkanes, amines and alcohols, and Burgess *et al.* (1976) study with more detail a series of 1,2 disubstituted ethanes. But these formulas seem especially interesting for dealing with macromolecules, since the large number of conformational parameters (even when restricting oneself to torsional angles) makes it compulsory to use the fastest methods available (it becomes even desirable to resort to statistical considerations, taking into account the entropic contribution to the free energy). The so-called 'conformational statistics' of macromolecules is treated in the books by Volkenstein (1963), Birshtein and Ptitsyn (1966), Flory (1969) and the reviews by Flory (1971) and Nagai (1971). Besides this subject, the recent review by Yamakawa (1974) covers the excluded volume effects in a single chain and in solutions (dilute or moderately concentrated), the use of the 'continuous chain' model, and various dynamic properties (hydrodynamics of polymers and Brownian motion theory). For these applications concerning biological macromolecules, especially polypeptides and proteins, we refer the reader to the two-volume book edited by Ramachandran (1967) especially Volume 1, and to the reviews by Ramachandran and Sasisekharan (1968) Scheraga (1968, 1971), Popov (1975) the book by Kitaigorodskii (1973, Ch. VIII), and the recent papers by Momany *et al.* (1975), Burgess and Scheraga (1975) and also Tanaka and Scheraga (1975).

A variety of new problems arise when macromolecules in solution or assemblies of macromolecules are considered. For example Gilbert (1968) and also Gilbert and Claverie (1968a,b) investigated the intercalation model for the binding of amino-acridine ions to DNA. The main problem was to explain the limit of the binding rate (about one-half the number of base pairs of the DNA molecule) and its variation with the ionic strength of the solution. It turned out that the interaction energies between some given amino-acridine ion and the various base pairs of the DNA were not sufficiently different for explaining the above mentioned experimental findings, while an explanation could be found in terms of the electrostatic repulsion between the intercalated amino-acridine ions: this repulsion increases rapidly when one tries to intercalate more than one ion per two base pairs, and actually varies with the ionic strength of the surrounding solvent.

The problem consists in solving the Poisson–Boltzmann equation (see Section D above) for a cylindrical cavity (simulating the DNA molecule) with charges disposed regularly along the axis of the cylinder, according to a model proposed by MacGillivray and McMullen (1966a,b, 1968).

As a final example, we mention the line of work concerning assemblies of large molecules (and eventually macromolecules) in two-dimensional structures (layers, double layers), as occur in membranes. Thus Salem (1962b) considered the role of long-range forces (electrostatic, polarization, dispersion) in the cohesion of lipoproteins (represented as a double layer of parallel lipid chains with their polar heads interacting with two external layers of proteins); more specifically, Salem considered the role of the dispersion energy, which appears dominant as concerns the interaction between the parallel lipid chains, and emphasized that the rapid variation of this term with distance makes the total energy rather sensitive to such changes as the addition of lateral groups (which prevent a close approach as the one possible for normal chains) or the occurrence of double bonds. Of course, the lack of a precise knowledge of the structure of such complex objects like membranes makes it difficult to perform detailed calculations of the interactions involved, and in the recent years, less 'microscopic' methods have been used: continuum-model calculations for the electrostatic energy, and the 'general theory of van der Waals forces' (see Section V.G and the beginning of V.E.3), which may be considered as a continuum model for the *electrodynamic* interactions (see the reviews by Parsegian, 1973; Israelachvili, 1974). It is possible that, in the future, some 'discrete-continuum' model combining 'microscopic' calculations (of interaction energies between neighbour molecules) and continuum-like ones (for taking into account the effect of the solvent and other distant molecules) will emerge as the best theoretical toll for dealing with such complex structures like membranes.

VII. CONCLUSION

The main topics considered in this chapter (theoretical foundations, derivation of simplified formulas, examples of applications) were conveniently surveyed in the Introduction, and the specific methods and difficulties of each topic were summarized in the introduction of conclusion of the corresponding parts: we shall therefore not repeat such considerations here. We would rather emphasize a leading idea underlying our presentation of the scientific material in this chapter, namely establishing, as far as possible, detailed relationships between the three main 'levels' of study above mentioned (basic theory, simplified formulas, applications): in other words, advances made at some level are used, if possible, for improving in a rational way the methods pertaining to the next level, and conversely the desire for improvement at some level may induce special interest for some peculiar lines of research among those pertaining to the previous level. Thus, performing applications concerning realistic (and therefore rather complex) systems requires the simplest possible formulas or computational methods, and in turn the derivation of such formulas or methods usually relies upon some type(s) of theoretical analysis among the various existing ones. For example, we recalled in the Introduction (Part I) that the search for simplified analytic formulas is a strong incentive to use perturbation methods at the basic theoretical level, and to decompose as much as possible the interaction energy into components varying in a simple way (as functions of the intermolecular distance and orientation, or more explicitly, as functions of the atom–atom, atom–bond and bond–bond distances between the two molecules).

What, now, about the reverse 'ascending' connection between a 'basic' level and a 'more applied' one? At the end of this chapter, we may conclude that this connection is certainly necessary for giving a firm basis to the 'upper level', but it is not sufficient for solving trivially the problems pertaining to this upper level. Thus, the thorough analysis of the inadequacy of the usual Rayleigh–Schrödinger treatment at short distance (Part II.A) shows that it would be of no use to extend this treatment beyond second-order, and that removing this inadequacy requires the use of a different treatment which will insure that the final (perturbed) state belongs to the right symmetry type (in our case the total antisymmetry with respect to space–spin variables). The often repeated statement that 'at short distance it becomes necessary to antisymmetrize the products of the eigenfunctions of the subsystems' thus receives its true justification. Of course, it may look somewhat disappointing that the principle of the most practical solution (namely the antisymmetrization at short distance and not at large distance) had been stated and applied more or less intuitively long before receiving a full justification: here, the theoretical advance legitimates an already existing practical procedure instead of opening a completely new way (which could have led to more performing procedures). But such a situation is by no means exceptional: in a field of research already explored to a non-negligible extent (as is the case for intermolecular interactions), it is not astonishing that many practically interesting lines of work have been entered upon, more or less intuitively, before a satisfactory theoretical analysis is given. Nevertheless, a firm theoretical basis is definitely necessary for deriving reliable and general simplified formulas expressing the intermolecular energy. First, as we saw in Section III.A, the theoretical analysis allows us to split the interaction energy into several contributions in a rational and systematic way, thus avoiding lacks or redundancies which may otherwise occur. For example, electrostatic and polarization terms were missing in empirical formulas initially designed for dealing with non-polar molecules; in the case of the so-called charge-transfer complexes, all long-range contributions to the interaction energy, and even the short-range repulsion, were neglected in early investigations because exclusive emphasis was put upon the charge-transfer contribution; examples of redundancy are displayed by various empirical treatments of the 'hydrogen bond' interaction, where a specific potential is attributed to this interaction, besides other usual contributions (electrostatic, polarization, dispersion, short-range repulsion): in actual fact, provided these four contributions are correctly evaluated at the hydrogen bond distance (which is shorter than the usual intermolecular equilibrium distances) and a charge-transfer contribution is properly included, the 'hydrogen bond' should be satisfactorily reproduced, without adding some *ad hoc* potential (of the Morse type, for example) which clearly involves a risk of redundancy (see Section VI.A.4).

After having allowed us to decompose the total interaction energy into a sum of simpler contributions, the theoretical analysis also gave us very valuable hints for deriving simplified formulas for these contributions. This is evident as concerns the electrostatic energy, since its theoretical expression is simple enough, so that sufficiently accurate simplified formulas may be easily derived even for complex molecules (namely electrostatic interactions between multipoles located at several centres in each molecule). The other contributions, are less easily simplified, but significant achievements are nevertheless reached as concerns the theoretical analysis of polarization, dispersion and short-range repulsion:

(1) the use of localized orbitals for representing the molecular state functions offers a possibility for deriving simplified expressions in terms of local 'interaction sites' (atoms or bonds) (example: use of bond or atom polarizabilities).

(2) as concerns the dispersion term, the methods based upon the use of the dynamic polarizability appear able to give accurate enough values of the interaction constants C_6 in the C_6/R^6 terms (and even the constants in the higher power terms C_n/R^n may be approximately evaluated if necessary).

(3) As concerns the short-range (exponentially varying) repulsion term, the electron gas model looks quite promising for computing this term with reasonable accuracy and much more easily than by performing an '*ab initio*' computation (all integrals between molecular orbitals computed exactly), especially as the number of electrons increases.

Moreover, the results obtained so far suggest that a simple exponential formula $B \exp(-bR)$ may be fitted with sufficient accuracy on the values obtained (for the short-range repulsion between rare gas atoms) in the region of main interest (van der Waals minimum): thus, a systematic application of this procedure to all pairs of atoms could possibly provide a set of constants (B, b) which would be theoretically founded instead of being mere empirical parameters.

At the present time, the main gap in the programme aiming to derive simple analytic formulas from basic theory seems to concern the charge-transfer contribution. This may look somewhat paradoxical, since a considerable amount of work has been devoted to the so-called charge-transfer complexes (see Section III.A.3); but the point is just that, in earlier times, the excessive emphasis put on the 'charge-transfer' aspect resulted in the absence of a simultaneous evaluation of all contributions to the intermolecular interaction energy; and later on, when accurate *ab initio* calculations were performed, the total interaction energy was obtained, but its analysis into various contributions remained a non-trivial task. And it is still a different task to derive for the charge-transfer contribution some formula(s) having the same kind of simplicity which has been achieved for the other contributions. As pointed out at the end of Section III.A.3, it is possible that the approximations presently used for expressing the other contributions partially compensate for the lack of the charge-transfer contribution; if this would be the case, special care should be exercised when adding an explicit charge-transfer contribution to the energy.

The lack of suitable simplified formulas for the charge-transfer term is one of the factors which prevent a theoretically founded simplified expression of the interaction potential in the hydrogen bond case; the other factor is the small value of the equilibrium distance (when compared with the usual Van der Waals minima), so that the approximations commonly used for expressing the other terms (electrostatic, polarization, dispersion, short-range repulsion) should eventually be reconsidered and improved in such a way as to cover this unusually short domain of interatomic distances. These shortcomings, of course, do not mean that the simplified potentials presently available for dealing with the hydrogen bond cases (see Section VI.A.4) are wrong in practice, but only that these potentials are theoretically less well founded for the short distances than for the larger ones.

Another kind of problem exists for aromatic molecules: since their π orbitals cannot be localized as strongly as other kinds of orbitals, the possibility of expressing the interaction energy in terms of local subunits may be more difficult to prove theoretically than in other cases.

The accurate *ab initio* quantum-mechanical calculations of intermolecular complexes ('supermolecules') offer the possibility of checking the simplified formulas, provided appropriate cautions are taken, for example, avoiding spurious 'basis extension effects', as it was recalled at the end of the Introduction (Part I). In recent years, it appeared an interesting tendency to analyse such *ab initio* results in terms of the usual contributions (electrostatic, polarization, dispersion, short-range repulsion, charge-transfer). Quite recently, Allen (1975a,b) thus proposed a simple representation of the properties of the hydrogen bond by using, among other data, the numerous *ab initio* calculations performed on hydrogen-bonded complexes; to be more precise, this model concerns the hydrogen bond at the equilibrium position: energy, charge redistribution, internuclear distance, directionality, force constants; but is does not consider the intermolecular energy as a function of distance and orientation. On the whole, we may conclude that, as concerns the interaction energy between two molecules only, the connection between the 'lower level' (basic theory) and the 'upper level' (simplified analytic formulas) is already presently in a satisfactory state.

The situation becomes less clear when we turn to the connection with a still 'higher level', namely the study of systems involving a large number of molecules, or macromolecules, or both: in all these cases, the large number of degrees of freedom becomes an essential problem, even if the interaction energy is assumed pair-wise additive and the basic pair potentials are assumed to be known (in actual fact, many-body terms exists, if only the polarization energy, and their effect should be estimated, at least in order to be sure that this effect is actually negligible in practice). Now, many (if not most) applications actually involve such systems with numerous degrees of freedom, and the corresponding problems cannot be avoided. If we remember the recent developments concerning the solute–solvent interactions (Section VI.D) or the interactions in crystals (Section VI.E), it appears that more and more concern is being rightly expressed about the environment effects, whether in the liquid or crystalline state, and it is likely that this effort (which was made possible by the simplified theoretical methods and computing facilities presently available) will be pursued. We observe a similar development of the studies concerning the conformations of macromolecules, based upon the evaluation of the intramolecular energy (Section VI.E).

Some of these problems (crystals, conformation of macromolecules, search of strong solvation sites) only require (at least as a first step) a minimization of the energy with respect to a moderate or large number of variables; this is a well-defined problem of applied mathematics, concerning which substantial advances have been made (for example, Kowalik and Osborne, 1968; Box *et al.*, 1969; Lootsma, 1972; Murray 1972; Gill and Murray, 1974). But most frequently some statistical treatment should be necessary (liquids and solutions). Here, two main approaches are used: the 'discrete' one and the 'continuum' one.

The discrete approach considers explicitly the interaction potential between all molecules and would therefore lead to rigorous results if a full statistical treatment was used; but the statistical theories for the liquids or solutions involving non-spherical molecules are still under development (see Part I and Section VI.D), and the direct 'computer experiments' (Monte-Carlo or Molecular Dynamics) are considerably time-consuming. Hence, at the present time (see Section VI.D), the 'discrete model' calculations limit themselves to the search of energy minima (e.g. for one solvent molecule or a shell of solvent molecules around the solute molecule).

Conversely, the continuum models are necessarily approximate by their very nature, their purpose being to get rather simple treatments and formulas at the expense of some loss in accuracy. The macroscopic theory of van der Waals forces (see Sections V.E.3 and V.G) uses macroscopic frequency-dependent dielectric constants for evaluating the interactions between macroscopic bodies eventually separated by some other dielectric medium. The continuum models for solute–solvent interactions (reviewed in Section VI.D) use the static dielectric constant for evaluating the electrostatic and polarization terms, and perform a direct evaluation of the dispersion and short-range repulsion terms directly based upon the interaction terms between two molecules. Clearly, a variety of continuum models are conceivable, according to the place (in the theory) at which the continuum assumption is applied.

For such problems like the solute-solvent one, it is of course tempting to consider a mixed 'discrete-continuum' model, with the solvent molecules close to the solute one being treated explicitly as discrete, while the rest of the solvent would be treated as a continuum medium.

For the time being, all approaches (discrete, continuum and, more recently, discrete-continuum) are being pursued, since none of them yet appears overwhelmingly superior to the others from the practical point of view. Actually, since there is a wide variety of statistical problems in the various applications involving intermolecular interactions, there is no reason why the same approach should appear superior in all cases, and it looks safer to be ready to use any of them, or even to use several of them on the same problem and to establish the proper relationships between the quantities calculated according to these different approaches: for example, in the solute–solvent problem, the discrete approach considers as its natural zero energy the situation where *all* interactions (solute–solvent *and* solvent–solvent) are zero, while the continuum approach will rather take as its zero of energy the situation where the solute–solvent interaction *only* is reduced to zero, the solvent–solvent interaction having its physical value: thus, when comparing results from the two approaches, energy differences (between two thermodynamic states) rather than the 'absolute' energy values themselves should be considered.

Thus, it appears clearly that the availability of satisfactory simplified formulas for the two-body interaction potential is of course necessary, but not yet sufficient for dealing successfully with most practical problems, to the extent that these problems involve a large number of degrees of freedom, and therefore suitable corresponding methods (especially statistical ones) must be used. Anyway, it must not be forgotten that many systems of interest are so complex (for example, the systems of biological interest) that the theoretical study cannot be self-sufficient, and may only contribute, together with the experimental investigation, to a complete understanding of the complex real system.

VIII. APPENDIX A: THE VARIOUS ANTISYMMETRIZATION OPERATORS

The usual antisymmetrizers are designed for being effective upon a completely arbitrary function of n particles, i.e. they involve *all* permutations P (whose number is $n!$) between these particles. We introduce notations for the following three different normalizations:

$$\mathcal{A} = \sum_{P} (-1)^P P \tag{2.376}$$

$$\mathfrak{A} = \mathscr{A}/\sqrt{n!} \tag{2.377}$$

$$A = \mathscr{A}/n! \tag{2.378}$$

As is well known, $\mathscr{A}$ generates a (Slater) determinant when applied to a (Hartree) product $\phi_1(1)\ldots\phi_n(n)$ of (spin) orbitals; if the (spin) orbitals are orthonormal, $\mathfrak{A}$ generates the normalized (Slater) determinant; and finally A is the projection operator (idempotence property: $A^2 = A$).

Now, let us consider products

$$W_a^{(1)}(1,\ldots,n_1)\,W_b^{(2)}(n_1+1,\ldots,n_1+n_2) \tag{2.379}$$

with $n_1+n_2=n$, *where the functions* $W_a^{(1)}$ *and* $W_b^{(2)}$ *are already totally antisymmetric with respect to their respective arguments* $(1,\ldots,n_1)$ and $(n_1+1,\ldots,n_1+n_2)$. It is clear that, in order to antisymmetrize such products with respect to all the n arguments, only the 'inter-system' permutations (i.e. those exchanging labels between the two subsets) are required, since the 'intra-system' permutations (inside a given subset) change the product into itself or into its opposite. We may therefore introduce an 'inter-system' antisymmetrizer, denoted $\mathscr{A}_{\text{int}}$, with the following properties (see Van Duijneveldt (1968), Appendix A and Chapter II, Section 1):

$$\mathscr{A}_{\text{int}} \equiv 1 - \mathscr{A}'_{\text{int}} = \sum_{P_{\text{int}}} (-1)^{P_{\text{int}}} P_{\text{int}} \tag{2.380}$$

$$= 1 - P_{(1)} + P_{(2)} + \ldots + (-1)^{\alpha} P_{(\alpha)} + \ldots + (-1)^{n_0} P(n_0) \tag{2.381}$$

where P_{int} denotes a purely *inter*-system permutation, and $P_{(\alpha)}$ more specifically denotes the sum of all the inter-system permutations which exchange exactly α labels between the two subsets $(1,\ldots,n_1)$ and $(n_1+1,\ldots,n_1+n_2)$: thus, the maximum value of α is $n_0 = \text{Min}\,(n_1, n_2)$. More explicitly we have

$$P_{(1)} = \sum_{i=1}^{n_1} \sum_{j=n_1+1}^{n_1+n_2} P_{ij} \tag{2.382}$$

$$P_{(2)} = \sum_{i=1}^{n_1-1} \sum_{i'>i}^{n_1} \sum_{j=n_1+1}^{n_1+n_2-1} \sum_{j'>j}^{n_1+n_2} P_{ij}P_{i'j'} \tag{2.383}$$

and so on for higher numbers of exchanges (here P_{ij} obviously denotes the permutation of the labels i and j); the restrictions $i'>i$, $j'>j$ are necessary for avoiding redundancies: indeed $P_{ij}P_{i'j'}$ exchanges the pairs (ii') and (jj') but $P_{ij'}P_{i'j}$ would also exchange these pairs, the only difference being in the internal order in each subset: but we want to count only once the exchange between two pairs (or more generally two blocks of α labels in each subset), irrespective of their internal orders, for the reason explained below. Let us note immediately that $P_{(1)}$ is the sum of n_1n_2 terms, $P_{(2)}$ contains $[n_1(n_1-1)/2]/[n_2(n_2-1)/2]$ terms, and quite generally

$$P_{(\alpha)} \text{ is the sum of } \binom{n_1}{\alpha}\binom{n_2}{\alpha} \text{ terms} \tag{2.384}$$

Now, with these definitions, we shall prove that:

All the permutations involved in the full antisymmetrizer $\mathscr{A}$ are generated once and only once by multiplying all the inter-system permutations of $\mathscr{A}_{\text{int}}$ by all the intra-system permutations (i.e. those exchanging between themselves the labels of each subset).

(1) all the permutations (in number $n_1!\ n_2!$) thus generated by multiplying a given inter-system permutation by all intra-system ones are obviously different.

(2) if we now consider two different inter-system permutations, the sets (each containing $n_1!\ n_2!$ permutations) generated from them are necessarily disjoint, since the labels exchanged between the subsystems $(1, \ldots, n_1)$ and $(n_1+1, \ldots, n_1+n_2)$ are the same in each of these sets but differ from one set to the other. Here is the point where the above mentioned restriction (concerning the summations in the expressions (2.383) for $P_{(\alpha)}$) plays an essential role: if this restriction was not made, the sets of permutations generated from two different inter-system permutations could contain common permutations (thus, if $(P_{ij'}P_{i'j})$ was allowed besides $(P_{ij}P_{i'j'})$, it is clear that we would generate $P_{ii'}P_{jj'}(P_{ij'}P_{i'j})=(P_{ij'}P_{i'j})P_{ii'}P_{jj'}$ which is nothing but $(P_{ij}P_{i'j'})$ itself).

(3) We have thus proved that all permutations generated in this way are different. It remains only to remark that any permutation of the n labels $(1, \ldots, n=n_1+n_2)$ may be obtained in this way: indeed, any such permutation exchanges a definite number (say α) of labels belonging to the subsystem $(1, \ldots, n_1)$ with an equal number of labels belonging to the other subsystem $(n_1+1, \ldots, n_1+n_2)$; we arrange these two sets of labels in increasing order: $(i_1<i_2 \ldots <i_\alpha)$ and $(j_1<j_2 \ldots <j_\alpha)$, and we consider the permutation $P_{i_1j_1}P_{i_2j_2} \ldots P_{i_\alpha j_\alpha}$ (which is one of the terms in $P_{(\alpha)}$; it is immediately seen that the initial permutation may be obtained from this 'standard' inter-system permutation by applying suitable *purely intra-system* permutations, and our proof is thus complete.

From this property, we may deduce the number N_{int} of the intersystem permutations involved in $\mathscr{A}_{\text{int}}$ as defined by (2.380–383): indeed, possible permutations (in number $n!$) are generated once and only once by multiplying each of the N_{int} inter-system permutations by all intra-system ones (which are in number $n_1!n_2!$), hence:

$$n! = N_{\text{int}}n_1!n_2! \quad \text{or} \quad N_{\text{int}} = \frac{n!}{n_1!n_2!} \equiv \frac{(n_1+n_2)!}{n_1!n_2!} \tag{2.385}$$

We may go further in this direction by establishing a relationship between the antisymmetrization operators ($\mathscr{A}$, $\mathscr{A}_{\text{int}}$, $\mathscr{A}^{(1)}$, $\mathscr{A}^{(2)}$) themselves. The previous results indeed suggest us to consider the product

$$\begin{aligned}\mathscr{A}_{\text{int}}\mathscr{A}^{(1)}\mathscr{A}^{(2)} &= \left(\sum_{P_{\text{int}}}(-1)^{P_{\text{int}}}P_{\text{int}}\right)\left(\sum_{P^{(1)}}(-1)^{P^{(1)}}P^{(1)}\right)\left(\sum_{P^{(2)}}(-1)^{P^{(2)}}P^{(2)}\right)\\ &= \sum_{P_{\text{int}}}\sum_{P^{(1)}}\sum_{P^{(2)}}(-1)^{P_{\text{int}}}(-1)^{P^{(1)}}(-1)^{P^{(2)}}P_{\text{int}}P^{(1)}P^{(2)} \qquad (2.386)\\ &= \sum_{P_{\text{int}}}\sum_{P^{(1)}}\sum_{P^{(2)}}(-1)^{(P_{\text{int}}P^{(1)}P^{(2)})}P_{\text{int}}P^{(1)}P^{(2)}\end{aligned}$$

since the signature of a product of permutations is equal to the product of the signatures of these factor permutations. But we saw that the set of all products $P_{\text{int}}P^{(1)}P^{(2)}$ is nothing but the set of all permutations P of the n labels $(1, \ldots, n)$, therefore the right-hand side of (2.386) is nothing but

$$\sum_P (-1)^P P = \mathscr{A}:$$

thus, $\mathscr{A}_{\text{int}}\mathscr{A}^{(1)}\mathscr{A}^{(2)} = \mathscr{A}$. The same result would be obtained by starting from $\mathscr{A}^{(1)}\mathscr{A}^{(2)}\mathscr{A}_{\text{int}}$, so that we may finally conclude:

$$\mathscr{A}_{\text{int}}\mathscr{A}^{(1)}\mathscr{A}^{(2)} = \mathscr{A}^{(1)}\mathscr{A}^{(2)}\mathscr{A}_{\text{int}} = \mathscr{A} \qquad (2.387)$$

$\mathscr{A}^{(1)}$ and $\mathscr{A}^{(2)}$ obviously commute, since they operate on disjoint sets of labels, and (2.387) shows that $\mathscr{A}_{\text{int}}$ commutes with $\mathscr{A}^{(1)}\mathscr{A}^{(2)} = \mathscr{A}^{(2)}\mathscr{A}^{(1)}$. It is interesting to look at these commutation properties in closer detail. Actually, *$\mathscr{A}^{(1)}$ and $\mathscr{A}^{(2)}$ both commute with each of the operators $P_{(\alpha)}$ introduced in* (2.381, 2.383, 2.383).

Proof: this is evident for $P_{(1)}$ since according to its definition (2.382) this operator is symmetric with respect to the labels $(1, \ldots, n_1)$ on one hand and $(n_1+1, \ldots, n_1+n_2)$ on the other hand. Consequently, $P_{(1)}$ commutes with every intra-system permutation. More explicitly, let Q be such a permutation for the subsystem 1 (labels $1, \ldots, n_1$). Then, it is easily seen that, for every inter-system permutation P_{ij} involved in $P_{(1)}$, we have

$$P_{ij}Q = QP_{kj} \quad \text{with} \quad k = Q^{-1}(i) \qquad (2.388)$$

Then

$$\left(\sum_i P_{ij}\right)Q = \sum_i QP_{k(i)j} = Q\sum_i P_{k(i)j}$$

but when i takes all values $1, \ldots, n_1$, $k(i) = Q^{-1}(i)$ also takes all these values exactly once, therefore

$$\sum_i P_{k(i)j} = \sum_k P_{kj}$$

hence

$$\left(\sum_i P_{ij}\right)Q = Q\left(\sum_k P_{kj}\right) \equiv Q\left(\sum_i P_{ij}\right) \qquad (2.389)$$

Thus Q commutes with

$$\sum_i P_{ij}$$

and therefore with

$$\sum_j \sum_i P_{ij} = P_{(1)}$$

The proof is similar when Q is a permutation concerning the subsystem 2 $(n_1+1, \ldots, n)$. Thus, $P_{(1)}$ commutes with any combination of intra-system permutations, and among them are $\mathscr{A}^{(1)}$ and $\mathscr{A}^{(2)}$.

For $P_{(\alpha)}$ with $\alpha > 1$, the proof is more difficult because these operators have no definite symmetry, owing to the restrictions in the summations defining them. It becomes therefore necessary to use the symmetry properties of the operators $\mathscr{A}^{(1)}$ and $\mathscr{A}^{(2)}$ themselves.

Let us consider for example the operator

$$P_{(2)} = \sum_{i}^{(1)} \sum_{< i'}^{(1)} \sum_{j}^{(2)} \sum_{< j'}^{(2)} P_{ij} P_{i'j'}$$

For each pair of doublets $(i<i')$ and $(j<j')$, there are 2! permutations possible: $P_{ij}P_{i'j}$ and $P_{ij'}P_{i'j}$ (in the general case, for each pair of α-uplets, there would be $\alpha!$ possible permutations): they are obtained by keeping the subscripts i, i' in increasing order, while using for j, j' the $2! = 2$ possible orderings: j, j' and j', j. We are therefore led to consider the operator:

$$\begin{aligned} \mathscr{P}_{(2)} = P_{(2)} + P'_{(2)} &= \sum_{i<} \sum_{i'} \sum_{j<} \sum_{j'} (P_{ij}P_{i'j'} + P_{ij'}P_{i'j}) \\ &= \tfrac{1}{2} \sum_{i \neq} \sum_{i'} \sum_{j \neq} \sum_{j'} P_{ij} P_{i'j'} \end{aligned} \tag{2.390}$$

which is obviously symmetrical with respect to the intra-system permutations, and which therefore commutes with any such permutation and consequently with $\mathscr{A}^{(1)}$ and $\mathscr{A}^{(2)}$ (an explicit proof could be given along the same line of reasoning previously used for $P_{(1)}$).

We shall now prove that

$$P'_{(2)}\mathscr{A}^{(1)}\mathscr{A}^{(2)} = P_{(2)}\mathscr{A}^{(1)}\mathscr{A}^{(2)} \quad \text{and} \quad \mathscr{A}^{(1)}\mathscr{A}^{(2)}P'_{(2)} = \mathscr{A}^{(1)}\mathscr{A}^{(2)}P_{(2)} \tag{2.391}$$

Indeed

$$P_{ij'}P_{i'j} = (P_{ij}P_{i'j'})P_{ii'}P_{jj'} \tag{2.392}$$

$$= P_{ii'}P_{jj'}(P_{ij}P_{i'j'}) \tag{2.393}$$

(2.392) gives:

$$\begin{aligned} (P_{ij'}P_{i'j})\mathscr{A}^{(1)}\mathscr{A}^{(2)} &= (P_{ij}P_{i'j'})P_{ii'}P_{jj'}\mathscr{A}^{(1)}\mathscr{A}^{(2)} \\ &= (P_{ij}P_{i'j'})(P_{ii'}\mathscr{A}^{(1)})(P_{jj'}\mathscr{A}^{(2)}) \\ &= (P_{ij}P_{i'j'})(-\mathscr{A}^{(1)})(-\mathscr{A}^{(2)}) \\ &= (P_{ij}P_{i'j'})\mathscr{A}^{(1)}\mathscr{A}^{(2)} \end{aligned} \tag{2.394}$$

and by summation of (2.394) over $i<i'$ and $j<j'$ we get the first relation (2.391); the second one would be proved in the same way. Note that the commutation property would not hold for $\mathcal{A}^{(1)}$ or $\mathcal{A}^{(2)}$ separately.

We then get from (2.390, 2.391):

$$\mathcal{P}_{(2)}\mathcal{A}^{(1)}\mathcal{A}^{(2)} = 2P_{(2)}\mathcal{A}^{(1)}\mathcal{A}^{(2)} \quad \text{and} \quad \mathcal{A}^{(1)}\mathcal{A}^{(2)}\mathcal{P}_{(2)} = 2\mathcal{A}^{(1)}\mathcal{A}^{(2)}P_{(2)} \tag{2.395}$$

and from the above mentioned commutation property of $\mathcal{P}_{(2)}$ with $\mathcal{A}^{(1)}$ and $\mathcal{A}^{(2)}$, we finally get the desired result:

$$P_{(2)}\mathcal{A}^{(1)}\mathcal{A}^{(2)} = \mathcal{A}^{(1)}\mathcal{A}^{(2)}P_{(2)} \tag{2.396}$$

(in the general case (operator $P_{(\alpha)}$), we would have the factor $\alpha!$ instead of 2 in the relations analogous to (2.390) and (2.395)).

This completes our study of the commutation properties of $\mathcal{A}_{\text{int}}$ with $\mathcal{A}^{(1)}\mathcal{A}^{(2)}$. The relations (2.385, 2.387) show that N_{int} plays with respect to $\mathcal{A}_{\text{int}}$ a role analogous to that of $n_1!$, $n_2!$ and $n!$ with respect to $\mathcal{A}^{(1)}$, $\mathcal{A}^{(2)}$ and $\mathcal{A}$ respectively. This suggests us to define the two new normalizations:

$$\mathfrak{A}_{\text{int}} = \mathcal{A}_{\text{int}}/\sqrt{N_{\text{int}}} \tag{2.397}$$

$$A_{\text{int}} = \mathcal{A}_{\text{int}}/N_{\text{int}} \tag{2.398}$$

but it must be emphasized that $\mathcal{A}_{\text{int}}$ is *not* a projection operator, because $(\mathcal{A}_{\text{int}})^2$ is *not* proportional to $\mathcal{A}_{\text{int}}$ (indeed, the product of two inter-system permutations may introduce intra-system permutations, while $\mathcal{A}_{\text{int}}$ did not contain such permutations).

From (2.387), we understand very clearly why $\mathcal{A}_{\text{int}}$ suffices for antisymmetrizing completely the product $W^{(1)}W^{(2)}$ of two functions already antisymmetric with respect to the intra-system permutations: we indeed have

$$A^{(1)}W^{(1)} = W^{(1)} \qquad A^{(2)}W^{(2)} = W^{(2)}$$

Hence

$$\begin{aligned}\mathcal{A}_{\text{int}}W^{(1)}W^{(2)} &= \mathcal{A}_{\text{int}}A^{(1)}A^{(2)}W^{(1)}W^{(2)} \\ &= \frac{1}{n_1!n_2!}\mathcal{A}_{\text{int}}\mathcal{A}^{(1)}\mathcal{A}^{(2)}W^{(1)}W^{(2)} \\ &= \frac{1}{n_1!n_2!}\mathcal{A}W^{(1)}W^{(2)}\end{aligned} \tag{2.399}$$

Let us now suppose that $W^{(1)}$ and $W^{(2)}$ are normalized Slater determinants built from orthonormal functions:

$$W^{(1)} = \mathfrak{A}^{(1)}(\phi_1 \ldots \phi_{n_1}) \qquad W^{(2)} = \mathfrak{A}^{(2)}(\phi_{n_1+1} \ldots \phi_n) \tag{2.400}$$

We obviously have

$$AA^{(1)} = AA^{(2)} = A(A^{(1)}A^{(2)}) = A \tag{2.401}$$

(A is a projection operator on a subspace of the eigenspaces of the projection operators $A^{(1)}$ and $A^{(2)}$), hence

$$\begin{aligned}\mathscr{A} W^{(1)} W^{(2)} &= n! A \mathfrak{A}^{(1)} \mathfrak{A}^{(2)} [\phi_1 \ldots \phi_{n_1} \phi_{n_1} +_1 \ldots \phi_n] \\ &= n! \sqrt{n_1! n_2!} A A^{(1)} A^{(2)} [\phi_1 \ldots \phi_n] \\ &= n! \sqrt{n_1! n_2!} A(\phi_1 \ldots \phi_n) = \sqrt{n!} \sqrt{n_1! n_2!} \mathfrak{A}[\phi_1 \ldots \phi_n]\end{aligned} \tag{2.402}$$

where we used the definitions (2.376, 2.377, 2.378), and we note that $\mathfrak{A}[\phi_1 \ldots \phi_n]$ is the Slater determinant normalized to unity. Then, (2.399) becomes:

$$\mathscr{A}_{\text{int}} W^{(1)} W^{(2)} = \sqrt{\frac{n!}{n_1! n_2!}} \mathfrak{A}(\phi_1 \ldots \phi_n) \tag{2.403}$$

or

$$\mathfrak{A}_{\text{int}} W^{(1)} W^{(2)} = \mathfrak{A}[\phi_1 \ldots \phi_n] \tag{2.404}$$

according to the definition (2.397). Thus $\mathfrak{A}_{\text{int}}$ generates the normalized n-particles Slater determinant provided that *all* mono-particle functions involved in $W^{(1)}$ and $W^{(2)}$ are orthonormal. In general, this condition is realized inside each subsystem, but not exactly between the subsystems. Nevertheless, in the intermolecular interaction problem, this requirement becomes more and more accurately fulfilled when the intermolecular distance increases to infinity, and consequently the functions $\mathfrak{A}_{\text{int}} W^{(1)} W^{(2)}$ become normalized in this limit of infinite separation.

According to the formulas (2.376, 2.377, 2.378) and the idempotence property

$$A^2 = A \tag{2.405}$$

we have

$$\mathfrak{A}^2 = \mathscr{A} \tag{2.406}$$

Analogous properties hold for A_{int}, $\mathfrak{A}_{\text{int}}$, $\mathscr{A}_{\text{int}}$ *provided we restrict these operators to the subspace* $\mathscr{H}^{\Lambda n_1} \otimes \mathscr{H}^{\Lambda n_2}$ *of the simple products* $W^{(1)} W^{(2)}$ *of (antisymmetric) functions pertaining to the subsystems* (see (2.379)).

Let us indeed introduce in (2.399) $\mathscr{A}_{\text{int}} = N_{\text{int}} A_{\text{int}}$ and $\mathscr{A} = n! A$, according to (2.398) and (2.378) respectively:

$$N_{\text{int}} A_{\text{int}} W^{(1)} W^{(2)} = \frac{n!}{n_1! n_2!} A W^{(1)} W^{(2)}$$

or, according to (2.385):

$$A_{\text{int}} W^{(1)} W^{(2)} = A W^{(1)} W^{(2)} \tag{2.407}$$

Then, let us consider $(A_{\text{int}})^2$:

$$\begin{aligned}(A_{\text{int}})^2 W^{(1)} W^{(2)} &= A_{\text{int}} [A_{\text{int}} W^{(1)} W^{(2)}] \\ &= A_{\text{int}} A W^{(1)} W^{(2)}\end{aligned} \tag{2.408}$$

according to (2.407). Now, it results from (2.387) that $\mathscr{A}_{\text{int}}$ and $\mathscr{A}$ commute

$$\begin{aligned}\mathscr{A}_{\text{int}}\mathscr{A} = \mathscr{A}_{\text{int}}[\mathscr{A}^{(1)}\mathscr{A}^{(2)}\mathscr{A}_{\text{int}}] = [\mathscr{A}_{\text{int}}\mathscr{A}^{(1)}\mathscr{A}^{(2)}]\mathscr{A}_{\text{int}} \\ = \mathscr{A}\mathscr{A}_{\text{int}}\end{aligned} \tag{2.409}$$

and this property also holds evidently for A_{int} and A. Consequently, (2.408) becomes

$$\begin{aligned}(A_{\text{int}})^2 W^{(1)} W^{(2)} &= AA_{\text{int}} W^{(1)} W^{(2)} \\ &= A(AW^{(1)} W^{(2)}) \quad \text{according to (2.407)} \\ &= A^2 W^{(1)} W^{(2)} = AW^{(1)} W^{(2)}\end{aligned}$$

or, using again (2.407):

$$(A_{\text{int}})^2 W^{(1)} W^2 = A_{\text{int}} W^{(1)} W^{(2)} \tag{2.410}$$

which is the *restricted idempotence property* for A_{int}. By introducing in (2.410) $(A_{\text{int}})^2 = (1/N_{\text{int}})(\mathfrak{A}_{\text{int}})^2$ and $A_{\text{int}} = (1/N_{\text{int}})\mathscr{A}_{\text{int}}$ according to the definitions (2.397, 2.398), and multiplying both sides by N_{int}, we get:

$$(\mathfrak{A}_{\text{int}})^2 W^{(1)} W^{(2)} = \mathscr{A}_{\text{int}} W^{(1)} W^{(2)} \tag{2.411}$$

which is the announced property analogous to (2.406). But it must be once more emphasized that the properties (2.410, 2.411) hold only when the operators are applied to functions belonging to the linear subspace (of the n-particle function space) generated by the simple product functions $W^{(1)} W^{(2)}$ (defined by (2.379)).

IX. APPENDIX B: INCLUSION OF THE TOTALLY ANTISYMMETRIC SUBSPACE INTO THE SPACE OF THE SIMPLE PRODUCTS

The totally antisymmetric subspace corresponds to the projection operator (total antisymmetrizer):

$$A = \mathscr{A}/n! = \frac{1}{n!}\sum_P (-1)^P P \tag{2.412}$$

The space of the simple products corresponds to the projection operator $A^{(1)}A^{(2)}$, product of the antisymmetrizers $A^{(1)}$ and $A^{(2)}$ associated with the respective sybsystems:

$$A^{(1)} = \frac{1}{n_1!}\sum_{P^1} (-1)^{P^{(1)}} P^{(1)} \qquad A^{(2)} = \frac{1}{n_2!}\sum_{P^{(2)}} (-1)^{P^{(2)}} P^{(2)} \tag{2.413}$$

It is easily seen that, given two projection operators A and B, the subspace corresponding to A is included in the subspace corresponding to B if and only if

$$AB = BA = A \tag{2.414}$$

Indeed, given a vector ψ_A of the subspace A, i.e. $A\psi_A=\psi_A$, we have $B\psi_A=B(A\psi_A)=(BA)\psi_A=A\psi_A=\psi_A$: hence ψ_A actually also belongs to the subspace associated with B.

Therefore, we have only to prove that (2.414) is fulfilled for A defined by (2.412) and for $B=A^{(1)}A^{(2)}$. We shall actually see that (2.414) holds for $B=A^{(1)}$ and $B=A^{(2)}$ separately.

For any permutation P, we have

$$PA=AP=(-1)^P A \tag{2.415}$$

Hence, taking for definiteness $B=A^{(1)}$:

$$\begin{aligned} A^{(1)}A &= 1/n_1!\sum_{P^{(1)}}(-1)^{P^{(1)}}P^{(1)}A \\ &= 1/n_1!\sum_{P^{(1)}}(-1)^{P^{(1)}}(-1)^{P^{(1)}}A \\ &= 1/n_1!\left(\sum_{P^{(1)}}1\right)A \\ &= 1/n_1!(n_1!)A=A \end{aligned} \tag{2.416}$$

and it would be seen in the same way that $AA^{(1)}=A$ and $A^{(2)}A=AA^{(2)}=A$. Thus finally we actually have $A(A^{(1)}A^{(2)})=(A^{(1)}A^{(2)})A=A$, and this completes the proof.

X. APPENDIX C: EXPLICIT FORMULAS FOR THE HEITLER–LONDON INTERACTION ENERGY IN THE CASE OF SLATER DETERMINANTS

(Claverie (1973), Ch. III, Section 3.A.3)

According to the formulas (2.158) and (2.164) of Section IV.A, the Heitler–London interaction energy $\varepsilon_1^{\mathrm{HL}}$ may be split into the electrostatic energy $\varepsilon_1^{\mathrm{RS}}$ and the 'exchange' term $\varepsilon_1^{\mathrm{HL-ex}}=\varepsilon_1^{\mathrm{ex}}+\varepsilon_1^{\mathrm{c-ex}}$:

$$\varepsilon_1^{\mathrm{HL}}=\varepsilon_1^{\mathrm{RS}}+(\varepsilon_1^{\mathrm{ex}}+\varepsilon_1^{\mathrm{c-ex}}) \tag{2.417}$$

$$\varepsilon_1^{\mathrm{RS}}=\langle\phi_0|\,V\,|\phi_0\rangle \tag{2.418}$$

$$\begin{aligned}\varepsilon_1^{\mathrm{ex}} &= \frac{\langle\phi_0|\,V\mathcal{A}_{\mathrm{int}}\,|\phi_0\rangle}{\langle\phi_0|\,\mathcal{A}_{\mathrm{int}}\,|\phi_0\rangle}-\varepsilon_1^{\mathrm{RS}} \\ &= \frac{-\langle\phi_0|\,V\mathcal{A}'_{\mathrm{int}}\,|\phi_0\rangle+\langle\phi_0|\,V\,|\phi_0\rangle\langle\phi_0|\,\mathcal{A}'_{\mathrm{int}}\,|\phi_0\rangle}{1-\langle\phi_0|\,\mathcal{A}'_{\mathrm{int}}\,|\phi_0\rangle}\end{aligned} \tag{2.419}$$

$$\varepsilon_1^{\mathrm{c-ex}}=\frac{\langle\phi_0|\,(\bar{E}_0^0-H_0)\mathcal{A}'_{\mathrm{int}}\,|\phi_0\rangle}{\langle\phi_0|\,\mathcal{A}_{\mathrm{int}}\,|\phi_0\rangle} \tag{2.420}$$

where

$$\bar{E}_0^0=\langle\phi_0|\,H_0\,|\phi_0\rangle$$

In this appendix we shall evaluate explicitly the two terms $\varepsilon_1^{\text{RS}}$ and $\varepsilon_1^{\text{HL-ex}}$ in the case of closed-shell subsystems, represented by Slater determinants built from doubly occupied molecular orbitals, denoted b_γ; $\{b_\gamma\}$ will denote the set of *all* occupied orbitals of the two systems; if it becomes necessary to indicate explicitly to which system (s) some orbital belongs, we shall denote $b_\gamma^{(s)}$ the orbital, and $\gamma^{(s)}$ the subscript if it must be used alone. Thus

$$W_0^{(s)} = \frac{1}{\sqrt{n_s!}} \det|\ldots b_\gamma^{(s)}\alpha b_\gamma^{(s)}\beta \ldots| \qquad (2.421)$$

$$\psi_0 = \mathfrak{A}_{\text{int}} W_0^{(1)} W_0^{(2)} = \frac{1}{\sqrt{(n_1+n_2)!}} \det|\ldots b_\gamma^{(1)}\alpha b_\gamma^{(1)}\beta \ldots b_\delta^{(2)}\alpha b_\delta^{(2)}\beta \ldots| \qquad (2.422)$$

In the energy mean value $\langle\psi_0|\, H\, |\psi_0\rangle/\langle\psi_0|\psi_0\rangle$ of the supersystem, the two matrix elements cannot be obtained from the usual Slater rules (for example, see Slater (1963) Appendix 9) since the set of orbitals $\{b_\gamma\}$ is not orthonormal (only the subsets $\{b_\gamma^{(1)}\}$ and $\{b_\delta^{(2)}\}$ are). It would be possible to use special procedures valid for non-orthonormal orbitals (Slater, 1963, Appendix 9; Prosser and Hagstrom, 1968), but we shall rather use the normalized determinant $|\psi_0'\rangle = |\psi_0\rangle/\langle\psi_0 \mid \psi_0\rangle^{1/2}$ built from the orbitals b_i obtained from the b_γ by some orthonormalization procedure:

$$b_i' = \sum_\alpha c_{i\alpha} b_\alpha \qquad (2.423)$$

We shall see that the final expression of the energy is independent, as it must, from the particular orthonormalization procedure used, and involves only the elements of the matrix $\mathbf{S}^{-1}$. If we choose, for definiteness, the procedure proposed by Löwdin (1950), the coefficients $c_{i\alpha}$ are the elements of the matrix $\mathbf{C}$:

$$\mathbf{C} = \mathbf{S}^{-1/2} \qquad (2.424)$$

where $\mathbf{S}$ is the overlap matrix of the set $\{b_\alpha\}$:

$$(\mathbf{S})_{\alpha\beta} = S_{\alpha\beta} = \langle b_\alpha \mid b_\beta\rangle \qquad (2.425)$$

Since the subsets $\{b_\alpha^{(1)}\}$ and $\{b_\beta^{(2)}\}$ are already orthonormal, the matrix $\mathbf{S}$ has the general form

$$\mathbf{S} = \begin{bmatrix} \mathbf{1} & \boldsymbol{\Sigma} \\ \boldsymbol{\Sigma}^+ & \mathbf{1} \end{bmatrix} = \mathbf{1} + \boldsymbol{\Delta} \quad \text{where} \quad \boldsymbol{\Delta} = \begin{bmatrix} \mathbf{0} & \boldsymbol{\Sigma} \\ \boldsymbol{\Sigma}^+ & \mathbf{0} \end{bmatrix} \qquad (2.426)$$

and $\boldsymbol{\Sigma}$ is the matrix of the *intermolecular* overlap integrals.

Then, the Heitler–London energy of the supersystem corresponds to the determinant

$$\psi_0' = \frac{1}{\sqrt{n!}} \det|b_1'\alpha(1)b_1'\beta(2)\ldots b_{n/2}'\alpha(n-1)b_{n/2}'\beta(n)|$$

and may be written upon separation of the nuclear repulsion:

$$\bar{E}_0^0 = \langle \psi_0' | \, H \, | \psi_0' \rangle = \sum_\mu \sum_\nu \frac{Z_\mu Z_\nu}{R_{\mu\nu}} + \langle \psi_0' | \, H_{\mathrm{el}} \, | \psi_0' \rangle \tag{2.427}$$

where (for example see Pilar (1968) Ch. 13):

$$\langle \psi_0' | \, H_{\mathrm{el}} \, | \psi_0' \rangle = 2 \sum_i I_i + \sum_i \sum_j (2J_{ij} - K_{ij}) \tag{2.428}$$

with

$$I_i = \langle b_i' | \, h \, | b_i' \rangle \qquad J_{ij} = \langle b_i'(1) b_j'(2) | \frac{1}{r_{12}} | b_i'(1) b_j'(2) \rangle$$

$$K_{ij} = \langle b_i'(1) b_j'(2) | \frac{1}{r_{12}} | b_j'(1) b_i'(2) \rangle \tag{2.429}$$

The monoelectronic operators h (for the super system) and $h^{(s)}$ (for the subsystems) being:

$$h = -\frac{\Delta}{2} + N^{(1)} + N^{(2)} \qquad h^{(s)} = -\frac{\Delta}{2} + N^{(s)} \tag{2.430}$$

with the nuclear attraction parts $N^{(s)}$:

$$N^{(s)}(\vec{r}) = -\sum_\mu^{(s)} \frac{Z_\mu}{|\vec{r} - \vec{r}_\mu|} \tag{2.431}$$

$\sum^{(s)}$ means, as usual, a summation restricted to the subscripts μ (corresponding here to the nuclei) pertaining to the subsystem (s).

Replacing in (2.428) the orthonormalized orbitals b_i' by their expression (2.423) in terms of the original orbitals b_α, we get:

$$\langle \psi_0' | \, H_{\mathrm{el}} \, | \psi_0' \rangle = 2 \sum_\alpha \sum_\beta I_{\alpha\beta} \sigma_{\alpha\beta}$$

$$+ \sum_\alpha \sum_\beta \sum_\gamma \sum_\delta (2M_{\alpha\beta}^{\gamma\delta} - M_{\alpha\beta}^{\delta\gamma}) \sigma_{\alpha\gamma} \sigma_{\beta\delta} \tag{2.432}$$

where we introduced the notations:

$$I_{\alpha\beta} = \langle b_\alpha | \, h \, | b_\beta \rangle \qquad M_{\alpha\beta}^{\gamma\delta} = \langle b_\alpha(1) b_\beta(2) | \frac{1}{r_{12}} | b_\gamma(1) b_\delta(2) \rangle \tag{2.433}$$

$$\sigma_{\alpha\beta} = \sum_i c_{i\alpha}^* c_{i\beta} \tag{2.434}$$

Note that in (2.432) as in (2.423) all summations run over the set of all orbitals $\{b_\alpha\} = \{b_\gamma^{(1)}, b_\delta^{(2)}\}$ of the supersystem.

We shall prove at the end of the appendix that the coefficients $\sigma_{\alpha\beta}$ are actually independent of the orthonormalization procedure, and are nothing but the elements of the matrix $\mathbf{S}^{-1}$ (inverse of the overlap matrix). At present, we pursue the analysis of the Heitler–London energy.

We note the symmetry properties: $I_{\alpha\beta}=I_{\beta\alpha}$; $M^{\gamma\delta}_{\alpha\beta}$ may be interpreted as the electrostatic interaction between the charge densities $(\alpha\gamma)=b_\alpha(1)b_\gamma(1)$ and $(\beta\delta)=b_\beta(2)b_\delta(2)$, and therefore remains unchanged if we exchange $\alpha\leftrightarrow\gamma$ and/or $\beta\leftrightarrow\delta$, or the pairs (α,γ) and (β,δ) themselves.

The second two-electron term

$$\sum_\alpha\sum_\beta\sum_\gamma\sum_\delta M^{\delta\gamma}_{\alpha\beta}\sigma_{\alpha\gamma}\sigma_{\beta\delta}$$

may be written as well, upon exchange of γ and δ:

$$\sum_\alpha\sum_\beta\sum_\delta\sum_\gamma M^{\gamma\delta}_{\alpha\beta}\sigma_{\alpha\delta}\sigma_{\beta\gamma}$$

so that (2.432) may also be written:

$$\langle\psi_0'|\,H_{\rm el}\,|\psi_0'\rangle=2\sum_\alpha\sum_\beta I_{\alpha\beta}\sigma_{\alpha\beta}+\sum_\alpha\sum_\beta\sum_\gamma\sum_\delta M^{\gamma\delta}_{\alpha\beta}(2\sigma_{\alpha\gamma}\alpha_{\beta\delta}-\sigma_{\alpha\delta}\sigma_{\beta\gamma}) \tag{2.435}$$

On the other hand, the energy $\bar{E}_0^{(s)}$ of the subsystem (s) is:

$$\bar{E}_0^{(s)}=\sum_\mu^{(s)}\sum_\nu^{(s)}\frac{Z_\mu Z_\nu}{R_{\mu\nu}}+\langle \mathbf{W}_0^{(s)}|\,H_{\rm el}^{(s)}\,|\mathbf{W}_0^{(s)}\rangle \tag{2.436}$$

$$\langle \mathbf{W}_0^{(s)}|\,H_{\rm el}\,|\mathbf{W}_0^{(s)}\rangle=2\sum_\alpha^{(s)}I_\alpha^{(s)}+\sum_\alpha^{(s)}\sum_\beta^{(s)}(2J_{\alpha\beta}-K_{\alpha\beta}) \tag{2.437}$$

$$I_\alpha^{(s)}=\langle b_\alpha|\,h^{(s)}\,|b_\alpha\rangle\ J_{\alpha\beta}=M^{\alpha\beta}_{\alpha\beta}\ K_{\alpha\beta}=M^{\beta\beta}_{\alpha\alpha} \tag{2.438}$$

We may now calculate the difference $\varepsilon_1^{\rm HL}=\bar{E}_0-(\bar{E}_0^{(1)}+\bar{E}_0^{(2)})$; since the difference of the nuclear repulsion parts makes no problem, we consider the difference of the electronic parts and we deal separately with the one- and two-electron terms $\varepsilon^{\rm HL}_{1({\rm mono})}$ and $\varepsilon^{\rm HL}_{1({\rm bi})}$:

(1) *The one-electron term* $\varepsilon^{\rm HL}_{1({\rm mono})}$

$$\varepsilon^{\rm HL}_{1({\rm mono})}=2\left[\sum_\alpha\sum_\beta I_{\alpha\beta}\sigma_{\alpha\beta}-\left(\sum_\alpha^{(1)}I^{(1)}_{\alpha\alpha}+\sum_\alpha^{(2)}I^{(2)}_{\alpha\alpha}\right)\right] \tag{2.439}$$

and, according to (2.430), $I^{(1)}_{\alpha\alpha}=I_{\alpha\alpha}-N^{(2)}_{\alpha\alpha}$ and $I^{(2)}_{\alpha\alpha}=I_{\alpha\alpha}-N^{(1)}_{\alpha\alpha}$ hence:

$$\varepsilon^{\rm HL}_{1({\rm mono})}=2\left[\sum_\alpha\sum_\beta I_{\alpha\beta}\sigma_{\alpha\beta}-\left(\sum_\alpha^{(1)}I_{\alpha\alpha}+\sum_\alpha^{(2)}I_{\alpha\alpha}\right)\right.$$
$$\left.+\sum_\alpha^{(1)}N^{(2)}_{\alpha\alpha}+\sum_\alpha^{(2)}N^{(1)}_{\alpha\alpha}\right]$$

Moreover

$$\sum_{\alpha}^{(1)} I_{\alpha\alpha} + \sum_{\alpha}^{(2)} I_{\alpha\alpha} = \sum_{\alpha} I_{\alpha\alpha} = \sum_{\alpha}\sum_{\beta} I_{\alpha\beta}\delta_{\alpha\beta}$$

(where $\delta_{\alpha\beta}$ is the Kronecker symbol, not to be confused with the subscript δ used together with α, β, γ for labelling the orbitals). Hence:

$$\varepsilon_{1(\text{mono})}^{\text{HL}} = 2\sum_{\alpha}\sum_{\beta} I_{\alpha\beta}\sigma'_{\alpha\beta} + 2\left(\sum_{\alpha}^{(1)} N_{\alpha}^{(2)} + \sum_{\alpha}^{(2)} N_{\alpha}^{(1)}\right) \tag{2.440}$$

$$\sigma'_{\alpha\beta} = \sigma_{\alpha\beta} - \delta_{\alpha\beta} \tag{2.441}$$

The second term in (2.440) is the one-electron part of the electrostatic interaction $\varepsilon_1^{\text{RS}}$, namely the attraction between the nuclei of (2) and the electrons of (1), and conversely (the factor 2 corresponds to the double occupation of the orbitals); we shall denote this term $\varepsilon_{1(\text{mono})}^{\text{RS}} = \varepsilon_{\text{el(mono)}}$. The first term is therefore the one-electron part $\varepsilon_{1(\text{mono})}^{\text{HL-ex}}$ of the exchange term $\varepsilon_1^{\text{HL-ex}}$

$$\varepsilon_{1(\text{mono})}^{\text{HL-ex}} = 2\sum_{\alpha}\sum_{\beta} I_{\alpha\beta}\sigma'_{\alpha\beta} \tag{2.442}$$

(2) *The two electron term* $\varepsilon_{1(\text{bi})}^{\text{HL}}$

$$\varepsilon_{1(\text{bi})}^{\text{HL}} = \sum_{\alpha}\sum_{\beta}\sum_{\gamma}\sum_{\delta}(2M_{\alpha\beta}^{\gamma\delta} - M_{\alpha\beta}^{\delta\gamma})\sigma_{\alpha\gamma}\sigma_{\beta\delta} - \sum_{s=1}^{2}\left[\sum_{\alpha}^{(s)}\sum_{\beta}^{(s)} 2M_{\alpha\beta}^{\alpha\beta} - M_{\alpha\beta}^{\beta\alpha}\right] \tag{2.443}$$

Now, in the first term $\sum_\alpha\sum_\beta\sum_\gamma\sum_\delta$, we may separate $\sum_\alpha\sum_\beta$ into

$$\left(\sum_{\alpha}^{(1)}\sum_{\beta}^{(1)} + \sum_{\alpha}^{(2)}\sum_{\beta}^{(2)}\right) + \left(\sum_{\alpha}^{(1)}\sum_{\beta}^{(2)} + \sum_{\alpha}^{(2)}\sum_{\beta}^{(1)}\right) = \sum_{s=1}^{2}\left(\sum_{\alpha}^{(s)}\sum_{\beta}^{(s)}\right) + \left(\sum_{\alpha}^{(1)}\sum_{\beta}^{(2)} + \sum_{\alpha}^{(2)}\sum_{\beta}^{(1)}\right)$$

and the second term may be rewritten:

$$\sum_{s=1}^{2}\left[\sum_{\alpha}^{(s)}\sum_{\beta}^{(s)}\sum_{\gamma}\sum_{\delta}(2M_{\alpha\beta}^{\gamma\delta} - M_{\alpha\beta}^{\delta\gamma})\delta_{\alpha\gamma}\delta_{\beta\delta}\right]$$

Hence (2.443) becomes:

$$\varepsilon_{1(\text{bi})}^{\text{HL}} = \sum_{s=1}^{2}\left[\sum_{\alpha}^{(s)}\sum_{\beta}^{(s)}\sum_{\gamma}\sum_{\delta}(2M_{\alpha\beta}^{\gamma\delta} - M_{\alpha\beta}^{\delta\gamma})(\sigma_{\alpha\gamma}\sigma_{\beta\delta} - \delta_{\alpha\gamma}\delta_{\beta\delta})\right] + \left(\sum_{\alpha}^{(1)}\sum_{\beta}^{(2)} + \sum_{\alpha}^{(2)}\sum_{\beta}^{(1)}\right)\sum_{\gamma}\sum_{\delta}(2M_{\alpha\beta}^{\gamma\delta} - M_{\alpha\beta}^{\delta\gamma})\sigma_{\alpha\gamma}\sigma_{\beta\delta} \tag{2.444}$$

It is now appropriate to isolate in $\varepsilon_{1(\text{bi})}^{\text{HL}}$ the two electron part $\varepsilon_{1(\text{bi})}^{\text{RS}} = \varepsilon_{\text{el(bi)}}$ of the electrostatic interaction $\varepsilon_1^{\text{RS}} = \varepsilon_{\text{el}}$, namely the repulsion of the electron densities of the subsystems:

$$\varepsilon_{\mathrm{el(bi)}} = 4 \sum_{\alpha}^{(1)} \sum_{\beta}^{(2)} M_{\alpha\beta}^{\alpha\beta} \tag{2.445}$$

(the factor $4 = 2 \times 2$ corresponds to the double occupation of each orbital b_α, b_β). Now, in order to match with the second term of (2.444), we rewrite (2.445) as:

$$\varepsilon_{\mathrm{el(bi)}} = \left(\sum_{\alpha}^{(1)} \sum_{\beta}^{(2)} + \sum_{\alpha}^{(2)} \sum_{\beta}^{(1)} \right) (2M_{\alpha\beta}^{\alpha\beta} - M_{\alpha\beta}^{\beta\alpha}) + 2 \sum_{\alpha}^{(1)} \sum_{\beta}^{(2)} M_{\alpha\alpha}^{\beta\beta} \tag{2.446}$$

where we used the relations $M_{\alpha\beta}^{\alpha\beta} = M_{\beta\alpha}^{\beta\alpha}$ and $M_{\alpha\beta}^{\beta\alpha} = M_{\beta\alpha}^{\alpha\beta} = M_{\alpha\alpha}^{\beta\beta}$. Then, subtracting (2.446) from (2.444), we get:

$$\begin{aligned}
\varepsilon_{1(\mathrm{bi})}^{\mathrm{HL-ex}} &= \varepsilon_{1(\mathrm{bi})}^{\mathrm{HL}} - \varepsilon_{\mathrm{el(bi)}} = \\
&= \sum_{s=1}^{2} \left[\sum_{\alpha}^{(s)} \sum_{\beta}^{(s)} \sum_{\gamma} \sum_{\delta} (2M_{\alpha\beta}^{\gamma\delta} - M_{\alpha\beta}^{\delta\gamma})(\sigma_{\alpha\gamma}\sigma_{\beta\delta} - \delta_{\alpha\gamma}\delta_{\beta\delta}) \right] \\
&+ \left(\sum_{\alpha}^{(1)} \sum_{\beta}^{(2)} + \sum_{\alpha}^{(2)} \sum_{\beta}^{(1)} \right) (2M_{\alpha\beta}^{\gamma\delta} - M_{\alpha\beta}^{\delta\gamma})(\sigma_{\alpha\gamma}\sigma_{\beta\delta} - \delta_{\alpha\gamma}\delta_{\beta\delta}) \\
&- 2 \sum_{\alpha}^{(1)} \sum_{\beta}^{(2)} M_{\alpha\alpha}^{\beta\beta}
\end{aligned}$$

and the summations of the first and second terms recombine into $\sum_\alpha \sum_\beta \sum_\gamma \sum_\delta$, hence:

$$\varepsilon_{1(\mathrm{bi})}^{\mathrm{HL-ex}} = \sum_{\alpha} \sum_{\beta} \sum_{\gamma} \sum_{\delta} (2M_{\alpha\beta}^{\gamma\delta} - M_{\alpha\beta}^{\delta\gamma}) \tau'_{\alpha\gamma;\beta\delta} - 2 \sum_{\alpha}^{(1)} \sum_{\beta}^{(2)} M_{\alpha\alpha}^{\beta\beta} \tag{2.447}$$

where

$$\tau'_{\alpha\gamma;\beta\delta} = \sigma_{\alpha\gamma}\sigma_{\beta\delta} - \delta_{\alpha\gamma}\delta_{\beta\delta} \tag{2.448}$$

Now, as we did previously for obtaining (2.435) from (2.432), we may exchange γ and δ in:

$$-\sum_{\alpha} \sum_{\beta} \sum_{\gamma} \sum_{\delta} M_{\alpha\beta}^{\delta\gamma} \tau'_{\alpha\gamma;\beta\delta}$$

thus getting

$$-\sum_{\alpha} \sum_{\beta} \sum_{\delta} \sum_{\gamma} M_{\alpha\beta}^{\gamma\delta} \tau'_{\alpha\delta;\beta\gamma}$$

and the alternative expression for (2.447):

$$\varepsilon_{1(\mathrm{bi})}^{\mathrm{HL-ex}} = \sum_{\alpha} \sum_{\beta} \sum_{\gamma} \sum_{\delta} M_{\alpha\beta}^{\gamma\delta} (2\tau'_{\alpha\gamma;\beta\delta} - \tau'_{\alpha\delta;\beta\gamma}) - 2 \sum_{\alpha}^{(1)} \sum_{\beta}^{(2)} M_{\alpha\alpha}^{\beta\beta} \tag{2.449}$$

Our evaluation of $\varepsilon_1^{\mathrm{HL}}$ is now achieved. In agreement with the separation (2.417, 2.418, 2.419, 2.420) of the electrostatic and exchange terms, we have:

$$\varepsilon_1^{\mathrm{RS}} = \varepsilon_{\mathrm{el}} = \sum_{\mu}^{(1)} \sum_{\nu}^{(2)} \frac{Z_\mu Z_\nu}{R_{\mu\nu}}$$

$$+2\left[\sum_{\alpha}^{(1)} N_{\alpha\alpha}^{(2)} + \sum_{\beta}^{(2)} N_{\beta\beta}^{(1)}\right] + 4\sum_{\alpha}^{(1)}\sum_{\beta}^{(2)} M_{\alpha\beta}^{\alpha\beta} \tag{2.450}$$

$$\varepsilon_1^{\mathrm{HL-ex}} = \varepsilon_{1(\mathrm{mono})}^{\mathrm{HL-ex}} + \varepsilon_{1(\mathrm{bi})}^{\mathrm{HL-ex}} \tag{2.451}$$

where $\varepsilon_{1(\mathrm{mono})}^{\mathrm{HL-ex}}$ is given by (2.442) and $\varepsilon_{1(\mathrm{bi})}^{\mathrm{HL-ex}}$ is given by (2.447, 2.448) or by (2.449).

Expression of the coefficients $\sigma_{\alpha\beta}$

It must be pointed out that the formulas obtained for $\varepsilon_1^{\mathrm{HL}}$ are quite general and do not assume that the intermolecular overlap integrals are small (it is only necessary that the set $\{b_\alpha\}$, reunion of the occupied orbital sets $\{b_\alpha^{(1)}\}$ and $\{b_\beta^{(2)}\}$, is not linearly dependent). These formulas actually express in an expanded form the mean energy for a system described by a set for non-orthonormal orbitals. We shall now examine more closely some properties of the coefficients $\sigma_{\alpha\beta}$ which appeared in the course of the derivation of the formulas.

According to the definition (2.434), we may write $\sigma_{\alpha\beta} = (\mathbf{C}^\dagger\mathbf{C})_{\alpha\beta}$ (the dagger† denotes as usual the hermitian conjugate matrix), i.e.

$$\boldsymbol{\sigma} = \mathbf{C}^\dagger\mathbf{C} \tag{2.452}$$

Then it is easily seen that $\boldsymbol{\sigma}$ does not depend on what particular orthonormalization procedure is used. Indeed, if we consider another set d_i' of orthonormal orbitals deduced from $\{b_\alpha\}$, their Slater determinant is always ψ_0', and the $\{d_i'\}$ are therefore unitarily equivalent to the $\{b_i'\}$, that is, the matrix which transforms the set $\{b_i'\}$ into the set $\{d_i'\}$ is unitary:

$$d_i' = \sum_j u_{ij} b_j' \quad \text{where} \quad u_{ij} = (\mathbf{U})_{ij} \quad \text{and} \quad \mathbf{U}^\dagger = \mathbf{U}^{-1}$$

Then the new coefficients $c_{i\alpha}'$ are easily obtained:

$$d_i' = \sum_j u_{ij}\left(\sum_\alpha c_{j\alpha} b\right) = \sum_\alpha\left(\sum_j u_{ij} c_{j\alpha}\right) b_\alpha$$

i.e.

$$c_{i\alpha}' = \sum_j u_{ij} c_{j\alpha} \quad \text{or} \quad \mathbf{C}' = \mathbf{U}\mathbf{C}$$

hence

$$\boldsymbol{\sigma}' = \mathbf{C}'^\dagger\mathbf{C}' = \mathbf{C}^\dagger\mathbf{U}^\dagger\mathbf{U}\mathbf{C} = \mathbf{C}^\dagger\mathbf{C} = \boldsymbol{\sigma} \qquad \text{q.e.d.}$$

The Löwdin orthonormalization procedure immediately gives this invariant expression of $\boldsymbol{\sigma}$: according to this procedure $\mathbf{C}=\mathbf{S}^{-1/2}$, and since $\mathbf{S}$ is hermitian, so is $\mathbf{S}^{-1/2}$, hence:

$$\boldsymbol{\sigma}=\mathbf{C}^{\dagger}\mathbf{C}=(\mathbf{S}^{-1/2})^{\dagger}\mathbf{S}^{-1/2}=\mathbf{S}^{-1/2}\mathbf{S}^{-1/2}=\mathbf{S}^{-1} \tag{2.453}$$

It is quite satisfactory to retrieve the matrix elements of $\mathbf{S}^{-1}$, since these elements were actually involved in the formulas given by Prosser and Hagstrom (1968) for the matrix elements between Slater determinants built from non-orthonormal orbitals.

If the matrix $\mathbf{S}$ is quasi-diagonal (i.e. has its non-diagonal elements small with respect to the diagonal ones), it is interesting to write $\mathbf{S}=\mathbf{1}+\boldsymbol{\Delta}$ (see (2.426) and to use the series expansion for $\mathbf{S}^{-1}$:

$$\boldsymbol{\sigma}=(\mathbf{1}+\boldsymbol{\Delta})^{-1}=\sum_{p=0}^{\infty}(-\boldsymbol{\Delta})^{p}=\mathbf{1}-\boldsymbol{\Delta}+\boldsymbol{\Delta}^{2}-\ldots \tag{2.454}$$

This situation occurs for the intermolecular interactions at medium and long-range: the matrix has the special form (2.426) and the elements of the non-diagonal block $\boldsymbol{\Sigma}$ may be as small as 0·01 even in the equilibrium region (in the helium-helium interaction, $\boldsymbol{\Sigma}$ reduces to a single overlap integral whose value is 0·007661 at $R=5{\cdot}5$ a.u. and 0·004249 at $R=6{\cdot}0$ a.u. for the four-component orbital considered in Section IV.C of the main text: see Table 2.2).

Under such circumstances, a good approximation will already be obtained by using $\mathbf{1}-\boldsymbol{\Delta}$ instead of the exact $\boldsymbol{\sigma}$, i.e. $-\boldsymbol{\Delta}$ instead of $\boldsymbol{\sigma}'=\boldsymbol{\sigma}-\mathbf{1}$ as introduced in (2.441). Note the exact expression $\boldsymbol{\sigma}'=\boldsymbol{\sigma}-\mathbf{1}=-\boldsymbol{\Delta}(\mathbf{1}-\boldsymbol{\Delta}+\boldsymbol{\Delta}^{2}-\ldots)=-\boldsymbol{\Delta}\boldsymbol{\sigma}$. Thus $\boldsymbol{\sigma}'$ is of order 1 with respect to intermolecular overlap and it is then easily seen that all terms involved in $\varepsilon_1^{\mathrm{HL-ex}}$ are at least of order 2. This may occur in three different ways: either we have intermolecular exchange integrals which are by themselves of order S^2, or we have hybrid integrals of order S multiplied by a coefficient σ itself of order S, or we have coulomb integrals multiplied by two coefficients σ of order S. The approximation of $\varepsilon_1^{\mathrm{HL-ex}}$ by its principal terms of order S^2 will be used for the study of the helium–helium interaction (see Appendix D and Section IV.C in the main text).

XI. APPENDIX D: APPROXIMATE ANALYTICAL FORMULAS FOR THE HEITLER–LONDON INTERACTION ENERGY (SHORT-RANGE) BETWEEN TWO HELIUM ATOMS

(atomic units are used throughout this Appendix)

Each helium atom ($s=1,2$) is represented by a Slater determinant $(1/\sqrt{2})$ det $|b^{(s)}\alpha b^{(s)}\beta|$. Since the multipolar part of the electrostatic energy $\varepsilon_{\mathrm{el}}$ vanishes for spherically symmetric subsystems, $\varepsilon_{\mathrm{el}}$ reduces to its penetration part $\varepsilon_1^{\mathrm{pen}}$, and consequently $\varepsilon_1^{\mathrm{HL}}=\varepsilon_{\mathrm{el}}+\varepsilon_1^{\mathrm{HL-ex}}=\varepsilon_1^{\mathrm{pen}}+\varepsilon_1^{\mathrm{HL-ex}}$ reduces to the short-range (exponentially varying) term of our approximate evaluation (2.124) of the total interaction energy. ε_{el} and $\varepsilon_1^{\text{HL-ex}}$ are given by the formulas derived in Appendix C (see also Griffing and Wehner (1955) or Slater (1963) Ch. 6 and Appendix 10). We only introduced a supplementary notation concerning the nuclear integrals by separating explicitly the nuclear charge $Z=2$:

$$\langle b_\alpha|\,N^{(s)}\,|b_\beta\rangle = N_{\alpha\beta}^{(s)} = -2n_{\alpha\beta}^{(s)} \tag{2.455}$$

where

$$n_{\alpha\beta}^{(s)} = \langle b_\alpha(\vec{r})|\frac{1}{|\vec{r}-\vec{r}_s|}|b_\beta(\vec{r})\rangle \tag{2.456}$$

($\vec{r}_s$ denotes the position of the nucleus s) and by denoting the kinetic integrals explicitly:

$$T_{\alpha\beta} = \langle b_\alpha|-\frac{\Delta}{2}|b_\beta\rangle \tag{2.457}$$

Then (2.450) gives (R denoting the internuclear distance):

$$\begin{aligned}\varepsilon_{el} &= \frac{4}{R} + 2[-2n_{11}^{(2)} - 2n_{22}^{(1)}] + 4M_{12}^{12}\\ &= 4\left[\frac{1}{R} - (n_{11}^{(2)} + n_{22}^{(1)}) + M_{12}^{12}\right]\end{aligned} \tag{2.458}$$

(note that, owing to the symmetry of the problem, we have $n_{11}^{(2)} = n_{22}^{(1)}$)

$$\begin{aligned}\varepsilon_{1(\text{mono})}^{\text{HL-ex}} &= 2[I_{11}(\sigma_{11}-1) + 2I_{12}\sigma_{12} + I_{22}(\sigma_{22}-1)]\\ &= 4[I_{11}(\sigma_{11}-1) + I_{12}\sigma_{12}]\end{aligned} \tag{2.459}$$

since $I_{11} = I_{22}$ and $\sigma_{11} = \sigma_{22}$. Finally (2.449) gives:

$$\begin{aligned}\varepsilon_{1(\text{bi})}^{\text{HL-ex}} &= 2M_{11}^{22}[4\sigma_{12}^2 - (\sigma_{11}\sigma_{22}-1) - \sigma_{12}^2]\\ &\quad + 4M_{11}^{12}(2\sigma_{11}\sigma_{12} - \sigma_{11}\sigma_{12}) + 4M_{12}^{22}(2\sigma_{22}\sigma_{12} - \sigma_{22}\sigma_{12})\\ &\quad + 2M_{12}^{12}[2(\sigma_{11}\sigma_{22}-1) - \sigma_{12}^2] + M_{11}^{11}(\sigma_{11}^2 - 1)\\ &\quad + M_{22}^{22}(\sigma_{22}^2 - 1) - 2M_{11}^{22}\\ &= 2M_{11}^{22}(3\sigma_{12}^2 - \sigma_{11}^2) + 8M_{11}^{12}\sigma_{11}\sigma_{12}\\ &\quad + 2M_{12}^{12}[2(\sigma_{11}^2 - 1) - \sigma_{12}^2] + 2M_{11}^{11}(\sigma_{11}^2 - 1)\end{aligned} \tag{2.460}$$

Denoting for brevity $S_{12} = S$, we obtain for the coefficient matrix $\boldsymbol{\sigma}$ given by (2.453):

$$\boldsymbol{\sigma} = \mathbf{S}^{-1} = \begin{bmatrix} 1 & S \\ S & 1 \end{bmatrix}^{-1} = \frac{1}{1-S^2}\begin{bmatrix} 1 & -S \\ -S & 1 \end{bmatrix} \tag{2.461}$$

hence

$$\sigma_{11}=\sigma_{22}=\frac{1}{1-S^2}=1+\frac{S^2}{1-S^2} \qquad \sigma_{12}=\sigma_{21}=-\frac{S}{1-S^2}$$

$$\sigma_{11}^2-1=(\sigma_{11}-1)(\sigma_{11}+1)=\frac{S^2}{1-S^2}\left(2+\frac{S^2}{1-S^2}\right)$$

It is easily verified that $\varepsilon_1^{\mathrm{HL}}$ contains terms of even order only with respect to S (the exchange integral M_{11}^{22} is of order S^2 and the hybrid integrals I_{12} and M_{11}^{12} are of order S). If we limit ourselves to the terms of order S^2, the error is of order S^4, and this is negligible for actual practice at medium- and long-range. We therefore study the S^2—approximation to $\varepsilon_1^{\mathrm{HL-ex}}$, which is easily obtained from (2.459, 2.460, 2.461) as:

$$\begin{aligned}\varepsilon_{1(S^2)}^{\mathrm{HL-ex}}=4[&I_{11}S^2-I_{12}S+M_{11}^{11}S^2\\&+\tfrac{3}{2}M_{12}^{12}S^2-2M_{11}^{12}S-\tfrac{1}{2}M_{11}^{22}]\end{aligned} \tag{2.462}$$

Now we decompose the monoelectronic integrals defined by (2.433) according to (2.430) and (2.455, 2.456, 2.457):

$$I_{\alpha\beta}=T_{\alpha\beta}+N_{\alpha\beta}^{(1)}+N_{\alpha\beta}^{(2)}=T_{\alpha\beta}-2(n_{\alpha\beta}^{(1)}+n_{\alpha\beta}^{(2)})$$

hence:

$$I_{11}=T_{11}-2(n_{11}^{(1)}+n_{11}^{(2)}) \tag{2.463}$$

$$I_{12}=T_{12}-2(n_{12}^{(1)}+n_{12}^{(2)})=T_{12}-4n_{12}^{(1)} \tag{2.464}$$

since $n_{12}^{(1)}=n_{12}^{(2)}$, and finally:

$$\begin{aligned}\varepsilon_{1(S^2)}^{\mathrm{HL-ex}}=4\{&S^2[(T_{11}-2n_{11}^{(1)}+M_{11}^{11})-(2n_{11}^{(2)}-\tfrac{3}{2}M_{12}^{12})]\\&+S[-T_{12}+4n_{12}^{(1)}-2M_{12}^{12}]-\tfrac{1}{2}M_{11}^{22}\}\end{aligned} \tag{2.465}$$

The sum of (2.458) and (2.465) gives a suitable approximation to $\varepsilon_1^{\mathrm{HL}}$ Upon insertion of analytic formulas for the various integrals involved, an explicit analytic expression would be obtained, but it would be rather complex even for the simple case of He . . . He. We shall therefore examine the possibility of approximating these integrals in order to obtain a final simplified formula. This will be done for the two choices of the orbitals $b^{(s)}(\vec{r})$ used by Murrell and Shaw (1967b, 1968b):

(1) $$b^{(s)}=1s(1{\cdot}6875) \tag{2.466}$$

(2) $$\begin{aligned}b^{(s)}=&0{\cdot}884315\,1s(1{\cdot}4-0{\cdot}051269\,2s(1.4)\\&+0{\cdot}175261\,1s(3{\cdot}0)+0{\cdot}010673\,2s(3{\cdot}0)\end{aligned} \tag{2.467}$$

(2.466) is the SCF Slater orbital, and (2.467) is the SCF orbital of Roothaan, Sachs, and Weiss (1960). In these formulas $ns(\zeta)$ denotes the normalized Slater orbital $C_n r^{n-1}e^{-\zeta r}$ with quantum number n and exponent ζ.

(1) *The one component orbital*

Then all analytic expressions for the integrals involved in (2.458) and (2.465) are rather simple. They are given in terms of R, ζ and $\rho=\zeta R$

$$T_{11}=\frac{\zeta^2}{2} \qquad n_{11}^{(1)}=\zeta \qquad M_{11}^{11}=0{\cdot}625\zeta \qquad (2.468)$$

Overlap integral:

$$S=\left(1+\rho+\frac{\rho^2}{3}\right)\mathrm{e}^{-\rho} \qquad (2.469)$$

Nuclear hybrid integral:

$$n_{12}^{(1)}=\zeta(1+\rho)\mathrm{e}^{-\rho} \qquad (2.470)$$

Kinetic hybrid integral:

$$T_{12}=-\frac{\zeta^2}{2}\left(S-\frac{2}{\zeta}n_{12}^{(1)}\right) \qquad (2.471)$$

Nuclear attraction integral:

$$n_{11}^{(2)}=\frac{1}{R}[1-(1+\rho)\mathrm{e}^{-\rho}] \qquad (2.472)$$

Coulomb integral:

$$M_{12}^{12}=\frac{1}{R}\left[1-(1+\frac{11}{8}\rho+\frac{3}{4}\rho^2+\frac{1}{6}\rho^3)\mathrm{e}^{-2\rho}\right] \qquad (2.473)$$

Hybrid integral:

$$M_{11}^{12}=\zeta\left[\left(\rho+\frac{1}{8}+\frac{5}{16\rho}\right)\mathrm{e}^{-\rho}-\left(\frac{1}{8}+\frac{5}{16\rho}\right)\mathrm{e}^{-3\rho}\right] \qquad (2.474)$$

Exchange integral:

$$M_{11}^{22}=\frac{\zeta}{5}\left\{\left(\frac{25}{8}-\frac{23}{4}\rho-3\rho^2-\frac{1}{3}\rho^3\right)\mathrm{e}^{-2\rho}\right.$$
$$\left.+\frac{6}{\rho}[S^2(C+\log\rho)+S'^2\,\mathrm{Ei}\,(-4\rho)-2SS'\,\mathrm{Ei}\,(-2\rho)]\right\} \qquad (2.415)$$

In (2.475), S is the overlap integral defined (2.469),

$$S'=\left(1-\rho+\frac{\rho^2}{3}\right)\mathrm{e}^{-\rho}, \qquad C=0{\cdot}577\,215\,664\,901\ldots$$

is the Euler Constant and

$$\mathrm{Ei}(-x)=-\int_x^{+\infty}\frac{\mathrm{e}^{-t}}{t}\,\mathrm{d}t$$

is the exponential integral (see, for example, Abramowitz and Stegun (1965) or Gradshteyn and Ryzhik (1965)). For (2.468, 2.469, 2.470, 2.471, 2.472, 2.473) see Roothaan (1951); for (2.474) see Hirschfelder and Linnet (1950), Appendix I, formula for $L(a_0, a_0, a_0, b_0)$; and for (2.475) see Sugiura (1927), or Gombas (1950) (Appendix 5, page 256), or Kotani *et al.* (1965).

Inserting (2.472, 2.473) in (2.458) we get the electrostatic (penetration) energy:

$$\varepsilon_{el} = \varepsilon_1^{pen} = \frac{4}{R}\left(1+\frac{5}{8}\rho-\frac{3}{4}\rho^2-\frac{1}{6}\rho^3\right)e^{-2\rho} \tag{2.476}$$

which is of the order of magnitude of S^2 owing to the factor $e^{-2\rho}$. This term is attractive (negative) at medium- and long-range and becomes repulsive at rather short-range only (the minimum occurs at about $\rho=2$); at the equilibrium distance $R=6{\cdot}75$ a.u., $\rho=9{\cdot}703125$ for $\zeta=1{\cdot}6875$ and $\varepsilon_1^{pen}=\varepsilon_{el}$ is negative.

Let us consider (2.465).

(a) In the bracketed expression multiplied by S^2, we have an intra-atomic term independent of R given by (2.468):

$$\begin{aligned} T_{11}-2n_{11}^{(1)}+M_{11}^{11} &= \frac{\zeta^2}{2}-2\zeta+0{\cdot}625\zeta \\ &= \zeta\left(\frac{\zeta}{2}-1{\cdot}375\right) \end{aligned} \tag{2.477}$$

and an interatomic term which may be evaluated with sufficient accuracy (at medium- and long-range) by reducing $n_{11}^{(2)}$ and M_{12}^{12} to their 'multipole part' $1/R$, thus obtaining $-2/R+3/(2R)=-1/(2R)$. The S^2 term therefore becomes:

$$S^2\left[\zeta\left(\frac{\zeta}{2}-1{\cdot}375\right)-\frac{1}{2R}\right] \tag{2.478}$$

(b) The bracket multiplied by S becomes, upon using (2.469, 2.470, 2.471) for T_{12}:

$$\frac{\zeta^2}{2}\left(S-\frac{2}{\zeta}n_{12}^{(1)}\right)+4n_{12}^{(1)}-2M_{11}^{12}=\frac{\zeta^2}{2}S+(4-\zeta)n_{12}^{(1)}-2M_{11}^{12} \tag{2.479}$$

In order to elaborate a simplified formula, we remark that, according to (2.474) and (2.469, 2.470, 2.471), the ratio $M_{11}^{12}/n_{12}^{(1)}\to 1$ as $R\to\infty$. The part containing $e^{-3\rho}$ in the formula (2.474) for M_{11}^{12} is actually negligible with respect to the part containing $e^{-\rho}$ for $\rho\simeq 10$ or larger, we shall therefore use the approximation:

$$M_{11}^{12}\simeq\zeta\left(\rho+\frac{1}{8}+\frac{5}{16\rho}\right)e^{-\rho}=n_{12}^{(1)}-\mathcal{R} \tag{2.480}$$

$$\mathcal{R}=\zeta\left(\frac{7}{8}-\frac{5}{16\rho}\right)e^{-\rho} \tag{2.481}$$

For example, for $\rho = 10$,

$$\mathcal{R}/n_{12}^{(1)} = \left(\frac{7}{8} - \frac{5}{16\rho}\right) \Big/ (\rho + 1) = 0{\cdot}0767$$

Then (2.479) becomes:

$$\frac{\zeta^2}{2} S + (2 - \zeta) n_{12}^{(1)} + 2\mathcal{R} \tag{2.482}$$

We now introduce the following approximations:

(α) we drop $2\mathcal{R}$ in (2.482) and simultaneously the penetration energy $\varepsilon_1^{\text{pen}} = \varepsilon_{\text{eI}}$. Since (2.482) is multiplied by $4S$ in (2.465), we have to compare $8S\mathcal{R}$ and $\varepsilon_1^{\text{pen}}$. Now they actually are of opposite sign (at medium- and long-range), and, according to (2.469, 2.470, 2.471, 2.476, 2.481), $8S\mathcal{R}/\varepsilon_1^{\text{pen}} \to 7/2$ as $R \to \infty$ (for $\rho = 10$, we already get the value $3{\cdot}18$ for this ratio). Thus, dropping $\mathcal{R}$ overcompensates the dropping of $\varepsilon_1^{\text{pen}}$: the resulting approximation would underestimate the total repulsion.

(β) in the remaining expression from (2.482), namely $S[\zeta^2/2 + (2-\zeta)n_{12}^{(1)}/S]$, we perform a new simplification: we replace $n_{12}^{(1)}/(\zeta S) = (1+\rho)/(1+\rho+\rho^2/3)$ by the ratio $\rho/(\rho^2/3) = 3/\rho$. This obviously overestimates $n_{12}^{(1)}/S$, but this overestimation may compensate for the underestimation due to the previous approximation (α). Actually, a complete calculation for $\rho = 10$ (and $\xi = 1{\cdot}6875$) shows that the replacement of

$$\varepsilon_1^{\text{pen}} + 4S\left[\frac{\zeta^2}{2} S + (2-\zeta) n_{12}^{(1)} + 2\mathcal{R}\right]$$

by

$$4S\left[\frac{\zeta^2}{2} S + (2-\zeta)\zeta S \frac{3}{\rho}\right] = 4S^2\left[\frac{\zeta^2}{2} + \zeta(2-\zeta)\frac{3}{\rho}\right] \tag{2.483}$$

results in a relative error of -1 per cent only.

(c) It remains only to consider the exchange integral M_{11}^{22}. Since $\text{Ei}(-x) \sim e^{-x}/x$ as $x \to \infty$, the most important terms in (2.475) are of the form $-\rho^3/3\, e^{-2\rho}$ and $6S^2(C + \log \rho)$. We are therefore led to check the simplified formula:

$$\frac{M_{11}^{22}}{\zeta} \simeq \frac{S^2}{\rho}(C + \log \rho) \tag{2.484}$$

The comparison between the approximate values given by (2.484) and the exact values of M_{11}^{22}/ζ (see Hirschfelder and Linnet, 1950, or Hirschfelder, Curtiss and Bird, 1954, Ch. 14, Appendix A) is presented in Table 2.6.

It appears that the accuracy of (2.484) is practically sufficient for $\rho > 8$ (which actually corresponds to the medium- and long-range region for $\zeta = 1{\cdot}6875$). A generalization of (2.484), namely

$$[\chi_a(1)\chi_b(1)\, |/r_{12}|\, \chi_{a'}(2)\chi_{b'}(2)]/\zeta \simeq S_{ab}S_{a'b'}(C + \log \rho)/\rho$$

Table 2.6. Comparison between the accurate values of the exchange integral $M_{11}^{22}=[1s_a(1)1s_b(1)\,|1/r_{12}|\,1s_a(2)1s_b(2)]$ and the approximate values given by (2.484). We note $\rho=\zeta R$, and, in the last column, $\Delta M=$(approximate value–accurate value), $M=$accurate value. $C=0{\cdot}577\ 215\ 664\ 901\ldots$ denotes the Euler constant

ρ	S	$S^2\dfrac{C+\log\rho}{\rho}$	M_{11}^{22}/ζ	$\dfrac{\Delta M}{M}\times 100$ (%)
1·0	0·858 385	0·425 307	0·436 651	−2·598
2·0	0·586 453	0·218 457	0·184 156	18·626
3·0	0·348 509	$0{\cdot}678\ 478\times10^{-1}$	$0{\cdot}585\ 08\times10^{-1}$	15·963
4·0	0·189 262	$0{\cdot}175\ 832\times10^{-1}$	$0{\cdot}156\ 27\times10^{-1}$	12·518
5·0	$0{\cdot}965\ 772\times10^{-1}$	$0{\cdot}407\ 906\times10^{-2}$	$0{\cdot}371\ 704\times10^{-2}$	9·739
6·0	$0{\cdot}470\ 963\times10^{-1}$	$0{\cdot}875\ 754\times10^{-3}$	$0{\cdot}814\ 027\times10^{-3}$	7·583
7·0	$0{\cdot}221\ 891\times10^{-1}$	$0{\cdot}177\ 468\times10^{-3}$	$0{\cdot}167\ 600\times10^{-3}$	5·888
8·0	$0{\cdot}101\ 757\times10^{-1}$	$0{\cdot}343\ 854\times10^{-4}$	$0{\cdot}328\ 960\times10^{-4}$	4·528
10·0	$0{\cdot}201\ 273\times10^{-2}$	$0{\cdot}116\ 663\times10^{-5}$	$0{\cdot}113\ 835\times10^{-3}$	2·484
12·0	$0{\cdot}374\ 797\times10^{-3}$	$0{\cdot}358\ 455\times10^{-7}$	$0{\cdot}354\ 839\times10^{-7}$	1·019

where the four Slater orbitals have the same exponent has been checked by Claverie (1973, Ch. III, Section 3.B.1, pp. 194–196) for all possible types of exchange integrals between orbitals with $n=2(2s, 2p_x, 2p_y$. For arbitrary exponents $(\zeta_a, \zeta_b, \zeta_{a'}, \zeta_{b'})$, the approximation

$$[\chi_a(1)\chi_b(1)\,|1/r_{12}|\,\chi_{a'}(2)\chi_{b'}(2)]\simeq S_{ab}S_{a'b'}\left(C+\log\frac{2\rho\rho'}{\rho+\rho'}\right)\Big/R$$

with $\rho=R(\zeta_a+\zeta_b)/2$ and $\rho'=R(\zeta_{a'}+\zeta_{b'})/2$ was proposed but not checked. Various proposals concerning the approximate evaluation of molecular electronic integrals may also be found in the papers by Čisek (1963), Kay and Silverstone (1970), Živković and Murrell (1971).

By adding (2.478), (2.483) and (2.484) we get the following approximate formula for $\varepsilon_1^{\rm HL}$ in the case of the one–component orbital:

$$\varepsilon_1^{\rm HL}\simeq 4S^2\left[\zeta\left(\frac{\zeta}{2}-1{\cdot}375\right)-\frac{1}{2R}+\frac{\zeta^2}{2}+(2-\zeta)\frac{3\zeta}{\rho}-\frac{\zeta(C+\log\rho)}{2\rho}\right]$$

$$=4\zeta S^2\left[(\zeta-1{\cdot}375)+\frac{11-6\zeta-C-\log\rho}{2\rho}\right] \tag{2.485}$$

$$=4\frac{S^2}{R}[(\zeta-1{\cdot}375)\rho+(11-6\zeta-C-\log\rho)/2] \tag{2.486}$$

Upon replacing ζ by the value 1·6875 of (2.466) and the Euler constant C by its value 0·57721566 . . . , we get:

$$\varepsilon_1^{\rm HL}\simeq 4\frac{S^2}{R}[0{\cdot}3125\,\rho+(0{\cdot}2977843-\log\rho)/2] \tag{2.487}$$

For the numerical results obtained with this formula and their discussion, see the main text (Section IV.C.1).

(2) *The four-component orbital*

There would be much more numerous elementary integrals between the basic Slater orbitals, and it would be very tedious to derive all corresponding explicit analytical formulas. We shall built directly approximate formulas for the various terms of (2.465).

(*a*) *The factor multiplied by* S^2. The intra-atomic term $(T_{11}-2n_{11}^{(1)}+M_{11}^{11})$, that we shall denote K, may be computed exactly according to Roothaan's (1951) formulas:

$$T_{11}=1{\cdot}430\ 798\ 48 \qquad n_{11}^{(1)}=1{\cdot}687\ 261\ 59 \qquad M_{11}^{11}=1{\cdot}025\ 779\ 21$$

$$K=T_{11}-2n_{11}^{11}+M_{11}^{11}=-0{\cdot}917\ 945\ 79 \tag{2.488}$$

As concerns the interatomic term $2n_{11}^{(2)}-\frac{3}{2}M_{12}^{12}$ it may be reduced to its non-zero multipolar part: $2/R-3/(2R)=1/(2R)$ as in the case of the one-component orbital (the penetration part involving $e^{-2\rho}$ may be neglected). Thus the complete S^2 term becomes:

$$4S^2\left(K-\frac{1}{2R}\right) \tag{2.489}$$

(*b*) *The factor multiplied by S.* Since the integrals between Slater orbitals with different exponents have much more cumbersome expresssions than the orbitals with equal exponents, and since the orbitals with the lowest exponent $\zeta_1=1{\cdot}4$ will give predominant contributions to the hybrid (or exchange) integrals at large distances, we shall use the first two components:

$$c_1 1s(\zeta_1)+c_2 2s(\zeta_1)=\chi_1$$

$$(\zeta_1=1{\cdot}4 \qquad c_1=0{\cdot}884\ 315 \qquad c_2=-0{\cdot}051\ 269).$$

We shall denote $S[\zeta_1]=\langle\chi_1\,|\,\chi_1\rangle$ the corresponding overlap integral:

$$S[\zeta_1]=(c_1^2P_{11}+c_1c_2\sqrt{3}P_{12}+c_2^2P_{22})e^{-\rho_1} \tag{2.490}$$

where $\rho_1=\zeta_1R$ and the polynomials $P_{nn'}(\rho_1)$ are those involved in the overlap integrals $\langle ns(\zeta_1)\,|\,n's(\zeta_1)\rangle$ as given by the formulas (25a) of Roothaan (1951) (the integral is $P_{nn'}(\rho_1)e^{-\rho_1}$ multiplied by a numerical coefficient when $n\neq n'$):

$$P_{11}(\rho)=1+\rho+\frac{\rho^2}{3} \qquad P_{12}(\rho)=1+\rho+\frac{4}{9}\rho^2+\frac{\rho^3}{9}$$

$$P_{22}(\rho)=1+\rho+\frac{4}{9}\rho^2+\frac{\rho^3}{9}+\frac{\rho^4}{15} \tag{2.491}$$

Now, as in the case of the one-component orbital we neglect the penetration energy $\varepsilon_1^{\text{pen}}$ and we replace M_{11}^{12} by $n_{12}^{(1)}$, so that $(-T_{12}+4n_{12}^{(1)}-2M_{11}^{12})$ becomes $(-T_{12}+2n_{12}^{(1)})$. Then Roothaan's formulas for T_{12} and $n_{12}^{(1)}$ give after some algebra (a detailed derivation is given by Claverie (1973) Ch. III, Section 3.B.2, pp. 200–201):

$$-T_{12}[\zeta_1]+2n_{12}^{(1)}[\zeta_1]=S[\zeta_1]\left[\frac{\zeta_1^2}{2}+(2-\zeta_1)\frac{n_{12}^{(1)}[\zeta_1]}{S[\zeta_1]}\right.$$
$$\left.-\frac{\zeta_1^2}{2}\frac{2c_1c_2\rho_1^2/\sqrt{3}+c_2^2(1+\rho_1+\rho_1^2/3)}{3S[\zeta_1]e^{\rho_1}}\right] \qquad (2.492)$$

We first simplify the third term of the bracketed expression. Since $|c_2|\ll c_1$, the term $2c_1c_2$ is dominant in the numerator, and, using (2.491), we see that the term $c_1^2\rho_1^2/3$ is dominant in $S[\zeta_1]$ provided ρ_1 is neither too small nor too large. We therefore reduce the fraction to $(2c_1c_2\rho_1^2/\sqrt{3})/(3c_1^2\rho_1^2/3)=2c_2/(c_1\sqrt{3})$. As concerns the ratio $n_{12}^{(1)}[\zeta_1]/(\zeta_1S[\zeta_1])$, it could be represented quite correctly by $3/(\rho_1+2)$; but, as in the case of the one-component orbital, we use the fraction $3/\rho_1$ instead of $3/(\rho_1+2)$ in order to compensate for the incomplete concellation of the two previous approximations (namely the neglect of $\varepsilon_1^{\text{pen}}$ and the replacement of the bielectronic hybrid integral M_{11}^{12} by the nuclear hybrid integral $n_{12}^{(1)}$). Finally, instead of using $S[\zeta_1]$ in front of the bracketed expression of (2.492), we shall use the complete S for making possible the factorization of S^2 in the final complete formula for $\varepsilon_1^{\text{HL}}$, and the S-term will therefore be approximated by:

$$S(-T_{12}+4n_{12}^{(1)}-2M_{11}^{12})$$
$$\simeq S^2\left[\frac{\zeta_1^2}{2}+(2-\zeta_1)\frac{3\zeta_1}{\rho_1}-\frac{\zeta^2}{2}\frac{2c_2}{c_1\sqrt{3}}\right] \qquad (2.493)$$

(c) *The exchange integral M_{11}^{22}*. We again take advantage of the dominant character of the orbitals with exponent $\zeta_1=1{\cdot}4$ at intermediate and large distances, and we introduce the approximation (where C is always the Euler constant):

$$M_{11}^{22}\simeq\zeta_1S^2\frac{C+\log\rho_1}{\rho_1} \qquad (2.494)$$

Putting together (2.489, 2.493, 2.494) we finally obtain the following approximate formula for $\varepsilon_1^{\text{HL}}$:

$$\varepsilon_1^{\text{HL}}\simeq 4S^2\left[\left(K-\frac{1}{2R}\right)+\frac{\zeta_1^2}{2}\left(1-\frac{2c_2}{\sqrt{3}c_1}\right)+(2-\zeta_1)\frac{3}{R}-\frac{C+\log R_1}{2R}\right]$$
$$\simeq 4\frac{S^2}{R}\left\{\left[K+\frac{\zeta_1^2}{2}\left(1-\frac{2c_2}{\sqrt{3}c_1}\right)\right]R+\tfrac{1}{2}(11-6\zeta_1-C-\log\rho_1)\right\} \qquad (2.495)$$

where $\rho_1 = \zeta_1 R$ and we recall the numerical values $\zeta_1 = 1{\cdot}4$

$c_1 = 0{\cdot}884315 \qquad c_2 = 0{\cdot}051269 \qquad K = 0{\cdot}91794579 \qquad$ (see (2.487))

$C = 0{\cdot}577215664\ldots,$

hence:

$$\varepsilon_1^{\mathrm{HL}} \simeq 4\frac{S^2}{R}[0{\cdot}1276601R + 0{\cdot}843156 - \tfrac{1}{2}\log R] \qquad (2.496)$$

The numerical results obtained with this formula are discussed in Section IV.C.2 of the main text; it will be seen that the multiplication by an *empirical* factor equal to 1·4 turns out to be necessary. The replacement of the complete overlap integral S by the simpler $S_1 = S[\zeta_1]$ (*see* (2.490)) is also considered there.

XI. APPENDIX E: SOME TENSORIAL IDENTITIES

(Rein *et al.*, 1968; (Claverie and Rein, 1969 Claverie, 1973 Appendix to Ch. III)

Let us first recall some definitions (for example, see Morse and Feshbach (1953, Ch. 1, and the Appendix to the chapter by Buckingham in the present book).

The *dyadic product* (or *tensor product*) of two vectors $\vec{U}$ and $\vec{V}$ is a tensor, denoted $\vec{U}\otimes\vec{V}$, which will essentially be used as an operator acting on other vectors; its action upon any vector $\vec{X}$ corresponds to tensor contracted product, and is written, in vector notations, as

$$(\vec{U}\otimes\vec{V})\vec{X} = \vec{U}(\vec{V}\,.\,\vec{X}) \qquad (2.497)$$

Introducing an orthonormal basis, we deduce from (2.497) the matrix elements of $\vec{U}\otimes\vec{V}$ in terms of the components of $\vec{U}$ and $\vec{V}$:

$$(\vec{U}\otimes\vec{V})_{ij} = U_i V_j^* \qquad (2.498)$$

(where the * denotes complex conjugation). (2.498) clearly corresponds to the definition of the tensor product in terms of explicit components. From (2.498) we get

$$\mathrm{Tr}(\vec{U}\otimes\vec{V}) = \sum_i (\vec{U}\otimes\vec{V})_{ii} = \sum_i U_i V_i^* = \vec{V}\,.\,\vec{U} \qquad (2.499)$$

If $\vec{U} = \vec{V} = \vec{e}$, where $\vec{e}$ denotes some vector with unit norm, we deduce from (2.497) that $\vec{e}\otimes\vec{e}$ is the *projection operator* corresponding to the direction $\vec{e}$, and (2.499) actually gives Tr $(\vec{e}\otimes\vec{e}) = 1$ (note that all these properties correspond to those of the expression $|\psi\rangle\langle\psi|$ which is used in quantum mechanics for representing the projection operator corresponding with some normalized function $|\psi\rangle$).

Explicit expression of a hermitian operator in terms of its eigenvalues and eigenvectors (in finite-dimensional linear space).

As is well-known, any hermitian operator $\bar{\bar{A}}$ in a finite-dimensional linear space has a complete set of orthonormal eigenvectors $\vec{e}_k$ (with corresponding real eigenvalues λ_k), and if we denote $\bar{\bar{P}}_k$ the projection operator corresponding with the eigenvector $\vec{e}_k$, the operator $\bar{\bar{A}}$ may be expressed by its spectral decomposition:

$$\bar{\bar{A}} = \sum_k \lambda_k \bar{\bar{P}}_k \tag{2.500}$$

and, as we saw above, $\bar{\bar{P}}_k = \vec{e}_k \otimes \vec{e}_k$, hence:

$$\bar{\bar{A}} = \sum_k \lambda_k \vec{e}_k \otimes \vec{e}_k \tag{2.501}$$

Product of Operators defined as dyadic products

Let us consider two dyadic products $\vec{U}_1 \otimes \vec{V}_1$ and $\vec{U}_2 \otimes \vec{V}_2$. We express according to (2.497) the action of their product onto an arbitrary vector $\vec{X}$:

$$\begin{aligned}(\vec{U}_2 \otimes \vec{V}_2)(\vec{U}_1 \otimes \vec{V}_1)\vec{X} &= (\vec{U}_2 \otimes \vec{V}_2)\vec{U}_1(\vec{V}_1 . \vec{X}) \\ &= \vec{U}_2(\vec{V}_2 . \vec{U}_1)(\vec{V}_1 . \vec{X}) \\ &= (\vec{V}_2 . \vec{U}_1)\ \vec{U}_2 \otimes \vec{V}_1)\vec{X}\end{aligned}$$

which proves that

$$(\vec{U}_2 \otimes \vec{V}_2)(\vec{U}_1 \otimes \vec{V}_1) = (\vec{V}_2 . \vec{U}_1)\vec{U}_2 \otimes \vec{V}_1 \tag{2.502}$$

Of course, when $\vec{e}$ is a unit vector, we check from (2.502) that the projection operator $\vec{e} \otimes \vec{e}$ is actually idempotent: $\vec{e} \otimes \vec{e})^2 = (\vec{e} . \vec{e})\vec{e} \otimes \vec{e} = \vec{e} \otimes \vec{e}$.

Some identities involving vectors and tensors

(1) Let $\vec{U}$ and $\vec{X}$ be two arbitrary vectors:

$$\begin{aligned}\vec{X}(\vec{U} \otimes \vec{U})\vec{X} &= \vec{X} . [(\vec{U} \otimes \vec{U})\vec{X}] = \vec{X} . [\vec{U}(\vec{U} . \vec{X})] \\ &= (\vec{X} . \vec{U})(\vec{U} . \vec{X}) = |\vec{U} . \vec{X}|^2\end{aligned} \tag{2.503}$$

(2) Let $\vec{X}$, $\vec{Y}$ be two arbitrary vectors and $\bar{\bar{B}}$ an arbitrary tensor. By introducing the coordinates of $\vec{X}$, $\vec{Y}$ and the matrix elements of $\bar{\bar{B}}$ in some orthonormal basis, we get:

$$\begin{aligned}\vec{X}\bar{\bar{B}}\vec{Y} &= \sum_i \sum_j X_i^* B_{ij} Y_j = \sum_i \sum_j B_{ij} Y_j X_i^* \\ &= \sum_i \sum_j B_{ij}(\vec{Y} \otimes \vec{X})_{ji} = \mathrm{Tr}\,[\bar{\bar{B}}(\vec{Y} \otimes \vec{X})]\end{aligned}$$

Thus

$$\vec{X}\bar{\bar{B}}\vec{Y} = \mathrm{Tr}\,[\bar{\bar{B}}(\vec{Y}\otimes\vec{X})] = \mathrm{Tr}\,[(\vec{Y}\otimes\vec{X})\bar{\bar{B}}] \tag{2.504}$$

(3) Let $\vec{X}$ and $\vec{Y}$ be two arbitrary vectors, $\bar{\bar{B}}$ and $\bar{\bar{C}}$ two arbitrary tensors. Introducing their coordinates and matrix elements as previously, we get:

$$\begin{aligned}\mathrm{Tr}\,[\bar{\bar{B}}(\vec{X}\otimes\vec{Y})\bar{\bar{C}}] &= \sum_i \sum_j \sum_k B_{ij} X_j Y_k^* C_{ki} \qquad (2.505)\\ &= \sum_k \sum_j Y_k^* \Big(\sum_i C_{ki} B_{ij}\Big) X_j \\ &= \sum_k \sum_j Y_k^* (\bar{\bar{C}}\bar{\bar{B}})_{kj} X_j\end{aligned}$$

or

$$\mathrm{Tr}\,[\bar{\bar{B}}(\vec{X}\otimes\vec{Y})\bar{\bar{C}}] = \vec{Y}(\bar{\bar{C}}\bar{\bar{B}})\vec{X} \tag{2.506}$$

We may also start again from the triple sum in the right-hand side of (2.505) and rearrange the terms in a different way:

$$\mathrm{Tr}\,[\bar{\bar{B}}(\vec{X}\otimes\vec{Y})\bar{\bar{C}}] = \sum_i \Big(\sum_k Y_k^* C_{ki})(\sum_j B_{ij} X_j)$$

Now $\sum_j B_{ij} X_j = (\bar{\bar{B}}\vec{X})_i$, and $\sum_k C_{ki} Y_k^* = (\sum_k C_{ki}^* Y_k)^* = (\sum_k (\bar{\bar{C}}^\dagger)_{ik} Y_k)^* = (\bar{\bar{C}}^\dagger \vec{Y})_i^*$ (where $\bar{\bar{C}}^\dagger$ denotes the hermitian conjugate of $\bar{\bar{C}}$. Hence the alternative identity:

$$\mathrm{Tr}[\bar{\bar{B}}(\vec{X}\otimes\vec{Y})\bar{\bar{C}}] = \sum_i (\bar{\bar{C}}^\dagger\vec{Y})_i^* (\bar{\bar{B}}\vec{X})_i = (\bar{\bar{C}}^\dagger\vec{Y})\,.\,(\bar{\bar{B}}\vec{X}) \tag{2.507}$$

(4) Let us finally consider four vectors: $\vec{X}$, $\vec{Y}$, $\vec{Z}$, $\vec{V}$ and two tensors: $\bar{\bar{B}}$, $\bar{\bar{C}}$. We want to transform the expression $(\vec{X}\bar{\bar{B}}\vec{Y})$ $(\vec{Z}\bar{\bar{C}}\vec{V})$, and, for that purpose, we introduce as previously the coordinates and matrix elements with respect to some arbitrary orthonormal basis:

$$\begin{aligned}(\vec{X}\bar{\bar{B}}\vec{Y})(\vec{Z}\bar{\bar{C}}\vec{V}) &= \Big(\sum_i \sum_j X_i^* B_{ij} Y_j\Big)\Big(\sum_k \sum_l Z_k^* C_{kl} V_l\Big) \\ &= \sum_i \sum_j \sum_k \sum_l X_i^* B_{ij} Y_j Z_k^* C_{kl} V_l\end{aligned}$$

Now, by isolating products of vector coordinates two by two, we recognize the matrix elements of dyadic products, according to (2.498) (e.g. $X_i^* Y_j = (\vec{Y}\otimes\vec{X})_{ji} = (\vec{X}\otimes\vec{Y})_{ij}^*$). If we perform these arrangements in such way as to have the same subscript in the first and last position, we may consider the sum as the Trace of the product of $\bar{\bar{B}}$, $\bar{\bar{C}}$ and two dyadic products. It must be emphasized that this may be done only by putting $\bar{\bar{B}}$ and $\bar{\bar{C}}$ in non-neighbour positions, and similarly for the two dyadic products. We may thus get various identities (special forms occur when $\bar{\bar{B}}$, or $\bar{\bar{C}}$, or both, are assumed to be hermitian):

$$\sum_i \sum_j \sum_k \sum_l X_i^* B_{ij} Y_j Z_k^* C_{kl} V_l = \sum_i \sum_j \sum_k \sum_l B_{ij} (Y_j Z_k^*) C_{kl} (V_l X_i^*)$$
$$= \sum_i \sum_j \sum_k \sum_l B_{ij} (\vec{Y} \otimes \vec{Z})_{jk} C_{kl} (\vec{V} \otimes \vec{X})_{li}$$
$$= \mathrm{Tr}\,[\bar{\bar{B}}(\vec{Y} \otimes \vec{Z})\bar{\bar{C}}(\vec{V} \otimes \vec{X})]$$

and, taking into account that $\mathrm{Tr}\,(\bar{\bar{A}}\bar{\bar{A}}') = \mathrm{Tr}\,(\bar{\bar{A}}'\bar{\bar{A}})$ for any pair of operators $\bar{\bar{A}}$, $\bar{\bar{A}}'$, we may write the identity:

$$(\vec{X}\bar{\bar{B}}\vec{Y})(\vec{Z}\bar{\bar{C}}\vec{V}) = \mathrm{Tr}\,[\bar{\bar{B}}(\vec{Y} \otimes \vec{Z})\bar{\bar{C}}(\vec{V} \otimes \vec{X})] \tag{2.508}$$
$$= \mathrm{Tr}\,[(\vec{V} \otimes \vec{X})\bar{\bar{B}}(\vec{Y} \otimes \vec{Z})\bar{\bar{C}}] \tag{2.509}$$
$$= \mathrm{Tr}\,[\bar{\bar{C}}(\vec{V} \otimes \vec{X})\bar{\bar{B}}(\vec{Y} \otimes \vec{Z})] \tag{2.510}$$
$$= \mathrm{Tr}\,[(\vec{Y} \otimes \vec{Z})\bar{\bar{C}}(\vec{V} \otimes \vec{X})\bar{\bar{B}}] \tag{2.511}$$

In the special case $\vec{V} = \vec{X}$, $\vec{Z} = \vec{Y}$, $\bar{\bar{B}} = \bar{\bar{C}}$, the identity (2.508) becomes

$$(\vec{X}\bar{\bar{B}}\vec{Y})(\vec{Y}\bar{\bar{B}}\vec{X}) = \mathrm{Tr}\,[\bar{\bar{B}}(\vec{Y} \otimes \vec{Y})\bar{\bar{B}}(\vec{X} \otimes \vec{X})] \tag{2.512}$$

Moreover, we obviously have (e.g. by using the explicit formula

$$\vec{X}\bar{\bar{B}}\vec{Y} = \sum_i \sum_j X_i^* B_{ij} Y_j :)$$

$$\vec{Y}\bar{\bar{B}}\vec{X} = (\vec{X}\bar{\bar{B}}^\dagger \vec{Y})^* \qquad \vec{X}\bar{\bar{B}}\vec{Y} = (\vec{Y}\bar{\bar{B}}^\dagger \vec{X})^* \tag{2.513}$$

Therefore, if $\bar{\bar{B}}$ is hermitian, $\vec{Y}\bar{\bar{B}}\vec{X}$ and $\vec{X}\bar{\bar{B}}\vec{Y}$ are complex conjugate, and (2.512) becomes

$$|\vec{X}\bar{\bar{B}}\vec{Y}|^2 = |\vec{Y}\bar{\bar{B}}\vec{X}|^2 = \mathrm{Tr}\,[\bar{\bar{B}}(\vec{Y} \otimes \vec{Y})\bar{\bar{B}}(\vec{X} \otimes \vec{X})]$$
$$= \mathrm{Tr}\,[(\vec{X} \otimes \vec{X})\bar{\bar{B}}(\vec{Y} \otimes \vec{Y})\bar{\bar{B}}] = \mathrm{Tr}\,[\bar{\bar{B}}(\vec{X} \otimes \vec{X})\bar{\bar{B}}(\vec{Y} \otimes \vec{Y})]$$
$$= \mathrm{Tr}\,[(\vec{Y} \otimes \vec{Y})\bar{\bar{B}}(\vec{X} \otimes \vec{X})\bar{\bar{B}}] \tag{2.514}$$

XIII. APPENDIX F: APPROXIMATION OF ATOMIC DIPOLES BY 'EFFECTIVE' ATOMIC CHARGES

Let us consider in a molecule an arbitrary atom (labelled 0) bearing a dipole moment $\vec{\mu}$ and let us label $\alpha = 1, \ldots, \nu$ the ν atoms bonded to it. We want to define 'effective charges' $q_0, q_1, \ldots, q_\alpha, \ldots, q_\nu$ located at these atoms, such that the total dipole moment associated with these charges is the closest possible to the given dipole $\vec{\mu}$; the sum of these charges is required to be zero:

$$q_0 + \sum_{\alpha=1}^{\nu} q_\alpha = 0 \tag{2.515}$$

For that purpose, we shall decompose (exactly or approximately) the total dipole $\vec{\mu}$ into bond dipoles $\vec{\mu}_\alpha$ directed along the bonds (0–α). Let us denote $\vec{\mu}_\alpha$ the unit vectors corresponding to these bonds (directed from 0 to α, say). These vectors span some linear subspace whose dimension will be denoted n (of course, $n \leqslant 3$, and $n \leqslant \nu$ since there are ν vectors $\vec{\mu}_\alpha$). As an example, we list the possibilities occurring in usual organic molecules:

$\nu = 1$: $n = 1$ (one bond)
$\nu = 2$: $n = 1$ (two colinear bonds); $n = 2$ (two non-colinear bonds)
$\nu = 3$: $n = 2$ (three coplanar bonds); $n = 3$ (three non-coplanar bonds)
$\nu = 4$: $n = 3$ (four non-coplanar bonds)

First, it must be emphasized that, for $n < 3$, the dipole $\vec{\mu}$ can be decomposed exactly along the bonds only if it belongs to the linear subspace generated by the basis $\vec{u}_\alpha$. Thus, we begin by considering the projection $\vec{\mu}'$ of $\vec{\mu}$ into this subspace. In actual practice, the error thus introduced is quite small (for coplanar bonds, the (hybridization) dipole of the central atom is almost exactly contained in the plane of these bonds).

Now, we must define bond dipoles $\mu_\alpha \vec{u}_\alpha$ whose sum is equal to $\vec{\mu}'$:

$$\sum_{\alpha=1}^{\nu} \mu_\alpha \vec{u}_\alpha = \vec{\mu}' \tag{2.516}$$

If $n = \nu$, the vectors $\vec{u}_\alpha$ are linearly independent, and (2.516) determines in a unique way the coefficients μ_α as the coordinates of $\vec{\mu}'$ with respect to the basis $\{\vec{u}_\alpha\}$ of the linear subspace defined by the bonds.

If $n < \nu$, the vectors $\vec{u}_\alpha$ are linearly dependent, and (2.516) does not determine the coefficients μ_α in a unique way. Some supplementary condition is necessary, and we choose, for simplicity, to require that

$$\sum_{\alpha=1}^{\nu} (\mu_\alpha)^2 \text{ is minimum} \tag{2.517}$$

If we introduce the coordinates $u_{\alpha i}$, μ_i' of the vectors $\vec{u}_\alpha$, $\vec{\mu}'$ in some basis of the physical space ($i = 1, 2, 3$), (2.516) appears as a set of 3 linear relations

$$\sum_{\alpha=1}^{\nu} \mu_\alpha u_{\alpha i} = \mu_i' \quad (i = 1, 2, 3) \tag{2.518}$$

but only n of them are linearly independent (the 3-n other are obtained as linear combinations of the n first, since both $\vec{\mu}'$ and the $\vec{u}_\alpha$s belong to the same n-dimensional subspace). The problem defined by n among the relations (2.518) ($i = 1, \ldots, n$), for definiteness) and (2.517) has now a unique solution: indeed, in an *abstract* space of dimension ν where the quantities μ_α may be considered as the coordinates of a point with respect to an orthonormal basis, the n relations (2.518) define a *linear* manifold of dimension ν-n (hyperplane, plane, straight line), since the relations (2.518) are linear with respect to the μ_αs, and this linear manifold does not contain the origin since the μ_i's are not all zero. Thus, the solution to our problem is given by the point in this linear manifold which is closest to the origin, and this point is unique.

For practical calculations we use the method of Lagrange multipliers: namely, we minimize with respect to the variables μ_α the function

$$F(\mu_1, \ldots, \mu_\alpha) = \sum_{\alpha=1}^{\nu} \mu_\alpha^2 + \sum_{i=1}^{n} \lambda_i \left(\sum_{\alpha=1}^{\nu} \mu_\alpha u_{\alpha i} - \mu_i' \right) \qquad (2.519)$$

Now

$$\frac{\partial F}{\partial \mu_\alpha} = 2\mu_\alpha + \sum_{i=1}^{n} \lambda_i u_{\alpha i} \qquad (\alpha = 1, \ldots, \nu) \qquad (2.520)$$

Hence $\partial F/\partial \mu_\alpha = 0$ gives:

$$\mu_\alpha = -\tfrac{1}{2} \sum_{i=1}^{n} \lambda_i u_{\alpha i} \qquad (\alpha = 1, \ldots, \nu) \qquad (2.521)$$

Thus the ν unknown quantities μ_α have been expressed as linear functions of n unknown Lagrange multipliers λ_i $(i = 1, \ldots, n;\ n < \nu)$, and these are determined by a linear system of n equations obtained by inserting the expressions (2.520) of μ_α in the first n relations (2.518):

$$-\tfrac{1}{2} \sum_{\alpha=1}^{\nu} \left(\sum_{j=1}^{n} \lambda_i u_{\alpha i} \right) u_{\alpha j} = \mu_j' \qquad (j = 1, \ldots, n)$$

or

$$\sum_{i=1}^{n} \lambda_i \left(\sum_{\alpha=1}^{\nu^*} u_{\alpha i} u_{\alpha j} \right) = -2\mu_j' \qquad (j = 1, \ldots, n) \qquad (2.522)$$

and the problem is now completely solved, since the components $u_{\alpha j}$ and μ_j' (of the vectors $\vec{u}_\alpha$ and $\vec{\mu}'$ respectively) are known quantities.

Practical remarks

In actual practice, the ν bond vectors $\vec{u}_\alpha$ are *not* an orthonormal basis of the linear subspace (if $n < \nu$, these vectors are even linearly dependent and therefore constitute a 'basis' which is not only non-orthogonal but even overcomplete). Thus, a direct calculation of $\vec{\mu}'$ (the projection of $\vec{\mu}$ onto the subspace spanned by the $\vec{u}_\alpha$s) through its usual (contravariant) components in the basis $(i = 1, 2, 3)$ may be somewhat involved. A convenient solution consists in expressing (2.516), namely $(\sum_{\alpha=1}^{\nu} \mu_\alpha \vec{u}_\alpha - \vec{\mu}') = 0$, by taking the scalar products with the vectors $\vec{u}_\beta$ (instead of the components over the basis $i = 1, 2, 3$), thus getting

$$\sum_{\alpha=1}^{\nu} \mu_\alpha (\vec{u}_\alpha \,.\, \vec{u}_\beta) - (\vec{\mu}' \,.\, \vec{u}_\beta) = 0 \qquad (\beta = 1, \ldots, \nu) \qquad (2.523)$$

but $\vec{\mu}' . \vec{u}_\beta = \vec{\mu} . \vec{u}_\beta$ since $(\vec{\mu} - \vec{\mu}') . \vec{u}_\beta = 0$ $(\beta = 1, \ldots, \nu)$ according to the very definition of $\vec{\mu}'$ ($\vec{\mu} - \vec{\mu}'$ is orthogonal to the subspace spanned by the bond vectors $\vec{u}_\beta$s). Therefore (2.523) becomes

$$\sum_{\alpha=1}^{\nu} \mu_\alpha (\vec{u}_\alpha . \vec{u}_\beta) = \vec{\mu} . \vec{u}_\beta \qquad (\beta = 1, \ldots, \nu) \tag{2.524}$$

but here also only n of these relations are linearly independent (since only n among the vectors $\vec{u}_\beta$ are independent). Thus, taking the n first relations (2.524) is sufficient, and these relations correspond to the n first relations (2.518).

Formally, using (2.524) amounts to replace $i = 1, \ldots, n$ by $\beta = 1, \ldots, n$, $u_{\alpha i}$ by $v_{\alpha\beta} = \vec{u}_\alpha . \vec{u}_\beta = \sum_{j=1}^{3} u_{\alpha j}$ and μ_i' by $m_\beta = \sum_{j=1}^{3} \mu_j u_{\beta j}$. The following developments (2.519–2.522) may be performed in a completely analogous way, thus giving

$$\mu_\alpha = -\tfrac{1}{2} \sum_{\beta=1}^{n} \lambda_\beta v_{\alpha\beta} \tag{2.525}$$

where the λ_βs are solution of the linear system

$$\sum_{\beta=1}^{n} \lambda_\beta \left(\sum_{\alpha=1}^{\nu} v_{\alpha\beta} v_{\alpha\gamma} \right) = -2m_\gamma \qquad (\gamma = 1, \ldots, n) \tag{2.526}$$

(2.525 and 2.526) replacing respectively (2.512 and 2.522) (of course, we may introduce the modified Lagrange multipliers $\lambda_\beta' = -\tfrac{1}{2}\lambda_\beta$: then the factors $=\tfrac{1}{2}$ in (2.525) and -2 in (2.526) are cancelled).

Definition of the effective charges

Once the decomposition of $\vec{\mu}'$ into bond dipoles $\mu_\alpha \mu_\alpha$ has been achieved, the definition of the effective charges is obvious: for every bond (0–α), we put the charge $q_\alpha = \mu_\alpha / l_\alpha$ (l_α = length of the bond (0–α)) on the atom α and $-q_\alpha$ on 0; when this has been done for all bonds, we have

$$q_0 = -\sum_{\alpha=1}^{\nu} q_\alpha.$$

The whole procedure is thus completed for the atom 0, and it must be applied to all the atoms of the molecule which bear a non-zero dipole moment.

XIV. REFERENCES

Abbott, J. A. Bolton, H. C. (1952). *Trans. Faraday Soc.*, **48,** 422.

Abrahamson, A. A. (1969). *Phys. Rev.*, **178,** 76.

Abramowitz, M. and Stegun, I. A. (1965). *Handbook of Mathematical Functions*, Dover, New York.

Ahlrichs, R. (1972a). *Phys. Rev. A.*, **5,** 605.

Ahlrichs, R. (1972b). *Chem. Phys. Lett.*, **15,** 609.

Ahlrichs, R. (1973a). *Chem. Phys. Lett.*, **18,** 67.

Ahlrichs, R. (1973b). *Chem. Phys. Lett.*, **18,** 521.
Ahlrichs, R. (1973c). *J. Math. Phys.*, **14,** 1860.
Ahlrichs, R. (1976). *Theoret. Chim. Acta*, **41,** 7.
Ahlrichs, R. and Claverie, P. (1972). *Int. J. Quant. Chem.*, **6,** 1001.
Ahmed, N. A. and Kitaigorodskii, A. I. (1972). *Acta Cryst.*, **B28,** 739.
Alexander, M. H. (1970a). *Phys. Rev. A.*, **1,** 1397.
Alexander, M. H. (1970b). *J. Chem. Phys.*, **52,** 3354.
Allen, L. C. (1975a). *Proc. Natl. Acad. Sci. U.S.A.*, **72,** 4701.
Allen, L. C. (1975b). *J. Amer. Chem. Soc.*, **97,** 6921.
Alonso, J. A. and Fonseca, M. P. (1974). *Anales de Fisica*, **70,** 371.
Alvarez-Rizzati, M. and Mason, E. A. (1975). *J. Chem. Phys.*, **63,** 5290.
Amos, A. T. (1970). *Chem. Phys. Letters*, **5,** 587.
Amos, A. T. (1971). *Intern. J. Quantum Chem.*, **5,** 699.
Amos, A. T. and Crispin, R. J. (1975). *J. Chem. Phys.*, **63,** 1890.
Amos, A. T. and Crispin, R. J. (1976a). *Mol. Phys.*, **31,** 147.
Amos, A. T. and Crispin, R. J. (1976b). *Mol. Phys.*, **31,** 159.
Amos, A. T. and Crispin, R. J. (1976c). p. 1 in *Theoretical Chemistry, Advances and Perspectives* Vol. 2 (eds. H. Eyring and D. Henderson). Academic Press, New York.
Amos, A. T., Crispin, R. J. and Smith, R. A. (1975). *Theoret. Chim. Acta*, **39,** 7.
Amos, A. T. and Musher, J. I. (1967). *Chem. Phys. Letters*, **1,** 149.
Amos, A. T. and Musher, J. I. (1968). *J. Chem. Phys.*, **49,** 2158.
Amos, A. T. and Musher, J. I. (1969). *Chem. Phys. Lett.*, **3,** 721.
Amos, A. T. and Yoffe, J. A. (1975a). *Chem. Phys. Lett.*, **31,** 53.
Amos, A. T. and Yoffe, J. A. (1975b). *Chem. Phys. Lett.*, **31,** 57.
Amos, A. T. and Yoffe, J. A. (1975c). *Theoret. Chim. Acta*, **40,** 221.
Amos, A. T. and Yoffe, J. A. (1975d). *J. Chem. Phys.*, **63,** 4723.
Andersen, A. M. (1975). *Acta Chem. Scand.*, **B29,** 239.
Andersen, H. C. (1975). *Ann. Rev. Phys. Chem.*, **26,** 145.
Andersen, H. C. and Chandler, D. (1972). *J. Chem. Phys.*, **57,** 1930.
Andersen, H. C., Weeks, J. O. and Chandler, D. (1971). *Phys. Rev. A*, **4,** 1597.
Andrews, L. J. and Keefer, R. M. (1964). *Molecular complexes in Organic Chemistry.* Holden-Day, San Francisco.
Applequist, J. and Carl, R. (1973). *J. Phys. Chem.*, **77,** 2090.
Applequist, J., Carl, R. and Fung, K. K. (1972). *J. Amer. Chem. Soc.*, **94,** 2952.
Arrighini, G. P., Biondi, F. and Guidotti, C. (1973). *Mol. Phys.*, **26,** 1137.
Atalay, B., Mann, A. and Peierls, R. (1973). *Proc. Roy. Soc. (London)*, **A335,** 251.
Axilrod, B. M. (1951a). *J. Chem. Phys.*, **19,** 719.
Axilrod, B. M. (1951b). *J. Chem. Phys.*, **19,** 724.
Axilrod, B. M. and Teller, E. (1943). *J. Chem. Phys.*, **11,** 299.
Babloyantz, A. (1959). *Mol. Phys.*, **2,** 39.
Bade, W. L. (1957). *J. Chem. Phys.*, **27,** 1280.
Bade, W. L. and Kirkwood, J. G. (1957). *J. Chem. Phys.*, **27,** 1284.
Bader, R. F. W., Novaro, O. A. and Beltran-Lopez, V. (1971). *Chem. Phys. Lett.*, **8,** 568.
Baker, Jr., G. A. (1965). *Advances in Theoretical Physics*, **1,** 1. Academic Press, New York.
Baker Jr., G. A. and Gammel, J. L., eds. (1970). *The Padé Approximant Method in Theoretical Physics.* Academic Press, New York.
Bakshiev, N. G., Girin, O. P. and Piterskaya, I. V. (1968). *Optika i Spectroskopiya*, **24,** 901. [*Optics and Spectroscopy*, **24,** 483 (1968)].
Banerjee, K. (1967). *Mol. Phys.*, **12,** 385.
Banerjee, K. and Salem, L. (1966). *Mol. Phys.*, **11,** 405.
Barker, J. A. and Henderson, D. (1967). *J. Chem. Phys.*, **47,** (a) 2856, (b) 4714.
Barker, J. A. and Henderson, D. (1968). *J. Chem. Educ.*, **45,** 2.

Barker, J. A. and Henderson, D. (1971). *Acc. Chem. Res.*, **4,** 303.
Barker, J. A. and Henderson, D. (1972). *Ann. Rev. Phys. Chem.*, **23,** 439.
Barker, J. A. and Leonard, P. J. (1964). *Phys. Letters*, **13,** 127.
Bartell, L. (1960). *J. Chem. Phys.*, **32,** 827.
Basilevsky, M. V. and Berenfeld, M. M. (1972a). *Intern. J. Quant. Chem.*, **6,** 23.
Basilevsky, M. V. and Berenfeld, M. M. (1972b). *Int. J. Quant. Chem.*, **6,** 555.
Battezzati, L., Pisani, C. and Ricca, F. (1975). *J. Chem. Soc. Faraday Trans. II*, **71,** 1629.
Bell, R. J. (1965). *Proc. Phys. Soc. (London)*, **86,** 17 and 239.
Bell, R. J. (1966). *Proc. Phys. Soc. (London)*, **87,** 594.
Bender, C. F., Schaefer, H. F. and Kollman. P. A. (1972). *Mol. Phys.*, **24,** 235.
Ben Naim, A. (1971a). *J. Chem. Phys.*, **54,** 1387.
Ben Naim, A. (1971b). *J. Chem. Phys.*, **54,** 3696.
Ben Naim, A. (1972a). *J. Chem. Phys.*, **57,** 5257.
Ben Naim, A. (1972b). *J. Chem. Phys.*, **57,** 5266.
Berrondo, M. (1973). *Mol. Phys.*, **26,** 329.
Berthod, H. and Pullman, A. (1964). *Biopolymers*, **2,** 483.
Berthod, H. and Pullman, A. (1965). *J. Chim. Phys.*, **62,** 942.
Berthod, H. and Pullman, A. (1966). *C. R. Acad. Sci. Paris*, **262,** 76.
Bertoncini, P. and Wahl, A. C. (1970). *Phys. Rev. Letters*, **25,** 991.
Beveridge, D. L., Kelly, M. M. and Radna, R. J. (1974). *J. Am. Chem. Soc.*, **96,** 3769.
Beveridge, D. L. and Schnuelle, G. W. (1974). *J. Phys. Chem.*, **78,** 2064.
Beveridge, D. L. and Schnuelle, G. W. (1975). *J. Phys. Chem.*, **79,** 2562.
Birshtein, T. and Ptitsyn, O. B. (1966). *Conformations of Macromolecules.* Interscience, New York.
Bleick, W. E. and Mayer, J. E. (1934). *J. Chem. Phys.*, **2,** 252.
Boehm, R. and Yaris, R. (1971). *J. Chem. Phys.*, **55,** 2620.
Bonamy, J. and Robert, D. (1970). *Chem. Phys. Letters*, **6,** 591.
Bondi, A. (1964). *J. Phys. Chem.*, **68,** 441.
Bonnor, W. B. (1951). *Trans. Farad. Soc.*, **47,** 1143.
Born, M. (1920). *Z. Phys.*, **1,** 45.
Box, M. J., Davies, D. and Swann, W. H. (1969). *Non Linear Optimization Techniques.* Oliver and Boyd, London.
Boyer, T. H. (1973). *Phys. Rev. A*, **7,** 1832.
Boyer, T. H. (1975). *Phys. Rev. D*, **11,** (a) 790, (b) 809.
Brändas, E. and Goscincki, O. (1970). *Intern. J. Quant. Chem. Symp.*, **3,** 383.
Bratoz, S. (1967). *Adv. Quant. Chem.*, **3,** 209.
Brieglieb, G. (1961). *Elektronen Donator-Acceptor Komplexe.* Springer-Verlag, Berlin.
Buckingham, A. D. (1965). *Disc. Faraday Soc.*, **40,** 232.
Buckingham, A. D. (1967). p. 107 in *'Intermolecular Forces', Advances in Chemical Physics, 12* (ed. J. O. Hirschfelder). Interscience, New York.
Buckingham, A. D. (1968). *J. Chem. Phys.*, **48,** 3827.
Buckingham, A. D., Disch, R. L. and Dunmur, D. A. (1968). *J. Amer. Chem. Soc.*, **90,** 3104.
Buckingham, A. D. and Utting, B. D. (1970). *Annual Review of Physical Chemistry*, **21,** 287. Annual Reviews, Inc., Palo-Alto, California.
Buckingham, R. A. (1938). *Proc. Roy. Soc. (London)*, **A168,** 264.
Buckley, F. (1958). *J. Res. Nat. Bur. Stand.*, **61,** 325.
Buckley, F. and Maryott, A. A. (1954). *J. Res. Nat. Bur. Stand.*, **53,** 229.
Buff, F. P., Goel, N. S. and Clay, J. R. (1975). *J. Chem. Phys.*, **63,** 1367.
Bugg, C. E., Thomas, J. M., Sundaralingam, M. and Rao, T. S. (1971). *Biopolymers*, **10,** 175.

Bukta, J. F. and Meath, W. J. (1972). *Intern. J. Quant. Chem.*, **6,** 1045.
Burgess, A. W. and Scheraga, H. A. (1975). *Proc. Natl. Acad. Sci. USA*, **72,** 1221.
Burgess, A. W., Shipman, L. L., Nemenoff, R. A. and Scheraga, H. A. (1976). *J. Am. Chem. Soc.*, **98,** 23.
Burgess, A. W., Shipman, L. L. and Scheraga, H. A. (1975). *Proc. Natl. Acad. Sci. USA*, **72,** 854.
Byers Brown, W. (1968a). *Chem. Phys. Letters*, **2,** 105.
Byers Brown, W. (1968b). *Chem. Phys. Letters*, **1,** 655.
Byers Brown, W. and Power, J. D. (1970). *Proc. Roy. Soc.* (*London*), **A317,** 545.
Caillet, J. (1972). '*Application de la méthode des interactions à l'étude de la stabilité de complexes intermoléculaires d'intérêt biochimique*'. Thesis (Doctorat d'Etat), C.N.R.S., Paris.
Caillet, J. and Claverie, P. (1974). *Biopolymers*, **13,** 601.
Caillet, J. and Claverie, P. (1975). *Acta Cryst.*, **A31,** 448.
Caillet, J., Claverie, P. and Pullman, B. (1976). *Acta Cryst.*, **B32,** 2740.
Califano, S., ed. (1975). '*Lattice Dynamics and Intermolecular Forces*', Proceedings of the International School of Physics 'Enrico Fermi', course 55 (July 1972). Academic Press, New York.
Carlström, D. (1973). *Acta Cryst.*, **B29,** 161.
Carlström, D., Bergin, A. and Falkenberg, G. (1973). *Quart. Rev. Biophys.*, **6,** 257.
Carr, W. J., Jr. (1963). *Phys. Rev.*, **131,** 1947.
Casanova, G., De Paz, M., Dondi, M. G., Klein, M. L. and Scoles, G. (1965). *Disc. Faraday Soc.*, **40,** 188.
Casimir, H. B. G. and Polder, D. (1948). *Phys. Rev.*, **73,** 360.
Certain, P. R. (1968). *J. Chem. Phys.*, **49,** 35.
Certain, P. R. (1971). *J. Chem. Phys.*, **55,** 3045.
Certain, P. R. and Bruch, L. W. (1972). Ch. 4, p. 113 in *Theoretical Chemistry* (ed. by W. Byers Brown), Vol. 1 in '*Physical Chemistry, series One*' of MTP (Medical and Technical Publishing Co.) International Review of Science. Butterworths, London and University Park Press, Baltimore.
Certain, P. R. and Byers Brown, W. (1972). *Intern. J. Quant. Chem.*, **6,** 131.
Certain, P. R., Dion, D. R. and Hirschfelder, J. O. (1970). *J. Chem. Phys.*, **52,** 5987.
Certain, P. R. and Hirschfelder, J. O. (1970a). *J. Chem. Phys.*, **52,** 5977.
Certain, P. R. and Hirschfelder, J. O. (1970b). *J. Chem. Phys.*, **52,** 5992.
Certain, P. R., Hirschfelder, J. O. and Dion, D. R. (1970). *J. Chem. Phys.*, **53,** 2992.
Certain, P. R., Hirschfelder, J. O. and Epstein, S. T. (1969). *Chem. Phys. Letters*, **4,** 401.
Certain, P. R., Hirschfelder, J. O., Kolos, W. and Wolniewicz, L. (1968). *J. Chem. Phys.*, **49,** 24.
Certain, P. R. and Merkel, R. R. (1975a). *J. Chem. Phys.*, **62,** 1402.
Certain, P. R. and Merkel, R. R. (1975b). *J. Chem. Phys.*, **62,** 2406.
Chalasinski, G. and Jeziorski, B. (1973). *Intern. J. Quantum Chem.*, **7,** 745.
Chalasinski, G. and Jeziorski, B. (1974). *Mol. Phys.*, **27,** 649.
Chalasinski, G., Jeziorski, B. and Szalewicz, K. (1977). *Intern. J. Quantum Chem.* **11,** 247.
Chalvet, O., Daudel, R., Diner, S. and Malrieu, J. P. (eds.) (1975, 1976). *Localization and Delocalization in Quantum Chemistry*, Vol. I (1975), Vol. II (1976) Reidel, Dordrecht.
Chandler, D. (1973). *J. Chem. Phys.*, **59,** 2742.
Chandler, D. (1974). *Acc. Chem. Res.*, **7,** 246.
Chandler, D. (1976). *Mol. Phys.*, **31,** 1213.
Chipman, D. M. (1972). '*Perturbation theories for the calculation of molecular interaction energies*'. Ph.D. Thesis, University of Wisconsin, Madison.

Chipman, D. M., Bowman, J. D. and Hirschfelder, J. O. (1973). *J. Chem. Phys.*, **59,** 2830.
Chipman, D. M. and Hirschfelder, J. O. (1973). *J. Chem. Phys.*, **59,** 2838.
Choi, D. S., Jhon, M. S. and Eyring, H. (1970). *J. Chem. Phys.*, **53,** 2608.
Çignitti, M. and Allen, T. L. (1965). *J. Chem. Phys.*, **43,** 4472.
Čižek, J. (1963). *Mol. Phys.*, **6,** 19.
Claverie, P. (1968a). p. 115 in *Molecular Associations in Biology* (ed. B. Pullman). Academic Press, New York.
Claverie, P. (1968b). *J. Chim. Phys.*, **65,** 57.
Claverie, P. (1969). *Intern. J. Quant. Chem.*, **3,** 349.
Claverie, P. (1970). *Studia Biophysica*, **24–25,** 161.
Claverie, P. (1971a). *Intern. J. Quant. Chem.*, **5,** 273.
Claverie, P. (1971b). *J. Mol. Biol.*, **56,** 75.
Claverie, P. (1973). *Contribution à l'Etude des Interactions Moléculaires.* Thesis, Paris, registered at the C.N.R.S. library under number A.O. 8214.
Claverie, P. (1976). p. 127 in *Localization and Delocalization in Quantum Chemistry*, Vol. II, (eds. O. Chalvet, R. Daudel, S. Diner and J. P. Malrieu). Reidel, Dordrecht.
Claverie, P. and Diner, S. (1976). p. 375 in *Localization and Delocalization in Quantum Chemistry*, Vol. II, (eds. O. Chalvet, R. Daudel, S. Diner and J. P. Malrieu). Reidel, Dordrecht.
Claverie, P., Diner, S. and Malrieu, J. P. (1967). *Intern. J. Quant. Chem.*, **1,** 751.
Claverie, P., Pullman, B. and Caillet, J. (1966). *J. Theoret. Biol.*, **12,** 419 (there is an error in the Table 4 of this paper: in the row

$$\left|\begin{matrix}\uparrow GC \\ CG\downarrow\end{matrix}\right|,$$

$E_{\rho\rho}$ should be $-1{\cdot}6$ instead of $1{\cdot}6$; consequently, in the column 'sum', $-4{\cdot}9$ must be replaced by $-8{\cdot}1$ and in the column 'total', $-24{\cdot}1$ must be replaced by $-27{\cdot}3$).
Claverie, P. and Rein, R. (1969). *Intern. J. Quant. Chem.*, **3,** 537.
Clementi, E. (1967). *J. Chem. Phys.*, **46,** 3851.
Coen, J. L. de, Elefante, G., Liquori, A. M. and Damiani, A. (1967). *Nature*, **216,** 910.
Cohan, N. V. (1969). *Mol. Phys.*, **17,** 307.
Cohen, J. S. and Pack, R. T. (1974). *J. Chem. Phys.*, **61,** 2372.
Coiro, V. M., Giglio, E., Lucano, A. and Puliti, R. (1973). *Acta Cryst.*, **B29,** 1404.
Coiro, V. M., Giglio, E. and Quagliata, C. (1972). *Acta Cryst.*, **B28,** 3601.
Conway, A. and Murrell, J. N. (1972). *Mol. Phys.*, **23,** 1143.
Conway, A. and Murrell, J. N. (1974). *Mol. Phys.*, **27,** 873.
Cook, D. and Rowlinson, J. S. (1953). *Proc. Roy. Soc.* (*London*) *A*, **219,** 405.
Cook, E. G. and Schug, J. C. (1970). *J. Chem. Phys.*, **53,** 723.
Corinaldesi, E. (1962). *Nuovo Cimento*, **25,** 1190.
Corinaldesi, E. (1963). *Nuovo Cimento*, **30,** 105.
Corinaldesi, E. (1964a). *Nuovo Cimento*, **31,** 1258.
Corinaldesi, E. (1964b). *Nuovo Cimento*, **33,** 1680.
Corinaldesi, E. (1964c). *Phys. Lett.*, **9,** 135.
Corinaldesi, E., Remiddi, E. and Strocchi, F. (1963). *Nuovo Cimento*, **30,** 1563.
Corinaldesi, E. and Lin, H. E. (1963). *Nuovo Cimento*, **28,** 105.
Coulson, C. A. (1953). *Valence.* Oxford University Press.
Coulson, C. A. and Davies, P. L. (1952). *Trans. Faraday Soc.*, **48,** 777.
Coulson, C. A. and Rogers, M. T. (1961). *J. Chem. Phys.*, **35,** 593.
Dacre, P. D. and McWeeny, R. (1970). *Proc. Roy. Soc.* (*London*) *A*, **317,** 435.
Dalgarno, A. (1962). *Adv. Phys.*, **11,** 281.

Dalgarno, A. (1967). p. 143, in '*Intermolecular Forces*', *Advances in Chemical Physics*, **12** (ed. J. O. Hirschfelder). Interscience, New York.
Dalgarno, A. and Davison, W. D. (1966). *Advances in Atomic and Molecular Physics*, **2,** 1 (eds. D. R. Bates and I. Estermann). Academic Press, New York.
Dalgarno, A. and Kingston, A. E. (1961). *Proc. Phys. Soc.* (*London*), **78,** 607.
Dalgarno, A. and Lewis, J. T. (1955). *Proc. Roy. Soc.* (*London*), **A233,** 70.
Dalgarno, A. and Lewis, J. T. (1956). *Proc. Phys. Soc.* (*London*), **A69,** 57.
Dalgarno, A. and Lynn, N. (1957). *Proc. Phys. Soc.* (*London*), **A70,** 802. (The notations used in this paper slightly differ from those used in the more recent literature (e.g. Dalgarno and Davison, 1966): thus, E_0 and E_n denote binding energies with *positive sign* (instead of the usual state energies with negative sign), and they are expressed in *rydberg units* instead of atomic units (1 rydberg $= \frac{1}{2}$ atomic unit); hence, $(E_0 - E_n)$ of Dalgarno and Lynn's paper should be replaced by $2(E_n - E_0)$ in order to recover the formulas corresponding to the more recent notations.)
Dalgarno, A., Morrison, I. H. and Pengelly, R. M. (1967). *Intern. J. Quant. Chem.*, **1,** 161.
Damburg, R. J. and Propin, R. Kh. (1971). *J. Chem. Phys.*, **55,** 612.
Das, G. and Wahl, A. C. (1971). *Phys. Rev.*, **A4,** 825.
Dashevskii, V. G. and Kitaigorodskii, A. I. (1967). *Theor. Eksp. Khim.*, **3,** 43 [*Theoret. Exp. Chem.*, **3,** 22 (1970)].
Dashevskii, V. G. and Sarkisov, G. N. (1974). *Mol. Phys.*, **27,** 1271.
Daudel, R. (1966). p. 21 in *Fondements Théoriques des Recherches sur les Actions Intermoléculaires*, 27è colloque national du Centre National de la Recherche Scientifique, éditions du Centre National de la Recherche Scientifique, Paris.
Daudel, R., Lafforgue, A. and Vroelant, C. (1952). *J. Chim. Phys.*, **49,** 545.
Daudey, J. P. (1974). *Int. J. Quant. Chem.*, **8,** 29.
Daudey, J. P., Claverie, P. and Malrieu, J. P. (1974a). *Int. J. Quant. Chem.*, **8,** 1.
Daudey, J. P., Malrieu, J. P. and Rojas, O. (1974b). *Int. J. Quant. Chem.*, **8,** 17.
Daudey, J. P., Malrieu, J. P. and Rojas, O. (1975). p. 155 in *Localization and Delocalization in Quantum Chemistry*, Vol. I, (eds. O. Chalvet, R. Daudel, S. Diner and J. P. Malrieu). Reidel, Dordrecht.
Davies, P. L. (1952). *Trans. Faraday Soc.*, **48,** 789.
Deal, W. J. and Young, R. H. (1976). *Int. J. Quantum Chem.*, **10,** 419.
Del Re, G. (1958). *J. Chem. Soc.*, p. 4031.
Denbigh, K. G. (1940). *Trans. Faraday Soc.*, **36,** 936.
Denis, A., Langlet, J. and Malrieu, J. P. (1973). *Theoret. Chim. Acta*, **29,** 117.
Denis, A., Langlet, J. and Malrieu, J. P. (1975). *Theoret. Chim. Acta*, **38,** 49.
Derjaguin, B. V., Dzyaloshinskii, I. E., Koptelova, M. M. and Pitaevskii, L. P. (1965). *Disc. Faraday Soc.*, **40,** 246.
Devaquet, A. (1970). *Mol. Phys.*, **18,** 233.
Dewar, M. J. S. and Thompson, Jr., C. C. (1966). *Tetrahedron Suppl.*, **7,** 97.
Diercksen, G. H. F. (1971). *Theoret. Chim. Acta*, **21,** 335.
Diercksen, G. H. F., Kraemer, W. P. and Roos, B. (1975). *Theoret. Chim. Acta*, **36,** 249.
Diner, S. (1967). '*Etude de la structure électronique des molécules organiques conjuguées en termes d'orbitales localisées sur les liaisons chimiques*'. Thesis (Doctorat d'Etat), C.N.R.S., Paris.
Diner, S. and Claverie, P. (1976). p. 441 in *Localization and Delocalization in Quantum Chemistry*, Vol. II, (eds. O. Chalvet, R. Daudel, S. Diner and J. P. Malrieu). Reidel, Dordrecht.
Diner, S., Claverie, P. and Malrieu, J. P. (1967). *C.R. Acad. Sc. Paris*, **264,** 1589.
Diner, S., Malrieu, J. P., Jordan, F. and Claverie, P. (1970). *Theoret. Chim. Acta*, **18,** 86.

Di Nola, A. and Giglio, E. (1970). *Acta Cryst.*, **A26,** 144.
Dreyfus, M. (1970). '*Etude non-empirique de la liaison hydrogène dans les dimères de la formamide*'. Thesis (Doctorat de 3è cycle), Paris.
Dreyfus, M. and Pullman, A. (1970). *Theoret. Chim Acta*, **19,** 20.
Dushman, S. and Seitz, F. (1937). *J. Phys. Chem.*, **41,** 233.
Dutta, N. C., Dutta, C. M. and Das, T. P. (1971). *Intern. J. Quant. Chem.*, **4,** 299.
Dutta, N. C., Ishihara, T., Matsubara, C. and Das, T. P. (1970). *Intern. J. Quant. Chem. Symp.*, **3,** 367.
Dymond, J. H., Rigby, M. and Smith, E. B. (1965). *J. Chem. Phys.*, **42,** 2801.
Dyson, F. J. (1956). *Phys. Rev.*, **102,** (a) 1217, (b) 1230.
Dzyaloshinskii, I. E., Lifschitz, E. M. and Pitaevskii, L. P. (1961). *Adv. Phys.*, **10,** 165.
Edmiston, C. and Ruedenberg, K. (1963a). *J. Chem. Phys.*, **43,** 597.
Edmiston, C. and Ruedenberg, K. (1963b). *Rev. Mod. Phys.*, **35,** 457.
Egelstaff, P. A. (1967). *An Introduction to the Liquid State.* Academic Press, London and New York.
Eisenberg, D. and Kauzmann, W. (1969). *The Structure and Properties of Water.* Oxford University Press, London.
Eisenschitz, R. and London, F. (1930). *Z. Phys.*, **60,** 491.
Eley, D. D. (1938). *Trans. Faraday Soc.*, **35,** (a) 1281, (b) 1421.
Ellis, R. L. (1976). *J. Chem. Phys.*, **64,** 342.
England, W. and Gordon, M. S. (1971). *J. Am. Chem. Soc.*, **93,** 4649.
England, W. and Gordon, M. S. (1972). *J. Am. Chem. Soc.*, **94,** (a) 4818, (b) 5168.
England, W. and Ruedenberg, K. (1971). *Theoret. Chim. Acta*, **22,** 196.
England, W., Salmon, L. S. and Ruedenberg, K. (1971). p. 31 in *Tropics in Current Chemistry*, Vol. **23.** Springer–Verlag, Berlin.
Epstein, S. T. (1968a). *Am. J. Phys.*, **36,** 165.
Epstein, S. T. (1968b). *J. Chem. Phys.*, **48,** 4716.
Epstein, S. T. and Johnson, R. E. (1968). *Chem. Phys. Letters*, **1,** 602.
Erdelyi, A. (1956). *Asymptotic Expansions.* Dover, New York.
Evans, D. J. and Watts, R. O. (1975). *Mol. Phys.*, **29,** 777.
Evans, D. J. and Watts, R. O. (1976). *Mol. Phys.*, **31,** 83.
Everett, D. H. (1965). *Disc. Faraday Soc.*, **40,** 177.
Ewing, G. E. (1972). *Angew. Chem. Intern. Edit.*, **11,** 486.
Eyring, H., Henderson, D. and Jost, J. ed. (1971). *Physical Chemistry, an Advanced Treatise*', Vol. VIII A and B *Liquid State.* Academic Press, New York.
Eyring, H., Henderson, D., Stover, B. J. and Eyring, E. M. (1964). *Statistical Mechanics and Dynamics.* Wiley, New York.
Eyring, H. and Jhon, M. S. (1969). *Significant Liquid Structures.* Wiley, New York.
Farberov, D. S., Mitrofanov, V. Ya. and Men, A. N. (1972). *Int. J. Quant. Chem.*, **6,** 1057.
Feinberg, G. and Sucher, J. (1970). *Phys. Rev.*, **A2,** 2395.
Fetter, A. L. and Walecka, J. D. (1971). *Quantum Theory of Many-Particle Systems.* McGraw–Hill, New York.
Fitts, D. D. (1966). *Ann. Rev. Phys. Chem.*, **17,** 59.
Fitts, D. D. (1972). Ch. 6, p. 197 in *Theoretical Chemistry* (ed. W. Byers Brown), Vol. 1 in *Physical Chemistry, Series One* of MTP (Medical and Technical Publishing Co.) International Review of Science. Butterworths, London and University Park Press, Baltimore.
Fleckinger, R., Gomes, A. and Soulet, Y. (1976). '*Introduction of the Ideal Formulation in Quantum Chemistry. I. Reformulation of some successful methods dealing with overcomplete basis*', preprint. (Université Paul Sabatier, Toulouse).
Fleckinger, R. and Soulet, Y. (1976). '*Introduction of the Ideal Formulation in*

Quantum Chemistry. II. Use of an overcomplete basis for the description of the chemical bonding. Example: the H_2^+ problem', preprint. (Université Paul Sabatier, Toulouse).

Fliszar, S., Goursot, A. and Dugas, H. (1974a). *J. Am. Chem. Soc.*, **96,** 4358.

Fliszar, S., Kean, G. and Macaulay, R. (1974b). *J. Am. Chem. Soc.*, **96,** 4353.

Flory, P. J. (1969). *Statistical Mechanics of Chain Molecules.* Interscience, New York.

Flory, P. J. (1971). *Pure Appl. Chem.*, **36,** 26.

Fontana, P. R. (1961). *Phys. Rev.*, **123,** 1865.

Foster, R. (1969). *Organic Charge-Transfer Complexes.* Academic Press, New York.

Foster, R., ed. (1973). *Molecular Complexes,* Vol. I., Paul Elek, London, and Crane, Russak and Co., New York.

Foster, R., ed. (1975). *Molecular Association,* Vol. 1, Academic Press, New York.

Fowler, R. M. (1936), *Statistical Mechanics.* Cambridge University Press, Cambridge.

Frank, H. S. and Evans, M. W. (1945). *J. Chem. Phys.*, **13,** 507.

Fukui, K. and Yamabee, T. (1968). *Intern. J. Quant. Chem.*, **2,** 359.

Gaydaenko, V. I. and Nikulin, V. K. (1970). *Chem. Phys. Letters,* **7,** 360.

Gentile, G. (1930). *Zeits. f. Phys.*, **63,** 795.

Gerratt, J. (1976). *Proc. Roy. Soc.* (*London*), **A350,** 363.

Geurts, P. J. M., Wormer, P. E. S. and Van der Avoird, A. (1975). *Chem. Phys. Letters,* **35,** 444.

Gianturco, F. A. and Dilonardo, M. (1975). *J. Chim. Phys.*, **72,** 315.

Giglio, E. (1969). *Nature,* **222,** 339.

Giglio, E. (1970). *Z. Kristallogr.*, **131,** 385.

Giglio, E. and Liquori, A. M. (1967). *Acta Cryst.*, **22,** 437.

Gilbert, M. (1968). '*Etude théorique de l'intercalement des amino-acridines dans l'ADN*', Thesis (Doctorat de 3è cycle), Paris.

Gilbert, M. and Claverie, P. (1968a). *J. Theoret. Biol.*, **18,** 330.

Gilbert, M. and Claverie, P. (1968b). p. 245 in *Molecular Associations in Biology* (ed. B. Pullman). Academic Press, New York.

Gilbert, T. L. (1964). p. 405 in *Molecular Orbitals in Chemistry, Physics and Biology,* ed. P. O. Löwdin, Academic Press, New York.

Gilbert, T. L. and Wahl, A. C. (1967). *J. Chem. Phys.*, **47,** 3425.

Gill, P. E. and Murray, W., eds. (1974). *Numerical Methods for Constrained Optimization.* Academic Press, New York.

Girardeau, M. D. (1963). *J. Math. Phys.*, **4,** 1096.

Girardeau, M. D. (1970). *J. Math. Phys.*, **11,** 681.

Girardet, C. and Robert, D. (1971). *J. Mol. Structure,* **7,** 31.

Girardet, C., Robert, D. and Galatry, L. (1970). *C. R. Acad. Sc. Paris,* **270,** 481.

Gombas, P. (1949). *Die Statistische Theorie des Atoms und ihre Anwendungen,* Springer–Verlag, Vienna.

Gombas, P. (1950). *Theorie und Lösungsmethoden des Mehrteilchenproblems der Wellenmechanik.* Birkhäuser Verlag, Basel.

Good, R. J. and Hope, C. J. (1970). *J. Chem. Phys.*, **53,** 540.

Gordon, R. G. (1968a). *Int. J. Quant. Chem. Symp.*, **2,** 151.

Gordon, R. G. (1968b). *J. Chem. Phys.*, **48,** 3929.

Gordon, R. G. and Kim, Y. S. (1972). *J. Chem. Phys.*, **56,** 3122.

Gor'kov, L. P. and Pitaevskii, L. P. (1964). *Sov. Phys. Doklady,* **8,** 788.

Goscinski, O. (1968). *Int. J. Quant. Chem.*, **2,** 761.

Goscinski, O. and Brändas, E. (1968). *Chem. Phys. Letters,* **2,** 299.

Goscinski, O. and Brändas, E. (1971). *Intern. J. Quant. Chem.*, **5,** 131.

Gouyet, J. F. (1973). *J. Chem. Phys.*, **59,** 4637.

Gouyet, J. F. (1974). *J. Chem. Phys.*, **60,** 3690.

Gradshteyn, I. S. and Ryzhik, I. M. (1965). *Table of Integrals, Series and Products.* Academic Press, New York.
Griffing, V. and Wehner, J. F. (1955). *J. Chem. Phys.*, **23,** 1024.
Groszek, A. J. (1970). *Proc. Roy. Soc.* (*London*), **A314,** 473.
Grundke, E. W., Henderson, D., Barker, J. A. and Leonard, P. J. (1973). *Mol. Phys.*, **25,** 883.
Günter, K. (1964). *Ann. Physik* (*7 Folge*), **14,** 296.
Gurikov, Yu. V. (1968). *Teoreticheskaya i Experimental'naya Khimiya*, **4,** (a) 607, (b) 616 [*Theoret. Exp. Chem.*, **4,** (a) 387, (b) 392 (1971)].
Hagler, A. T., Huler, E. and Lifson, S. (1974). *J. Am. Chem. Soc.*, **96,** 5319.
Hagler, A. T. and Lifson, S. (1974). *J. Am. Chem. Soc.*, **96,** 5327.
Hall, G. G. (1973). *Chem. Phys. Lett.*, **6,** 501.
Halicioglu, T. and Sinanoglu, O. (1969). *Ann. N. Y. Acad. Sci.*, **158,** 308.
Hamilton, W. C. (1957). *J. Chem. Phys.*, **26,** 1018. (In this paper, two formulas of Table IV must be corrected: for $(2s_a|\,x\,|2p\pi_b)$ and $(2p\pi_a|\,x\,|2s_b)$, the numerical factor before the bracketed expression [. . .] must be divided by $\sqrt{3}$; moreover, for $(2s_a|\,x\,|2p\pi_b)$, one should read R^6 instead of R^5, as for $(2p\pi_a|\,x\,|2s_b)$).
Hanna, M. W. (1968). *J. Am. Chem. Soc.*, **90,** 285.
Hanna, M. W. and Lippert, J. L. (1973). pp. 1–48 in *Molecular Complexes*, Vol. I, ed. R. Foster. Paul Elek (London), Crane Russak and Co. (New York).
Harris, S. J., Janda, K. C., Novick, S. E. and Klemperer, W. (1975). *J. Chem. Phys.*, **63,** 881.
Helfand, E., Reiss, H., Frisch, H. L. and Lebowitz, J. L. (1960). *J. Chem. Phys.*, **33,** 1379.
Henderson, D. (1974). *Ann. Rev. Phys. Chem.*, **25,** 461.
Henderson, D. and Davison, S. G. (1967). p. 339, Ch. 7 in *Statistical Mechanics*, Vol. II of *Physical Chemistry, an Advanced Treatise*, (eds. H. Eyring, D. Henderson and W. Jost). Academic Press, New York.
Herman, R. M. (1963). *Phys. Rev.*, **132,** 262.
Herman, R. M. (1966). *J. Chem. Phys.*, **44,** 1346.
Hermann, R. B. (1971). *J. Phys. Chem.*, **75,** 363.
Hermann, R. B. (1972). *J. Phys. Chem.*, **76,** 2754.
Hermann, R. B. (1975a). *J. Phys. Chem.*, **79,** 163.
Hermann, R. B. (1975b). p. 441 in *Molecular and Quantum Pharmacology.* (*The Jerusalem Symposia on Quantum Chemistry and Biochemistry*, Vol. 7) eds. E. D. Bergmann and B. Pullman. Reidel, Dordrecht.
Herndon, W. C. and Feuer, J. (1968). *J. Am. Chem. Soc.*, **90,** 5914.
Herring, C. and Flicker, M. (1964). *Phys. Rev.*, **134,** A362.
Hildebrand, J. H., Prausnitz, J. M. and Scott, R. L. (1970). *Regular and Related Solutions—The Solubility of Gases, Liquids and Solids.* Van Nostrand–Reinhold, Princeton, New Jersey.
Hildebrand, J. H. and Scott, R. L. (1962). *Regular Solutions.* Prentice–Hall, Englewood Cliffs, New Jersey.
Hill, T. L. (1944). *J. Chem. Phys.*, **12,** 147.
Hill, T. L. (1948). *J. Chem. Phys.*, **16,** 399.
Hill, T. L. (1956). *Statistical Mechanics.* McGraw–Hill, New York.
Hill, T. L. (1960). *Introduction to Statistical Thermodynamics.* Addison–Wesley, Reading, Massachusetts.
Hirschfelder, J. O. (1965). p. 325 in *Molecular Biophysics*, B. Pullman and M. Weissbluth eds., Academic Press, New York.
Hirschfelder, J. O. (1966). p. 3 in *Perturbation Theory and its applications in Quantum Mechanics* (ed. C. H. Wilcox). John Wiley, New York.
Hirschfelder, J. O. (1967a). *Chem. Phys. Letters*, **1,** 325.
Hirschfelder, J. O. (1967b). *Chem. Phys. Letters*, **1,** 363.

Hirschfelder, J. O., ed. (1967c). '*Intermolecular forces*', *Advances in Chemical Physics*, **12.** Interscience, New York.
Hirschfelder, J. O. (1971). *Intern. J. Quant. Chem. Symp.*, **4,** 257.
Hirschfelder, J. O., Byers Brown, W. and Epstein, S. T. (1964). *Advances in Quantum Chemistry*, **1,** 255. Academic Press, New York.
Hirschfelder, J. O. and Certain, P. R. (1968a). *Intern. J. Quant. Chem. Symp.*, **2,** 125.
Hirschfelder, J. O. and Certain, P. R. (1968b). *Chem. Phys. Letters*, **2,** 539.
Hirschfelder, J. O., Curtiss, C. F. and Bird, R. B. (1954). *Molecular Theory of Gases and Liquids.* John Wiley, New York (second printing, corrected and with notes added: 1964).
Hirschfelder, J. O. and Linnett, J. W. (1950). *J. Chem. Phys.*, **18,** 130.
Hirschfelder, J. O. and Meath, W. J. (1967). p. 3 in 'Intermolecular forces', *Advances in Chemical Physics*, **12** (ed. J. O. Hirschfelder). Interscience, New York.
Hirschfelder, J. O. and Silbey, R. (1966). *J. Chem. Phys.*, **45,** 2188.
Howard, B. B. (1963). *J. Chem. Phys.*, **39,** 2524.
Huler, E. and Warshel, A. (1974). *Acta Cryst.*, **B30,** 1822.
Huler, E. and Warshel, A. (1975) *Chem. Phys.*, **8,** 239.
Huron, M. J. and Claverie, P. (1969). *Chem. Phys. Letters*, **4,** 429. (erratum: *Chem. Phys. Letters*, **11,** 152 (1971)).
Huron, M. J. and Claverie, P. (1971). *Chem. Phys. Letters*, **9,** 194.
Huron, M. J. and Claverie, P. (1972). *J. Phys. Chem.*, **76,** 2123.
Huron, M. J. and Claverie, P. (1974a). *J. Phys. Chem.*, **78,** 1853.
Huron, M. J. and Claverie, P. (1974b). *J. Phys. Chem.*, **78,** 1862.
Isnard, P., Robert, D. and Galatry, L. (1976). *Mol. Phys.*, **31,** 1789.
Israelachvili, J. N. (1974). *Quart. Rev. Biophys.*, **6,** 341.
Jansen, L. (1962). *Phys. Rev.*, **125,** 1798.
Jansen, L. (1965). *Adv. Quant. Chem.*, **2,** 119.
Jansen, L. (1967). *Phys. Rev.*, **162,** 63.
Jansen, L. and Lombardi, E. (1965). *Disc. Faraday Soc.*, **40,** 78.
Jansen, L. and Lombardi, E. (1967). *Chem. Phys. Letters*, **1,** 417.
Jehle, H. (1965). *Adv. Quant. Chem.*, **2,** 195.
Jehle, H. (1969). *Ann. N.Y. Acad. Sci.*, **158,** 240.
Jeziorski, B., Bulski, M. and Piela, L. (1976). *Int. J. Quantum Chem.*, **10,** 281.
Jeziorski, B. and Piela, L. (1972). *Acta Physica Polonica*, **A42,** 177.
Jeziorski, B. and Van Hemert, M. (1976). *Mol. Phys.*, **31,** 713.
Jhon, M. S., Eyring, H. and Sung, Y. K. (1972). *Chem. Phys. Lett.*, **13,** 36.
Joesten, M. D. and Schaad, L. J. (1974). *Hydrogen Bonding.* Marcel Dekker, New York (this book contains a complete bibliography from 1960 to 1973 inclusive).
Johansen, H. (1975). *Int. J. Quant. Chem.*, **9,** 459.
Johnson, R. E. and Epstein, S. T. (1968). *Chem. Phys. Letters*, **1,** 599.
Jug, K. (1973). *Theoret. Chim. Acta*, **31,** 63.
Jug, K. (1975). *Theoret. Chim. Acta*, **39,** 301.
Junker, B. R. and Menendez, M. G. (1975). *J. Chem. Phys.*, **63,** 1243.
Karplus, M. and Kolker, H. J. (1964). *J. Chem. Phys.*, **41,** 3955.
Kauzmann, W. (1957). *Quantum Chemistry.* Academic Press, New York.
Kauzmann, W. (1959). *Adv. Protein Chem.*, **14,** 34.
Kay, K. G. and Silverstone, H. J. (1970). *J. Chem. Phys.*, **53,** 4269.
Kestner, N. R. (1966). *J. Chem. Phys.*, **45,** (a) 208 (b) 213.
Kestner, N. R. and Sinanoglu, O. (1963). *J. Chem. Phys.*, **38,** 1730.
Kestner, N. R. and Sinanoglu, O. (1966). *J. Chem. Phys.*, **45,** 194.
Kihara, T. (1953). *Rev. Mod. Phys.*, **25,** 831.
Kihara, T. (1955). *Rev. Mod. Phys.*, **27,** 412.

Kihara, T. (1958). *Advances in Chemical Physics*, **1,** 267. Interscience, New York.
Kihara, T. (1967). *Suppl. Progr. Theoret. Phys.*, **40,** 177.
Kihara, T. (1970). p. 663, Ch. 12 in *Valency*, Vol. V of *Physical Chemistry, and Advanced Treatise*, (eds. H. Eyring, D. Henderson and W. Jost). Academic Press, New York.
Kihara, T. and Jhon, M. S. (1970). *Chem. Phys. Lett.*, **7,** 559.
Kihara, T. and Yamazaki, K. (1971). *Chem. Phys. Lett.*, **11,** 62.
Kihara, T., Yamazaki, K., Jhon, M. S. and Kim, U. R. (1971). *Chem. Phys. Lett.*, **9,** 62.
Kim, Y. S. and Gordon, R. G. (1974a). *J. Chem. Phys.*, **60,** 1842.
Kim, Y. S. and Gordon, R. G. (1974b). *J. Chem. Phys.*, **60,** 4323.
Kim, Y. S. and Gordon, R. G. (1974c). *J. Chem. Phys.*, **60,** 4332.
Kim, Y. S. and Gordon, R. G. (1974d). *J. Chem. Phys.*, **61,** 1.
Kingston, A. E. (1964). *Phys. Rev.*, **135,** A1018.
Kirkwood, J. G. (1934). *J. Chem. Phys.*, **2,** 351.
Kirkwood, J. G. (1968). *Theory of Solutions* (a reprint volume). Gordon and Breach, New York.
Kirkwood, J. G. and Westheimer, F. H. (1938). *J. Chem. Phys.*, **6,** 506.
Kiselev, A. V. (1965). *Disc. Faraday Soc.*, **40,** 205.
Kitaigorodskii, A. I. (1961a). *Tetrahedron*, **14,** 230.
Kitaigorodskii, A. I. (1961b). *Doklady Akademii Nauk SSSR*, **137,** 116 [*Proc. Acad. Sc. USSR, Physical Chemistry Section*, **137,** 231)].
Kitaigorodskii, A. I. (1961c). *Organic Chemical Crystallography* (transl. from Russian). Consultants Bureau, New York.
Kitaigorodskii, A. I. (1965). *Acta Cryst.*, **18,** 585.
Kitaigorosdkii, A. I. (1966). *J. Chim. Phys.*, **63,** 9.
Kitaigorodskii, A. I. (1970). p. 173 in *Advances in Structure Research by Diffraction Methods*, Vol. **3**, ed. R. Brill and R. Mason. Viewieg, Braunschweig.
Kitaigorodskii, A. I. (1973). *Molecular Crystals and Molecules*, Academic Press, New York.
Kitaigorodskii, A. I. and Ahmed (1972). *Acta Cryst.*, **A28,** 207.
Kitaigorodskii, A. I. and Dashevskii, V. G. (1967). *Theoret. Eksp. Khim.*, **3,** 35. [*Theoret. Exp. Chem.*, **3,** 18 (1970)].
Kitaigorodskii, A. I. and Dashevskii, V. G. (1968). *Tetrahedron*, **24,** 5917.
Kitaigorodskii, A. I. and Mirskaya, K. V. (1961). *Kristallografiya*, **6,** 507. [*Soviet Physics-Crystallography*, **6,** 408 (1962)].
Kitaigorodskii, A. I. and Mirskaya, K. V. (1964a). *Kristallografiya*, **9,** 174. [*Soviet Physics-Crystallography*, **9,** 137 (1964)].
Kitaigorodskii, A. I. and Mirskaya, K. V. (1964b). *Kristallografiya*, **9,** 634. [*Soviet Physics-Crystallography*, **9,** 539 (1965)].
Kitaigorodskii, A. I. and Mirskaya, K. V. (1965). *Kristallografiya*, **10,** 162. [*Soviet Physics-Crystallography*, **10,** 121 (1965)].
Kitaigorodskii, A. I. and Mirskaya, K. V. (1970). p. 339 in *Molecular Crystals and Liquid Crystals*, Vol. **6,** Gordon and Breach, New York.
Kitaigorodskii, A. I., Mirskaya, K. V. and Nauchitel, V. V. (1969). *Kristallografiya*, **14,** 900 [*Soviet Physics-Crystallography*, **14,** 769 (1970)].
Klein, D. J. (1971a). *Intern. J. Quantum Chem. Symp.*, **4,** 271.
Klein, D. J. (1972). *Perturbational Scheme for Multistate Kets*, preprint, unpublished.
Klein, D. J. and Seitz, W. A. (1973). *Phys. Rev. B*, **8,** 2236 (see especially references 5, 6, 7, 8).
Kochanski, E. (1974). *Int. J. Quant. Chem. Symp.*, **8,** 219.
Kochanski, E. (1975). *Theoret. Chim. Acta*, **39,** 339.
Kochanski, E. and Gouyet, J. F. (1975a). *Mol. Phys.*, **29,** 693.
Kochanski, E. and Gouyet, J. F. (1975b). *Theoret. Chim. Acta*, **39,** 329.

Kollman, P. A. and Allen, L. C. (1970). *Theoret. Chim. Acta*, **18,** 399.
Kollman, P. A. and Allen, L. C. (1972). *Chem. Rev.*, **72,** 283.
Kolos, W. (1967). *Int. J. Quantum Chem.*, **1,** 169.
Kolos, W. (1974). *Int. J. Quantum Chem. Symp.*, **8,** 241.
Kolos, W. (1975). *Int. J. Quantum Chem.*, **9,** 133.
Kolos, W. and Les, A. (1972). *Int. J. Quantum Chem.*, **6,** 1101.
Kolos, W. and Wolniewicz, L. (1974). *Chem. Phys. Lett.*, **24,** 457.
Kolos, W. and Wolniewicz, L. (1975). *J. Mol. Spectry.*, **54,** 303.
Kotani, M., Amemiya, A., Ishiguro, E. and Kimura, T. (1955). *Table of Molecular Integrals.* Maruzen, Tokyo (second edition: 1963).
Kouznetsov, A. V. and Guiochon, G. (1969). *J. Chim. Phys.*, **66,** 257.
Kowalik, J. and Osborne, M. R. (1968). *Methods for Unconstrained Optimization Problems.* American Elsevier, New York, London, Amsterdam.
Kramer, H. L. (1970). *J. Chem. Phys.*, **53,** 2783.
Kramer, H. L. and Herschbach, D. R. (1970). *J. Chem. Phys.*, **53,** 2792.
Kreek, H. and Meath, W. J. (1969). *J. Chem. Phys.*, **50,** 2289.
Kreek, H. and Meath, W. J. (1971). *Mol. Phys.*, **22,** 915.
Kreek, H., Pan, Y. H. and Meath, W. J. (1970). *Mol. Phys.*, **19,** 513.
Kutzelnigg, W. (1966). *Theoret. Chim. Acta*, **4,** 417.
Kutzelnigg, W. (1968). 'Natural states of interacting systems and the calculation of long-range intermolecular forces'. Technical Report.
Kutzelnigg, W. (1969). *Chem. Phys. Letters*, **4,** 435.
Kutzelnigg, W. and Gelus, M. (1970). *Chem. Phys. Letters*, **7,** 296.
Kutzelnigg, W., Staemmler, V. and Gelus, M. (1972). *Chem. Phys. Letters*, **13,** 496.
Kvasnička, V., Laurinč, V. and Hubač, I. (1974). *Phys. Rev.*, **A10,** 2016.
Ladanyi, B. M. and Chandler, D. (1975). *J. Chem. Phys.*, **62,** 4308.
Laidlaw, W. G., Lekkerkerker, H. N. W. and Wieser, H. (1971). *Trans. Faraday Soc.* **67,** 2819.
Lamotte, M. (1973). '*Contribution á l'étude des interactions moléculaires faibles par spectroscopie ultra-violette. Application à l'interprétation des phénomènes liès aux interactions electrostatiques localisées et aux interactions orientées*'. Thesis (Doctorat d'Etat), Université de Bordeaux I. (C.N.R.S. registration number: A.O. 8074).
Lamotte, M. and Joussot-Dubien, J. (1973). *Chem. Phys.*, **2,** 245.
Lamotte, M., Joussot-Dubien, J., Mantione, M. J. and Claverie, P. (1974). *Chem. Phys. Letters*, **27,** 515.
Landau, L. D. and Lifshitz, E. M. (1958). *Statistical Physics.* Pergamon Press, London.
Langbein, D. (1973). *J. Phys. A: Math., Nucl., Gen.*, **6,** 1149.
Langhoff, P. W., Gordon, R. G. and Karplus, M. (1971). *J. Chem. Phys.*, **55,** 2126.
Langhoff, P. W. and Karplus, M. (1970a). p. 41 in '*The Padé Approximant in Theoretical Physics*' (G. A. Baker, Jr. and J. L. Gammel eds.). Academic Press, New York.
Langhoff, P. W. and Karplus, M. (1970b). *J. Chem. Phys.*, **53,** 233.
Langhoff, P. W. and Karplus, M. (1970c). *J. Chem. Phys.*, **52,** 1435.
Laughlin, C. and Amos, A. T. (1971). *J. Chem. Phys.*, **55,** 4837.
Laughlin, C., Woodward, M. R. and Amos, A. T. (1974). *Intern. J. Quantum Chem.*, **8,** 491.
Lawley, K. P., ed. (1975). '*Molecular scattering. Physical and chemical applications*', *Adv. Chem. Phys.*, **30.** Wiley–Interscience, New York.
Le Fevre, C. G. and Le Fevre, R. J. W. (1955). *Rev. Pure Appl. Chem.*, **5,** 261.
Le Fevre, R. J. W. (1965). *Advances in Physical Organic Chemistry*, **3,** 1. Academic Press, New York.
Lekkerkerker, H. N. W. and Laidlaw, W. G. (1970a). *J. Chem. Phys.*, **52,** 2953.
Lekkerkerker, H. N. W. and Laidlaw, W. G. (1970b). *Trans. Far. Soc.*, **66,** 1830.

Lennard-Jones, J. E. (1924). *Proc. Roy. Soc.* (*London*), **A106,** 463.
Lennard-Jones, J. E. and Pople, J. A. (1950). *Proc. Roy. Soc.* (*London*), **A210,** 446.
Lennard-Jones, J. E. and Pople, J. A. (1951). *Proc. Roy. Soc.* (*London*), **A220,** 190.
Leonard, P. J., Henderson, D. and Barker, J. A. (1970). *Trans. Faraday Soc.*, **66,** 2469.
Leonas, V. B. (1972). *Usp. Fiz. Nauk*, **107,** 29 [*Sov. Phys. Uspekhi*, **15,** 266 (1973)].
Les, A. (1976). *Chem. Phys. Lett.*, **37,** 399.
Lewin, S. (1974). *Displacement of Water and its Control in Biochemical Reactions.* Academic Press, New York.
Linder, B. (1960). *J. Chem. Phys.*, **33,** 666.
Linder, B. (1961). *J. Chem. Phys.*, **35,** 371.
Linder, B. (1962). *J. Chem. Phys.*, **37,** 963.
Linder, B. (1965). *Disc. Faraday Soc.*, **40,** 164.
Linder, B. (1967). p. 225 in '*Intermolecular forces*', *Adv. Chem. Phys.*, **12** (J. O. Hirschfelder ed.). Wiley–Interscience, New York.
Linder, B. and Hoernschemeyer, D. (1964). *J. Chem. Phys.*, **40,** 622.
Linder, B. and Rabenold, D. A. (1972). *Adv. Quantum Chem.*, **6,** 203.
Lippert, J. L., Hanna, M. W. and Trotter, P. J. (1969). *J. Am. Chem. Soc.*, **91,** 4035.
Liquori, A. M. (1966). *J. Polymer Sc.*, **C12,** 209.
Liquori, A. M. (1969). *Quart. Rev. Biophys.*, **2,** 65.
Liu, B. and McLean, A. D. (1973). *J. Chem. Phys.*, **59,** 4557.
London, F. (1930a). *Z. Phys.*, **63,** 245.
London, F. (1930b). *Z. Physik. Chemie* (*B*), **11,** 222.
London, F. (1937). *Trans. Faraday Soc.*, **33,** 8.
London, F. (1942). *J. Phys. Chem.*, **46,** 305.
Longman, I. M. (1971). *Intern. J. Computer Math.*, **B3,** 53.
Longuet-Higgins, H. C. (1956). *Proc. Roy. Soc.* (*London*), **A235,** 537.
Longuet-Higgins, H. C. and Salem, L. (1961). *Proc. Roy. Soc.* (*London*), **A259,** 433.
Lootsma, F. A. ed. (1972). *Numerical Methods for Non-linear Optimization.* Academic Press, New York.
Lowden, L. J. and Chandler, J. P. (1973). *J. Chem. Phys.*, **59,** 6587 (erratum and addenda: *J. Chem. Phys.*, **62,** 4246 (1975)).
Lowden, L. J. and Chandler, J. P. (1974). *J. Chem. Phys.*, **61,** 5228.
Löwdin, P. O. (1950). *J. Chem. Phys.*, **18,** 365.
Löwdin, P. O. (1953). *J. Chem. Phys.*, **21,** 374.
Löwdin, P. O. (1956). *Adv. Phys.*, **5,** 22.
Löwdin, P. O. (1962). *J. Math. Phys.*, **3,** 969.
Löwdin, P. O. (1963). *J. Mol. Spectry.*, **10,** 12.
Löwdin, P. O. (1965). *Phys. Rev.*, **139,** A357.
Löwdin, P. O. (1966). p. 255 in *Perturbation Theory and its Applications in Quantum Mechanics* (ed. C. H. Wilcox). John Wiley, New York.
Löwdin, P. O. (1967). *Intern. J. Quantum Chem. Symp.*, **1,** 811.
Löwdin, P. O. (1968a). *Intern. J. Quantum Chem. Symp.*, **2,** 137.
Löwdin, P. O. (1968b). *Intern. J. Quantum Chem.*, **2,** 867.
Löwdin, P. O. (1971). *Intern. J. Quantum Chem. Symp.*, **5,** 685.
MacGillivray, A. D. and McMullen, A. I. (1966). *J. Theoret. Biol.*, **12,** (a) 75, (b) 260.
MacGillivray, A. D. and McMullen, A. I. (1968). *J. Theoret. Biol.*, **19,** 159.
Maeder, F. (1972). '*Application de la méthode des pseudo-potentiels au calcul des forces de Van der Waals, entre atomes à un et deux électrons de valence.*' Thesis (Doctorat de 3è cycle), Paris.
Magnasco, V. (1971). *Theoret. Chim. Acta*, **21,** 267.
Magnasco, V., Battezzati, M. and Figari, G. (1976). *J. Chem. Soc. Faraday Trans., II*, **72,** 22.

Magnasco, V. and Musso, G. F. (1971a). *Theoret. Chim. Acta*, **21,** 280.
Magnasco, V. and Musso, G. F. (1971b). *J. Chem. Phys.*, **54,** 2925.
Mahanty, J. and Ninham, B. W. (1973). *J. Phys. A*, **6,** 1140.
Malrieu, J. P. (1971a). *Intern. J. Quantum Chem.*, **5,** 435.
Malrieu, J. P. (1971b). *Intern. J. Quantum Chem.*, **5,** 455.
Malrieu, J. P. and Claverie, P. (1968). *J. Chim. Phys.*, **65,** 735.
Mantione, M. J. (1968a). p. 411 in '*Molecular associations in Biology*' (ed. B. Pullman). Academic Press, New York.
Mantione, M. J. (1968b). *Theoret. Chim. Acta*, **11,** 119.
Mantione, M. J. (1969a). *Intern. J. Quantum Chem.*, **3,** 185.
Mantione, M. J. (1969b). *Theoret. Chim. Acta*, **15,** 141.
Mantione, M. J. (1971). '*Etude de quelques domaines d'application de la théorie des interactions moléculaires*'. Thesis (Doctorat d'Etat), C.N.R.S., Paris
Mantione, M. J. and Daudey, J. P. (1970). *Chem. Phys. Letters*, **6,** 93.
March, N. H. (1957). *Adv. Phys.* **6,** 1.
March, N. H., Young, W. H. and Sampanthar, S. (1967). *The Many-Body Problem in Quantum Mechanics.* Cambridge University Press, Cambridge.
Marchand, D. (1970). '*Contribution à l'étude de l'interaction de deux atomes d'hélium*'. Thesis (Doctorat de 3è cycle, Chimie Structurale), Paris.
Margenau, H. (1939a). *Rev. Mod. Phys.* **11,** 1.
Margenau, H. (1939b). *Phys. Rev.* **56,** 1000.
Margenau, H. and Kestner, N. R. (1969). *Theory of Intermolecular Forces.* Pergamon Press, Oxford.
Margenau, H. and Rosen, P. (1953). *J. Chem. Phys.*, **21,** 394.
Margenau, H. and Stamper, J. (1966). *Advances in Quantum Chemistry*, **3,** 129. Academic Press, New York.
Mason, E. A. and Hirschfelder, J. O. (1957). *J. Chem. Phys.*, **26,** 173.
Mason, E. A. and Rice, W. E. (1954). *J. Chem. Phys.*, **22,** 843.
Mason, E. A. and Spurling, T. M. (1969). *The Virial Equation of State.* Pergamon Press, Oxford.
Matcha, R. Z. and Nesbet, R. K. (1967). *Phys. Rev.*, **160,** 72.
Matsen, F. A. and Junker, B. R. (1971). *J. Phys. Chem.*, **75,** 1878.
Mavroyannis, C. and Stephen, R. J. (1962). *Mol. Phys.*, **5,** 629.
Mayer, S. W. (1963). *J. Chem. Phys.*, **38,** 1803.
McClellan, A. L. (1963). *Tables of Experimental Dipole Moments.* Freeman, San Francisco.
McDonald, J. K. L. (1933). *Phys. Rev.*, **43,** 830.
McGuire, F., Momany, A. and Scheraga, H. A. (1972). *J. Phys. Chem.*, **76,** 375.
McLachlan, A. D. (1963a). *Proc. Roy. Soc.* (*London*), **A271,** 387.
McLachlan, A. D. (1963b). *Mol. Phys.*, **6,** 423.
McLachlan, A. D. (1963c). *Proc. Roy. Soc.* (*London*), **A274,** 80.
McLachlan, A. D. (1965). *Disc. Faraday Soc.*, **40,** 239.
McLaughlin, D. R. and Schaeffer III, H. F. (1971). *Chem. Phys. Lett.*, **12,** 244.
McQuarrie, D. A., Terebey, J. and Shire, S. J. (1969). *J. Chem. Phys.*, **51,** 4683.
McWeeny, R. (1960). *Rev. Mod. Phys.*, **32,** 335.
McWeeny, R. and Sutcliffe, B. T. (1969). *Molecular Quantum Mechanics.* Academic Press, London and New York.
McWeeny, R. and Yonezawa, F. (1965). *J. Chem. Phys.*, **43,** S 120.
Messiah, A. (1959–1960). *Mécanique Quantique*, Dunod, paris (part I: 1959; part II: 1960) (English translation: *Quantum Mechanics*, North Holland, Amsterdam (1961)).
Micha, D. A. (1968). *J. Chem. Phys.*, **48,** 3639.
Michels, M. A. J. and Suttorp, L. G. (1972). *Physica*, (a) **59,** 601; (b) **61,** 481; (c) **61,** 506; (d) **61,** 517.

Michels, M. A. J. and Suttorp, L. G. (1973). *Physica*, **67,** 137.
Midzuno, Y. and Kihara, T. (1956). *J. Phys. Soc. Japan*, **11,** 1045.
Mies, F. H. (1965). *J. Chem. Phys.*, **42,** 2709.
Millie, P., Levy, B. and Berthier, G. (1975). p. 59 in *Localization and Delocalization in Quantum Chemistry* (eds. O. Chalvet, R. Daudel, S. Diner and J. P. Malrieu). Reidel, Dordrecht.
Milonni, P. W. (1976). *Physics Reports* (Section *C* of *Phys. Lett.*), **25,** 1.
Mitchell, D. J. and Ninham, B. W. (1972). *J. Chem. Phys.*, **56,** 1117.
Mitchell, D. J., Ninham, B. W. and Richmond, P. (1971). *J. Phys. B.*, **4,** L81.
Mitchell, D. J., Ninham, B. W. and Richmond, P. (1972). *Austral. J. Phys.*, **25,** 33.
Momany, F. A., Carruthers, L. M., McGuire, R. F. and Scheraga, H. A. (1974a). *J. Phys. Chem.*, **78,** 1595.
Momany, F. A., Carruthers, L. M. and Scheraga, H. A. (1974b). *J. Phys. Chem.*, **78,** 1621.
Momany, F. A., McGuire, R. F., Burgess, A. W. and Scheraga, H. A. (1975). *J. Phys. Chem.*, **79,** 2361.
Morris, J. M. (1973). *Austral. J. Chem.*, **26,** 649.
Morris, J. M. (1974). *Mol. Phys.*, **28,** 1167.
Morokuma, K. (1971). *J. Chem. Phys.*, **55,** 1236.
Morokuma, K. and Pedersen, L. (1968). *J. Chem. Phys.*, **48,** 3275.
Morse, P. M. and Feshbach, H. (1953). *Methods of Theoretical Physics*. McGraw-Hill, New York.
Mortensen, E. M. (1968). *J. Chem. Phys.*, **49,** 3732.
Motherwell, W. D. S. and Isaacs, N. W. (1972). *J. Mol. Biol.*, **71,** 231.
Mulliken, R. S. (1949). *J. Chim. Phys.*, **46,** 497 (see p. 521).
Mulliken, R. S. (1950). *J. Am. Chem. Soc.*, **72,** 600.
Mulliken, R. S. (1952). *J. Am. Chem. Soc.*, **74,** 811.
Mulliken, R. S. (1955). *J. Chem. Phys.*, **23,** 1833.
Mulliken, R. S. (1964). *J. Chim. Phys.*, **61,** 20.
Mulliken, R. S. and Person, W. B. (1962). *Ann. Rev. Phys. Chem.*, **13,** 107.
Mulliken, R. S. and Person, W. B. (1969). *Molecular Complexes: a Lecture and Reprint Volume*. Wiley, New York.
Murray, W. ed. (1972). *Numerical Techniques for Non-linear Optimization*. Academic Press, New York.
Murrell, J. N. (1974). p. 311 in *Orbital Theories of Molecules and Solids*, ed. N. H. March. Clarendon Press, Oxford.
Murrell, J. N. and Harget, A. J. (1972). *Semi-empirical Self-consistent-field Molecular-orbital Theory of Molecules*. Wiley–Interscience, London, New York.
Murrell, J. N., Kettle, S. F. A. and Tedder, J. M. (1965). *Valence Theory*, Wiley, London.
Murrell, J. N., Randić, M. and Williams, D. R. (1965). *Proc. Roy. Soc.* (*London*), **A284,** 566.
Murrell, J. N. and Shaw, G. (1967a). *J. Chem. Phys.*, **46,** 1768.
Murrell, J. N. and Shaw, G. (1967b). *Mol. Phys.*, **12,** 475.
Murrell, J. N. and Shaw, G. (1968a). *Theoret. Chim. Acta*, **11,** 434.
Murrell, J. N. and Shaw, G. (1968b). *Mol. Phys.*, **15,** 325.
Murrell, J. N. and Shaw, G. (1968c). *J. Chem. Phys.*, **49,** 4731.
Murrell, J. N. and Teixeira-Dias, J. J. C. (1970a). *Mol. Phys.*, **19,** 521.
Murrell, J. N. and Teixeira-Dias, J. J. C. (1970b). *Mol. Phys.*, **19,** 329.
Murrell, J. N. and Teixeira-Dias, J. J. C. (1971). *Mol. Phys.*, **22,** 535.
Murrell, J. N. and Varandas, A. J. C. (1975). *Mol. Phys.*, **30,** 223.
Musher, J. I. (1963). *J. Chem. Phys.*, **39,** 2409.
Musher, J. I. (1967). *Rev. Mod. Phys.*, **39,** 203.
Musher, J. I. and Amos, A. T. (1967). *Phys. Rev.*, **164,** 31.

Musher, J. I. and Salem, L. (1966). *J. Chem. Phys.*, **44,** 2943.
Muto, Y. (1943). *Proc. Phys. Math. Soc. Japan*, **17,** 629.
Nagai, K. (1971). *Progr. Polym. Sci. Japan* (ed. M. Imoto and S. Onogi), **1,** 215.
Nemethy, G. (1967). *Angew. Chem.*, **6,** 195.
Nemethy, G. and Scheraga, H. A. (1962a). *J. Chem. Phys.*, **36,** 3382.
Nemethy, G. and Scheraga, H. A. (1962b). *J. Phys. Chem.*, **66,** 1773.
Nikulin, V. K. (1971). *Zh. Tekh. Fiz.*, **41,** 41 [*Sov. Phys.–Tech. Phys.*, **16,** 28 (1971)].
Ninham, B. W. and Parsegian, V. A. (1970). *J. Chem. Phys.*, **53,** 3398.
Ninham, B. W., Parsegian, V. A. and Weiss, G. (1970). *J. Stat. Phys.*, **2,** 323.
Ninio, J. (1971). *J. Mol. Biol.*, **56,** 63.
Nir, S., Adams, S. and Rein, R. (1973). *J. Chem. Phys.*, **59,** 3341.
Novaro, O. A. and Beltran-Lopez, V. (1972). *J. Chem. Phys.*, **56,** 815.
O'Connor (1973). *Comm. Math. Phys.*, **32,** 319.
O'Leary, B., Duke, B. J. and Eilers, J. E. (1975). *Adv. Quantum Chem.*, **9,** 1.
Onsager, L. (1936). *J. Am. Chem. Soc.*, **58,** 1486.
Pack, G. R., Wang, H. Y. and Rein, R. (1972). *Chem. Phys. Lett.*, **17,** 381.
Pade, H. (1892). Thesis: *Ann. Sci. Ecole Normale Supérieure* (*Paris*), **9,** *Suppl.* pp. 1–92.
Pan, Y. H. and Meath, W. J. (1971). *Mol. Phys.*, **20,** 873.
Parsegian, V. A. (1973). *Ann. Rev. Biophys. Bioeng.*, **2,** 221.
Parsonage, N. G. and Pemberton, R. C. (1967). *Trans. Faraday Soc.*, **63,** 311.
Peierls, R. (1973). *Proc. Roy. Soc.* (*London*), **A333,** 157.
Piela, L. (1970). *Chem. Phys. Lett.*, **7,** 591.
Piela, L. (1971). *Intern. J. Quantum Chem.*, **5,** 85.
Piela, L. and Jeziorski, B. (1972). *Acta Physics Polonica*, **A42,** 185.
Pierotti, R. A. (1963). *J. Phys. Chem.*, **67,** 1840.
Pierotti, R. A. (1965). *J. Phys. Chem.*, **69,** 281.
Pilar, F. L. (1968). *Elementary Quantum Chemistry*, McGraw-Hill, New York.
Pisani, C., Ricca, F. and Dovesi, R. (1974). *Japan. J. Appl. Phys. Suppl.*, **2, Pt. 2,** 269 (Proceedings of the 2nd International Conference on Solid Surfaces).
Pitzer, K. S. (1959). *Advances in Chemical Physics*, **2,** 59. Interscience, New York.
Politzer, P. (1971). *Theoret. Chim. Acta*, **23,** 203.
Politzer, P. and Harris, R. R. (1970). *J. Am. Chem. Soc.*, **92,** 6451.
Politzer, P. and Mulliken, R. S. (1971). *J. Chem. Phys.*, **55,** 5135.
Politzer, P. and Politzer, A. (1973). *J. Am. Chem. Soc.*, **95,** 5450.
Politzer, P. and Reggio, P. H. (1972). *J. Am. Chem. Soc.*, **94,** 8308.
Politzer, P. and Stout Jr., E. W. (1971). *Chem. Phys. Lett.*, **8,** 519.
Pollak, M. and Rein, R. (1967). *J. Chem. Phys.*, **47,** 2045.
Pollak, M. and Rein, R. (1968). *J. Theoret. Biol.*, **19,** 241.
Poltev, V. I. and Sukhorukov, B. I. (1967). *Biofizika*, **12,** 763 [*Biophysics*, **12,** 879 (1967)].
Poltev, V. I. and Sukhorukov, B. I. (1968a). *Biofizika*, **13,** 941 [*Biophysics*, **13,** 1103 (1968)].
Poltev, V. I. and Sukhorukov, B. I. (1968b). *Zh. Strukt. Khim.*, **9,** 298 [*J. Struct. Chem.*, **9,** 234 (1968)].
Pople, J. A. and Beveridge, D. L. (1970). *Approximate Molecular Orbital Theory*. McGraw-Hill, New York.
Popov, E. M. (1975). *Molekulyarnaya Biologiya*, **9,** 578 [*Mol. Biol.*, **9,** 468 (1975)].
Popov, E. M., Dashevskii, V. G., Lipkind, G. M. and Arkhipova, S. F. (1968). *Molekulyarnaya Biologiya*, **2,** 612 [*Mol. Biol.*, **2,** 498 (1968)].
Port, G. N. J. and Pullman, B. (1974). *Theoret. Chim. Acta*, **33,** 275.
Poshkus, D. P. (1965). *Disc. Faraday Soc.*, **40,** 195.
Power, E. A. (1967). p. 167 in '*Intermolecular Forces*', *Advances in Chemical Physics*, **12** (ed. J. O. Hirschfelder). Wiley-Interscience, New York.

Powers, J., Keedy, D. A. and Stein, R. S. (1961). *J. Chem. Phys.*, **35,** 376.
Prosser, F. and Hagstrom, S. (1968). *Intern. J. Quantum Chem.*, **2,** 89.
Pullman, A. (1975). p. 457 in *Structure and Conformation of Nucleic Acids and Protein-Nucleic Acid Interactions* (ed. by M. Sundaralingham and S. T. Rao). University Park Press, Baltimore.
Pullman, A. and Pullman, B. (1968). *Advances in Quantum Chemistry*, **4,** 267. Academic Press, New York.
Pullman, A. and Pullman, B. (1975). *Quart. Rev. Biophys.*, **7,** 505.
Pullman, B. (1977). in *Adv. Quantum Chem.*, **10,** in press.
Pullman, B., Berthod, H. and Courriere, Ph. (1974). *Int. J. Quant. Chem. Biol. Symp.*, **1,** 93.
Pullman, B., Berthod, H. and Gresh, N. (1975). *C. R. Acad. Sc. Paris*, **280D,** 1741.
Pullman, B., Claverie, P. and Caillet, J. (1966a). *Proc. Natl. Acad. Sci. U.S.A.*, **55,** 904.
Pullman, B., Claverie, P. and Caillet, J. (1966b). *J. Mol. Biol.*, **22,** 373.
Pullman, B., Claverie, P. and Caillet, J. (1967). *Proc. Natl. Acad. Sci. U.S.A.*, **57,** 1663.
Pullman, B., Coubeils, J. L., Courriere, Ph. and Gervois, J. P. (1972). *J. Medic. Chem.*, **15,** 17.
Pullman, B., Courriere, Ph. and Berthod, H. (1974). *J. Medic. Chem.*, **17,** 439.
Pullman, B. and Pullman, A. (1974). *Adv. Protein Chem.*, **28,** 347.
Rabenold, D. A. and Linder, B. (1975). *J. Chem. Phys.*, **63,** 4079.
Rabinowitz, J. R. and Rein, R. (1972). *Intern. J. Quantum Chem.*, **6,** 669.
Rae, A. I. M. (1973). *Chem. Phys. Letters*, **18,** 574.
Rae, A. I. M. (1975). *Mol. Phys.*, **29,** 467.
Rae, A. I. M. and Mason, R. (1968). *Proc. Roy. Soc.* (*London*), **A304,** (a) 487, (b) 501.
Ramachandran, G. N., editor (1967). *Conformations of Biopolymers*, Vol. 1 and 2. Academic Press, New York.
Ramachandran, G. N. and Sasisekharan, V. (1968). *Adv. Prot. Chem.*, **23,** 283.
Razi Naqvi, K. and Byers Brown, W. (1972). *Intern. J. Quantum Chem.*, **6,** 271.
Rein, R. (1973). *Adv. Quantum Chem.*, **7,** 335.
Rein, R., Claverie, P. and Pollak, M. (1968). *Intern. J. Quantum Chem.*, **2,** 129.
Rein, R., Rabinowitz, J. R. and Swissler, T. J. (1972). *J. Theoret. Biol.*, **34,** 215.
Reiss, H., Frisch, H. L., Helfand, E. and Lebowitz, J. L. (1960). *J. Chem. Phys.*, **32,** 119.
Reiss, H., Frisch, H. L. and Lebowitz, J. L. (1959). *J. Chem. Phys.*, **31,** 369.
Reiss, H. and Mayer, S. W. (1961). *J. Chem. Phys.*, **35,** 1513.
Richmond, P. and Ninham, B. W. (1971a). *J. Phys. C.*, **4,** 1988.
Richmond, P., and Ninham, B. W. (1971b). *J. Low Temp. Phys.*, **5,** 177.
Riera, A. and Meath, W. J. (1972). *Mol. Phys.*, **24,** 1407.
Ritchie, A. B. (1968a). *Phys. Rev.*, **171,** 125.
Ritchie, A. B. (1968b). *J. Chem. Phys.*, **49,** 2167.
Robert, D., Girardet, C. and Galatry, L. (1969). *Chem. Phys. Lett.*, **3,** 102.
Robinson, G., March, N. H. and Perrin, R. C. (1971). *Intern. J. Quantum Chem.*, **5,** 261.
Robinson, P. D. (1961). *Proc. Phys. Soc.*, **78,** 537.
Rojas, O. (1976). '*Contribution à l'étude du découpage quantique de l'énergie de l'état fondamental des molécules en termes de liaisons. Application à l'analyse conformationnelle*'. Thesis (Doctorat d'Etat). C.N.R.S., Paris.
Roothaan, C. C. J. (1951). *J. Chem. Phys.*, **19,** 1445.
Roothaan, C. C. J., Sachs, L. M. and Weiss, A. W. (1960). *Rev. Mod. Phys.*, **32,** 186.
Rose, J. (1967). *Molecular Complexes*. Pergamon Press, Oxford.
Rosen, N. (1931). *Phys. Rev.*, **38,** 255.

Rosen, P. (1950). *J. Chem. Phys.*, **18,** 1182.
Rothstein, E. V. (1970). *Chem. Phys. Letters*, **7,** 232.
Rowe, D. J. (1969). *J. Math. Phys.*, **10,** 1774.
Rowell, R. L. and Stein, R. S. (1967). *J. Chem. Phys.*, **47,** 2985.
Rowlinson, J. S. (1969). *Liquids and Liquid Mixtures.* Butterworth, London (first edition: 1959).
Rowlinson, J. S. (1971). *Trans. Faraday Soc.*, **67,** 576.
Rubinstejn, F. and Yaris, R. (1971). *J. Chem. Phys.*, **55,** 4731.
Ruedenberg, K. (1951). *J. Chem. Phys.*, **19,** 1433.
Ruedenberg, K. (1965). p. 85 in *Modern Quantum Chemistry*, Vol. I, ed. O. Sinanoğlu, Academic Press, New York.
Rumyantsev, A. A. (1973). *Zh. Eksp. Teor. Fiz.*, **65,** 926 [*Sov. Phys. JETP*, **38,** 459 (1974)].
Salahub, D. R. (1974). *Mol. Phys.*, **28,** 243.
Salem, L. (1960). *Mol. Phys.*, **3,** 441.
Salem, L. (1961). *Proc. Roy. Soc.* (*London*), **A264,** 379.
Salem, L. (1962a). *J. Chem. Phys.*, **37,** 2100.
Salem, L. (1962b). *Canad. J. Biochem. Physiol.*, **40,** 1287.
Salem, L. (1964). p. 293 in *Electronic Aspects of Biochemistry* (ed. B. Pullman). Academic Press, New York.
Salem, L. (1965). *Disc. Faraday Soc.*, **40,** 150.
Sanders, W. A. (1969). *J. Chem. Phys.*, **51,** 491.
Santos, E. and Villagra, A. (1972). *Phys. Rev. B*, **6,** 3134.
Schaeffer III, H. F., McLaughlin, D. R., Harris, F. E. and Alder, B. J. (1970). *Phys. Rev. Lett.*, **25,** 988.
Scheludko, A., Platikanov, D. and Manev, E. (1965). *Disc. Faraday Soc.*, **40,** 253.
Scheraga, H. A. (1968). *Advances in Physical Organic Chemistry*, **6,** 103. Academic Press, New York.
Scheraga, H. A. (1971). *Chem Rev.*, **71,** 195.
Schlier, C. (1969). *Ann. Rev. Phys. Chem.* **20,** 191.
Schnuelle, G. W. and Beveridge, D. L. (1975). *J. Phys. Chem.*, **79,** 2566.
Schulman, J. M. (1972). *J. Chem. Phys.*, **57,** 4413.
Schwartz, C. (1959). *Ann. Phys.* (*New York*), **6,** 156.
Scott, R. A. and Scheraga, H. A. (1966). *J. Chem. Phys.*, **45,** 2091.
Sen, P. and Basu, S. (1968). *J. Chem. Phys.*, **48,** 4075.
Shipman, L. L. (1975). *Chem. Phys. Lett.*, **31,** 361.
Shipman, L. L., Burgess, A. W. and Scheraga, H. A. (1975). *Proc. Natl. Acad. Sci. U.S.A.*, **72,** 543.
Shipman, L. L., Burgess, A. W. and Scheraga, H. A. (1976). *J. Phys. Chem.*, **80,** 52.
Sinanoğlu, O. (1962a). *J. Chem. Phys.*, **37,** 191.
Sinanoğlu, O. (1962b). *J. Phys. Chem.*, **66,** 2283.
Sinanoğlu, O. (1964). *Advances in Chemical Physics*, **6,** 315. Interscience, New York.
Sinanoğlu, O. (1965). p. 221 in *Modern Quantum Chemistry, Istanbul Lectures, Part II*: *Interactions* (ed. O. Sinanoğlu) Academic Press, New York.
Sinanoğlu, O. (1967a). p. 283 in '*Intermolecular forces*', *Advances in Chemical Physics*, **12,** (ed. J. O. Hirschfelder). Interscience, New York.
Sinanoğlu, O. (1967b). *Chem. Phys. Letters*, **1,** 340.
Sinanoğlu, O. (1968). p. 427, in *Molecular Associations in Biology* (ed. B. Pullman). Academic Press, New York.
Sinanoğlu, O. and Abdulnur, S. (1964). *Photochem. Photobiol.*, **3,** 333.
Sinanoğlu, O. and Abdulnur, S. (1965). *Fed. Proc. Fed. Am. Soc. Exp. Biol.*, **24,** S12.
Singh, T. R., Kreek, H. and Meath, W. J. (1970). *J. Chem. Phys.*, **52,** 5565 (erratum: ibid. **53,** 4121 (1970)).

Slater, J. C. (1928). *Phys. Rev.*, **32,** 349.
Slater, J. C. (1960). *Quantum Theory of Atomic Structure* Vol. I., McGraw-Hill, New York.
Slater, J. C. (1963). *Quantum Theory of Molecules and Solids* Vol. I., McGraw-Hill, New York.
Slater, J. C. and Kirkwood, J. G. (1931). *Phys. Rev.*, **36,** 682.
Slifkin, M. (1971). *Charge Transfer Interactions of Biomolecules.* Academic Press, London and New York.
Snook, I. K. and Spurling, T. M. (1970). *Austral. J. Chem.*, **23,** 819.
Snook, I. K. and Spurling, T. H. (1975). *J. Chem. Soc. Farad. Trans. II*, **71,** 852.
Soos, Z. G. (1974). *Ann. Rev. Phys. Chem.*, **25,** 121.
Soos, Z. G. and Klein, D. J. (1975). pp. 1–109 in *Molecular Association*, Vol. **1** (ed. R. Foster). Academic Press, New York.
Stamatiadou, M. N., Swissler, T. J., Rabinowitz, J. R. and Rein, R. (1972). *Biopolymers*, **11,** 1217.
Starkschall, G. and Gordon, R. G. (1971). *J. Chem. Phys.*, **54,** 663.
Starkschall, G. and Gordon, R. G. (1972a). *J. Chem. Phys.*, **56,** 2102.
Starkschall, G. and Gordon, R. G. (1972b). *J. Chem. Phys.*, **56,** 2801.
Steele, W. A. (1973). *Surface Sci.*, **36,** 317.
Sternheimer, R. M. (1954). *Phys. Rev.*, **96,** 951.
Sternlicht, H. (1964). *J. Chem. Phys.*, **40,** 1175.
Stogryn, D. E. (1970). *Phys. Rev. Lett.*, **24,** 971.
Stogryn, D. E. (1971). *Mol. Phys.*, **22,** 81.
Stogryn, D. E. and Stogryn, A. P. (1966). *Mol. Phys.*, **11,** 371.
Sugiura, Y. (1927). *Z. Phys.*, **45,** 484.
Sun Chia-Chung and Su Chan (1976). *Scientia Sinica*, **19,** 91.
Surel-Pée, C. (1974). '*Etude de la conformation d'agrégats moléculaires et de son influence sur les données spectroscopiques dans une matrice amorphe*'. Thesis (Doctorat 3è cycle), Université de Bordeaux I (Nr. 1189).
Sustmann, R. and Binsch, G. (1971a). *Mol. Phys.*, **20,** 1.
Sustmann, R. and Binsch, G. (1971b). *Mol. Phys.*, **20,** 9.
Suttorp, L. G. (1973). *Physica*, **64,** 217.
Swissler, T. J. and Rein, R. (1972). *Chem. Phys. Lett.*, **15,** 617.
Tait, A. D. and Hall, G. G. (1973). *Theoret. Chim. Acta*, **31,** 311.
Tanaka, S. and Scheraga, H. A. (1975). *Proc. Natl. Acad. Sci. USA*, **72,** 3802.
Tang, K. T. (1968). *J. Chem. Phys.*, **49,** 4727.
Tang, K. T. (1969). *Phys. Rev.*, **177,** 108.
Tang, K. T. (1970). *Phys. Rev.*, **A1,** 1033.
Tang, K. T. (1971). *J. Chem. Phys.*, **55,** 1064.
Tang, K. T. and Karplus, M. (1968). *Phys. Rev.*, **171,** 70.
Toennies, J. P. (1973). *Chem. Phys. Lett.*, **20,** 239.
Uhlig, H. H. (1937). *J. Phys. Chem.*, **41,** 1215.
Unsöld, A. (1927). *Z. Phys.*, **43,** 563.
Van der Avoird, A. (1967a). *Chem. Phys. Letters*, **1,** 24.
Van der Avoird, A. (1967b). *J. Chem. Phys.*, **47,** 3649.
Van der Avoird, A. (1967c). *Chem. Phys. Letters*, **1,** 411.
Van der Avoird, A. (1968). '*Perturbation theory for intermolecular forces*'. Thesis, Eindhoven.
Van Duijneveldt, F. B. (1968). *J. Chem. Phys.*, **49,** 1424.
Van Duijneveldt, F. B. (1969). '*Intermolecular Forces and the Hydrogen Bond*'. Thesis, Utrecht University.
Van Duijneveldt, F. B. and Murrell, J. N. (1967). *J. Chem. Phys.*, **46,** 1759.
Venkatachalam, C. M. and Ramachandran, G. N. (1967). p. 83 in *Conformations of Biopolymers*, Vol. 1 (ed. G. N. Ramachandran). Academic Press, New York.

Vickery, B. C. and Denbigh, K. G. (1949). *Trans. Farad. Soc.*, **45,** 61.
Victor, G. A. and Dalgarno, A. (1970). *J. Chem. Phys.*, **53,** 1316.
Volkenstein, M. V. (1963). *Configurational Statistics of Polymeric Chains.* Interscience, New York.
Wada, A. (1954). *J. Chem. Phys.*, **22,** 198.
Wagner, A. F., Das, G. and Wahl, A. C. (1974). *J. Chem. Phys.*, **60,** 1885.
Warshel, A. (1973). *Israel J. Chem.*, **11,** 709.
Warshel, A. and Karplus, M. (1972). *J. Am. Chem. Soc.*, **94,** 5612.
Weeks, J. D., Chandler, D. and Andersen, H. C. (1971). *J. Chem. Phys.*, **54,** 5237.
Weintraub, H. J. R. and Hopfinger, A. J. (1975). p. 131 in *Molecular and Quantum Biology* (*The Jerusalem Symposia on Quantum Chemistry Biochemistry,* Vol. 7), eds. E. D. Bergmann and B. Pullman, Reidel, Dordrecht.
Westheimer, F. H. and Kirkwood, J. G. (1938). *J. Chem. Phys.*, **6,** 513.
Whitton, W. N. and Byers Brown, W. (1976). *Int. J. Quantum Chem.*, **10,** 71.
Wilhelm, E. and Battino, R. (1972). *J. Chem. Phys.*, **56,** 563.
Williams, D. E. (1965). *J. Chem. Phys.*, **43,** 4424.
Williams, D. E. (1966). *J. Chem. Phys.*, **45,** 3770.
Williams, D. E. (1967). *J. Chem. Phys.*, **47,** 4680.
Williams, D. E. (1972a). *Acta Cryst.*, **A28,** 84.
Williams, D. E. (1972b). *Acta Cryst.*, **A28,** 629.
Williams, D. R., Schaad, L. J. and Murrell, J. N. (1967). *J. Chem. Phys.*, **47,** 4916.
Williams, J. E., Stang, P. J. and Schleyer, P. von R. (1968). *Ann. Rev. Phys. Chem.*, **19,** 531.
Wojtala, J. (1964). *Acta Physics Polonica,* **25,** 27.
Wormer, P. E. S. (1975). '*Intermolecular forces and the group theory of many-body systems*'. Thesis, Nijmegen.
Wormer, P. E. S., Van Berkel, T. and Van der Avoird, A. (1975). *Mol. Phys.*, 1181.
Wormer, P. E. S. and Van der Avoird, A. (1975). *J. Chem. Phys.*, **62,** 3326.
Yamakawa, H. (1974). *Ann. Rev. Phys. Chem.*, **25,** 179.
Young, R. H. (1975). *Int. J. Quantum Chem.*, **9,** 47.
Yorke, E. D. and De Rocco, A. G. (1970). *J. Chem. Phys.*, **53,** 764.
Živković, T. and Murrell, J. N. (1971). *Theoret. Chim. Acta,* **21,** 301.
Zorkii, P. M., Lazareva, S. G. and Zefirov, Yu. V. (1970). *Kristallografiya,* **15,** 698 [*Sov. Phys.-Crystallogr.*, **15,** 602 (1971)].
Zwanzig, R. (1963). *J. Chem. Phys.*, **39,** 2251.

Note 1. For two helium atoms, $(E_{\text{coul}}+E_{\text{exch}})$, as computed by Murrell and Shaw (1968b), is practically equal to ε_{HF} as computed by McLaughlin and Schaeffer III (1971), and $\varepsilon_1^{\text{HL}}$, as computed by Murrell and Shaw (1967b: see $\bar{E}$ for the 4-terms orbital), is only slightly larger than $(E_{\text{coul}}+E_{\text{exch}})$.

CHAPTER 3

Studies of biomolecular interactions: principles of nucleic acid structure and function from the point of view of constituent interactions

R. Rein

Roswell Park Memorial Institute, Buffalo, New York, U.S.A.

Abstract

Various refinements, introduced in the perturbation theory of intermolecular interactions during the last decade, are surveyed. The predictive value of the theory is tested by a comprehensive comparison with experimental results for the case of nucleic acid base interactions. The theory predicts, within reason, base pairing and stacking stabilities and patterns. Forces responsible for double helical structure of DNA are analysed in terms of hydrogen bonding, stacking, solvent effect and conformational adjustment energies. Thermodynamic stability of helix formation, expressed in terms of these components, compares somewhat favourably with experimental values.

Two new concepts emerge from these considerations: (a) The quantitative role of conformation constraint in maintaining the exclusiveness of Watson–Crick pairs in the double helix; (b) a sequence dependent fine structure of the helix which arises as a consequence of the interplays of stacking orientational preferences and backbone constraints. The predicted sequence, polymorphism, has been observed experimentally. The possible biological significance of this polymorphism is that it may give rise to recognizable markers for enzymes and regulator molecules.

I. INTRODUCTION

Nucleic acids and proteins are the principal macromolecular components in the living cell. These two classes of macromolecules are fundamental in the most important biological functions of the living process: heredity and metabolic control. Both, nucleic acids and proteins, are chainlike polymers. In the former the repeat units are nucleotides and in the latter, the repeat units are amino acid residues. The residues are connected by covalent bonds, conferring a primary structure to the nucleotide or amino acid sequence. In their native form, these molecules possess specific three-dimensional structures, such as the double helical structure of DNA and the higher order structure of proteins. This structural specialization is often the first clue in understanding their biological function. X-ray crystallographic studies, during the past few decades, have contributed greatly to the understanding of the structure of many molecules in their crystalline state. However, these molecules, comprised of flexible chains, undergo structural transitions to alternative and random structures under the influence of a variety of environmental conditions. This happens because the structural features of nucleic acids and proteins are not conferred exclusively by the primary covalent bond structure, but also by weak intermolecular interactions, exerted between the constituent molecules on the one hand, and by interactions with a variety of substrates, ions, and solvent environments on the other hand.

In this broader context, scientific interest turns also toward the question; why is a structure developing, under what conditions, and due to the interplay of what molecular forces? To answer these questions the characterization of the key molecular interactions playing a role in conferring structure-function relationships have to be investigated.

In the present paper a review of the theoretical and experimental studies of intermolecular interactions between nucleic acid components, and the structural consequences of these interactions is presented. The discussion is designed to enlighten the relation between molecular forces, structure and biological functions.

The following, Section II, introduces the relationship between nucleic acid component interactions and structure, and an examination of the role of these interactions in the biosynthesis of nucleic acids.

Section III includes a brief review of the quantum mechanical analysis of hydrogen bonding. This is followed by a discussion of the methods for studying intermolecular interactions in large systems, with a particular emphasis on the expansion techniques of the interaction operator and the approximation of the arising matrix elements.

In Section IV, experimental results on hydrogen bonding of nucleic acid bases in solution, crystals, and polymers are reviewed and compared with theoretical results. Aspects of base stacking such as directionality, nature of contributing forces, compositional and orientational dependence of helical configuration and thermodynamic stability, are discussed in view of existing experimental and theoretical results.

In Section V.a, the theory of solvent effects on molecular associations and helix formation is discussed.

In Section V.b, the conformational aspects of the backbone in relation to nucleic acid helix formation are considered. Also point mutations are examined from the point of view of energetics of mispaired replication complexes.

Section VI reviews further application to the analysis of genetic mechanisms. The topics discussed are: rare tautomers and their possible role in mutagensis; induced mutations in view of nucleotide base analogue interactions; interactions of heterocyclics with DNA in the context of their mutagenic activity, and the structure of wobble pairs.

In the final section, the validity of the theoretical model is discussed. Relevance of the theory for structure interpretation and prediction is further evaluated. The usefulness of the analysis for gaining insight into the molecular aspects of biological mechanisms is also examined.

II. BIOSYNTHESIS OF NUCLEIC ACIDS AND POINT MUTATIONS IN VIEW OF NUCLEOTIDE INTERACTIONS

The biosynthesis of nucleic acids underlies heredity and is thus perhaps, the most fundamental process in cellular biology. Transcription of the base

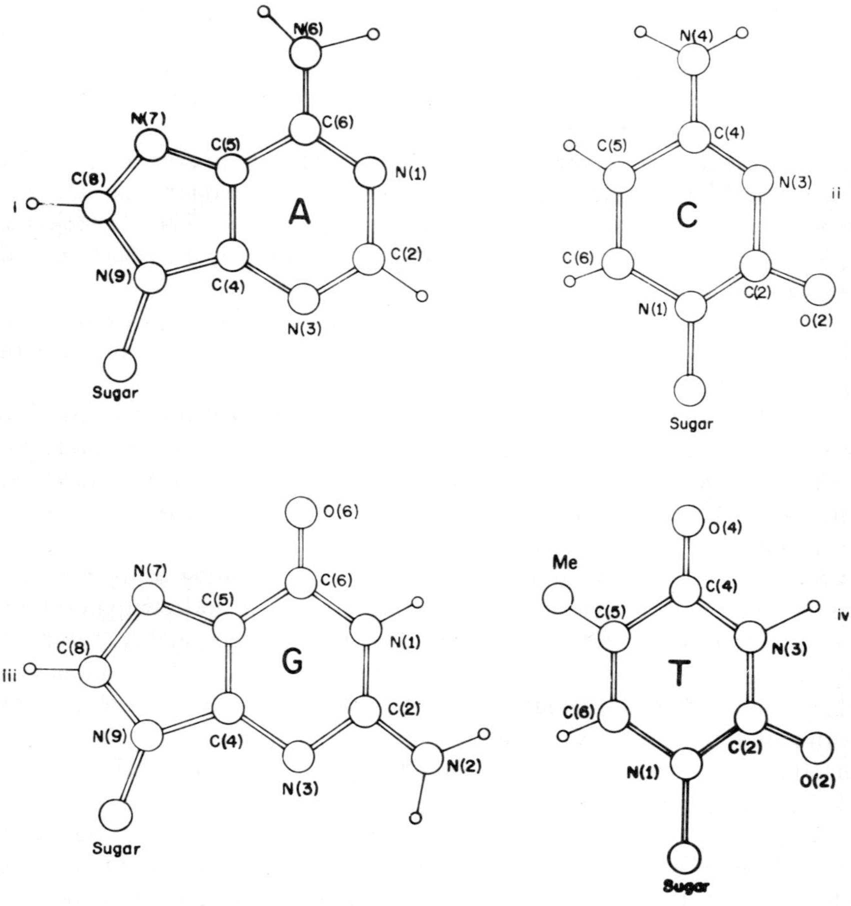

Figure 3.1. The numbering on DNA bases

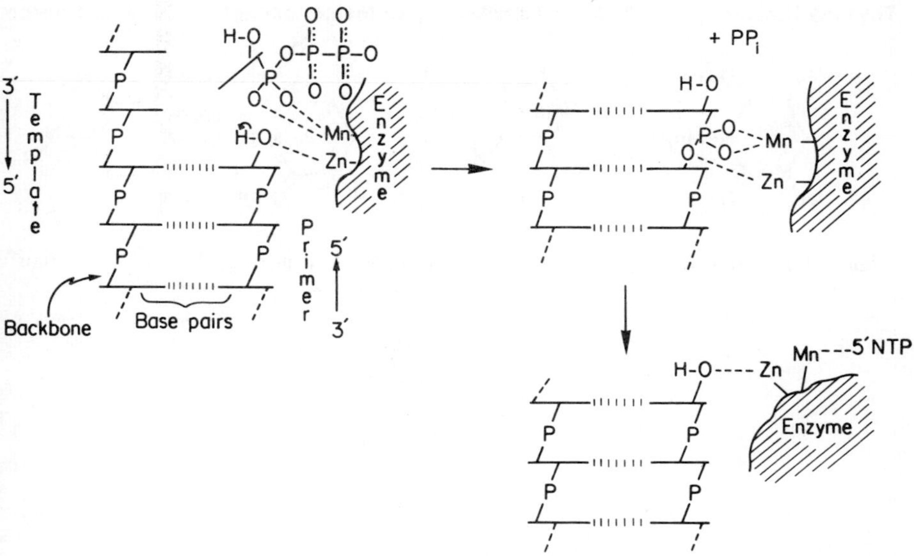

Figure 3.2. Mechanism for polynucleotide synthesis

sequence of a gene into the base sequence of the messenger RNA is the first step in the biosynthesis of proteins and enzymes necessary for the metabolic activities of the cell. Numerous studies by Kornberg and coworkers (see for example Kornberg, 1969) have demonstrated the general aspects of nucleic acid biosynthesis. An essential step in nucleic acid synthesis involves phosphoester bond formation, which connects two successive nucleotide units. This step is enzyme catalysed by polymerase; the substrate is an activated nucleotide in the form of a nucleoside tri-phosphate which is added to the 3-OH terminus of the preceding nucleotide unit. This step also involves divalent ions such as Mn^{++} and Mg^{++}. The major steps of the synthetic process are summarized in Figure 3.2 (Figure 3.1, 3.3 and 3.4 showing DNA components.)

The heart of the problem is to understand the nature of template activity, that is, how a particular nucleotide sequence complementary to the parental strand is assembled. Implicit in the Watson–Crick double helical model of DNA structure is the explanation of the preferential hydrogen-bonding stability between complementary bases, i.e. adenine with thymine and guanine with cytosine, as the basis of template selection. It is important to recognize, however, that these preferences are not absolute; thus, mismatched pairs are occasionally formed. For example, Freese (1959) explains point mutations, one of the common mechanisms of heredity changes, by assuming incorporation of mismatched pairs as an intermediate step. The nature and strength of various bonding schemes between nucleotide bases have been characterized by X-ray crystallographic (Sobell *et al.*, 1963; Hoogsteen, 1959; Davies, 1967; Haschemeyer and Rich, 1967; Mazza *et al.*,

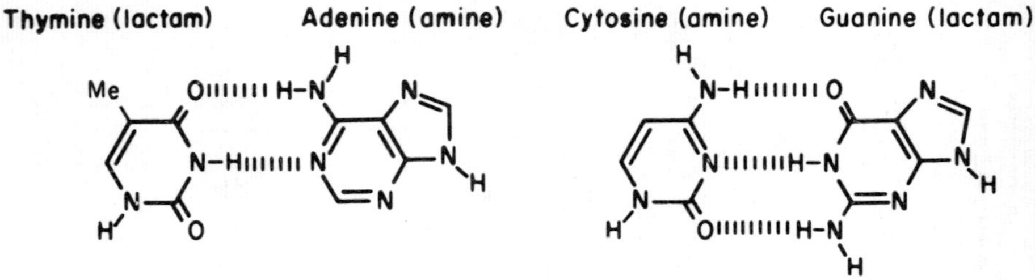

Figure 3.3. The usual forms of the DNA bases and their normal Watson–Crick pairing scheme

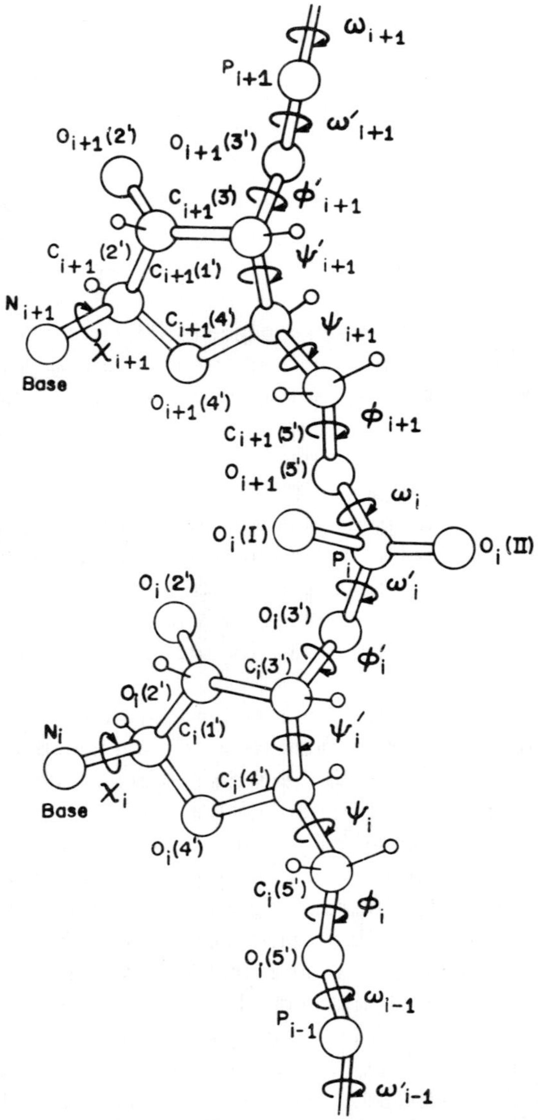

Figure 3.4. Torsional angles for DNA backbone

1969) and other experimental techniques, (Ts'o 1964, 1970; Ts'o *et al.*, 1964; Huang, 1966).

The experimental elucidation of the energetics of these bonding schemes still remains a difficult matter. These difficulties stem from the fact that thermodynamic quantities obtained from melting experiments with complexes such as synthetic double stranded oligo and polynucleotides (Von Hippel and Wong, 1964, 1965; Hamaguchi and Geiduschek, 1962) pertain to the overall energetics of formation which involves a variety of interactions other than hydrogen-bonding interactions. This includes stacking energies, strand conformational energies, ionic and solvent interactions and chain entropies. The decomposition of the measured thermodynamic quantities into these terms is not a trivial matter and often rests on a multitude of assumptions (Goel *et al.*, 1968).

This has prompted theoretical chemists since the early sixties to attempt to calculate these interactions by quantum mechanical methods. Some of the more prominent of these earlier calculations are due to: DeVoe and Tinoco (1962), Pullman, Claverie and Caillet (1905a, 1905b, 1966), and Rein, Claverie and Pollack (1969). Most of these calculations consist of an electrostatic term and a second order interaction term in the form of polarizability approximations.

The calculation of these interactions and the development of the nature of the methodology have been discussed in detail in several recent reviews (Pullman 1968; Claverie, 1968; Rein, 1973, 1975). These are also the subject of two contributions in the present volume (Buckingham and Claverie). In view of this, we have kept the discussion of these developments short and refer the reader for further details to the above sources.

The first nucleotide interaction calculation, which has dealt explicitly with the biological problem of nucleic acid replication is due to Nash and Bradley (1965, 1966). These authors have been concerned qualitatively with the question of what planar hydrogen bonded complexes can be formed between pairs of nucleotide bases. In order to answer this question, they have set up a computational procedure, involving electrostatic hard sphere potential energies, in a high density sampling of the planar configurational space of pairs of bases, to locate the minima i.e., stable base pair configurations. Figure 3.5 and 3.6 are reproduced from Bradley and Nash and illustrate the searching procedure and stable base pair configurations.

Table 3.1 reproduced from the same paper, summarizes the base pair combinations for which stable complexes are predicted. The principal conclusion, arrived at from this study, is that no specificity resides in the hydrogen bonding schemes. In their search to find what confirms the base selection specificity in replication, these authors suggest the factors of steric constraint. They consider the geometrical problem arising from the addition of a nucleotide into the double helical conformation of a template. They showed that depending on the hydrogen bonding scheme between the entering base and the template base, the phosphate end of the new nucleotide may be displaced from its normal position. Nash and Bradley's

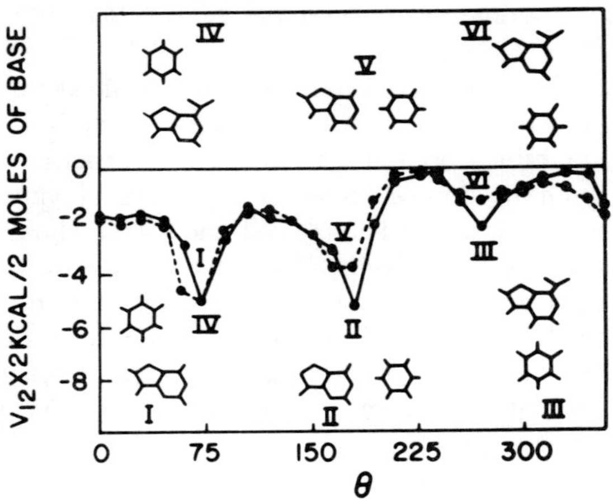

Figure 3.5. Potential energy curves of the adenine–uracil (A–U) base pair computed from the electrostatic hard-sphere model (after Nash and Bradley, 1966)

displacement data is shown in Table 3.2. For each of the various pair combinations, it shows the distance from the proximal phosphate of the entering base to the normal position it would assume in Watson–Crick pairing. This table demonstrates that non-complementary pairs deviate significantly from the normal double helical configuration. To explain specificity in replication, these authors postulate a steric-filter action by the polymerase enzyme, due to its fixed position (relative to the template) and its small active site.

The work of Bradley and Nash has undoubtedly made a very significant conceptual contribution to the problem of replication by stressing and

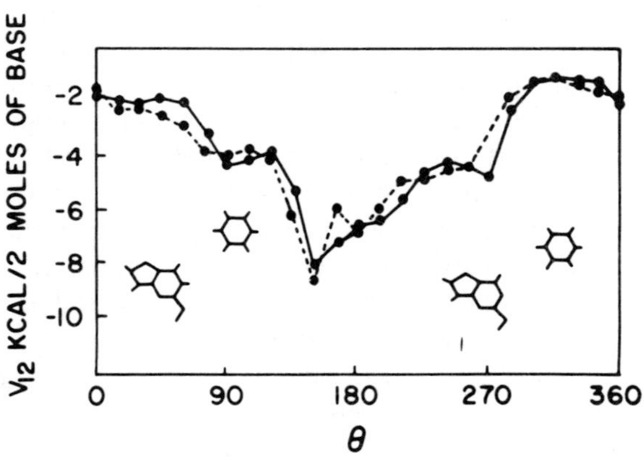

Figure 3.6. Potential energy curves for the guanine–uracil (G–U) base pair computed from the electrostatic hard-sphere model (after Nash and Bradley, 1966)

Table 3.1. Comparison of $V_{ij(\min)}$ (upper values) with ΔH (lower values) in $CHCl_3$ for various combinations of nucleotide bases (after Bradley and Nash, 1968).

Base	A	G	C	U	I
A	−3·55 −4·0±0·8	−6·61	−5·06	−6·07 −6·2±0·6	−6·65 (<−4)
G		−14·38 (−8·5 to −10)	−15·03 (−10 to −11·5)	−9·81	−13·48
C			−9·84 (−6·3±0·6)	−4·77	−11·61 (−9)
U				−7·03 −4·3±0·4	−9·41
I					−12·64 (−8)

quantitatively illustrating the interplay of hydrogen bonding and steric factors in base selection specificity. This study, however, does not arrive at a final conclusion with regard to base selection specificity, nor does it elaborate on the role of the enzyme in this process. A more comprehensive study of this problem should consider not only the hydrogen bonding energies for non-complementary base pairs, but also the entire energy of formation of a DNA fragment (mini-helix) comprised of two successive base pairs. From the preceding discussion it is clear that when a non-complementary pair is formed, the geometrical relation between the successive base pairs is altered. Thus, a change in stacking energies will also appear. Furthermore, since hydration energies differ depending on the base pair, different solvent effects may be anticipated for the various schemes. As a result of the alteration of the geometry in the position of a non-complementary nucleotide base there is a different steric requirement on the phosphate terminus for juxtapositioning with the 3′-OH group of the primer terminus which is necessary for the condensation step to take place. Both the geometrical and energetic aspects of strand deformation should be considered. The question of the positioning of a non-complementary pair on top of a Watson–Crick pair and subject to the strand constraint is not obvious. Assuming that the four bases occupying the two planes assume a minimum

Table 3.2. Deviations from normal in the position of the proximal phosphate of the entering base (after Bradley and Nash, 1968)

	Entering base			
Parental base	A	G	C	U
A	12·67	4·59	14·39	0·39
G	3·32	4·52	0·51	2·34
C	8·90	0·51	11·57	2·53
U	0·51	3·87	2·06	3·48

energy configuration, it would seem that the non-complementary pairing in general will lead to a strain on both of the connecting strands. This has to be relaxed by a configurational adjustment of the strands. The hydrogen bonding, stacking and strand conformational energies have to be simultaneously optimized.

To predict selection preferences, the formation energies of normal and mismatched mini-helices in their minimum energy configuration are required.

III. THEORETICAL APPROACH

A. Theoretical models of hydrogen bonding

As with other non-covalent intermolecular interactions, a theoretical analysis of hydrogen bonds can be accomplished either by a supermolecular approach or by perturbation theory. In the former the interacting molecules are considered as a single system. A calculation of the energies of the separated constituents follows and the difference of these quantities is taken to be the interaction energy.

In the context of hydrogen bond energies of large biomolecular systems, the supermolecule approach is not very practical. First, the interaction energy is a small quantity appearing as a difference of two large quantities. A meaningful calculation therefore requires a high degree of accuracy in the calculation of the combined and separate molecule energies which exceeds the practical limits, except for the most sophisticated *ab initio* calculations. In order to construct an *ab initio* potential function, it is necessary to determine the appropriate energies from among geometrical configurations of the biomolecular complex. This procedure is of limited value for biomolecules because of the prohibitive cost involved. Secondly, there is a more fundamental obstacle imposed by the nature of the wave function. A single determinantal SCF wave function may be a good description of a molecule at or near its equilibrium configuration. Whether such a description remains optimal for a biomolecular complex, particularly at longer separations, is not yet satisfactorily determined. Experience indicates that such situations could be handled only by wave functions constructed from a large number of electronic configurations. With present day computational technology such calculations are practical only for the smallest systems such as two hydrogen atoms and are thus not applicable in the biomolecular context.

A major alternative to the supermolecular approach for treatment of intermolecular interactions is based on the assumption that the individual interacting molecules preserve their individuality. That is, the molecular orbitals of the individual molecules can be utilized for construction of good zeroth order basis functions for a description of the perturbation of the interacting system. When the interacting molecules are sufficiently apart their electronic systems are not significantly overlapping. The above assumption is indeed a very good approximation for the interacting system.

At close approach such as Van der Waals contacts, the assumption of the overlap between the constituents is not valid any more and the additional effects of exchange, non-orthogonality (Lowdin 1950, 1956) and charge transfer come into play.

Recently, studies concerned with small hydrogen-bonded systems, notably the studies of Dreyfus and Pullman (1970), of Morokuma (1971) and of Kollman and Allen (1970), have provided an insight into the nature of these close range interaction components, as well as to how these components change with increasing separation distance. These studies are of interest here not only for their conceptual application, but also for their practical value as guide lines to what components should be included in an interaction scheme at a given separation distance, and what empirical corrections can be made for the more computationally difficult short range terms.

Morokuma has applied his decomposition scheme to the study of components contributing to hydrogen bonding in water dimers, and in formamide–water dimers. The contribution to hydrogen bonding in the case of water dimers has been studied also by Coulson and Danielson (1954). Further results are available on the contributions to hydrogen bonding in the case of the linear dimer of formamide from the work of Dreyfus and Pullman (1970). These results are summarized in Table 3.3.

Table 3.3. Decomposition of hydrogen bond energies (-kcal/mol dimer)

System	E_{Total}	$E_{Electst.}$	$E_{Exch.Rep.}$	$E_{Char.Tranf.}$	E_{Pol}	$E_{Disp.}$	Reference
$(H_2O)_2$	8·09	8·0	−9·86	8·16	0·2	1·54	Kollman and Allen, *J. Chem. Phys.*, **52,** 5085 (1970)
$(H_2O)_2$	8·6	6·0	−8·4	8·0		3·0	Morokuma, *J. Chem. Phys.*, **55,** 1236 (1971)
$H_2CO{\cdot}H_2O$	3·45	4·64	−6·71	5·34	0·18		Dreyfus and Pullman, *Theor. Chim. Acta*, **17,** 109 (1970)
$(H_2CO)_2$	8·0	10·0	−7·0	5·0			Coulson and Danielson, *Arkiv. Fysik.*, **8,** 239 (1954)

The most salient feature emerging from these results is that the electrostatic contribution to hydrogen bonding at the equilibrium geometry seems to approximate fairly well the total interaction energy. This appears to be a consequence of the fact that exchange repulsion in this domain is nearly compensated by the attractive terms contributing in this range, i.e. charge transfer and polarization.

Another point of great interest is to observe how the various components change with increasing separation between interacting systems. Both Morokuma, (1971) and Dreyfus and Pullman (1970) have published curves representing the magnitude of the various components for a range of distances from the equilibrium configuration to about 10 Å. From these results, it is apparent that all but the electrostatic terms are negligible for distances larger than 5 Å. This in essence confirms the earlier results of Coulson and Danielson (1954) based on valence bond theory. According to these workers the contributions to the hydrogen-bonding energy are electrostatic, dispersion, polarization, charge transfer, and short range repulsion energy. The later terms are of similar order of magnitudes and nearly cancel each other. For long bonds, the H-bond interaction is totally electrostatic.

These results form the rationale for use of the perturbation model for large hydrogen bonded systems and the justification for the assumptions that the stability of the hydrogen bonded complexes is essentially governed by the electrostatic interaction.

The perturbation model originated in the thirties. When London (1937) developed his theory of intermolecular forces, wave functions for larger molecules were not yet available. To overcome this difficulty, the perturbation operator to the Hamiltonian ($1/r_{12}$) was expanded in a Taylor series. Depending on the centre of expansion, two schemes have been derived. In the first one the expansion is over the entire molecule, and in the second the molecule is first divided into segments and then the expansion is made for each segment. For both cases the interaction energies can be expressed in terms of interactions between multipoles of the two molecules and transition multipoles of the two molecules for first and second order interactions. The expansions are terminated after the first non-zero term. That means a dipole–dipole interaction is calculated for the first method and an atomic point charge–atomic point charge interaction for the second. The experimental knowledge of dipole moments and polarizabilities for the first and second order, respectively, thus circumvents the need for explicit knowledge of wave functions. However, it has been shown recently (Rein, Rabinowitz and Swissler, 1972) that truncation of the interaction series to one term is generally not valid below separation distances of the order of ten angstroms, thus excluding most biological problems of interest.

The approach taken was to carry the expansion to higher terms until convergence is assured.

While dipole and quadrupole moments for a collection of polyatomic molecules is now available (Flygare, W. H. and Benson R. C., 1971; Rabinowitz and Rein, 1972), the segmental moments as well as the higher moments are derivable only from wave functions. This introduces the requirement for reliable wave functions. The reliability of the wave function in the particular context can be best established by calculation of moments which have been experimentally observed.

The question of dispersion interactions is of interest, not only from the practical point of view, but also conceptually. There are at least three different ways to look at dispersion forces. In a static view they appear as the consequence of interactions of induced charges of one molecule with those of another molecule, where the magnitude of the changes are calculated from the first order perturbation correction to the respective wave function. Another way to look at dispersion forces is to consider them as arising from electron correlations (Sinanoglu, 1964). A third way is to obtain the dispersion forces from an electrodynamic point of view (Lifshitz, 1955; Nir, Rein and Weiss, 1972; Nir, 1974). The interesting point is that whichever of these descriptions is adopted by a suitable set of approximations, they can be brought into an equal form in which the dispersion interactions will appear in terms of polarizabilities.

The differences in applications of this theory to large hydrogen bonded systems during the last two decades are reflected principally in the way the operator expansion and the representation (approximation) of matrix elements is handled. These applications together with the outline of the approximations used are summarized in Table 3.4.

B. Methods and analysis of forces

The interaction model includes the following components: $E_{tot} = E_{el} + E_{disp} + E_{pol} + E_{rep}$ where the energies are; (I) electrostatic, polarization and dispersion energies and (II) short range overlap repulsive energies.

A detailed description of the electrostatic interactions between biomolecules has been given in a series of papers from this laboratory (Rabinowitz and Rein, 1972; Pack *et al.*, 1972; Rabinowitz, Swissler and Rein, 1972; Rein, 1973). The method for calculation of polarization and dispersion interactions has been presented by Rein, Claverie and Pollack (1968) and Claverie and Rein (1969).

Fundamental to these calculations is the representation of the electrostatic potential at biological distances of importance in terms of a convergent series of successive multipole interactions. Prior studies by Stamatiadou *et al.* (1972) have been made on nucleotide base interactions using terms as high as quadrupole interactions. The results indicated that while the electrostatic interactions tended to converge at hydrogen bonded distances, terms higher than quadrupole might be needed to assure complete convergence. Pack, Wang and Rein (1972) justified this approach by showing by comparing with *ab initio* calculations that the electrostatic series of pyridine tended to converge at or about Van der Waal's distances. The behaviour of the electrostatic and polarization energies with respect to inverse powers of R is presented in Table 3.5.

The more recent study (Egan, Swissler, Rein, 1975) i.e. the inclusion of terms up to octopole interactions and subsequent decomposition into inverse powers of R, gives a better insight into the convergence problem than the previous study (Stamatiadou *et al.*, 1972). With respect to electrostatic interactions it was observed that after some oscillations, the energy approaches a convergent value. The mean energy for all interactions was calculated to serve as a reference $E = \Sigma E(1/R^{-n})/n$. For all cases studied, the highest term in the series came within one standard deviation of this value. This is a powerful indicator that convergence has been reached. It is

Table 3.4. Nucleotide interaction calculation

Introduced improvements (authors) \ Contributions to the total interaction and stacking	Electrostatic	Polarization	Dispersion	Repulsion (short-range)	Reference
DeVoe and Tinoco (1962)	*M, DD,* μ_e, *W(H)*	*m, Me,* αe, *DD*	*m,* αe	W_{emp}	*J. Mol. Biol.*, **4,** 500 (1962).
Bradley *et al.* (1964)	*a, m, DR,*	*m,* μe, αe, *tm(DR) DD*			In *Electronic Aspects of Biochem.*, Pullman, ed., (1964) p. 77.
Nash and Bradley (1965)	*a, m, W(H)*	*m, MM*			*Biopolymers*, **3,** 261 (1965).
Claverie, Pullman, Caillet (1966)	*a, m, W(H)*	*a, m, Mc,* αc *MM, MD*	*a, m, Mc,* αc		*J. Theor. Biol.*, **12,** 419 (1966).
Rein and Pollack (1967)	*a, m, W(PPP)*	*a, m, tm(PPP)*	*a, m, tm(PPP) Mc,* αc, *Ie*		*J. Chem. Phys.*, **47,** 2039 (1967).
Pollack and Rein (1967)	*a, m, W(PPP)*	*a, m, tm(PPP)*	*a, m, tm(PPP) Mc,* αc, *Ie*		*J. Chem. Phys.*, **47,** 2045 (1967).
Rein, Claverie, Pollack (1968)	*a, m, W(PPP)*	*a, M, tm(PPP)* μe, αe, *MM MD, DD*	*a, M, tm(PPP)* μe, αe, *MM DD*		*Int. J. Q. Chem.*, **2,** 129 (1968).
Rein, Goel, Fukuda (1969)	*a, m, W(PPP)*	*a, m, tm(PPP) Me,* αe	*a, m, tm(PPP) Me,* αe, *Ie MM, DD*		*Ann. N.Y. Ac. Sci.*, **153**(3), 805 (1969).
Claverie (1968)	*a, m, LCAO*	*a, m,* μ, αe	*a, m,* μc, αe	W_{emp}	In *Molec. Assoc in Biology* (1968) p. 130.
Stamatiadou *et al.* (1972)	*a, W(IEHT), W(CNDO)*				*Biopolymers*, **11,** 1217 (1972).
Egan *et al.* (1974)	*a, W(IEHT)* $-\Omega$	*a,* $-\Omega$, μc, αe	*a, DD,* αe	W_{emp}	*Int. J. Q. Chem., Q. B. Sym.*, **1,** 71 (1974).
Rein *et al.* (1975)	α, *M, W(IEHT), W(CNDO)* $-Q$	*a,* $-\Omega$, μc, αe *MD*	*a, DD,* αe *M(BP)*	W_{emp}	*Int. J. Q. Chem., Q. B. Sym.*, **2,** 145 (1975).
Kudritskaya, Danilov (1976)	*a, m,* μ, *W(CNDO)*	*a, m, MD, DD*	*a, m, MD, DD* α		*J. theor. Biol.*, In press (1977).

a	atomic multipole expansion
α	atomic polarizability
αe	bond polarizability
DD	dipole–dipole approximation
DR	DelRe method
Ie	experimental ionization potential
m	monopole approximation
M	molecular multipole expansion
μc	Calculated dipole moment
μe	experimental dipole moment
LCAO	LCAO method
MD	monopole–dipole approximation
MM	monopole–monopole approximation
M(BP)	monopole–bond polarizability
$-\Omega$	octopole approximation
$-Q$	quadrupole approximation
tm(PPP)	π electron transition monopole calculated by Pariser–Parr–Pople method
W(CNDO)	CNDO wave functions
W(H)	Hückel wave functions
W(IEHT)	LEHT wave functions
W(PPP)	Pariser–Parr–Pople wave functions

interesting to note that terms carried only as far as R^{-4} which contained moments up to quadrupole, approached to within 90 per cent of the final value. If monopole approximations are used the difference between monopole and convergent values in some cases observed is as great as 40 per cent. These results lead us to believe that the use of terms as high as R^{-7} are justified in investigating the role of electrostatic interactions in DNA type complexes.

Table 3.5. Contributions to electrostatic and polarization energies in inverse powers of R (kcal/mol)

				Electrostatic					
Pair	R^{-1}	R^{-2}	R^{-3}	R^{-4}	R^{-5}	R^{-6}	R^{-7}	R^{-8}	Mean
A–T	−6·081	−7·21	−10·11	−10·96	−9·88	−10·01	−8·94	—	−9·03
G–C	−18·05	−20·20	−23·56	−24·03	−23·35	−23·40	−22·48	—	−22·52
				Polarization					Total
G–C	—	−3·422	—	−3·786	—	−3·841	—	−3·841	−3·841
A–T	—	−0·763	—	−1·028	—	−1·080	—	−1·090	−1·090

The use of full multipole representation of the electric field in the polarization calculations, however, was not seen to be as necessary. Convergence occurred quickly after the second term in the series. This indicates that the expansion of the electric field to include R^{-4} terms should be sufficient for this convergence. Since the R^{-4} terms contain dipole and quadrupole expressions, this implies that the inclusion of octopole interactions is not as necessary for the convergence of the polarization energy as it was for the electrostatic energy. Though no special refinements were made on the dispersion calculations, the quick convergence of the polarization series indicated that the dispersion energy with R^{-6} dependence is also convergent.

It becomes clear from this table that the repulsion energy tends to cancel the second order energies such as polarization and dispersion; this leaves only the electrostatic forces as the predominant one in hydrogen bonding. This result supports the work of Nash and Bradley (1965) where the electrostatic forces are mainly responsible for H-bonding stability.

The components of the hydrogen bond interactions for A–T and G–C pairs, are listed in Table 3.6.

Table 3.6. Decomposition of energies in hydrogen bonded Watson–Crick base pairs (kcal/mol)

Author	Pair	E_{El}	E_{Pol}	E_{Disp}	E_{Rep}	E_{Total}
Kudritskaya and Danilov (1976)	A–T	−5·70	−0·57	−0·73	—	−7·00
Egan *et al.* (1975)	A–T	−8·99	−1·09	−1·06	+3·81	−8·04
Claverie (1968)	A–T	−6·67	—	—	+3·81	−2·86
Kudritskaya and Danilov (1976)	A–U	−5·68	−0·66	−0·87	—	−7·21
Egan *et al.* (1975)	A–U	−9·70	−1·50	−0·12	+5·29	−6·03
Pullman *et al.* (1969)	A–U	−4·64	−0·25	−0·69	—	−5·58
Kudritskaya and Danilov (1976)	G–C	−14·10	−1·90	−0·79	—	−16·79
Egan *et. al.* (1975)	G–C	−22·48	−3·84	−1·49	+5·28	−22·53
Pullman *et. al.* (1969)	G–C	−15·91	−2·02	−1·25	—	−19·18
Claverie (1968)	G–C	−21·30	—	—	+5·28	−16·02

IV. INTERACTION OF NUCLEIC ACID BASES

A. Configuration of hydrogen-bonded base pairs

Donohue (1956) presented what was probably the first systematic discussion of the possible hydrogen-bonded based pair arrangements. Donohue began by considering the bases guanine, adenine, cytosine and thymine in their normal tautomeric form and assumed that these bases can form planar N–H . . . N and N–H . . . O type hydrogen bonded pairs with 3·0 and 2·8 Å bond distances respectively with a deviation from linearity not exceeding 15°. Under these assumptions Donohue has shown that it is possible to form as many as 25 different pair configurations. In a later paper Donohue and Trueblood (1956) extended this compilation to include 29 configurations.

The suggested base pair configurations can be classified broadly into (a) complementary pairing, (b) crossed pairing, and (c) self associating pairs. In complementary pairing G pairs with C, while A pairs with T or U. In the crossed pairing category, the combination of purine and pyrimidines are reversed such that A–C, and G–U or G–T pairs may be formed. In the latter category, purines pair with purines and pyrimidines with pyrimidines.

Attempts have been made to compute the stability of some of these complexes. DeVoe and Tinoco (1962) computed the London dispersion, dipole and induced dipole forces in A–T and G–C complexes of fixed geometry. Since then substantial progress has been made in the study of the electronic properties of the bases of nucleic acids. A clear and informative review on the quantum mechanical approach has been presented by Pullman and Pullman (1969). In this work these investigators also address themselves to the problem of abnormal pairing of bases. They perform studies on Hoogsteen pairs (1959) of different arrangements as well as normal Watson–Crick pairs. They reinvestigated the results of DeVoe and Tinoco using more refined approximations.

Using formal charges residing at the centre of each atom rather than dipoles located at the centre of each nucleotide base, Bradley, Lifson and Honig (1964) have also performed theoretical calculations and computed the electrostatic contribution for some cases.

Nash and Bradley (1965) extended this monopole method to the 10 complexes which can be formed from the four bases if each is held in a single orientation.

In still a later study, Nash and Bradley (1966), using as a point of departure that the principal force acting between two nucleotide bases is the electrostatic force, computed the energy of interaction between bases in a sampling of all possible coplanar orientations and determined the value and geometry of the minima in electrostatic potential energy. The results are discussed in relation to the customary hydrogen bond criteria used to predict relative stabilities and geometries of nucleotide base pairs.

Kudritskaya and Danilov (1976) extended the study of hydrogen bonding to include 34 hydrogen bonded configurations using a monopole approximation for the electrostatic and a transition monopole approximation for the electrostatic and a transition monopole approximation for the polarization and dispersion energies. Also included in this analysis are short range repulsions and solvent effects which are calculated from reaction field formalism. Rein *et al.* (1975) have calculated the hydrogen bond energies for complementary pairs and a number of mismatched pairs. These calculations include electrostatic terms up to octopole. The polarization and dispersion terms were calculated by the bond polarizability method and the repulsion term according to Claverie's (1968) method. The solvent effect in this calculation has been estimated from a discrete dehydration model constituting a simplification of the method of Poland and Scheraga (1965). The number of sites and the hydration energies have been assigned according to the water nucleotide base interaction energies calculated by Port and Pullman (1973).

In the next section, experimental results on the stability of various hydrogen bonding schemes are briefly reviewed. The experimental results will be compared with the theoretically predicted stabilities in order to assess the reliability of the theoretical calculations.

B. Stability of hydrogen-bonded base pairs in solution

Within the past decade much experimental data has accumulated concerning the interaction of bases in solvents. The experimental NMR and IR investigations of the interactions between bases in organic solvents (Katz and Penman, 1966; Pitha, Jones and Pithova, 1966; Shoup, Miles and Becker, 1966) indicates that Watson–Crick type pairing are favoured between base monomers or their derivatives in various organic solvents. These investigations also show that the formation constants for complementary pairing are much higher than those for corresponding homo-associates (Katz and Penman, 1966; Kyogoku, Lord and Rich, 1966; Shoup *et al.*, 1966).

Concerning the behaviour of interacting nucleotides in an aqueous media, it would appear that planar complementary complexes of different monomers are not favoured. Ts'o (1974) indicates that partially overlapping stacked complexes appeared to be favoured in water. Similar results were obtained by Nakano and Igarashi (1969), Gratzer, (1969) and Solie and Schellman (1968). More recent studies by Razka and Kaplan (1972) using pulse NMR techniques, however, have detected the presence of planar type hydrogen bonding schemes between nucleotide monophosphates which appear to follow Watson–Crick pairing rules. Even more recently D'albis, Wickens and Gratzer (1975) have studied the association of chloroform-soluble derivatives of uracil and adenine in the presence of water. Analysis by infrared spectroscopy shows that complex formation still occurs in these conditions, and that the extent of association is substantially unchanged by the presence of water. These workers conclude that not only is the hydrogen bonding between base and base stronger than that between the bases and water, but that a large excess of water would be needed to cause dissociation of the base pairs. Ts'o (1974) has compiled a table of association constants for crossed and self-associated nucleotide derivatives in organic solutions and represents an informative tabulation of organic solvent results up to that time. We refer the reader to that work.

We have presented in Table 3.7 a qualitative summary of the relative stability of hydrogen bonding between nucleotide pairs.

Table 3.7. Experimental results on the relative strengths of hydrogen bonding

Result	Solvent	Method	References
A–U > A–A or U–U	$CDCl_3$	IR	Hamlin *et al.* (1965)
	$CDCl_3$	IR	Kyogoku *et al.* (1967)
	$CDCl_3$	IR	Miller and Sobell (1962)
G–C > G–G or C–C	$CDCl_3$	IR	Katz and Penman (1966)
G–C > A–T or A–U	Me_2SO	IR	Katz and Penman (1966)
	$CHCl_3$	IR	Pitha *et al.* (1966)
	Me_2SO	NMR	Shoup *et al.* (1966)
G–G > C–C	$CHCl_3$	IR	Kyogoku *et al.* (1966)
U–U > A–A	$CHCl_3$	IR	Kyogoku *et al.* (1966)
G–C ≃ A–U > G–I ≫ G–A*	H_2O	PMR	Razka and Kaplan (1972)
A–U ≃ A–C ≫ A–I ≧ A–G			
G–G ≃ G–C > G–U > G–X ≧ G–A ≧ G–I			

* These hydrogen bond stabilities were observed on nucleotide monophosphates.

From this table it is possible to deduce the following orders of stability between II–bonded base pairs

I: A–U > U–U > A–A
II: G–C > G–G > C–C
III: G–C > A–T ≃ A–U
IV: G–C > G–U ≫ G–A
V: A–U > A–C ≫ A–G
VI: C–G ≃ C–C > C–U > C–A

Calculated hydrogen bond energies of various base pair configuration are listed in Table 3.8. The values calculated by Nash and Bradley (N), Kudritskaya and Danilov (K), Pullman and Pullman (P), Claverie (C), and Rein (R) are listed for each type of base pairing studied. Upon scrutiny of this table, several generalizations can be made.

For cases where several configurations of a given pair were taken into consideration by more than one author, there is a general agreement as to which scheme is more stable, except most notably for the case of AU. It can be seen that the Watson–Crick forms of hydrogen bonding are favoured for both G–C and A–T configurations. Also, all authors listed G–C as the most stable base pair. This certainly conforms to experimental findings.

This table also shows that although Watson–Crick configurations are usually favoured for each pair, there does not seem to be sufficient differences in the energetics between non-complementary and complementary pairing schemes. This implies that there is insufficient recognition potency in

Table 3.8. Calculated energies of various H-bonding schemes

Pair	Configuration	No.	E_{total} (kcal/mole) N	K	P	C	R
AA	N6(H) . . . N7 N7 . . . (H)N6	1	−3·48				
		2		+0·35			
	N6(H) . . . N1 N1 . . . (H)N6	3	−3·25				
		4		−5·25			
	N6(H) . . . N7 N1 . . . (H)N6	5	−3·55				
		6		−5·60			
	not specified (one of the above)	7			−5·8		
	not specified (H-bond w/glyc. N)	8			−8·13		
GG	O6 . . . (H)N1 N1(H) . . . O6	9	−14·38				
		10		−16·04			
	N2(H) . . . N7 N2(H) . . . O6 N1(H) . . . O6	11	−11·26				
	N2(H) . . . N7 N1(H) . . . O6	12		−10·98			
	N2(H) . . . N3 N3 . . . (H)N2	13	−3·15				
		14		−6·36			
	N2(H) . . . O6 N1(H) . . . N7	15	−11·72				
	not specified	16			−14·5		
CC	N3 . . . (H)N4 O2 . . . (H)N4 N4(H) . . . N3 N4(H) . . . O2	17	−9·84				
	N3 . . . (H)N4 N4(H) . . . N3	18		−10·73			
	N3 . . . (H)N1 N4(H) . . . O2	19		−13·72			
	not specified	20			−13·0		
TT	O2 . . . (H)N3 N3(H) . . . O2	21		−3·73			
	N3(H) . . . O4 O4 . . . (H)N3	22		−5·21			
	not specified	23			−5·2		
UU	N3(H) . . . O4 O4 . . . (H)N3	24	−7·03				
		25		−5·42			
		26					−6·1
	O2 . . . (H)N3 N3(H) . . . O2	27	−6·56				
		28		−3·89			
	N3(H) . . . O2 O4 . . . (H)N3	29	−6·61				

Table 3.8. (*Continued*)

Pair	Configuration	No.	E_{total} (kcal/mole) N	K	P	C	R
AG	N6(H) . . . O6 N1 . . . (H)N1	30	−6·61				
		31		−9·40			
		32			−7·5		
		33					−13·41
	N6(H) . . . O6 N7 . . . (H)N1	34	−5·45				
		35		−5·86			
	N1 . . . (H)N2 N6(H) . . . N3	36	−4·05				
AC	N7 . . . (H)N4 N6(H) . . . N3	37	−5·06				
		38		−8·30			
	N1 . . . (H)N4 N6(H) . . . N3	39	−4·89				
		40		−7·32			
	not specified	41			−7·8		
AT	N6(H) . . . O4 N1 . . . (H)N3 (Watson–Crick)	42		−7·00			
		43			−5·65		
		44				−2·86	
		45					−8·04
	N6(H) . . . O2 N1 . . . (H)N3	46		−6·78			
	N6(H) . . . O4 N7 . . . (H)N3	47		−6·44			
	N6(H) . . . O2 N7 . . . (H)N3	48	−6·67				
AU	N6(H) . . . O4 N1 . . . (H)N3 (Watson–Crick)	49	−5·65				
		50		−7·21			
		51			−5·58		
		52					−6·03
	N6(H) . . . O2 N1 . . . (H)N3 (reversed Watson–Crick)	53	−5·22				
		54		−6·97			
		55			−4·94		
	N6(H) . . . O4 N7 . . . (H)N3 (Hoogsteen)	56	−5·59				
		57		−6·61			
		58			−6·96		
	N6(H) . . . O2 N7 . . . (H)N3 (reversed Hoogsteen)	59	−6·07				
		60		−6·85			
		61			−6·74		

GC	O6 . . . (H)N4 N1(H) . . . N3 N2(H) . . . O2 (Watson–Crick)	62	−15·03				
		63		−16·79			
		64			−19·18		
		65				−16·02	
		66					−22·53
	O6 . . . (H)N4	67		+1·51			
	N1(H) . . . O2 N2(H) . . . N3	68			−12·85		
	N1(H) . . . O2 N2(H) . . . O2 N3(H) . . . N3	69	−10·08				
	N7 . . . (H)N3 O6 . . . (H)N4	70			−5·75		
	N7 . . . (H)N4 O6 . . . (H)N3	71			−3·67		
GT	N1(H) . . . O2 O6 . . . (H)N3	72		−7·74			
	N1(H) . . . O4 O6 . . . (H)N3	73		−9·86			
	not specified	74			−7·4		
GU	N1(H) . . . O2 O6 . . . (H)N3	75	−8·47				
		76		−7·78			
		77					−14·26
	N1(H) . . . O4 O6 . . . (H)N3	78	−9·81				
		79		−9·84			
CT	N3 . . . (H)N3 N4(H) . . . O4	80		−3·67			
	N3 . . . (H)N3 N4(H) . . . O2	81		−4·40			
	not specified	82			−6·5		
CU	N3 . . . (H)N3 N4(H) . . . O4	83	−4·77				
		84		−3·85			
	O2 . . . (H)N3 N3 . . . (H)N3 N4(H) . . . O2	85	−4·51				
	N3 . . . (H)N3 N4(H) . . . O2	86		−4·57			

In summary of the table, here are the relative orders of stability given by the various authors

N Nash and Bradley (1966)

$$GC > GG \gg CC \cong GU > UU > AG > AU > AC > CU > AA$$

K Kudriatskaya and Danilov (1976)

$$GC > GG > CC \gg GT \cong GU > AG > AC > AU > AT > AA > UU > TT > CU > CT$$

P Pullman and Pullman (1969)

$$GC > GG > CC \gg GT > AA > AC > AG > AU > CT > AT > TT$$

R Rein, Coeckelenbergh, Egan (1975)

$$GC \gg GU > AG > AT > AU > UU$$

the nature of the hydrogen bond itself to account for the efficiency of the replication process in DNA.

An attempt was made to correlate the relative stability of the theoretical energies with those of experiment. We must be cognizant of the fact, however, that in so doing we are comparing *in vacuo* interaction energies and their concomitant stabilities with those involved in solvent experiments. To this end we have constructed a table consisting of the six different orders of relative experimental stability. We then extracted the Watson–Crick pairs from each author's list and for non-complementary base pairs we selected the most stable configuration for comparison where more than one such scheme was provided. Table 3.9 provides a qualitative measure of the

Table 3.9. Table of correlation between experiment and theory

Experimental stability	N	K	Author P	R
A–U > A–A > U–U	NO	YES	NO	YES
G–C > G–U > C–C	YES	YES	YES	YES
G–C > A–T ≃ A–U	YES	YES	YES	YES
G–C > G–U ≫ A–G	YES	YES	YES	YES
A–U > A–C > A–G	YES	NO	NO	NO
G–C ≃ C–C > C–U > A–C	YES	NO	YES	insufficient data

N = Nash and Bradley (1966)
K = Kudritskaya and Danilov (1976)
P = Pullman and Pullman (1969)
R = Rein, Coeckelenbergh, Egan (1975)

The first row indicates poor performance in two out of four cases. Nash and Bradley data, because their most stable U–U configuration is stronger than their A–U. For Pullman data, because their A–A appears stronger than their A–U.

The next row of poor conformance is indicated by row 5. This has occurred for the following reasons: the most stable A–C pair for Kudritskaya is lower in energy than his A–U. This is a true case also for Pullman data, in case of Rein data on the basis that his G–A energies are lower than either A–U or A–T; Nash and Bradley are in conformance.

conformance between theory and experiment for these cases. If the ranking conforms a 'yes' is entered into the table, where there is no conformance or partial conformance a 'no' is entered. Claverie's data was not included in the Table 3.9 because of too few entries.

If one grants the validity of this qualitative exercise and if one can trust the results of experiment concerning preferred stabilities, then the results of Nash and Bradley seem slightly better in correlation with experiment. If this is so, then the electrostatic model of hydrogen bonding is further promoted.

The lack of conformance with experiment in several of the cases may point to a need for further consideration of solvent effects.

C. Base stacking

Overlapping arrangements of parallel planar molecules at Van der Waals distances of separation are generally referred to as stacking. Base stacking is an important feature of the DNA double helix. It is also observed in polynucleotides, in most crystals of purines, pyrimidines, nucleosides and nucleotides, as well as in aqueous aggregates of these molecules. Stacking interactions are among the factors influencing the secondary structure of nucleic acids and conferring stability to these structures. In view of the intimate relation between secondary structure and biological activity of nucleic acid, the nature and specificity of stacking forces is likely to exert an influence on the mechanism of many important biochemical and genetic processes. During the last decade this rationale has prompted a long series of investigations of the phenomena of stacking of nitrogen bases. Stacking patterns in the crystalline state have been studied extensively by X-ray crystallographic methods. Bugg *et al.* (1971) have reviewed 70 published crystal structures and analysed the pattern of base stacking for reoccurring features. NMR studies of base stacking in aggregates and complexes of bases and oligo and polynucleotides in aqueous solution have been carried out by Ts'o and coworkers, and reviewed by Ts'o (1970). A review of the theoretical literature of stacking until 1968 has been presented by Pullman (1968). Papers by Claverie (1968) and by Rein (1973) review some aspects of the theory of stacking interactions. A most informative paper on stacking has been published very recently by Polozov *et al.* (1975). There are a number of questions concerned with stacking, which are of general interest, and which many of the studies and reviews have aimed to enlighten.

The most interesting among these problems are: (a) the characterization of the bonding pattern or directionality of stacking specificity; (b) the elucidation of the relative stabilities or compositional dependencies of the stacking of nucleotide bases and base analogues; (c) a comparison of the magnitude of stacking forces with the magnitude of other forces influencing nucleic acid structure; (d) the thermodynamic parameters of stacking; (e) the elucidation of the nature of forces responsible for stacking; (f) the explanation of environmental effects on stacking in terms of the bonding concepts; (g) the illustration of the direct role which stacking interactions may play in effecting fundamental biological processes. This section is designed to discuss in further depth the above and related questions in view of the existing literature of stacking and to summarize the conclusions which can be reached at the present stage.

D. Bonding patterns and directional specificity of stacking interactions

The conclusions of Bugg *et al.* (1971) in the analysis of the base stacking patterns of over 70 crystal structures are summarized as follows: Among the examined structures there are several re-occurring patterns which persist in various crystal environments. A typical pattern is composed of a partial overlap of the hetero substituent of one of the bases, such as the carboxyl or the amino group, with the imidazole ring of the other base. Bugg *et al.* (1971) interpret the interactions in these patterns as arising from induction (polarization) forces between the electro-negative substituent and the aromatic system of the respective bases. To support this interpretation these authors invoke several related experimental observations. It has been observed (Ts'o *et al.*, 1963; Ts'o, 1968) that the degrees of association between bases (and between nucleosides), in aqueous solution are, in order of decreasing strength: purine–purine, purine–pyrimidine, and pyrimidine–pyrimidine. Purines are more polarizable than pyrimidines. These facts are considered by the authors as an indication of a correlation of stacking energies with base polarizabilities. Based on this correlation the authors conclude that the dominant forces responsible for stacking are the inductive (polarization forces). Convincing arguments can be put forward against this interpretation, both on general theoretical grounds and in view of specific results. Inductive forces even on the level of the molecular dipole approximation should be correlated not only with the polarizabilities, but also with the respective molecular dipole moments and polarizabilities of the interacting molecules. In fact, the polarizabilities would be more indicative of dispersive forces rather than of inductive forces. Moreover, it has been clearly demonstrated in several studies that the dipole approximation is not applicable for stacking distances, for the calculation of electrostatic terms, nor for the second order interaction terms (Rein, Claverie and Pollack, 1968; Rein, 1973; and references there in). Stacking interactions have been studied extensively by more refined methods than the dipole approximation. This includes the whole range of methods such as the monopole atom polarizability multipole bond polarizability methods and method based on non-empirical perturbation theory of Murrell, Randic and Williams (1965) (Claverie, 1968; Pullman, 1968; Rein *et al.* 1969, 1975; Kudritskaya and

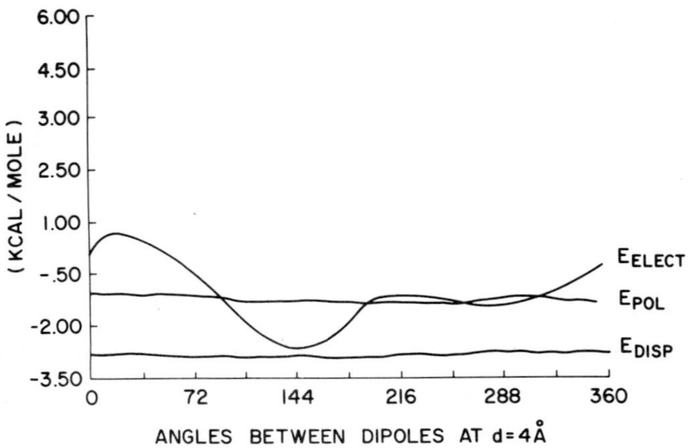

Figure 3.7. Electrostatic, polarization, and dispersion energy behaviour in stacked [A, T] rotations

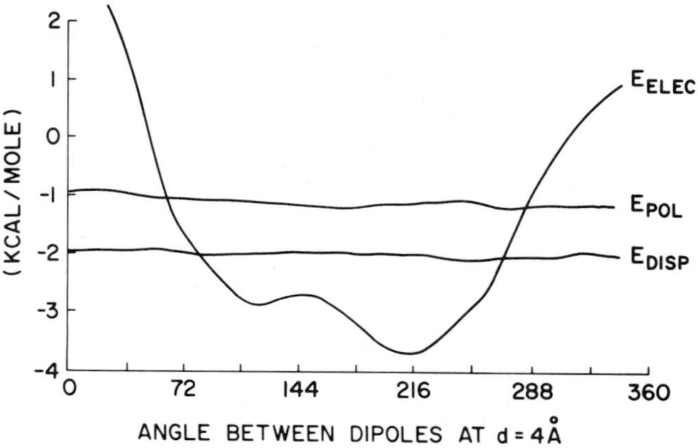

Figure 3.8. Electrostatic, polarization, and dispersion energy behaviour in stacked [G, C] rotations at (O, O)

Danilov, 1976; Polozov *et al.* 1975; and Fugita, Imamura and Nagata, 1974). Without exception these studies indicate that the inductive forces are of lesser importance than the corresponding electrostatic and dispersion forces. Results obtained in several studies of the dependence of stacking forces on the configuration of the interacting bases i.e., base pair separation and gliding as well as orientational angle (Pullman *et al.* 1969; Egan *et al.* 1975), and the decomposition of the stacking interactions into their components, have consistently demonstrated that the position of the energy minima is dominated by the electrostatic component of stacking with the polarization and dispersion term having only a very weak dependence on orientation. This is illustrated in Figures 3.7 and 3.8.

E. Configurational dependence of stacking in the double helix

The three most important degrees of freedom in the configuration of stacked base pairs in the double helix are: rotation about the helix axis of the far pair relative to the vicinal, vertical translations, and horizontal translations. The first two degrees of freedom can be combined into a helical movement of the pairs relative to each other. Such a movement would occur if the length of the sugar phosphate strand remains invariant under rotation. Claverie (1968) has calculated the stacking potential for helical movements in ten distinguishable combinations of pairs. Some of the results of Claverie's calculations are reproduced in Figure 3.9.

The minima of these potentials appear to correctly predict both the base pair separation distance and the helical pitch corresponding to DNA in the B form. This may lead one to assume that stacking interactions are of dominant importance in establishing the secondary structure of DNA. However, it may be somewhat misleading to accept these calculations as evidence to this effect, because they do not allow for any variation in the length of the backbone. The minima appear to be partly a consequence of

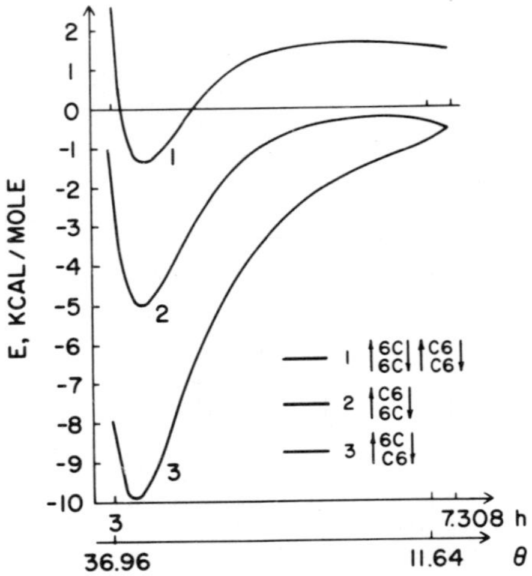

Figure 3.9. The interaction energy between two consecutive base pairs in a 'helical' movement (after Claverie, 1968)

the coupling of two modes of motion. That is, a negative (counterclockwise) effects a vertical separation of the bases, thereby contributing to the attractive energy due to the vertical mode. A rotation in the positive direction compresses the base pairs, leading to an exponential increase of the repulsion in the short range domain. Thus altogether the calculated minima may be an artifact due to the assumed fixed length of the backbone. More recent studies by Fugita, Imamura and Nagata (1974) and by Polozov *et al.* (1976) are more informative in this respect; these authors have studied the effect of the rotation and vertical movement separately. Indeed for the rotational movement, several minima are observed of which only one is in the neighbourhood of the one occurring in the DNA helix (Figures 3.10, 3.11)

Some of the energy minima positions are not compatible at all with the backbone constraint, such as the one occurring in the vicinity of 140° of rotation of one AU pair over another. These results illustrate very clearly the interplay of the various force fields in the establishment of the double helical structure. In this particular case, it is safe to assume that this interplay is between stacking and backbone conformational preferences.

F. Nature of the forces involved in base stacking

The elucidation of the nature of the forces involved in base stacking is interesting not only from the theoretical point of view, but also for an understanding of the sequence dependent structural features of nucleic acids.

A detailed analysis of data obtained from theoretical calculations seems to support the following generalizations: (1) The largest attractive contribution

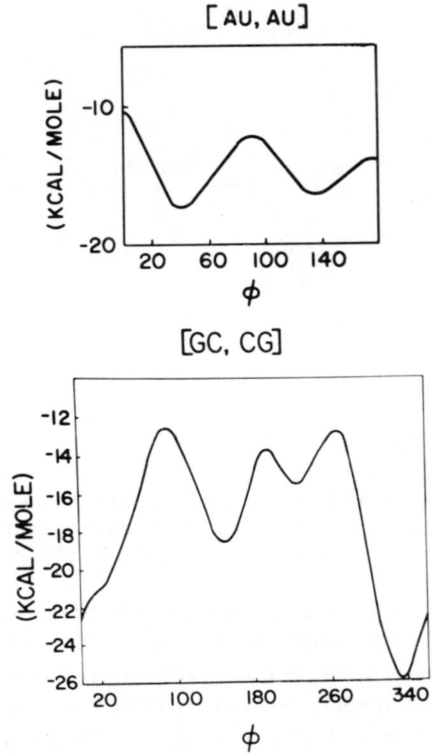

Figure 3.10. The dependence of stacking energy on angle of rotation, ϕ (after Polozov *et al.*, 1975)

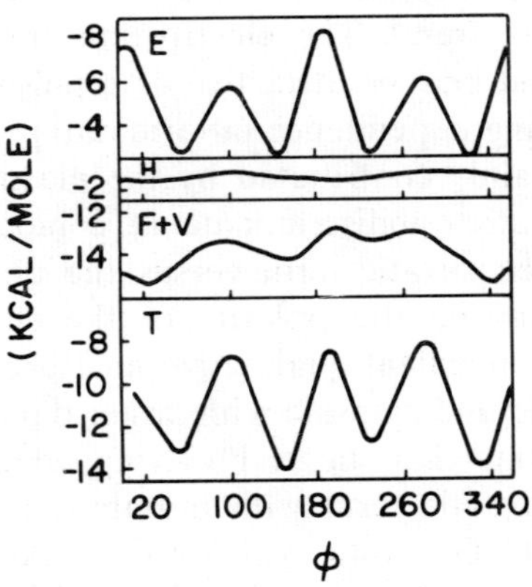

Figure 3.11. The dependence of the total interaction energy between base pairs and its terms on ϕ (combination of pairs [AU, UA], $d = 3{\cdot}2$ Å) (after Polozov *et al.*, 1975)

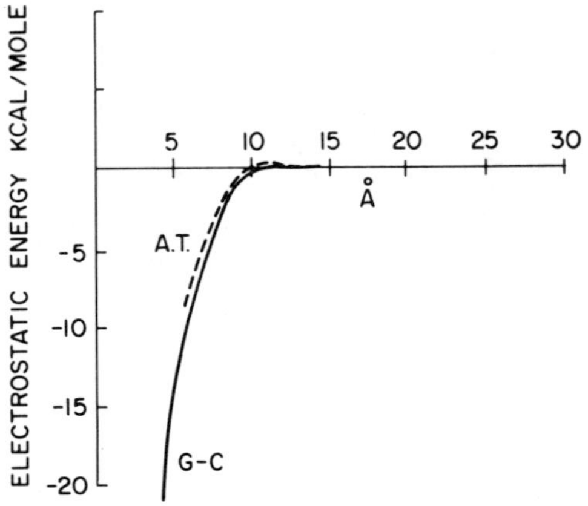

Figure 3.12. Electrostatic energy behaviour of Watson–Crick pairs with respect to centre-of-mass separation along the bond axis

to stacking arises in most of the cases from dispersion. The next attractive term in importance is the polarization term. (2) The dispersion and polarization contributions vary monotonically with the rotational angle in a helical arrangement. (3) The dispersion term is dependent on the extent of base overlap. Given that the geometry is consistent, the purine–purine configurations reveal the largest dispersion term with purine–pyrimidine, pyrimidine–pyrimidine following. This trend is reflected in the total stacking energy between two units if we do not include those configurations where the electrostatic term is the largest. (4) The short range repulsion term follows more or less the same trend. The electrostatic term changes sign several times with rotation and may be attractive or repulsive in the mini-helix. (5) Thus the major stacking dependence on orientation and composition of the respective bases is mainly contributed by the electrostatic term, the other contributions being rather indiscriminate in comparison. (6) The relative importance of the electrostatic term versus the uniform and monotonous attractive part depends on the polarity of the charge distributions. The dipole moments are convenient qualitiative measures for this. Thus interactions involving guanine and cytosine with larger dipole moments give rise to large electrostatic terms than those involving adenine and thymine. The changes in the electrostatic term with orientation would therefore reflect itself more for the G–C, G–C dimers than for the A–T, A–T dimers with the mixed cases somewhere in between (Figure 3.12). The structural consequences of the above generalization will be discussed in Section VII.

G. Composition dependence of stacking

Differences in the pattern of stacking and stabilization for various base compositions, may be the source of some variability in the secondary structure of nucleic acids. It has been suggested that this structural variation may provide the recognizable features of nucleic acids necessary to proteins for genetic regulation and enzyme specificity (Bugg *et al.* 1971; Carter and Kraut, 1974; Rein *et al.* 1970; Singer *et al.* 1963). Due to this fundamental biological significance the question of the relative stability for various combinations of bases in the stacked pairs is of great interest.

There are several sources of experimental information relevant to this question: the phenomena of aggregation of purines, pyrimidines and a variety of bases in aqueous solutions has been well investigated by NMR and thermodynamic methods (Ts'o, 1964; and references therein). These studies have shown that the relative stabilities follow the order: purine–purine, purine–pyrimidine, pyrimidine–pyrimidine. Stacking interactions have been studied further in single stranded oligo and polynucleotides (Warshaw and Tinoco, 1965; Cantor and Tinoco, 1965; Fasman *et al.* 1964; Witz and Luzzati, 1965). Claverie, Pullman and Caillet (1966) have calculated the respective stacking interactions; these values are in agreement with the experimental observed order. In the case of cytosine–cytosine and uracil–uracil, the experimentally observed order is reproduced only for the syn configurations (rotation of a base by 180° around the glycosidic linkage).

Melting temperature studies of double stranded di-oligo and polynucleotides, as well as nucleic acids (Marmur and Doty, 1959, 1962; Radding *et al.* 1962; Bollum, 1965; Wells *et al.* 1905; Michelson *et al.* 1967; Von Hippel and Wong, 1964; Hamaguchi and Geiduscek, 1962 and others) have been conducted by various investigators also to assess the relative stabilities, thermodynamic parameters, and force components contributing to the stability of double helices. A variety of statistical mechanical models have been formulated to facilitate the discussion of these results (Zimm, 1960; Crothers and Zimm, 1960; Montroll and Goel, 1966; Lehman, 1964; Strassler, 1967; Reiss *et al*, 1966; Scheraga and Poland, 1970). In spite of this a systematic and detailed interpretation in terms of component interactions remains elusive. Efforts to extract stacking enthalpies for the various possible base pair combinations from melting curves of polymers of known sequences based on an extended Ising model (Montroll and Goel, 1966) have been only partially successful. The analysis (Goel *et al.* 1968; Rein *et al.* 1969) gave stacking energies for the eight base pairs (four of them combined as averages) which were almost independent of the nature of the bases and were between 5–6 kcal/mol base pair (for m base pairs m-1 stacking energies). The values obtained for the GC and AT bond energies are 1·7 and 1·4 kcal/mol. It is to be noted that according to the Ising model definition of bond energies, these latter values contain in addition to the intrinsic hydrogen bond energies, solvent effects and possibly conformation adjustment energies. Thus they should not be directly compared with calculated hydrogen bond energies. However, according to the Ising model formulation the vertical energies should correspond to the intrinsic stacking interactions and thus a comparison with calculated values is warranted.

The uniformity of the obtained stacking energies can be interpreted as an average stacking energy. It appears that the extended Ising model with the limited experimental information (available melting curves for polymers of known sequences) is not capable of resolving the sequence dependence of these interactions. Calculation of the average of the theoretical stacking energies (Rein *et al.*, 1969) gave a value of 7·3 kcal/mol. This compared moderately with the above experimental values. Considering the approximations involved both in the theoretical procedure and the Ising statistical anlysis of the experimental data, the obtained agreement within about 20 per cent of the two values seems quite satisfactory. Although the quantum mechanical calculations of 1968 could account for the general melting behaviour of DNA-like helices, with regard to the question of sequence dependence effects, the crudeness of the analysis of the experimental results did not permit clear conclusions.

Since 1968 there have been significant advances in the methodology of

Table 3.10 Stacking energies (kcal/mol)

Stacking / Mini-helix	Rein *et al.* (1968)		Coeckelenbergh (1975)		Claverie (1968)		Pullman *et al.* (1969)		Kudritskaya *et al.* (1975)		Standard deviation of ranking
	Energy	Rank	Energy	Rank	Energy	Rank	Energy	Rank	Energy	Rank	
↑ T–A × A T ↓	−5·03	2	−9·3	2	−2·5	3	−5·0	1	+0.39	1	0.84
↑ T–A × G C ↓	−4·9	1	−11·6	6	−2·4	2	−7·0	3	−0.97	3	1·87
↑ A–T × T A ↓	−5·04	3	−10·29	4	−4.5	6	−6·1	2	−2.71	7	2.07
↑ G–C × G C ↓	−5·20	4	−12·6	8	−1.26	1	−7·7	7	−0·66	4	2·77
↑ C–G × T A ↓	−6·1	5	−10·2	3	−3·5	4	−9·9	9	−0·1	2	2·70
↑ T–A × T A ↓	−6·33	6	−10·8	5	−4·0	5	−7·4	6	−1·53	6	0·55
↑ T–A × C G ↓	−9·35	8	−12·5	7	−6·0	8	−7·0	4	−2.86	8	1·73
↑ C–G × G C ↓	−9·76	9	−12·9	9	−1·99	7	−11·3	10	−0·99	5	2·00
↑ G–C × C G ↓	−12·9	10	−8·6	1	−9·8	10	−8·5	8	−6·66	10	3·90
↑ G–C × T A ↓	−9·0	7	−18·4	10	−7·3	9	−7.2	5	−4·75	9	2·00

stacking calculations. Claverie (1968) has introduced the short range repulsion term in the theory. Rein *et al.* have developed the segmental multipole series calculation of the electrostatic term. Progress has been made to account for solvent effects. In view of these developments stacking energies have been recalculated and further studied by several workers (Kudritskaya and Danilov, 1976; Rein *et al.*, 1975). In Table 3.10 we list the results of several of these calculations.

A first inspection of this table would indicate only a limited correlation between the various calculations. This is reflected both in the value of stacking and in the stability ordering (the stabilities are ranked from 1–10 as obtained by each author). However, a more careful analysis shows that there is a moderately good agreement between the various calculations for the order of stability for the ten different stacking cases.The mini-helices appear in the table in order of increasing stabilites according to the average ranking of the various authors. The last column of the table lists the standard deviation of the rank for each stacking case. It can be seen that this is approximately two. Considering the differences in the methods used by the authors and particularly the differences in the charge distributions employed in the calculation of the electrostatic terms as well as the relatively small energy differences required for a correct ordering within the types, this result is not too discouraging.

H. Thermodynamic stability of nucleic acid fragments (mini-helices)

Borer *et al.* (1974) has studied the melting behaviour of 19 double stranded oligo nucleotides of known base composition. From an analysis of this data the authors have obtained the thermodynamic parameters for mini-helix formation, i.e. the process of forming a base pair at the end of a helix. The authors present a list of free energies, enthalpies and entropies for this process for six different sequences (two of them averages). The enthalpies together with theoretical estimates (Rein, Coeckelenbergh and Egan, 1975, 1976) are listed in Table 3.11.

Borer enthalpies fall, broadly speaking, into three groups: (a) the most stable are the fragments involving GC pairs; (b) AT dimers are intermediate in stability; (c) while mixed AT, GC dimers are by about 7–8 kcal/mol less stable than those of (a). The interesting aspect of the observed stability order is that it doesn't quite parallel the hydrogen bond stabilities which would dictate that the mixed dimers with one stronger GC type hydrogen bond should be more stable than those in group (b). Also the stacking order would normally favour group (c) over group (b). Thus, neither hydrogen bonding nor stacking can provide an explanation for the preference of group (b) over group (c).

From this one may conclude that the observed thermodynamic stability must depend on additional factors. The two most likely additional factors contributing to stability are the solvent effect, and conformational energies.

Table 3.11. Enthalpies of mini-helix formation (kcal/mol)[a]

Group	Mini-helix \ Stacking	Borer *et al.* (1974)	Rein *et al.* (1975)
c	G–C ↑ × ↓ T A C–G ↑ × ↓ T A T–A ↑ × ↓ G C T–A ↑ × ↓ C G	−5·9	−9·1
b	A–T ↑ × ↓ T A T–A ↑ × ↓ A T	−6·5	−5·6
	T–A ↑ × ↓ T A	−8·2	−4·2
	G–C ↑ × ↓ C G	−13·0	−7·2
a	G–C ↑ × ↓ G C	−13·7	−11·0
	C–G ↑ × ↓ G C	−14·7	−11·5

[a] T and U are considered equivalent for this calculation.

Using our earlier estimates for these two quantities (Rein, Coeckelenbergh, Egan, 1975; Coeckelenbergh 1976, see also Section 6) and stacking and hydrogen bond energies for the various cases (Coeckelenbergh 1976; Egan, 1976) we have calculated the mini-helix energies (column 3 of Table 3.11). The agreement of the thermodynamic values with our calculations is acceptable, however groups (b) and (c) remain out of order also in this case. What this indicates to us that there must be structural differences between the various sequences with associated energy changes. It appears that with the assumption of a uniform secondary structure for all sequences (basic to our calculations) the trend of the observed order cannot be fully reproduced. Our analysis of the components and orientation dependence of stacking has indicated that AT type stacking is likely to be least orientation dependent (smallest electrostatic term contributing to stacking), thus it is plausible that this type of stacking may put the minimum strain on the backbone with a concomitant lowering of the conformational energy for these cases. We are then suggesting that the way to explain the inversion of stability order for case (b) and (c) rests on the differences of backbone strain induced by possible torques between stacked paris, which may be larger for the more polarized GC, AT cases than for the poor AT, AT cases.

V. ENVIRONMENTAL EFFECTS

A. Solvent effects on nucleic acid stability

Most organic solvents denature DNA. Water seemed to be unique in stabilizing the double helical form. Sinanoglu and Abdulnur (1964) were the first to analyse the effect of solvent on helix coil transition. These authors have considered four terms to contribute to the unitary free energy of helix coil transition. The first term is the difference between the interaction of the double helix with the solvent and the single helix with the solvent. These terms were calculated according to continuum reaction field theory (See Sinanoglu, Abdulnur, Kestner (1964) and the references there in, Sinanoglu (1965, 1967a,b); Halicioglu and Sinanoglu (1969). The second term is a reduction of the intrinsic interactions in the helical state obtained also from reaction field considerations. Claverie (1968) has pointed out that this term may be counted twice in the original formulation. The third term is the difference in the free energies of creating the cavities for the double and the single helices in the given solvent. This term is approximated from the estimate in the change in the surface area in going from coils to helix and the surface tension of the solvent. These three terms constitute the solvent effect. The fourth term is the intrinsic free energy of the helix, i.e. the base–base interaction energies less the configurational entropy change.

The results of Sinanoglu and Abdulnur were successful in qualitatively explaining the observed solvent effect for water and a number of organic solvents. However, the solvent effect values in water of 14 kcal/mol for GC dimer and −19 kcal/mol for AT dimer formation are much too large to be reconcilable with thermodynamic values for these processes. A continuum approach which avoids the use of the surface tension and thus starts for the case of pure liquids from first principles was developed by Linder and coworkers (Linder, 1960, 1961, 1962, 1964, 1967; Linder and Rabenold (1972) with some further developments and application by Huron and Claverie (1972) and Nir (1975, 1976).

An application of the continuum model to the question of preferred conformations of solute molecules in various solvents has been given by Beveridge *et al.* (1974), Rein *et al.* (1975) and Renugopalakrishnan *et al.* (1976). In the last two studies use has been made of the detailed procedure of Hermann (1972) for calculating molecular surface areas.

A second approach emphasizes the importance of specific sites on the solute molecule, where properly oriented solvent molecules can be preferentially attached. In this approach the basic entity investigated is the supermolecule, i.e., the complex of solute molecule with the solvent molecules attached to it. Alagona *et al.* (1973) developed this approach for the hydration of formamide, and Port and Pullman (1973) performed a similar study for adenine, guanine, cytosine and thymine. The approach of the supermolecule has been presented prior to the above studies in numerous *ab initio* self consistent field calculations by the French group: Dreyfus Maigret and A. Pullman (1970), Dreyfus and A. Pullman (1970a,b), Berthold and Pullman (1972).

In order to reduce the computational efforts involved in *ab initio* calculations Alagona *et al.* (1973) carried out the computations in two steps. First, in order to search for the most probable geometries of the associations, the electrostatic contribution to the energy of the two interacting molecules was calculated. In the second stage the SCF procedure was used for a refinement. The study of Port and A. Pullman (1973) on the DNA bases included only the first stage. The justification for this short-cut was based on the previous studies of B. Pullman and A. Pullman (1969), Dreyfus and A. Pullman (1970a,b), Bonaccorsi *et al.* (1971), in which it has been shown that the electrostatic component of the interaction energy plays the dominant role in hydrogen bonding, particularly in determining the relative orientations of the two interacting molecules at equilibrium distance. However, it must be cautioned that the absence of short range repulsive interactions in exclusively electrostatic interactions gave a poorer agreement for cases involving overlapping of Van der Waals spheres (Bonaccorsi *et al.*, 1971), e.g. a situation in which water acts as a proton donor, such as the close approach of a water molecule to the carbonyl oxygen of formamide (Alagona *et al.* (1973)). Consequently, in the study of Port and Pullman (1973) a constant distance of 2.85 Å was maintained between the oxygen of water and the atoms of the base participating in the particular hydrogen bonding, whereas for other atoms a closer approach than the sum of their Van der Waals radii was forbidden.

Port and Pullman (1973) concluded that for the four bases the most favourable positions for hydration are around the base peripheries with the oxygen of water in the molecular plane. Out of plane monohydration appeared unlikely in the case of the bases. However, Port and Pullman (1973) emphasized that their calculations do not exclude the possibility of out of plane hydration bridging two molecules.

In evaluating the above studies we must appreciate the potential of the detailed knowledge of the hydration sites. The above studies on the DNA bases have not yet considered the question of whether the stacked or hydrogen bonded configuration is preferred in given solvent, e.g. water.

To calculate the solvent effect on base pair formation, Rein *et al.* (1975) have used several additional assumptions. It was assumed that the solvent effect arises principally due to the dehydration of the bases accompanying pair formation. The hydration sites on the bases are the same that will enter base–base hydrogen bonding. Thus the number of water molecules displaced are three for G and C and two for A and T, respectively. The dehydration energies are assigned values by judicial choice from the hydration energies calculated by Port and Pullman. The solvent effect is calculated according to the formula below which can be considered a simplified version of the method of Poland and Scheraga: (1965) $E_{\text{solv eff}} = E_{\text{base pair}} \ldots nH_2O$-- $(E_{\text{base 1}} \ldots n'H_2O + E_{\text{base 2}} \ldots n''H_2O)$. These calculations gave a solvent effect on the GC pair formation of 7kCal/mol and −2kCal/mol for the AT pair formation. Using these values together with calculated hydrogen bond, stacking and conformational adjustment energies, the authors have obtained a qualitiative agreement with the experimentally determined thermodynamic data of Borer *et al.* (1975) (see also Section IV).

Based on these results it would seem that the dehydration model is useful for explaining the solvent effect in helix formation.

Obvious improvements in this approach would be to consider the complete supermolecule including the interaction of water molecules with each other, instead each hydration region in isolation. Although a consideration of the above mentioned factors may require a significant increase in the complexity of the calculations such an extension is feasible. A more comprehensive solvent effect theory would have also to account for both dehydration and continuum effects.

B. Backbone conformational adjustment to helical structure

An analysis of the energetics of the backbone must include the interactions with the other units of the double helix in order to account for the actual helical conformation.

We propose, therefore, that the relative position of the stacked bases results from the compromise between stacking orientation and backbone conformation preferences. One of the immediate consequences is related to the investigation of mismatched base pairs that would depend on similar simultaneously applied forces.

We have seen previously that the double-helical stacked conformation is by no means the only stable arrangement between two superimposed bases. We will now review the results of several investigations performed on the backbone conformation in view of its relative stability.

1. Experimental investigation of polynucleotide stability

The conformational study of the monomeric unit of nucleic acids may be divided in three parts: the sugar conformation, the glycosyl bond, and the ribose phosphate backbone. The following generalizations result from the compilation by Sundaralingam (Sundaralingam, 1968) of 80 different nucleoside, nucleotide and polynucleotide observed structures:

(a) The majority of the observed conformations of furanose rings fall into two narrow sectors around the C2′ endo (or C3′ exo) and the C3′ endo forms. As a rule, ribose systems show a preference for the C3′ endo, while deoxyribose systems show a preference for the C2′ endo conformation.

(b) The conformation of the base relative to the sugar is expressed in terms of *chi* angles defining two domains: *syn* and *anti.* In polynucleotides, only *anti* conformations are observed; small *chi* values being generally related to C3′ endo, larger *chi* values being related to C2′ endo.

(c) Two angle maps may be expressed. They show the highly populated regions and they are used as reference for the theoretical calculations involving two successive angles.

2. Molecular orbital calculation on the conformation of polynucleotides

Hard-sphere calculations (Olson, 1972; Sasisekharan, 1973) have been performed to determine the angle of rotation around selected bonds. Only first-order overlap is considered by Sasisekharan and the disallowed ranges correspond therefore to the case when the different rotations are treated independently. Second-order overlap are included in the data from Olson.

Several quantum mechanical calculations were performed on molecules representing a simplified system (only part of the polynucleotide chain). The results are presented as potential energy functions of a torsional angle for a given bond, the other angles remaining unchanged, or as conformational energy maps describing the potential energy variation resulting from the similtaneous scanning of two successive torsion angles. The investigation of simultaneous variation of two torsion angles lead to the drawing of conformational energy maps showing energy minima corresponding to the most probable angle. The other angles remain rigid and their value is assigned from experimental occurrence and energy considerations.

The results from several authors (Haschemeyer, 1967; Wilson, 1969, 19ul; Lakshminaryayanan (1967, 1969, 1973; Jordan, 1968; Chatkoff, 1971; Kang, 1971; Berthod, 1971a,b, 1973; Pullman, 1972a,b; DeCoen, 1970; Govil, 1971a,b; Saran, 1971) are presented in Table 3.12. They are presented in the form of the angles limiting the energy minima domains. In order to be meaningful they should however be compared with the original conformation maps.

Olson, (1972) performed partitioned potential function calculations aimed at the reproduction of the experimentally derived characteristic ratio of polynucleotide chains. This characteristic ratio was calculated on the basis of the energy description of the successive torsion angles. Upon satisfactory agreement with the experiment, two angles energy maps were proposed and served as a basis for the proposition of stacked conformations. Little agreement was, however, obtained between these maps and crystallographic data.

3. Mini-helix configuration

The synthesis of mispaired DNA was simulated (Rein, 1975) with an accompanying search for a ribose phosphate backbone fit of the new stereochemical requirement defined by the position and direction of the glycosyl bond. Backbone conformations were obtained for two possible mismatched pairs: G–U and U–U. The obtained torsion angles fall in the observed crystallographic ranges with one exception (30° difference). The distance matrix was, however, calculated so that no steric overlap was

Table 3.12. Comparison of the minima from two angles rotational maps

Variable angles	Rigid angles	Minima	Method
ω', ω	$\phi' = 240°, \phi = 180°, \psi = 60°$	(300°, 300°), (270°, 30)	PCILO
ω', ω	$\phi' = 240°, \phi = 180°, \psi = 180°$	(300°,270°),(30°,30°)	PCILO
ω', ω	$\phi' = 240°, \phi = 180°, \psi = 180°$	(60°, 60°), (60°, 180°), (180°, 60°) (180°, 180°), (180°, 300°) (300°, 180°)(300°, 300°)	EHT
ω', ω	$\phi = 260°, \phi = 180°, \psi = 180°$	(60°, 60°), (300, 300°)	Empirical[a]
ω', ω	$\phi' = 260°, \phi = 180°, \psi = 180°$	(180°, 180°)	Empirical[b]
ω', ω	$\phi' = 260°, \phi = 180°, \psi = 180°$	(60°, 60°), (180°, 180°)	Empirical[c]
ϕ', ω'	$\omega = 300°, \phi = 180°, \psi = 60°$	(180°, 270°), (180°–240°, 270°)	PCILO
ϕ', ω'	$\omega = 300°, \phi = 180°, \psi = 180°$	(240°, 300°), (240°, 180°)	EHT
ϕ', ω'	$\omega = 180°, \phi = 180°, \psi = 180°$	$\phi' = 180°, \phi' = 300°$	Empirical
ω', ϕ	$\phi' = 240°, \omega' = 300°, \omega = 60°$	(270°–300°, 180°)	PCILO
ω, ϕ	$\phi' = 240°, \omega' - 300°, \omega = 180°$	(180°, 120°), (180°, 180°) (300°, 180°)	EHT
ω, ϕ	$\phi' = 260°, \omega' = 180°, \omega = 180°$	(60°, 60°), (60°, 180°), (60°, 300°) (180°, 60°), (180°, 180°) (180°, 300°), (300°, 60°), (300°, 180°), (300°, 300°)	Empirical
ϕ, ψ	$\phi' = 240°, \omega' = 300°, \omega = 300°$	(180°, 60°), (180°, 180°)	PCILO
ϕ, ψ	$\phi' = 240°, \omega' = 300°, \omega = 300°$	(180°, 60°), (180°, 180°) (180°, 300°)	EHT
ϕ, ψ	$\phi' = 260°, \omega' = 180°, \omega = 180°$	(180°, 60°), (180°, 180°) (180°, 300°), (300°, 300°)	Empirical

For references to the methods, see text.
[a] empirical—non-bonded and torsional energy.
[b] empirical including electrostatic interaction.
[c] empirical including electrostatic interaction.

tolerated. The difference between the conformational energy in the mispaired case versus the minimum conformational energy of the backbone is called energy of conformational adjustment.

4. Interpretation of the conformation energy maps

The methods of calculation are different and the obtained results are often different, if not contradictory. Qualitative interpretation is, therefore, the best that may be obtained at the present time. One way to test the conformational results obtained from quantum mechanical calculations is to compare them with the conformations observed by crystallography. Such a satisfactory agreement is obtained from the PCILO calculations (Perahia, Saran, Pullman, 1973). On the bases of two angle conformational maps, Pullman proposes a set of most stable torsion angles. $\omega, \omega' = 300, 300$; $\psi = 60$; $\phi = 180$; $\phi' = 240$. Assuming that two angle conformation energies are additive, and limiting our calculation to the couples ω, ω'; ϕ', ω'; ϕ, ψ; ω, ϕ; we would obtain a conformation energy of 1·0 kcal for the most stable set of torsion angles.

The angles found in normal polynucleotides are different, and it seems therefore that the conformation of, say DNA B, is not exactly the most stable as far as only the backbone is concerned.

On the basis of the two angle conformational energy maps, an additional, but somewhat questionable, approximation must be made in order to compare a given backbone conformation to the proposed most stable one. This approximation is to neglect the influence of the remaining angles on the conformational maps, and use those conformational maps having the closest fixed angles to those of the conformation that we investigate. Limiting ourselves to the same four coupples, ω, ω'; ϕ', ω'; ϕ, ψ'; ω, ϕ, we obtain a conformational adjustment energy of 3·0 kcal for DNA B, 6·0 kcal for the mismatched pair G–U, and 7·0 kcal for the mismatched pair U–U. The contribution from each available couple is presented in Table 3.13.

Therefore, if we want to know the conformation adjustment energy of DNA B versus the reference minimum energy we have to do the following operations:

(1) (Total angle-couples contribution for DNA B)-(Total from reference)

$$3{\cdot}0\ \text{kcal} - 1{\cdot}0\ \text{kcal} = 2{\cdot}0\ \text{kcal}$$

(2) Add 4·5 kcal for the x angle conformation energy in C2′-ENDO sugar conformation.

(3) Multiply $2{\cdot}0\ \text{kcal} + 4{\cdot}5\ \text{kcal} = 6{\cdot}5\ \text{kcal}$ by 2 in order to obtain the conformation adjustment energy for the two strands = 13·0 kcal.

This result, because of the approximations previously mentioned, has however to be considered more for qualitative than quantitative purposes. This would, on a qualitative basis, emphasize the initial proposition that the actual polynucleotide conformation results from the compromise between a series of forces that include the backbone conformational stability.

Two other conclusions may be drawn from the preceding. First, even if the preferred conformations of the backbone and the preferred orientation of two stacked bases are not exactly compatible, they are at least not drastically different, and this may constitute an essential factor for the remarkable conservational stability of the DNA double helix. Secondly, it

Table 3.13. Conformation energy for angle couples in kcal/mol for their respective values in proposed mini-helix conformations. Energy values taken from Perahia, Saran and Pullman, 1973.

Angles	Reference	DNA–B	G–U	U–U
ω, ω'	0·0	0·0	1·0	1·0
ϕ', ω'	1·0	2·0	3·0	3·0
ϕ, ψ	0·0	0·5	1·0	2·0
ω, ϕ	0·0	0·5	1·0	1·0
Total	1·0	3·0	6·0	7·0

would seem more efficient to investigate the detailed conformation pertaining to a specific biological situation and then calculate the corresponding energy rather than depend on very drastic approximations in the use of existing conformational energy maps. Such calculations would give us a quantitative description of the forces involved in the double helix stability.

VI. INTERACTIONS AND MECHANISMS OF BIOLOGICAL PROCESSES

A. Nucleic acid base tautomerism

The Watson–Crick model of DNA (Watson and Crick, 1953a,b) depends on the unique pairing of complementary purine and pyrimidine bases in their usual tautomeric forms (Fig. 3.3). It was also suggested by Watson and Crick (1953b,c) that the existence of bases in their rare tautomeric forms may lead to miscouplings of the bases (Figure 3.13) and thus to perturbation of the genetic code. Although tautomeric forms of the bases can arise either (a) from prototrophy involving an extra-ring atom, or (b) protrophy involving only ring atoms, the latter type do not directly alter the normal pairing scheme (except, however, for the ring nitrogens in positions 1 and 4 of the purines and pyrimidines, respectively) and are thus not further considered in our discussion.

It is generally believed that rare tautomers may result either from (a) chemical rearrangement, or (b) proton tunnelling in normal base pairs. In order to establish the significance of tautomerism as a mode of mutagenesis, we must therefore establish the probability of the nucleic acid bases to exist in their rare forms.

1. Experimental and theoretical basis of tautomerism

The specific pairing of adenine with thymine and of guanine with cytosine, characteristic of the Watson–Crick model of DNA, is dependent on the predominance of the lactam and amino tautomers of the bases over lactim and imine forms (Pullman and Pullman, 1959). That this is actually the case

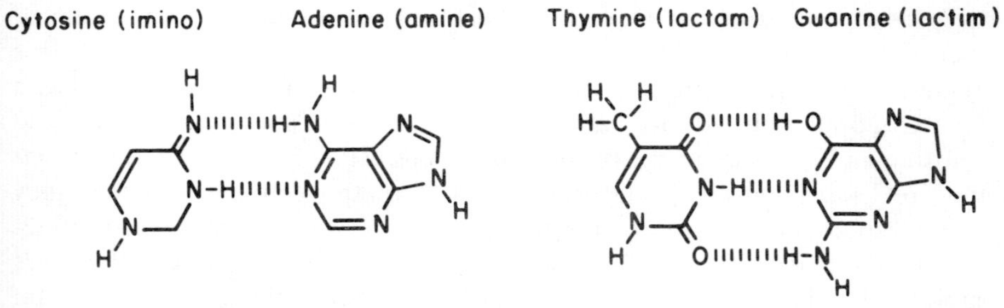

Figure 3.13. Examples of rare tautomeric forms involved in miscoupling of DNA bases

has been substantiated by numerous physico-chemical techniques, such as infrared, Raman, and ultraviolet spectroscopies, NMR, measurements of ionization constants and X-ray crystallography (Pullman and Pullman, 1971; Kwiatkowski, Pullman (1975) and references therein).

The tendency of the bases to exist in their rare forms has been shown to be quite limited by numerous experimental techniques. For instance, the lactam–amine form of cytidine seems to prevail over its lactam–imine tautomer with a factor $\sim 10^5$ according to the evidence presented in the above review. There were a few conflicting reports of the higher abundance of rare forms, however these reports have since been re-interpreted or retracted. For example, Morita and Nagakura (1968) have suggested that observed electronic absorption spectra of cytidine implies the existence of an imino tautomer with an equilibrium constant of 0·03 in aqueous solution at 30°C, but Jonnson *et al.*, 1971 suggest that these spectral characteristics are due to the existence of solvent–solute complexes in solvents capable of donating a proton. On the basis of NMR studies it has been suggested that the imino tautomer of cytosine (Lee *et al.*, 1972) and the lactim tautomer of guanosine 2′-monphosphate (Lee and Chan, 1972) are each present to an extent of about 15 per cent at room temperature in aqueous solutions. However, in a later publication this interpretation has been retracted as erroneous and the observed results have been attributed to the presence of paramagnetic impurities (Pieber *et al.*, 1973). Also on the basis of IR studies, Miles (1961) has concluded that the lactim tautomer of guanine is negligible under ordinary conditions. Further support for the dominant tautomer of cytosine (the lactam form) comes from analysis of ^{14}N NQR (nuclear quadrupole resonance), UV absorption, ionization constant, and other methods reviewed by Kwiatkowski and Pullman (1975).

The predominance of the diketo form of thymine and uracil according to the reviewed experimental evidence also seems to be firmly established. The diketo form for example is more abundant then the lactam form by a factor of $\sim 10^4$ (Katrizky and Waring).

The relative stability of the lactam and amino forms of the bases have also been confirmed by quantum mechanical calculations. In the earlier calculation the molecular energies of the different tautomeric forms of the bases have been approximated by adding their resonance energies to the sum of their bond energies (Pullman and Pullman 1962, 1963; Pullman, 1964). This analysis concerned itself also with the question of which of the bases has a greater population of rare imino and lactim forms the results predicted cytosine and guanine, respectively. This is interesting in view of the fact that the G–C pairs constitute the unstable part of the genome and that they mutate spontaneously more frequently than the A–T pairs (Freese, 1963, Drake, 1966). Danilov's (1967) calculation for tautomeric forms of the bases showed also that the equilibrium is greatly in favour of the amine and lactam forms, although cytosine and guanine are more likely than adenine and thymine to exist partly in the imine and lactim forms respectively.

Molecular orbital studies by Pullman and Pullman (1962) and Lowdin (1965) have shown that the resonance energy of the imino form of cytosine is about 2 kcal/mol higher than that of the amino tautomer, while Bodor *et al.* (1970) predicted on the basis of SCF–MO calculations that the difference in the heats of atomization between the two tautomers of cytosine is also about 2 kcal/mol higher in the amino tautomer. Numerous other theoretical calculations (reviewed in Kwiatkowski and Pullman, 1975) also suggest that the lactam tautomer of cytosine is the most stable. The CNDO/2 calculations of Breen and Flurry (1971) is the only exception in indicating that the lactim tautomer of cytosine as more stable than the lactam form; however, because of uncertainties in the geometry their SCF–CI calculations were unable to draw a clear distinction between these two tautomers.

It was pointed out by Lowdin (Lowdin, 1963, 1964a,b, 1965) that proton tunnelling may be a mechanism for tautomeric interconversion and that the A–T pair is considerably more stable than the G–C pair to proton tunnelling. Thus in an attempt to understand the relationship between tunnelling, tautomeric interconversion, and mutagenesis Ladik (1964a,b), Rein and Ladik (1964), Rein and Harris (1964a,b, 1965a,b, 1966), and Lunell and Sperber (1967) studied various molecular orbital properties of the hydrogen bonds in the base pairs. These studies demonstrated that tautomeric interconversion resulting from proton tunnelling in ground state base pairs is very rare. In effect these studies confirmed further the exceedingly low population of the rare tautomeric forms.

In summary it seems that both the existing theoretical and experimental evidence indicates that the population of rare tautomeric forms of the bases is likely to be at least by a factor of 10^4 less than the normal forms. This fact should be borne in mind in any speculations about the implication of rare forms in mutational mechanisms.

2. Interaction studies and biological consequences of tautomerism

Interaction energy calculations for various tautomeric forms in different base pairing schemes have produced interesting results. For instance, Pullman and Caillet (1967) have shown, based on Van der Waals–London interaction energies that miscouplings containing the rare forms of guanine or cytosine should be appreciably less stable than those containing the rare forms of adenine or thymine (Table 3.14) and thus the former ones may therefore introduce local elements of instability into the DNA. Rein and coworkers (Stamatiadou, unpublished data) have determined the interaction energy for tautomers in Watson–Crick pairs, miscoupled pairs, and Hoogsteen pairing configurations. Their results (Table 3.15) are qualitatively similar to those of Pullman *et al.* However because of the very low (perhaps negligible) abundance of the rare tautomers participating in these pairing schemes, great caution should be exercised before attributing biological role to these interactions. Rein et al. (1975) have suggested recently, instead mismatched

Table 3.14. Interaction energies in the Watson–Crick and miscoupled base pairs (from Pullman and Caillet, 1967)

Base pair	Total energy (kcal/mol)
A–T	−5·65
G–C	−19·18
A*–C	−19·44
C*–A	−9·24
G*–T	−7·99
T*–G	−17·45

pairs of normal tautomers as point mutational intermediates. The involvement of such pairs in processes leading to mutation would depend especially on their adaptability to the double helical configurations as discussed in detail in other sections of this paper.

Most recently it has been shown that the mutagenic effect of the base analogue 2-aminopurine, can be explained as incorporational and replicational errors involving only normal tautomeric form of the analogue and not its rare form (Rein and Garduno, 1976).

Thus, in summary, tautomerism may neither be ruled out completely, from involvement in certain aspects of mutagenesis, nor does present data permit one to satisfactorily confirm its involvement.

B. Aromatic binding to nucleic acids and their biological significance

Introduction. The complexes between DNA and various aromatic molecules have been widely studied in order to obtain information on the mechanism

Table 3.15. Electrostatic energy of various base pair types (from Stamatiadou *et al.*, unpublished data)

Base pair	Energy (kcal/mol)
A–T (WC)	−9·03
G–C (WC)	−22·20
A–G* (WC)	−9·06
G–T* (WC)	−11·90
G–C (Hoogsteen)	−7·29
A–T (Hoogsteen)	−10·75
G–C* (Hoogsteen)	−18·30
A–A* (WC)	−6·75
C–C (WC)	−7·20
T–T* (WC)	−2·29
A–C (WC)	−2·87

(WC) indicates normal Watson–Crick type pairing, while (Hoogsteen) indicates base pairing according to the Hoogsteen system.
* Rare tautomer.

of the interaction which may initiate various biological consequences including mutagenesis and chemical carcinogensis. The first stage of biologically significant complex formation involves physical interactions which apparently facilitate the formation of specific, chemical linkages.

Since the extensive studies of Brock and coworkers (1938), and Weil-Malherbe (1946), it has been known that aromatic hydrocarbons may interact with purines and pyrimidines and be solubilized by them. More recently, the extension to the solubilization or intercalation of aromatic dyes (Lerman, 1961, 1963) and hydrocarbons (Liquori, 1962; Boyland and Green, 1962) by nucleic acids has been proposed. For instance, the combination in solution of DNA with small amounts of acridines, which intercalate between adjacent nucleotide-pair layers by extensive unwinding of the deoxyribose–phosphate backbone, result in enhanced viscosity and diminution of the sedimentation coefficient of the DNA (Lerman, 1961, 1963). (Although aromatics can apparently bind to DNA by an external mechanism, this usually occurs only at higher concentrations of the aromatic (reviewed in Liquori *et al.*, 1969).)

Numerous studies (reviewed in Leng *et al.*, 1973) show that aromatic molecules may bind to single-stranded polymers of nucleotides in a fashion similar to that observed in double-stranded DNA molecules. Detailed X-ray analysis of co-crystallized complexes confirms the existence of both types of intercalation models. The first has been demonstrated between solutions of a fully methylated purine, tetramethyluric acid (TMU) and 3,4-benzpyrene (Damiani *et al.* 1966, 1967). This complex consists of parallel stacks of a benzpyrene molecule and two TMU molecules with each adjacent ring system 3·45 Å apart. This model is consistent with hypotheses (Blake and Peacock, 1968; Pritchard *et al.*, 1966) suggesting that aromatic dyes can intercalate in a single-stranded DNA, but that electrostatic attraction between positive centres on the dye and negative centres of the backbone may also be involved.

The second type of intercalation model has been observed in co-crystallized complexes with 9-aminoacridine (Seeman *et al.*, 1975) or ethidium bromide (Tsai *et al.*, 1975). For instance, the complex between 9-aminoacridine and dinucleoside phosphate adenylyl-3′, 5′-uridine (ApU) consists of adjacent layers of hydrogen bonded adenine and uracil with a 9-aminoacridine molecule sandwiched between the base pairs and stacked parallel to them. This model supports earlier hypotheses suggesting the aromatic interacts almost equally with both bases of the adjacent base pair.

The nature of the intermolecular forces operating to solubilize aromatic molecules by purines and pyrimidines has been the subject of numerous theoretical investigations. Pullman and co-workers (1965a,b) have calculated the interaction energies for numerous hydrocarbons and have shown the Van der Waals–London interaction energies (E_L) between a series of bases accounts for the relative order of the solubilizing power of the bases.

Table 3.16. Interaction energies of TMU with aromatic hydrocarbons (after Caillet and Pullman, 1968)

Hydrocarbon	$E_{Elec.}$	$E_{Pol.}$	$E_{V.V. London Int.}$	$E_{Rep.}$	E_{Total}
1,2,3,4-Dibenzopyrene	−1·93	−1·11	−10·37	4·95	−8·46
Anthanthrene	−1·34	−0·93	−8·72	4·79	−6·20
Perylene	−1·35	−0·86	−7·80	4·25	−5·76
1,2-Benzopyrene	−1·77	−0·96	−9·23	4·51	−7·45
3,4-Benzopyrene	−1·38	−0·86	−8·64	4·48	−6·40
Pyrene	−1·46	−0·78	−7·76	4·12	−5·88
1,2,5,6-Dibenzan thracene	−1·09	−0·79	−8·23	3·88	−6·23
Chrysene	−1·43	−0·77	−8·11	4·16	−6·15
1,2-Benzanthracene	−1·18	−0·80	−7·92	4·15	−5·75
Phenanthrene	−1·57	−0·65	−7·13	3·77	−5·58
Anthracene	−1·27	−0·78	−6·84	3·67	−5·22

More recent calculations of the interaction energies of TMU with hydrocarbons, displayed in Table 3.16 confirm the non-specificity of this type of interaction with respect to carcinogenic activity and, thus indicate that this interaction probably has no direct significance for the mechanism of carcinogenesis by this type of molecule (Caillet and Pullman, 1968).

Biological role and chemical interactions. It has been suggested (Brenner *et al.*, 1961; Lerman, 1963; Ames, 1968) that aromatic molecules which can intercalate in the DNA may be mutagens which cause additions of deletions of nucleotides. The mutagenicity of these molecules is, however, strongly dependent on the presence of a reactive side group that can covalently interact with DNA (Ames and Whitfield, 1966; Ames *et al.*, 1973a,b; Creech, 1972). It is well known (Miller, 1970) that most aromatic carcinogens must be metabolically converted, *in vivo*, to electrophilic reactants which are then capable of interacting chemically with the DNA. These observations are consistent with physical chemical studies (Kuroki and Heidelberger, 1971; Schreiber and Daune, 1974; Tada *et al.*, 1974), indicating the presence of covalent linkages between aromatics and the DNA.

Ts'o and coworkers (1969) have demonstrated that 3,4-benzpyrene which is carcinogenic can intercalate and bind covalently with DNA, while 1,2-benzpyrene shows no such covalent reaction and is non-carcinogenic. As the radical cations of these hydrocarbons are likely intermediates, and as the presence of aromatic hydrocarbons linked to DNA have been detected *in vivo* (Brookes and Lawley, 1964), Pullman and coworkers (1969) have investigated the nature of the interactions of the radical cations of 3,4- and 1,2-benzpyrene with a G–C base pair. The association pattern of the neutral molecules with the G–C pair at maximum interactions (see Figures 1 and 2 in original reference) are very similar and the major contributant to the interaction energy, as expected, comes from its dispersion component as shown in Table 3.16. Several interesting points emerge with respect to the association with the radical cations. In the first place, these interactions are energetically more favourable than the associations with the neutral molecules, largely due to the appearance of significant electrostatic components. As a result of the electrostatic contribution, the geometrical patterns of the association are modified (see Figures 3–6 in original reference). In addition, there is a significant difference between the association patterns of the two different radical ions. It thus appears that only one of the two most probable geometries calculated for the complex between 1,2-benzpyrene cation with the G–C base pair involves forces favourable for the possible arylation by the radical at N_7 of the guanine, while both of the probable geometries for 3,4-benzpyrene cation display this favourable balance of forces (Pullman *et al.*, 1969).

As indicated by mutational studies (Ames *et al.*, 1972; Fahmy and Fahmy, 1973), stereochemical factors are critically involved in determining which reactive aromatics can induce genetic damage. The formation of biologically significant associations between aromatics and base pairs is strongly dependent, thus, on the specific nature of the physical and chemical interactions as well as on favourable stereochemical factors. Furthermore, it appears that the intercalation process has an important indirect consequence in the induction of genetic damage by facilitating the close approach of reactive centres on the activated aromatics and the DNA constituents.

C. Wobble pairs

Because of its biological significance and because it belongs rightfully in the category of hydrogen bonding, we now address ourselves to the topic of the 'wobble pairs'. One interesting feature of the genetic code is that although the codon for each amino acid is not unique, almost invariably the codon triplets which code for a given amino acid differ only in the last member of the triplet codon. To explain these observations, the wobble concept was devised. It postulates that the base at the 5′ end of the anticodon is not as sterically confined as the other two, allowing it to form hydrogen bonds with any of several bases located at the 3′ end of a codon. Not all combinations are feasible, with pairing restricted to those listed here (Crick, 1966).

Anticodon base	Codon base
G	U or C
C	G
A	U
U	A or G
I	A, U or C

As a result some pairs which differ from the standard Watson–Crick pairs can occur in the third member of the triplet. The inclusion of such pairs would presumably hinge upon the nature of the bonding forces between the constituent bases which thus require that the steric departure from standard pairs is not too drastic.

Pollak and Rein (1966) have studied the structure of the wobble pairs. They calculated the interaction energy between the corresponding bases which make up the pairs suggested by Crick. The configuration was varied, and that configuration which corresponds to the minimum energy was found. All the 'wobble' pairs postulated by Crick were examined except those which involve inosine. The latter could not be studied because the wavefunctions of inosine were not available.

The calculated interaction energy includes the electrostatic dispersion and polarization energies. At the time the added refinements mentioned previously were not developed and the terms in the interaction series were only carried as far as monopole. The procedure for the calculations is described in another paper (Rein, Claverie and Pollak, 1968). The configurations with energy minima which correspond to the postulated 'wobble' configurations are illustrated in Figure 3.14 below.

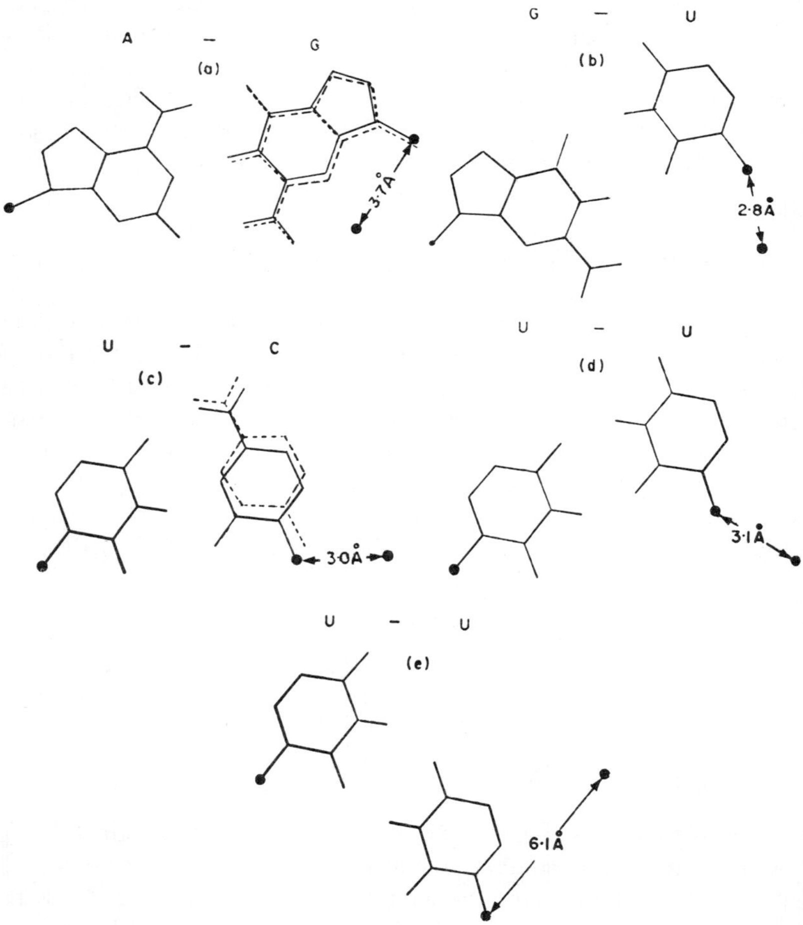

Figure 3.14. Configurations for some 'wobble' pairs

The first interesting result to be noticed is that all the configurations of the 'wobble' type which were postulated by Crick and which were examined are, indeed, stable. Furthermore all but one of the configurations, namely U–U (a), have a distance (d) approximately as expected by Crick (see Figure 8, Crick, 1965). The distance *d* for U–U (e) seems to be outside a reasonable expectation allowed by the steric constraint. This however, does not alter any of Crick's conclusions with respect to coding because triplets coded with U–U (d) would be identical to those coded with U–U (e). Since *d* for U–U (d) is definitely in the range of Crick's hypothesis, the existence of U–U (e) is not essential.

The total binding energies of the various 'wobble' configurations are all in the same range, and furthermore, are in between the binding energies of the Watson–Crick base pairs, A–T and G–C (Pollak and Rein, 1966). Thus also from this point of view, the calculations support the feasibility of the 'wobble' pairs proposed by Crick.

VII. DISCUSSION

A. Validity of the theoretical model

This review has focused on the analysis of forces underlying the structure of biopolymers, specifically nucleic acids. We inquired also into the causative relation between 'forces, structure and mechanism' for a number of fundamental biological processes. It remains in this section, to summarize and discuss in some further details the conclusions which can be arrived from the analysis of the presented material.

Biopolymer constituent interactions are generally treated according to a refined version of the London theory of intermolecular interactions.

This model is conceptually very attractive. It permits the dissection of the interactions into components such as electrostatic polarization, dispersion and short range effects. These components can, through a suitable set of approximations, be related to molecular properties such as electrostatic moments, polarizabilities and ionization potentials. Thus, causative relations can be sought between these molecular properties and the manifested interactions. The force laws for these component interactions are well understood. Thus in the scope of this model, inferences can be made regarding the effects of configurational or compositional variables; for example, how are changes in base pair composition or mutual orientations effecting the interaction energies.

The underlying interaction model is derived from quantum mechanics. However, it contains many arbitrary elements: This includes the assumption of electronic separability of the constituents; inclusion of an empirical term for accounting short range effects, is somewhat contradictive to the former assumption; the truncation of the operator expansion and approximation of its matrix elements; the variable quality of charge distributions (wavefunctions) used in the calculations. Attempts were made to improve the crudeness of these approximations. However, evaluation of the effect of these improvements is not always obvious. Clarification is hindered by the difficulties in comparisons arising because of the arbitrariness in parameters. If comparison with experiment is used as criterion of evaluation, it may happen that a simpler approximation with a more judiciously chosen set of parameters may lead to better agreement than an improved approximation but with less optimized parameters. This is particularly obvious if one examines the agreement of some of the older calculations with experimental results.

In view of these ambiguities the fundamental question one would like to answer is: How well can the validity of the model be ascertained from the comprehensive comparison with experimental results?

In support of the model one can list: (a) the general (not without exception) agreement of the predicted and observed orders of base pair stability in solution; (b) the agreement of the mean stacking energies with the value obtained from the statistical mechanical analysis of melting curves; (c) magnitude and partial order of stability agreement with the thermodynamic values of helix formation; (d) the moderate agreement of the theoretical stability ordering among the various calculations (in spite of differences in approximations); (e) correct prediction of stacking patterns.

It seems, thus, that the refined London theory of intermolecular forces is capable for rationalizing the gross features of the existing experimental data on the stability of base complexes. There are exceptions and occasional disagreements between different sets of calculations, however, these are more likely to be the consequences of different approximations and parameters than of principal deficiency of the model. It appears that the most serious source of the remaining disagreements between various sets of calculations is the electrostatic term. From the analysis of the convergence properties of the interaction series, it is established clearly that segmental multipole representation is preferable and limitation of the series to the monopole terms limits substantially the accuracy. However the accuracy of the results may also be affected by the quality of the wave function used in these calculations. Thus differences arising from the use of a variety of semiempirical wave functions like CNDO and IEHT may lead to considerable disagreement. Altogether it would seem most desirable to replace these wave functions in future calculations by *ab initio* wave functions. In this case one could expect much more uniformity between calculations by various groups.

B. Sources of helix stability

Perhaps the most valuable aspect of the study of intermolecular interactions between DNA components is the insight gained in sources of helix stability.

From these studies it appears that the double helical structure of nucleic acids arises as a consequence of at least four different energy factors. Hydrogen bonding contributes of the order of 5–15 kcal/mol base pair to the stability of the helix (electronic or intrinsic energy). This contribution is selective, i.e., there is a preferential stability of the Watson–Crick G–C pair relative to all other pairs. However this is not the case for the A–T pair where Hoogsteen pairs are of comparable stability. In unconstrained cases also, many other pairing schemes can be realized. The solvent effect which is another factor of stability seems to decrease somewhat further bonding scheme preferences. For example in the model we used to estimate solvent effect we find that the G–C pair formation is destabilized strongly in water, while the A–T pair formation is mildly enhanced. The continuum model of solvent effect leads to similar qualitative conclusions. Thus it seems fairly clear that the hydrogen bonding preferences are not enough to explain the exclusiveness of the complementary pairs in the DNA double-helix. To explain the exclusiveness, the backbone constraint has to be taken into account in interplay with the pairing preference. Studies of backbone conformation indicate that only a limited range of conformational states are comparable with the helical configuration. The Watson–Crick pairs with their equal dimensions can bridge such strands. Deviation in the pair geometry from this standard value introduces strain in the backbone. First estimates of these conformational adjustment energies indicate that they are of sufficient magnitude to eliminate the incorporation of most mispairs into the helix.

The fourth component of DNA stability arises from stacking interactions. These interactions contribute approximately of the same order as hydrogen bonding. Perhaps the most interesting aspect of the stacking interactions which emerges from the theoretical analysis is the fact that the stacking maxima are not necessarily at the angles the successive base pair plans assume in a regular double helix. An analysis of the component forces of stacking and their dependence on orientation unravels that for fragments involving the interaction of more polar bases (G, C) the importance of the electrostatic component will be accentuated and consequently a much stronger orientational preference may arise than in the case for an A–T/A–T type stacking in which case interaction is dominated by the second order terms depending much less on orientation. From existing conformational maps we have estimated that the backbone is under some strain even in the case of regular double helix (DNA B).

It is possible that this strain can be relaxed by some deformation or rotation of the connecting pairs. This of course could occur with greater ease if the potentials for the pair positioning (stacking) as in the case of (AT, AT) is less steep. Consequently some sequence dependent structure peculiarities may arise. That is, the double helix may have a fine structure contingent on the nature of the subsequent base pairs. Indeed such sequence dependent polymorphism has been reported in the recent literature (Bram S., 1971a, 1971b).

In conclusion, it seems that theoretical studies of the interaction between DNA components have led to a general comprehension of the factors of DNA stability, and are in the stage that important structural inferences can be drawn based on the results of such analysis.

C. Biological inferences

We have reviewed in Section VI a few applications of the analysis of interactions, while considering their role in biological mechanisms.

Perhaps the most interesting are the conclusions which can be deduced from these studies regarding the mechanism of point mutations. In earlier considerations a good deal of weight has been given to the possible role of unusual tautomeric forms in induction of base substitutional errors. Based on the reviewed material the role of these forms seems somewhat less likely. It appears that their abundance is not unambiguously demonstrated, theoretically or experimentally. On the other hand, we have explained the mechanism of base analogue induced mutations purely on account of the energetic differences of the alternative pairing schemes of normal tautomeric form of 2-aminopurine (Rein and Garduno 1976).

Regarding the mechanism of spontaneous mutations the concept of conformational strain proved to be fruitful. Estimates of the conformational deformation energies, necessary for incorporation of mismatched bases, have shown that the probabilities of incorporation of such erroneous pairs would be by several orders of magnitude reduced relative to the incorporation of normal pairs. This, in most cases, would be sufficient to maintain a moderate fidelity of the transcription process with error frequencies somewhere of the order of that observed *in vitro* polymerization experiments (1:10000). The estimates of the mutational frequencies at the present are at best illustrative, and more accurate modelling of the strand deformation corresponding to various mispairing as well as more accurate calculation of the corresponding conformational deformation energies will be necessary in the future.

However, irrespective of the necessary refinements, what seems important is that the analysis of component interactions has led to new structural concepts for explanation of the detailed mechanism of a fundamental genetic process.

The likeliness which emerged from the analysis that the DNA double helix has a fine structure depending on the sequence (confirmed also by experiment) is also of potentially great biological significance. It should not escape one's notice that such structural peculiarities could act as recognizable features on the landscape around the double helix for specific interactions with enzymes and regulator molecules.

VIII. ACKNOWLEDGEMENTS

My deepest appreciation goes to John Egan, Ramón Garduño, Yves Coeckelenbergh and Ricky Ornstein for their contribution to the discussion in sections IV and VI. Special thanks are due to Karen Haydock for carefully checking the typescript.

I wish to express thanks to Dr. Shlomo Nir and Ms. Renee LeBoeuf for useful comments.

I am indebted to Mrs. Ruth Harvey and Miss Patt Slatery for technical help in the preparation of the typescript.

This work was supported partially by NASA–AMES University Consortium Interchange NCA2-OR635-701 and Grant NSG-3705 from the National Aeronautics Space Administration and by core Grant CA-17609 by the U.S. Public Health Service.

IX. REFERENCES

Alagona, G., Pullman, A., Scrocco, E. and Tomasi, F. (1943). *Int. J. Peptide Prot. Res.* **5,** 251.

Ames, B. N. and Whitfield, H. F. (1966). *Cold Spring Harbor Symp. in Quantitative Biology*, **31,** 221.

Ames, B. N. *et al.* (1972). *Science*, **176,** 47.

Ames, B. N. *et al.* (1973a). *Proc. Natl. Acad. Sci. US*, **70,** 782.

Ames, B. N. *et al.* (1973b). *Proc. Natl. Acad. Sci. US*, **70,** 2281.

Berthold, H. and Pullman, B. (1971a). *Biochim. Biophys. Acta*, **232,** 595.

Berthold, H. and Pullman, B. (1971b). *Biochim. Biophys. Acta*, **246,** 359.

Berthold, H. and Pullman, B. (1973). *FEBS Lett.*, **33,** (5), 147.

Blake, A. and Peacock, A. R. (1968). *Biopolymers*, **6,** 1225.

Bodor, N. *et al.* (1970). *J. Am. Chem. Soc.*, **92,** 2929.

Bollum, F. S. (1965). *Bull. Biophys. Soc. Abst.*, **T62,** p. 103.

Bonaccorsi, R., Petrongolo, E., Scrocco, E. and Tomasi, J. (1970). *Quantum Aspects of Heterocyclic Compounds in Chemistry and Biochemistry* (*Jerusalem Symposia on Quantum Chemistry and Biochemistry, II*). (Eds. E. D. Bergmann and B. Pullman) The Israel Academy of Sciences and Humanities, Jerusalem p. 181.

Bonaccorsi, R., Petrongolo, E., Scrocco, E. and Tomasi, J. (1971). *Theoret. Chim. Acta*, **20,** 331.

Bonaccorsi, R., Pullman, A., Scrocco, E. and Tomasi, J. (1972). *Chem. Phys. Lett.*, **12,** 622.

Boyland, E. and Green, B. (1962). *Brit. J. Cancer*, **16,** 587.

Bradley, D. F., Lifson, S. and Honig, B. (1964). In *Electronic Aspects of Biochemistry* (Pullman, B., ed.) p. 77 (Academic Press).

Bradley, D. F. and Nash, H. A. (1968). *Mol. Ass. Biol. Proc. Int. Symp.* 1967 (B. Pullman, ed.) pp. 137–144.

Bram, S. (1971a). *Nature New Biol.*, **232,** 174.

Bram, S. (1971b). *Nature New Biol.*, **233,** 161.

Brenner, S. *et al.* (1961). *J. Mol. Biol.*, **3,** 121.

Brock, N. *et al.* (1938). *Arch. Expt. Pathol. Pharmakol.*, **189,** 709.

Brook, R. and Lawley, P. D. (1964). *Nature*, **202,** 781.

Bugg, C. E., Thomas, J. M., Sunderalingam, M. and Rao, S. T. (1971). *Biopolymers*, **10,** 175.

Caillet, J. and Pullman, B. (1968). *Molecular Associations in Biology* (Pullman, B., ed.) p. 217 (Academic Press).

Cantor, C. and Tinoco, I. (1965). *J. Mol. Biol.*, **13,** 65.

Carter, C. W. and Kraut, J. (1974). *Proc. Nat. Acad. Sci. U.S.*, **71,** (2) 283.
Chan, S. I. and Lee, G. C. Y. (1974). *The Purines–Theory and Experiment* (Ed. E. D. Bergmann and B. Pullman) Israel Acad.
Chatkoff, M. and Beckel, C. C. (1971). *Chem. Phys. Lett.*, **12,** 185.
Claverie, P., Pullman, B. and Caillet, J. (1966). *J. Theor. Biol.*, **12,** 419.
Claverie P. (1968). In *Molecular Associations in Biology*, pp. 115–135. (Academic Press).
Claverie, P. and Rein, R. (1969). *Int. J. Quantum Chem.*, **3,** 537.
Coeckelenbergh, Y. (1976). Ph.D. Thesis, Suny at Buffalo.
Coulson, C. A. and Danielson, V. (1954). *Arkiv. Fysik*, **8,** 239.
Creech, H. J. *et al.* (1972). *J. Med. Chem.*, **15,** 739.
Crick, F. H. (1965). *Information Exchange Group No. 17,* Scientific Memo., No. 14.
Crowthers, O. and Zimm, B. H. (1974). *J. Mol. Biol.* **9,** 1.
D'Albis, A., Wickens, M. P. and Gratzer, W. P. (1975). *Biopolymer*, **14,** 1425.
Damiani, A. *et al.* (1966). *J. Mol. Biol.*, **20,** 211.
Damiani, A. *et al.* (1967). *J. Mol. Biol.*, **23,** 113.
Danilov, V. T. (1967). *Biofizika*, **12,** 540.
Davis, R. R. (1967). *A. Rev. Biochem.*, **36,** 321.
Decoen, J. L. and Kerck, J. (1970). *FEBS Lett.*, **11,** 241.
Devoe, H. and Tinoco, I. Jr. (1962). *J. Mol. Biol.*, **4,** 500.
Donohue, J. (1956). *Proc. Natl. Acad. Sci. U.S.*, **42,** 60.
Donohue, J. and Trueblood, P. (1956). *J. Mol. Biol.*, **2,** 363.
Drake, J. W. (1966). *Proc. Natl. Acad. Sci. U.S.*, **55,** 738.
Dreyfus, M. and Pullman, A. (1970). *Theor. Chim. Acta*, **17,** 109.
Dreyfus, M., Maigret, B. and Pullman, A. (1970). *Theor. Chim. Acta*, **17,** 109.
Dreyfus, M. and Pullman, A. (1970). *Compt. Rend.*, **271,** 457.
Egan, J. T., Swissler, T. J. and Rein, R. (1974). *Int. J. Quant. Chem. Q. B. Symp.*, **1,** 71.
Egan, J. T. (1976). Unpublished data (Ph.D. Thesis).
Fahmy, O. G. and Fahmy, M. J. (1973). *Cancer Res.*, **33,** 302.
Fasman, G. D., Lindalow, C. and Grossman, L. (1964). *Biochem.*, **3,** 1015.
Flurry, R. L. Jr, and Breen, D. (1970). *Nature* (*London*), **225,** (5234), 729–730.
Flygare, W. H. and Benson, R. C. (1971). *Mol. Phys.*, **20,** 225.
Freese E. B. (1959). *Brookhaven Symp. Biol.*, **12,** 63.
Freese, E. B. (1963). In *Molecular Genetics, Part I.* (J. H. Taylor, ed.) p. 135 (Academic Press).
Fugita, H., Imamura, A. and Nagata, C. (1974). *J. Theor. Biol.*, **45,** 411.
Goel, N., Fukuda, N. and Rein, R. (1967). *J. Theor. Biol.*, **18,** 350.
Govil, G. and Saran, A. (1971a). *J. Theor. Biol.*, **30,** 407.
Govil, G. and Saran, A. (1971b). *J. Theor. Biol.*, **30,** 691.
Gratzer, N. (1969). *Eur. J. Biochem.*, **10,** 184.
Halicioglu, T. and Sinanoglu, O. (1969). *Ann. N. Y. Acad. Sci.*, **158,** 308.
Hamaguchi, K. and Geiduschek, E. P. (1962). *J. Am. Chem. Soc.*, **84,** 1329.
Hamlin, R. M., Lord, R. C. and Rich, A. (1965). *Science*, **148,** 1734.
Haschmeyer, A. E. and Rich, A. (1967). *J. Mol. Biol.*, **46,** 369.
Hoogsteen, K. (1959). *Acta Cryst.*, **12,** 822.
Huang, W. M. and Ts'o, P. O. P. (1966). *J. Mol. Biol.*, **16,** 523.
Huron, M. J. and Claverie, P. (1972). *Journal of Physical Chem.*, **76,** No. 15, 2123.
Jordan, F. and Pullman, B. (1968). *Theor. Chim. Acta*, **9,** 242.
Johnson, J. *et al.* (1971). *Biopolymers*, **10,** 923.
Kang, S. (1971). *J. Mol. Biol.*, **58,** 297.
Katritzky, A. R. and Waring, A. J. (1963). *J. Chem. Soc.*, 3046.
Katz, K. and Penman, S. (1966). *J. Mol. Biol.*, **15,** 220.
Kim, S. H. and Rich, A. (1968). *Proc. Natl. Acad. Sci. U.S.*, **60,** 402.

Kollman, P. A. and Allen, L. C. (1970). *J. Chem. Phys.*, **52,** 5085.
Kornberg, A. (1969). *Science*, **163,** 1410.
Kudritskaya, Z. G. and Danilov, V. I. (1976). *J. Theor. Biol.* In press.
Kuroki, T. and Heidelberger, C. (1971). *Cancer Res.*, **31,** 2168.
Kwiatkowski, J. S. and Pullman, B. (1975). *Advances in Heterocyclic Chem.*, **18,** 199.
Kyoguku, Y., Lord, R. D. and Rich, A. (1966). *Science*, **154,** 518.
Ladik, J. (1964a). *J. Theor. Biol.*, **6,** 201.
Ladik, J. (1964b). In *Electronic Aspects of Biochemistry* (B. Pullman, ed.) p. 203 (Academic Press).
Lakshninarayanan, V. and Sasieskharan, V. (1969). *Biopolym.*, **8,** 475.
Lakshninarayanan, V. and Sasieskharan, V. (1970). *Biochim. Biophys. Acta*, **204,** 49.
Lee, G. C. Y., Prestegard, J. H. and Chan, S. I. (1971). *Biochem. Biophys. Res. Commun.*, **43,** 435.
Lee, G. C. Y. and Chan, S. I. (1972a). *J. Am. Chem. Soc.*, **94,** 3218.
Lee, G. C. Y. *et al.* (1972b). *J. Am. Chem. Soc.*, **94,** 951.
Lehman, G. (1967). In *Statistical Mechanics: Foundations and Applications* (T. A. Back, ed.) pp. 204–236 (Benjamin Inc).
Leng, M. *et al.* (1973). *Physico-Chemical Properties of Nucleic Acids*, Vol. **3** (Dcuhesne, ed.) p. 20 (Academic Press).
Lerman, L. S. (1961). *J. Mol. Biol.*, **3,** 18.
Lerman, L. S. (1963). *Proc. Natl. Acad. Sci. U.S.*, **49,** 94.
Lifshitz, E. M. (1955). *Zn. Eksperime. Theor. Fiz.*, **29,** 94.
Linder, B. (1960). *J. Chem. Phys.*, **33,** 668.
Linder, B. (1961). *J. Chem. Phys.*, **35,** 37.
Linder, B. (1962). *J. Chem. Phys.*, **37,** 963.
Linder, B. (1964). *J. Chem. Phys.*, **40,** 2003.
Linder, B. (1967). *Adv. Chem. Phys.*, **12,** 225.
Linder, B. and Rabenold, D. A. (1972). *Adv. Quantum Chem.*, **6,** 203.
Liquori, A. M. *et al.* (1962). *J. Mol. Biol.*, **5,** 521.
Liquori, A. M. *et al.* (1969). *The Jerusalem Symposium on Quantum Chemistry and Biochemistry: Physico-Chemical Mechanisms of Carcinogenesis* (Bergmann, F. and Pullman, B., eds.) The Israel Academy of Sciences, Jerusalem, p. 159.
London, F. (1937). *Trans. Faraday Soc.*, **33,** 8.
Lowdin, P. O. (1950). *J. Chem. Phys.*, **18,** 363.
Lowdin, P. O. (1956). *Adv. Phys.*, **5,** (17), 153.
Lowdin, P. O. (1963). *Rev. Mod. Phys.*, **35,** 724.
Lowdin, P. O. (1964a). *Biopolym. Symp.*, **1,** 161, 293.
Lowdin, P. O. (1964b). In *Electronic Aspects of Biochemistry* (B. Pullman ed.) p. 167 (Academic Press).
Lowdin, P. O. (1965). *Adv. Quant. Chem.*, **2,** 213.
Lunell, S. and Sperber, G. (1967). *J. Chem. Phys.*, **40,** 2119.
Marmur, J. and Doty, P. (1939). *Nature* (*London*), **183,** 1427.
Marmur, J. and Doty, P. (1962). *J. Mol. Biol.*, **5,** 109.
Mazza, F., Sobell, H. M. and Kartha, G. (1969). *J. Mol. Biol.*, **43,** 407.
Michelson, A., Massoulie, J. and Guschlbauer, W. (1967). In *Progress in Nucleic Acid Research* (Davison, ed.) p. 85 (Academic Press).
Miles, H. T. (1961). *Proc. Natl. Acad. Sci. U.S.*, **47,** 791.
Miller, J. A. (1970). *Cancer Res.*, **30,** 559.
Montroll, E. and Goel, N. S. (1966). *Biopolym.*, **1,** 885.
Morita, H. and Nagakura, S. (1968). *Theor. Chim. Acta* (*Berl*), **11,** 279.
Morokuma, K. (1971). *J. Chem. Phys.*, **55,** 1236.
Murrell, H. N., Randic, M. and Williams, D. R. (1965). *Proc. R. Soc.*, **A284,** 566.
Nakano, N. I. and Igarashi, S. J. (1970). *Biochem.*, **9,** 577.
Nash, H. A. and Bradley, D. F. (1965). *Biopolymers*, **3,** 261.

Nash, H. A. and Bradley, D. F. (1966). *J. Chem. Phys.*, **45,** 1380.
Nir, S., Rein, R. and Weiss, L. (1973). *J. Theor. Biol.*, **41,** 561.
Nir, S. (1974). *J. Chem. Phys.*, **61,** 2316.
Nir, S. (1975). *Int. J. Quantum Chem.*: *Quantum Biology Symp. No. 2,* 263.
Nir, S. (1976). *Biophysical Journal,* **16,** 59.
Olson, W. K. and Flory, P. S. (1972). *Biopolymers,* **11,** 1.
Perahia, D., Saran, A. and Pullman, B. (1969). *The Jerusalem Symposia on Quantum Chemistry and Biochemistry* (Bergmann, E. and Pullman, B., eds.) p. 255.
Pieber, M., Kroon, P. A., Prestegard, J. H. and Chan, S. I. (1973). *J. Amer. Chem. Soc.*, **95,** 3408.
Pitaa, J., Jones, R. N. and Pithova, P. (1969). *Canad. J. Chem.*, **4,** 1015.
Poland, D. C. and Scheraga, H. S. (1965). *Biopolymers,* **3,** 315.
Pollack, M. and Rein, R. (1966). *J. Theor. Biol.*, **19,** 241.
Polozov, R. V., Poltev, V. I. and Sukhovukov, B. I. (1975). *J. Theor. Biol.*, **55,** 491.
Port, G. N. J. and Pullman, A. (1973). *Theoret. Chim. Acta (Berl.),* **31,** 231.
Port, G. N. J. and Pullman, A. (1973). *FEBS Letters, No. 1,* **31,** 70.
Pritchard, N. J. *et al.* (1966). *Nature,* **212,** 1360.
Pullman, A., Alagona, G. and Tomasi, J. (1974). *Theor. Chim. Acta,* **33,** 87.
Pullman, A. (1974). In *The World of Quantum Chemistry* (Daudel, R. and Pullman, B., eds.) pp. 239–251 (Reidel Publishing Co.).
Pullman, B. and Pullman, A. (1959). *Biochim. Biophys. Acta,* **36,** 346.
Pullman, B. and Pullman, A. (1962). *Biochim. Biophys, Acta,* **64,** 403.
Pullman, B. and Pullman, A. (1963). *Nature,* **196,** 1137.
Pullman, B. (1964). In *Electronic Aspects of Biochemistry* (B. Pullman, ed. p. 135 (Academic Press).
Pullman, B. *et al.* (1965). Science, **147,** 1305.
Pullman, B., Claverie, T. and Caillet, J. (1965a). *Ct. R. Acad. Sci. (Paris),* **260,** 5387.
Pullman, B., Claverie, T. and Caillet, J. (1965b). *Ct. R. Acad. Sci. (Paris),* **260,** 5915.
Pullman, B., Claverie, T. and Caillet, J. (1966). *Proc. Natl. Acad. Sci. U.S.*, **55,** 904.
Pullman, B. and Caillet, J. (1967). *Ct. R. Acad. Sci. (Paris),* **264,** 1900.
Pullman, B. (1968). In *Molecular Associations in Biology* (Academic Press). pp. 1–18.
Pullman, B. *et al.* (1969). *The Jerusalem Symposia on Quantum Chemistry and Biochemistry* (Bergmann, E. and Pullman, B., eds.) p. 325.
Pullman, B. and Pullman, A. (1969). *Prog. Nucl. Acid. Res. Mol. Biol.*, **9,** 257.
Pullman, B. and Pullman, A. (1971). *Adv. Heterocyclic Chem.*, **13,** 77.
Pullman, B. and Berthod, H. (1972a). *FEBS Lett.*, **20,** 341.
Pullman, B., Perahia, D. and Saran, A. (1972b). *Biochim. Biophys. Acta,* **269,** 1.
Radding, C. M., Fosse, J. and Kornberg, A. (1962). *J. Biol. Chem.*, **237,** 2869.
Rabinowitz, J. and Rein, R. (1972). *Int. J. Quant. Mech.*, **VI,** 669.
Rabinowitz, J. R., Swissler, T. and Rein, R. (1972). *Int. J. Quant. Chem.*, **6,** 353.
Razka, M. and Kaplan, N. O. (1972). *Proc. Natl. Acad. Sci. U.S.*, **69,** 2025.
Rein, R. and Ladik, J. (1964). *J. Chem. Phys.*, **40,** 2466.
Rein, R. and Harris, F. E. (1964a). *Science,* **146,** 649.
Rein, R. and Harris, F. E. (1964b). *J. Chem. Phys.*, **41,** 3393.
Rein, R. and Harris, F. E. (1965a). *J. Chem. Phys.*, **42,** 2177.
Rein, R. and Harris, F. E. (1965b). *J. Chem. Phys.*, **43,** 4415.
Rein, R. and Harris, F. E. (1966). *J. Chem. Phys.*, **45,** 1797.
Rein, R., Claverie, P. and Pollack, M. (1968). *Int. J. Quant. Chem.*, **2,** 129.
Rein, R., Goel, N. S., Fukuda, N., Pollack, M. and Claverie, P. (1969). *Ann. N.Y. Acad. Sci.*, **153,** 805.
Rein, R., Rendell, M. S. and Harlos, J. P. (1970). In *Molecular Orbital Studies in Chemical Pharmacology* (Kier, L., ed.) p. 191 (Springer-Verlag).

Rein, R. Rabinowitz, J. R. and Swissler, T. J. (1972). *J. Theor. Biol.*, **34,** 215.
Rein, R. (1973). *Adv. Quant. Chem.*, **7,** 335.
Rein, R. and Pack, G. R. (1973). *Nature*, **242,** (539a), 256.
Rein, R. (1975). *NATO Symposium in Quantum Chemistry* (*Electronic Structure of Polymers and Molecular Crystals*). (Ed. J. M. Andre and J. Landik) Plenum Press, New York p. 505.
Rein, R., Coeckelenbergh, Y. and Egan, J. T. (1975a) *Int. J. Quantum Chem. Q.B. Symp.*, **2,** 145.
Rein, R., Renugopalakrishnan, V., Nir, S. and Swissler, T. J. (1975b). *Int. J. Quant. Chem. Q.B. Symp.*, **2,** 99.
Rein, R. and Garduno, R. (1976). In a volume honouring P. Lowdin. In press.
Reiss, H., McQuarrie, D., McTague, J. P. and Cohen, E. (1966). *J. Chem. Phys.*, **44,** 9367.
Saran, A. and Govil, G. (1971). *J. Theor. Biol.*, **33,** 399.
Sasisekharan, V. Lakshminarayan, A. V. and Ramachandran, G. N. (1967). In *Conformation of Biopolymers*, Vol. II (Ramachandran, G. N., ed.) p. 641 (Interscience, N.Y.).
Sasisekharan, V. and Lakshminarayan, A. V. (1969). *Biopolymers*, **8,** 505.
Sasisekharan, V. (1973). *Conformation of Biological Molecules and Polymers. The Jerusalem Symposia on Quantum Chemistry and Biochemistry*, **V** (Bergman, E. and Pullman, B., eds.) p. 247.
Scheiber, J. P. and Daune, M. P. (1974). *J. Mol. Biol.*, **83,** 487.
Seeman, N. C. *et al.* (1975). *Nature*, **253,** 324.
Shoup, R. R., Miles, H. T. and Becker, E. D. (1966). *Biochem. Biophy. Res. Comm.*, **23,** 194.
Sinanoglu, O. (1964). *Adv. Chem. Phys.*, **6,** 315.
Sinanoglu, O. and Abdulnur, S. (1964a). *J. Photochem. Photobiol.*, **3,** 333.
Sinanoglu, O., Abdulnur, S. and Kesther, N. R. (1964b). In *Electronic Aspects of Biochemistry* (Pullman, B., ed.) p. 301 (Academic Press).
Sinanoglu, O. and Abdulnur, S. (1965). *Fed. Proc. Fed. Am. Soc. Exp. Biol.*, **24,** 5.
Sinanoglu, O. (1967a). *Adv. Chem. Phys.*, **12,** 283.
Sinanoglu, O. (1967b). *Chem. Phys. Letters*, **1,** 340.
Singer, M., Jones, O. and Risenberg, M. (1963). *Proc. Natl. Acad. Sci. U.S.*, **49,** 392.
Sobell, H. M., Tomita, K. and Rich, A. (1963). *Proc. Natl. Acad. Sci. U.S.*, **49,** 885.
Solie, T. N. and Shellman, J. A. (1968). *J. Mol. Biol.*, **33,** 61.
Stamatiadou, M. (unpublished data).
Stamatiadou, M., Swissler, T. J. and Rein, R. (1972). *Biopolymers*, **15,** 1217.
Strassler, S. (1967). *J. Chem. Phys.*, **46,** 1037.
Sundaralingam, M. (1969). *Biopolymers*, **7,** 821.
Swissler, T. and Rein, R. (1972). *Chem. Phys. Lett.*, **15,** (4) 617.
Tada, M. *et al.* (1974). *Biochem. Biophys. Res. Comm.*, **29,** 469.
Tinoco, I. Jr., Borer, P. N., Dengler, B., Levine, M., Ulenbergh, O. C., Crothers, D. and Groella, J. (1973). *Nature New Biol.*, **246,** 40.
Tsai, C. C. *et al.* (1975). *Proc. Natl. Acad. Sci. U.S.*, **72,** 628.
Ts'o, P. O. P. (1963). *J. Am. Chem. Soc.*, **81,** 5241.
Ts'o, P. O. P. and Chang, S. I. (1964a). *J. Am. Chem. Soc.*, **86,** 4176.
Ts'o, P. O. P. (1964b). In *Nucleo Histones* (Bonner, J. and Ts'o, P., eds.).
Ts'o, P. O. P. (1968). In *Molecular Associations in Biology* (Pullman, B., ed.) pp. 39–75 (Academic Press).
Ts'o, P. O. P. *et al.* (1969). *The Jerusalem Symposia on Quantum Chemistry and Biochemistry* (Bergmann, E. and Pullman, B., eds.) p. 106.
Ts'o, P. O. P. (1970). In *Fine Structure of Proteins and Nucleic Acids* (Fasman, G. and Timisheff, eds.) Marcel Dekker.

Ts'o, P. O. P. (1974). *Basic Principles in Nucleic Acid Chemistry*, Vol. I (Academic Press).
Von Hippel, P. H. and Wong, K. Y. (1964). *Science*, **145,** 577.
Von Hippel, P. H. and Wong, K. Y. (1965). *J. Biol. Chem.*, **240,** 3909.
Weil-Malherbe, H. (1946). *Biochem. J.*, **40,** 351.
Watson, J. D. and Crick, F. H. (1953a). *Nature*, **171,** 757.
Watson, J. D. and Crick, F. H. (1953b). *Nature*, **171,** 964.
Watson, J. D. and Crick, F. H. (1953c). *Cold Spring Harbor Symp. Q.B.*, **18,** 123.
Warshaw. M. and Tinoco, I. (1965). *J. Mol. Biol.*, **13,** 54.
Wells, R. O., Ohtsuka, E. and Khorama, H. (1965). *J. Mol. Biol.*, **14,** 221.
Witz, J. and Luzzatti, F. (1965). *J. Mol. Biol.*, **11,** 62.
Wilson, H. R., Rahman, A. and Tollin, P. (1969). *J. Mol. Biol.*, **46,** 585.
Wilson, H. R. and Rahman, A. (1971). *J. Mol. Biol.*, **86,** 129.
Zimm, B. M. (1966). *J. Chem. Phys.*, **33,** 1349.

CHAPTER 4

The fine structure of the hydrogen bond

P. Schuster

Institut für Theoretische Chemie und Strahlenchemie der Universität Wien, Wien, Austria

I. INTRODUCTION

The 'hydrogen bond' is now more than fifty years old. For an authentic reference see the monograph by Lewis (1923) and the historic and autobiographical survey by Huggins (1971). The expression has also been created by Huggins (1919, 1922) and was used almost simultaneously by Latimer and

This chapter is dedicated to Prof. Dr. H. Nowotny

Rodebush (1920) in order to explain a number of structural regularities and unusual properties of compounds with polar XH bonds. In the meantime a huge amount of experimental data has been accumulated, many theoretical approaches to explain the phenomenon 'hydrogen bond' have been proposed, but still hydrogen bonding represents a subject of interest which is worth to be treated in reviews summarizing most recent results. Moreover, it turned out, that nowadays a single person is unable to cover the whole field in a competent way and one of the most recent, comprehensive treatises appeared as a large collective value of about forty authors (Schuster, Zundel and Sandorfy, 1976). A number of other recent monographs and extended reviews in this field have been published as well: Hadži and Thompson (1959), Pimentel and McClellan (1960), Sokolov and Chulanovskii (1964), Covington and Jones (1968), Hamilton and Ibers (1968), Vinogradov and Linell (1971), Pimentel and McClellan (1971), Joesten and Schaad (1974).

Reviews on quantum mechanical calculations dealing with equilibrium properties and IR spectra of hydrogen-bonded structures have been presented as well in the past. Some representative examples are mentioned here (Bratož, 1967; Murthy and Rao, 1970; Kollman and Allen, 1972; Schuster 1973; Morokuma, Iwata and Lathan, 1974; Schuster *et al.*, 1974; Schuster, 1976a; Janoschek, 1976). New theoretical results have been reported during the last two years, calculations more accurate than the previous ones have been performed and therefore it seems well justified to present another summary on this subject. What we intend to present here is the quantum chemist's view on hydrogen bonding as it appears in the recent literature. Theoretical results will be supported and completed by experimental facts as far as they are concerned directly with a deeper understanding of this phenomenon. We will concentrate here mainly on the results of extensive *ab initio* LCAO MO calculations, and perturbation theory only as far as large distances are concerned.

Semi-empirical MO calculations were found to be highly useful for theoretical investigations on large molecules and their interactions (as far as hydrogen bonding and conformational equilibria are concerned: see for example Murthy and Rao (1970), Schuster (1973), Pullman and Pullman (1974). Provided the parameters have been determined carefully, fairly correct predictions on geometries of molecular associations and energies of interaction can be obtained. In case of the most frequently applied methods based on the CNDO assumption partitioning of intermolecular energies has shown that this agreement is merely a result of error compensation (for example, see Kollman and Allen, 1970; Schuster, Jakubetz and Marius, 1975; Schuster, 1976a,b). Therefore, we cannot expect to gain more insight into principles and details of intermolecular interactions from semi-empirical MO calculations and consequently they have been excluded from this summary.

The term hydrogen bond or hydrogen bridge is used for two different phenomena. In the language of valence bond theory they may be classified as (1) electron deficient three centre two electron bonds and (2) three-centre four-electron interactions (**I** and **II**).

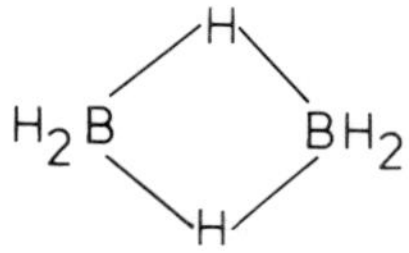

I

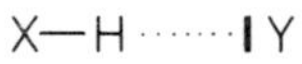

II

The first type of bond is common in beryllium hydride (BeH_2) boron hydrides and related compounds. Here we will restrict our discussion exclusively to the latter case which covers a wide range of intermolecular interactions as far as energetic aspects are concerned. In order to characterize weak interactions like those between polar molecules like HF or H_2O and noble gas atoms, or the binding in hydrogen bifluoride anion people use the same expression. From the theorists' point of view these two extreme cases have not much more in common than the fact that a hydrogen nucleus is placed in the region between the interacting molecules or ions. In the first class of examples energies of interaction are in the range of a few tenths of a kcal/mol and intermolecular overlap is more or less negligible whereas the second example resembles more closely a particular kind of chemical bond than an intermolecular complex. In $[FHF]^-$ overlap is large and the energy of interaction amounts roughly 50 kcal/mol.

Finally, before going into further details, we shall try to explain how 'hydrogen bonding' will be understood in this chapter. Instead of presenting a rigorous definition we propose a number of characteristic criteria for hydrogen bonds, which have been observed experimentally and which have to be explained by any serious modern theoretical approach to this subject (Figure 4.1):

(1) Distances between the neighbouring atoms of the two functional groups (XH . . . Y) involved in hydrogen bonds are substantially smaller than the sum of their Van der Waals radii.

(2) XH Stretching modes are shifted towards lower wave numbers on hydrogen bond formation.

(3) Polarities of XH bonds increase on hydrogen bond formation, in general leading to larger dipole moments of complexes than expected from vectorial addition. Furthermore, largely enhanced IR intensities indicate an increase in dipole moment derivatives on hydrogen bond formation too.

(4) NMR chemical shifts of protons in hydrogen bonds are substantially smaller than those observed in the corresponding isolated molecules. The deshielding effect observed is a result of reduced electron densities at protons involved in hydrogen bonds.

Moreover, another interesting regularity has been observed in hydrogen bonded systems: all the effects, (1) to (4) mentioned above are less pronounced in isolated dimers than in hydrogen bonded liquids or crystals, thus

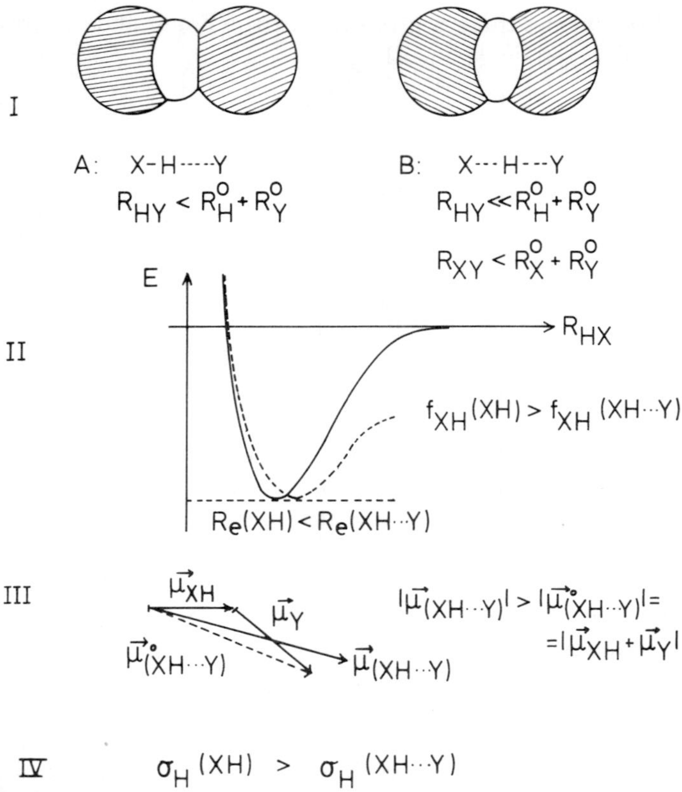

Figure 4.1. Characteristic properties of hydrogen bonded systems. (I) intermolecular geometries, (A) ordinary, (B) strong hydrogen bonds; (II) vibrational spectra, XH stretching frequencies; (III) increase in polarity on complex formation and (IV) NMR deshielding effect observed at protons involved in hydrogen bonds)

indicating the importance of long range interactions through chains or networks of hydrogen bonds.

In the following sections we will start with a discussion of the present stage of quantum mechanical predictions of molecular properties. Then, energies and wave functions of hydrogen bonded complexes and crystals will be discussed. The usefulness of energy partitioning is the subject of another section and at the end we present two brief remarks on the actual stage of our knowledge in this field and an estimate on possible developments in the near future.

II. PROPERTIES OF ISOLATED MOLECULES AND IONS

Within the frame of numerical calculations based on the stationary electronic Schrödinger equation LC–MO methods were found to be most appropriate and economic. The self-consistent field theory (SCF) of Hartree, Fock and Roothaan (Roothaan, 1951) combined with efficient procedures to calculate effects of electron correlation explicitly by configuration interaction (CI) is generally expected, nowadays, to produce the most accurate theoretical data on small molecules. Other SCF methods based on consistent approximations of exchange potentials like multiple scattering-Xα theory (for example, see Slater, 1965; Weinberger and Schwarz, 1975) have not yet reached a comparable stage, when applied to small molecular systems. The reason for that inferiority with respect to accurate results may be due to lack of experience with these fairly new methods or to more fundamental methodological deficiencies. Extensive calculations based on valence bond theory (VB) are rare (for example, see Raimondi, Tantardini and Simonetta, 1975) and have not been pushed forward to high accuracy so far. Consequently, no reliable comparison with respect to numerical efforts is possible. From a theoretical point of view the VB approach is equivalent to MO calculations including CI (Longuett-Higgins, 1948).

Ab initio LC–MO calculations are classified appropriately with respect to the limits they approach on basis set extension:

(1) the independent particle or Hartree–Fock (HF) limit and
(2) the non-relativistic (NR) limit.

Many calculations coming very close to the HF limit have been performed on diatomic and a few polyatomic molecules. In a considerable number of examples electron correlation effects have been studied as well. Mainly two strategies have been applied: direct CI expansion using high numbers of conformations and refined numerical procedures to calculate matrix elements very efficiently (for example, see Roos, 1972; Diercksen, Kraemer and Roos, 1975; Rosenberg and Shavitt, 1975; Matsuoka, Clementi and Yoshimine, 1976) and CI calculations with limited numbers of configurations after a suitable transformation of the SCF orbitals in order to achieve better convergence of the expansion series (for a recent review on electron pair theories applied to the calculation of correlation energies see Kutzelnigg, 1973). Among all approaches of the second kind we will mention here only those which have been applied also to intermolecular interactions. Two closely related procedures have been used in this context: the independent electron pair approximation (IEPA) and the coupled electron pair approximation (CEPA). Both methods apply pair natural orbitals (PNO) in the CI calculations (for further details see Meyer, 1971, 1973, and Kutzelnigg, 1973).

Considering the principal aim of this article we will restrict ourselves here to a few characteristic results concerning energies and other molecular properties which are relevant for studies on intermolecular interactions, especially for studies on hydrogen bonded systems. Primarily, we will compare here proton affinities (ΔE_{PA}), force constants, electric multipole moments and

$$\Delta E_{PA}(M) = E_{MH^+} - \{E_M + E_{H^+}\} = E_{MH^+} - E_M \tag{4.1}$$

polarizabilities as well as NMR chemical shifts.

Proton affinities have been chosen here as representative energetic quantities because they are more closely related to energies of interaction than the total energies of molecules or ions. Total energies as the minimized quantities of the variational procedure reflect all improvements in the wavefunctions, which need not necessarily be of advantage for calculations on intermolecular interactions. A better description of inner shell electrons roughly improving total energies only may serve as a representative example. In a way ΔE_{PA} values reflect the extreme case of the interaction with proton donors, whereby the donor has degenerated to a naked proton. For a long time proton affinities have been an enormous challenge for theoreticians, since these quantities are difficult to measure and therefore accurate experimental data are rare. In Table 4.1 a survey of the most reliable theoretical and experimental ΔE_{PA} values is presented. We realize that the

Table 4.1. Proton affinities in the vapour phase (energies in kcal/mol)

Reaction	ΔE_{PA} [1]	Reference	ΔH_{PA} [1]	ΔW_0^0 [1]	Reference
$He+H^+$	−46·0	*a*	≤−41·5		*b*
			−42·4		*c*
$Kr+H^+$			−85±7		*b*
			≤78·3		*c*
H_2+H^+	−102	*d*	≤91·4		*e*
$HF+H^+$	−121	*f*			
CH_4+H^+	−129	*g*	−122±7		*b*
$CO+H^+$	−120	*h*	−131	5·0	*h*
$C_2H_4+H^+$	−169	*i*	−157±8		*b*
H_2CO+H^+	−172	*h*	−161±3		*h*
H_2O+H^+	−174	*f*	−169±2	5·3	*b*
	−168	*m*	−164±4		*j*
			−168±3		*k,l*
H_2S+H^+			−178±2		*j*
CH_3OH+H^+			−180±3		*b*
$C_2H_5OH+H^+$			−193±9		*b*
NH_3+H^+	−213	*h*	−216		*b*
$H_3PCH_2+H^+$	−268	*o*			
$e_0^-+H^+$	−314	2	−314		2
I^-+H^+			−308	3·2	*n*
Br^-+H^+			−318	3·7	*n*
Cl^-+H^+			−324	4·1	*n*
F^-+H^+	−373	*f*	−366	5·8	*n*
HO^-+H^+	−396	*f*	−411	7·5	*n*
$H_2N^-+H^+$	−427	*h*	−380	8·8	*n*
			(−422)		

[1] Among calculated ΔE values the results of the most extensive treatments are presented. ΔH stands for the experimental enthalpy of the reaction $B+H^+ \rightarrow [HB]^+$. Zero point energy corrections, ΔW_0^0, are derived from spectroscopic data: $\Delta E \sim \Delta H - \Delta W_0^0$

[2] Ionization potential of *H*

References

[a] Wolniewicz (1965); [b] Vedeneyev *et al.* (1966); [c] Chupka and Russel (1968); [d] Kutzelnigg *et al.* (1967); [e] Chupka, Russel and Refaye (1968); [f] Lischka (1973); [g] Dyczmons and Kutzelnigg (1974); [h] Hopkinson *et al.* (1968); [i] Zurawski, Ahlrichs and Kutzelnigg (1973); [j] Beauchamp and Buttrill (1968); [k] Chong, Myers and Franklin (1972); [l] Cotter and Koski (1973); [m] Diercksen, Kraemer and Roos (1975); [n] Herzberg (1966); [o] Lischka (1976).

naked proton is bound to every neutral molecule or anion in a highly exothermic reaction. Additionally, there is an interesting gap in the series of proton affinities between neutral and negatively charged species indicating that the electrostatic contribution which results from charge neutralization overshadows molecular individualities. Not even highly basic and reactive species like methylene phosphorane, which may be characterized best by a zwitterionic structure, $H_3P^+{-}CH_2^-$, containing a carbanion, come close to the proton affinity for the simplest 'anion' possible (e_0^-). The series of ΔE_{PA} values shown in Table 4.1 can be regarded also as an absolute acidity scale in the vapor phase, which provides explanations for some features of chemistry in extremely acidic or basic media. We give here just one example: hydrogen fluoride has a $|\Delta E_{PA}|$ value which is slightly smaller than that of methane and the other hydrocarbons. Carbocations can be studied therefore in the presence of hydrogen fluoride (for example, see Olah, 1973). Provided specific solvation effects are not important qualitative predictions can be made, which are very useful in the extreme ranges of acidities, where no chemical experience is accessible and intuition usually fails.

Some years ago already proton affinities were found to be largely basis-set dependent (Hopkinson *et al.*, 1968; see also Schuster, 1970). Throughout in this paper we will use the following short notation for basis sets: STO and GTO stand for Slater and Gaussian atomic orbitals respectively. (α, β, γ, θ/ε, ϕ) denotes that αs-type, βp-type, γd-type and θf-type functions have been used for second row atoms and εs-type and ϕp-type functions for hydrogen atoms. Contractions are shown by square brackets: $[\alpha', \beta', \gamma', \theta'/\varepsilon', \phi']$.

Especially, minimum basis sets are very weak also in reproducing relative ΔE_{PA} values only (Table 4.2). The main reason for these failures comes from a side effect of the enlargement of calculated systems. Due to the lack in flexibility of small basis sets addition of further orbitals to the original basis lowers the total energies substantially. The calculated proton affinity might be overshadowed therefore by this kind of artificial stabilization energy, which we will have to discuss more extensively later on. In case of fluoride anion or hydrogen fluoride this effect is largest since the minimum basis set, (31/1), as applied by Hopkinson *et al.* (1968), has almost no flexibility. In ammonia we have a somewhat better ratio of electrons to basis orbitals and the error is somewhat smaller therefore. Nevertheless large basis sets are inevitable if reliable numerical results are desired. On the other hand proton affinities are not extremely sensitive to further basis set extension provided reasonably balanced basis sets are used (Table 4.2). The influence of electron correlation on ΔE_{PA} is not very large. In all examples studied so far, proton affinities calculated by IEPA and CEPA method were a few kcal/mol smaller than the corresponding SCF values. An other calculation based on the conventional large scale CI approach (Diercksen, Kraemer and Roos, 1975) gave essentially the same results for the water molecule.

Table 4.2. Influence of basis sets and correlation energies on calculated proton affinities for some selected molecules or ions (energies in kcal/mol)

Molecule or anion	Basis set (GTO's)	Method	ΔE_{PA}	ΔH_{PA}(exp.)	Reference
HF	(3, 1/1)	SCF	−401		a
	(5, 2/2)	SCF	−157		a
	(7, 3/3)	SCF	−104		a
	(10, 6/4)	SCF	−108		a
	(11, 7, 1/6, 1) → [5, 4, 1/3, 1]	SCF	−120·1		b
	(11, 7, 1/6, 1) → [7, 4, 1/4, 1]	SCF	−121·8		c
		IEPA	−122·5		c
	(11, 7, 2/6, 2) → [7, 4, 2/4, 2]	SCF	−121·8		c
		IEPA	−122·5		c
H_2O	(3, 1/1)	SCF	−345	−168	a
	(5, 2/2)	SCF	−209		a
	(7, 3/3)	SCF	−174		a
	(10, 6/4)	SCF	−181		a
	(11, 7, 1/6, 1) → [5, 4, 1/3, 1]	SCF	−174·3 (−169·0)[1]		d
		SCF–CI	−172·8 (−167·5)[1]		d
	(11, 7, 1/6, 1) → [7, 4, 1/4, 1]	SCF	−176·6		c
		IEPA	−176·6		c
	(11, 7, 2/6, 2) → [7, 4, 2/4, 2]	SCF	−174·0		c
H_3N	(3, 1/1)	SCF	−302		a
	STO-3G [2]	SCF	−252	e	
	(5, 2/2)	SCF	−234		a
	(7, 3/3)	SCF	−213		a
	(10, 6/4)	SCF	−220	−216	a
F^-	(3, 1/1)	SCF	−1022	−389	a
	(5, 2/2)	SCF	−504		a
	(7, 3/3)	SCF	−403	a	
	(10, 6/4)	SCF	−387		a
	HF-limit	SCF	−383·5		c
	(11, 7, 1/6, 1) → [7, 4, 1/4, 1]	SCF	−380·9		c
		IEPA	−382·1		c
	(11, 7, 2/6, 2) → [7, 4, 2/4, 2]	SCF	−374·3		c
		IEPA	−372·9		c
OH^-	(3, 1/1)	SCF	−785		a
	(5, 2/2)	SCF	−496		a
	(7, 3/3)	SCF	−425		a
	(10, 6/4)	SCF	−403		a
	HF-limit	SCF	−408·0		c
	(11, 7, 1/6, 1) → [7, 4, 1/4, 1]	SCF	−406·4		c
		IEPA	−407·5		c
	(11, 7, 2/6, 2) → [7, 4, 2/4, 2]	SCF	−398·2		c
		IEPA	−396·4		c

[1] The values including zero point energy corrections are given in parenthesis.
[2] For a definition of STO–NG basis sets see: Hehre, Stewart and Pople (1969) and Ditchfield, Hehre and Pople (1971).

References

[a] Hopkinson *et al.* (1968); [b] Diercksen, von Nessen and Kraemer (1973); [c] Lischka (1973); [d] Diercksen, Kraemer and Roos (1975); [e] Pullman and Brochen (1975).

Fairly accurate molecular geometries are obtained already by *ab initio* MO calculations at the SCF level. In most cases even small basis sets will serve very well for this purpose. Generally, inclusion of electron correlation effects still improves the HF results. As we can conclude from some selected examples in Table 4.3 the values calculated by CEPA method come very close to the experimental data.

Force constants are sensitive to both, basis set extension and electron correlation effects. Because of the incorrect dissociative behaviour of single determinant wavefunctions force constants for stretching vibrations are obtained too large at the HF limit. Some recent results are summarized in Table 4.3 as representative examples. In some cases, for example, in molecules with triple bonds like N_2 or HCN, the SCF results are particularly bad. Considering the most accurate calculations including electron correlation, the experimental values of stretching vibrational force constants are reproduced reasonably well. The errors amount to a few per cent only. Due to error compensation medium size basis sets sometimes are able to reproduce force constants correctly too (for example, see Karpfen and Schuster, 1976).

Electric properties of molecules are somewhat more difficult to discuss, since experimental data are available only for dipole and quadrupole moments. In most cases information on the latter is rather uncertain. With the exception of dipole moments and a few quadrupole moments we are forced therefore to compare the theoretical data among themselves. Calculated electric moments—for the definition used here see equation (4.2)—in general, depend strongly on both basis

$$\left.\begin{aligned} Q_m^1 &= e_0\left\{\sum_k Z_k R_k^1 Y_{1,m}(\theta_k \Phi_k) - \int \Psi^* \sum_i r_i^1 Y_{1,m}(\delta_i \phi_i)\Psi \, d\tau\right\} \\ Q_{x^\xi y^\eta z^\zeta}^l &= e_0\left\{\sum_k Z_k x_k^\xi y_k^\eta z_k^\zeta - \int \Psi^* \sum_i x_i^\xi y_i^\eta z_i^\zeta \Psi \, d\tau\right\}; \qquad 1 = \xi + \eta + \zeta \end{aligned}\right\} \quad (4.2)$$

set extension and inclusion of electron correlation effects (Table 4.4). Moreover, very flat polarization functions which have almost no influence on total energies are important for correct predictions of molecular multipole moments. The reason for this particular sensitivity of multipole moments simply originates from the different relative importance of outer and inner regions of the wavefunctions in calculations of different expectation values. Evidently, the inner parts are more important for energies, whereas the weighting factor r^1 appearing in the equations of the multipole moments (4.2) shifts the range of predominant contributions outwards. Accordingly, higher multipole moments depend more strongly on basis set extension than dipole moments do. In a number of examples like those shown here electron pair methods which account for the major part of correlation contributions, for example, like IEPA or CEPA, are able to predict dipole moments in good agreement with experiment. There are some other examples, like CO

Table 4.3. Molecular geometries and force constants of some small molecules

Molecule	Basis set (GTO's)	Method	Equilibrium geometry (bond distances (Å) and bond angles)	Force constants (in dyn/Å)	Reference
N_2	STO-3G [1]	SCF	$R_e = 1{\cdot}134$		*a*
	(9, 5, 1/5, 1) → [5, 3, 1/3, 1]	SCF	1·069	31·0	*b*
		PNO–CI	1·090	26·3	*b*
		IEPA	1·147	14·0	*b*
		CEPA	1·100	24·1	*b*
		EXP	1·098	22·94	*b*
F_2	STO-3G [1]	SCF	$R_e = 1{\cdot}315$		*a*
	(9, 5, 1/5, 1) → [5, 3, 1/3, 1]	SCF	1·336	8·70	*b*
		PNO–CI	1·379	7·40	*b*
		IEPA	1·472	3·55	*b*
		CEPA	1·411	5·00	*b*
		EXP	1·42	4·78	*b*
HF	STO-3G [1]	SCF	$R_e = 0{\cdot}956$	11·3	*a*
	(8, 4/4) → [4, 3/3]	SCF	0·917	9·4	*c*
	(11, 7, 2/6, 1) → [7, 4, 2/4, 1]	SCF	0·900	11·2	*d*
		PNO–CI	0·913	10·2	*d*
		IEPA	0·927	9·27	*d*
		CEPA	0·917	9·92	*d*
		EXP	0·917	9·66	*d*

HCN[2]	STO-3G[1]	SCF	$R_e^{HC} = 1{\cdot}070$	$R_e^{CN} = 1{\cdot}153$	$f_{HC} = 8{\cdot}0$	$f_{CN} = 26{\cdot}9$	a
	(11, 6/6) → [6, 3/3]	SCF	1·054	1·136	6·7	23·6	e
	(11, 6, 1/6, 1) → [6, 3, 1/3, 1]	SCF	1·059	1·127	6·9	24·3	e
		SCF–CI	1·065	1·150	6.54	20·7	e
		EXP	1·066	1·153	6·24	18·8	e
H_2O[2]	STO-3G[1]	SCF	$R_e^{OH} = 0{\cdot}990$	$\alpha_e^{HOH} = 100{\cdot}0°$	$f_{OH} = 9{\cdot}7$	$R_e^2 \cdot f_{HOH} = 1{\cdot}32$	a
	(9, 5/4) → [4, 3/2]	SCF	0·951	112·6°	9·4	0·83	f
	(9, 5, 2/4, 1) → [4, 3, 2/2, 1]	SCF	0·941	106·1°	9·9	0·90	f
	(11, 7, 2/5, 1) → [6, 5, 2/3, 1]	SCF	0·941	106·6°	9·83	0·867	f
	(11, 7, 1/6, 1) → [7, 4, 1/4, 1]	SCF	0·942	106·5°	9·31	0·821	g
		IEPA	0·973	103·3°	8·37	0·722	g
	(11, 7, 1/6, 1) → [5, 4, 1/3, 1]	SCF	0·944	105·3°	9·50	0·816	h
		SCF–CI	0·960	103·8°	8·44	0·752	h
		EXP	0·957	104·5°	8·35	0·76	h

[1] See Hehre, Stewart and Pople (1969) and Ditchfield, Hehre and Pople (1971).
[2] $C_{\infty v}$ and C_{2v} symmetry was obtained for HCN and H_2O at the equilibrium geometry in all calculations reported here.

References
[a] Newton *et al.* (1970); [b] Ahlrichs *et al.* (1976); [c] Karpfen and Schuster (1976); [d] Lischka (1974); [e] Wahlgren, Pacansky and Bagus (1975); [f] Dunning, Pitzer and Aung (1972); [g] Lischka (1973); [h] Diercksen, Kraemer and Roos (1975).

Table 4.4. Electric multipole moments obtained by *ab initio* calculations on HF and H_2O

Molecule	Basis set	Method	Dipole moment [1] $\mu_z (D)$	Quadrupole moments [1] (10^{-26} esu cm^2) Θ_{xx}	Θ_{yy}	Θ_{zz}	Octopole moments [1] (10^{-34} esu cm^3) Ω_{xxz}	Ω_{yyz}	Ω_{zzz}	Reference
HF	(9, 5, 1/4, 1) → [5, 4, 1/3, 1]	SCF	2·06	−1·09	−1·09	2·19				a
	(11, 7, 3/6, 1) → [7, 4, 3/4, 1]	SCF	1·93							a
		PNO–CI	1·84							a
		IEPA	1·78							a
		CEPA	1·82							a
		EXP	1·82							a
H_2O	STO(5, 4, 1/3, 1)	SCF	2·05	2·61	−2·54	−0·07	2·39	−0·98	−1·44	b
	STO-4G [2]	SCF	1·93	1·38	−1·50	0·13	1·35	−0·61	−0·75	c,d
	(9, 5/4) → [5, 3/3]	SCF	2·64	2·52	−2·33	−0·19	1·81	−0·74	−1·07	
	(9, 5, 2/4, 1) → [4, 3, 2/2, 1]	SCF	1·95	2·47	−2·39	−0·09	2·30	−0·99	−1·32	c,d
	(11, 7, 2/5, 1) → [6, 5, 2/3, 1]	SCF	2·03	2·66	−2·47	−0·20	2·19	−0·84	−1·35	b
	(11, 7, 2/6, 2) → [7, 5, 2/4, 2]	SCF	1·98	2·57	−2·45	−0·12	2·28	−0·92	−1·35	c,d
	(11, 7, 3/6, 2) → [7, 5, 3/4, 2]	SCF	2·00	2·57	−2·41	−0·10	2·29	−0·93	−1·35	c,d
	(11, 7, 2, 1/6, 2) → [7, 5, 2, 1/4, 2]	SCF	2·00	2·55	−2·43	−0·14	2·29	−0·93	−1·35	c,d
	STO(5, 4, 2/3, 1)	SCF	2·00	2·62	−2·52	−0·10				e
		SCF–CI	1·92	2·61	−2·52	−0·09				e
		EXP	1·85	2·63	−2·50	−0·13				e

[1] The coordinate system is chosen in such a way that the z-axis coincides with the C_∞ or C_2 symmetry axis of the molecule, the H_2O molecule lies in the xz plane. Quadrupole and octopole tensors are calculated relative to the centre of mass in H_2O^{16}.

[2] See Hehre, Stewart and Pople (1969) and Ditchfield, Hehre and Pople (1971).

References. [a] Lischka (1974); [b] Dunning, Pitzer and Aung (1972); [c] Beyer, Lischka and Schuster (1976); [d] Schuster, Lischka and Beyer (1976); [e] Rosenberg and Shavitt (1975).

(Grimaldi, Lecourt and Moser, 1967) or LiH (Bender and Davidson, 1968), where the contribution of singly excited states was found to be of particular importance.

Great efforts have been made in order to calculate reliable molecular polarizabilities. Apart from some older work based on uncoupled HF perturbation theory (Liebmann and Moskowitz, 1971) polarizabilities were obtained substantially too small (for example, see Arrighini, Maestro and Moccia, 1967) as long as no flat polarization functions were included in the basis set. Recently, a numerical simulation of the electric field in the Hamiltonian equivalent to the coupled HF perturbation theory has been used to calculate atomic and molecular polarizabilities (Werner and Meyer, 1976). The authors investigated also the influence of electron correlation, which again tends to increase calculated polarizabilities. Typical results are summarized in Table 4.5. Besides excellent agreement with the available

Table 4.5. Electric polarizabilities obtained by *ab initio* calculations on HF, H_2O and NH_3

Molecule	Basis set (GTO's)	Method	Polarizabilities (10^{-24} cm^3 = Å^3)[1] α_{xx}	α_{yy}	α_{zz}	$\bar{\alpha}$	Reference
HF	(8, 5, 1/4, 1) → [5, 4, 1/3, 1]	SCF	0·0261	0·0261	0·653	0·0391	a
		IEPA	0·0296	0·0296	0·714	0·436	a
		CEPA	0·267	0·267	0·685	0·406	a
	(11, 7, 3/6, 1) → [7, 4, 3/4, 1]	SCF	0·662	0·662	0·848	0·725	a
		IEPA	0·824	0·824	0·950	0·865	a
		CEPA	0·725	0·725	0·907	0·785	a
	(11, 6, 3/5, 2)	SCF	0·663	0·663	0·853	0·726	b
		PNO–CI	0.740	0·740	0·930	0·803	b
		CEPA	0·766	0·766	0·954	0·828	b
		EXP	0·720	0·720	0·960	0·800	a
H_2O	STO-4G[2]	SCF	0·774	0·005	0·360	0·380	c,d
	(9, 5/4) → [5, 3/3]	SCF	1·026	0·350	0·728	0·701	c,d
	(9, 5, 2/4, 1) → [4, 3, 2/2, 1]	SCF	1·224	0·769	1·054	1·016	c,d
	(11, 7, 2/6, 2) → [7, 5, 2/4, 2]	SCF	1·341	1·157	1·230	1·233	c,d
	(11, 7, 3/6, 2) → [7, 5, 3/4, 2]	SCF	1·355	1·157	1·246	1·253	c,d
	(11, 6, 3/5, 2)	SCF	1·340	1·184	1·255	1·260	b
		PNO–CI	1·425	1·348	1·378	1·384	b
		CEPA	1·454	1·421	1·428	1·434	b
		EXP				1·455	b
H_3N	(11, 6, 3/5, 2)	SCF	1·931	1·931	2·072	1·978	b
		PNO–CI	2·027	2·027	2·298	2·117	b
		CEPA	2·060	2·060	2·414	2·178	b
		EXP				2·196	b

[1] Polarizabilities calculated at the equilibrium geometries. The coordinate system is chosen in such a way, that the z-axis coincides with the C_∞, C_2 or C_3 axis in HF, H_2O and H_3N respectively. The H_2O molecule lies in the xz plane.
[2] See Hehre, Stewart and Pople (1969) and Ditchfield, Hehre and Pople (1971).

References.
[a] Lischka (1974); [b] Werner and Meyer (1976b); [c] Beyer, Lischka and Schuster (1976); [d] Schuster, Lischka and Beyer (1976).

experimental data we notice that in some cases, e.g. H_2O, the calculations represent the only information of the anisotropy of the polarizability tensor. As in case of electric multipole moments we find that very flat polarization functions are necessary for numerically correct predictions. It seems worth mentioning here, that the basis sets applied in the most extensive calculations on hydrogen bonding and ion–molecule interactions reported so far are not flexible enough to reproduce polarizabilities correctly. A new approach to calculate molecular polarizabilities has been presented very recently by Sadlej (1977). He applied wavefunctions which are changed appropriately in the electric field. As reference systems the exact solutions for the harmonic oscillator in the electric field were used. The results obtained thereby are promising: using medium-size basis sets Sadlej (1977) was able to calculate reasonably correct molecular polarizabilities (errors $<15\%$).

Ab initio calculations of chemical shifts for nuclear magnetic resonance have been tried by several groups of authors. Due to a number of difficulties, however, the numerical results obtained have not yet reached a high degree of precision which is comparable to other molecular quantities. Lack of gauge invariance and basis set dependence of calculated results represent the major problems. We will refer here only to two particular approaches which have been applied also to hydrogen bonded complexes. Jaszunski and

Table 4.6. Proton magnetic shielding constants in some small molecules

Molecule	Basis set (GTO's)	Method	Shielding constants (ppm) σ_{xx}	σ_{yy}	σ_{zz}	$\bar{\sigma}$	$\bar{\sigma}-\bar{\sigma}\,(CH_4)$	Reference
CH_4	STO-5G [1]	SCF, GIAO [2]				32·40		[a]
	4–31G [1]	SCF, GIAO [2]				32·73		[a]
H_3N	STO-5G [1]	SCF, GIAO [2]				33·41	1·01	[a]
	4–31G [1]	SCF, GIAO [2]				33·59	0·86	[a]
		EXP					0·05	[a]
H_2O	STO-5G [1]	SCF, GIAO [2]				33·77	1·37	[a]
	4–31G [1]	SCF, GIAO [2]				32·79	0·06	[a]
	(11, 7, 1/6, 1) → [5, 4, 1/3, 1]	UCHF [3], EC	26·9	20·3	37·7	28·3		[b]
		UCHF [3], H	24·3	11·9	36·7	24·3		[b]
		EXP				30·2	−0·60	[a,b]
HF	STO-5G [1]	SCF, GIAO [2]				34·60	2·20	[a]
	4–31G [1]	SCF, GIAO [2]				30·61	−2·12	[a]
	(11, 7, 1/6, 1) → [5, 4, 1/3, 1]	CHF, F [3]				27·3		[c]
		CHF, EC [3]				28·7		[c]
		CHF, H [3]				54·2		[c]
		EXP				29·2	−2·50	[a,c]

[1] See Hehre, Stewart and Pople (1969), Ditchfield, Hehre and Pople (1971), Hehre, Ditchfield and Pople (1972) and Harihanan and Pople (1972).
[2] GIAO stands for 'Gauge Invariant Atomic Orbitals'.
[3] UCHF = uncoupled Hartree Fock perturbation theory; CHF = coupled Hartree Fock perturbation theory. The origin of the gauge is given as well: EC = centred at the electronic centroid, H = at the hydrogen nucleus and F = at the fluorine nucleus.

References. [a] Ditchfield (1974); [b] Jaszunski and Sadlej (1972); [c] Sadlej and Raynes (1975).

Sadlej (1972a,b; 1973) used the uncoupled HF procedure of Karplus and Kolker (1963) to calculate proton chemical shifts in the water molecule and in water dimer. More recently, Ditchfield (1974) proposed an SCF perturbation method which leads to gauge invariant chemical shifts. As expected, the numerical results obtained depend strongly on the basis set applied. Most of the relative changes in nuclear magnetic shielding are reproduced correctly by the largest basis sets he had used. Some representative examples are presented in Table 4.6.

Concluding this section we might characterize the present situation of *ab initio* calculation of molecular properties as follows. Due to the modern facilities for numerical computations, methods which previously were restricted to systems with a few electrons can now be extended to structures with up to roughly 30 electrons. For almost all important quantities of small molecules theoretical results can be obtained with an accuracy which can compete favourably with the experimental error limits. In some fortunate cases, for example, in case of anisotropies of molecular polarizabilities, quadrupole, or higher electric moments, etc., the calculated values are more reliable than experimental data or even represent the only source of information.

III. ENERGIES OF INTERACTIONS AND VIBRATIONAL FREQUENCIES OF HYDROGEN-BONDED COMPLEXES

Within the frame of MO theory intermolecular interactions can be studied in a straightforward way by the 'supermolecule' approach. A Hamiltonian is constructed for the complete system which consists of two or more atoms, molecules or ions and the Schrödinger equation is solved by the standard approximation techniques. For a binary complex, AB we obtain:

$$\mathbf{H}\Psi_{AB} = E_{AB}\Psi_{AB}; \qquad \mathbf{H} = \mathbf{H}_A^0 + \mathbf{H}_B^0 + \mathbf{V}_{AB} \tag{4.3}$$

H consists of the Hamiltonians of the isolated subsystems $\mathbf{H}_A^0$ and $\mathbf{H}_B^0$ and the interaction term $\mathbf{V}_{AB}$. The energies of interaction are obtained as differences:

$$\Delta E = E_{AB} - (E_A^0 + E_B^0) \tag{4.4}$$

E_A^0 and E_B^0 are total energies obtained by independent calculations of the isolated subsystems. In order to be consistent the same approximation method as used in the calculation of the complex has to be applied again. The major problem of the supermolecule approach to intermolecular forces can be visualized immediately from equation (4.3). E_{AB} and $E_A^0 + E_B^0$ are huge numbers compared to their difference. Consequently, no MO calculation on intermolecular interactions could be successful, before the numerical techniques had reached sufficiently high accuracy.

A. SCF results

The results of SCF calculations on energy surfaces of hydrogen-bonded complexes have been reviewed recently (Kollman and Allen, 1972; Schuster, 1976a). Therefore, we will discuss here only a few representative examples showing all important and characteristic features of hydrogen bonds: $(HF)_2$, $(H_2O)_2$, $(H_5O_2)^+$, $(H_3O_2)^-$, $(HF_2)^-$ and $F^-.H_2O$. In order to be able to recognize the differences, in case there are any, between hydrogen-bonded complexes and other strong intermolecular associations we will compare some results also with those obtained on a typical ion-molecule complex, $Li^+.OH_2$ (see also Schuster *et al.* (1975), Schuster, Lischka and Beyer (1976). Two very recent and extensive studies on hydrogen-bonded complexes should be mentioned here too: calculations on a large variety of complexes including hydrogen-bonded structures involving π-electron systems and compounds of third row elements (HCl, H_2S, PH_3, HCP and H_2CS) too (Kollman *et al.*, 1975) and a systematic study on basis set effects using various medium size basis sets (Dill *et al.*, 1975). Finally, we will refer the interested reader also to a recent study on symmetric intramolecular hydrogen bonds (Isaacson and Morokuma, 1975).

$(HF)_2$ is the simplest example for a hydrogen-bonded complex between two neutral molecules. Since it consists of two linear molecules the whole energy surface is four dimensional (Figure 4.2) if we assume frozen geometries of both HF molecules. A representative calculation covering the energetically important parts is within the range of present computational possibilities. Furthermore the most stable linear arrangement of all four atoms has almost the same energy as the tilted true equilibrium geometry. Linear geometry can be assumed therefore for model calculations without loosing all relevance for the real system. Due to higher symmetry the computation on linear arrangements are less time consuming and more accurate treatments are accessible, Additionally, $(HF)_2$ and $(H_2O)_2$ are the only two, small hydrogen-bonded complexes for which direct experimental data are available.

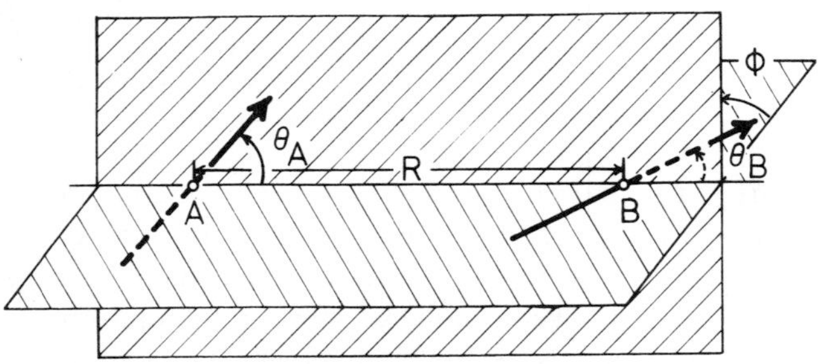

Figure 4.2. Four intermolecular coordinates (R, θ_A, θ_B and ϕ) characterizing the geometry of a complex formed from two linear molecules A and B

The most extensive calculations on linear $(HF)_2$ have been performed by Lischka (1974). The most complete investigations of the energy surface have been presented by Yarkony *et al.* (1974) and Diercksen and Kraemer (1970). The equilibrium structure obtained in the latter two investigations is roughly the same (Table 4.7). Variation of the HF bond length (R_{HF}) in the proton donor molecule leads to very small elongation only and therefore the rigid molecule approach of Yarkony *et al.* (1974) seems to be well justified. The calculated FF distance (R_{FF}) comes very close to the experimental value obtained by Dyke, Howard and Klemperer (1972) using molecule beam electric resonance technique: $R_{FF} = 2{\cdot}79$ Å. Variation of the angle θ_B lead to a very flat minimum around $\theta_B = 50°$ which is only 0·1 kcal/mol more stable than the completely linear structure. Variation of θ_A finally shows that the equilibrium geometry of the hydrogen bond is slightly tilted: $\theta_A \sim 355°$. As far as structural data for $(HF)_2$ are availabe the agreement with the SCF results is excellent indeed.

Table 4.7. Results of *ab initio* calculations on $(HF)_2$

Basic set (GTO's)	Method	Equilibrium geometry[1] R_{FF}(Å)	ΔR_{HF}(Å)	Θ_A	Θ_B	ΔE (kcal/mol)	Reference
(9, 5, 1/4, 1)→[4, 2, 1/2, 1]	SCF	2·80		355°	52°	−4·63	[a]
(11, 7, 1/6, 1)→[5, 4, 1/3, 1]	SCF	2·85		0°[2]	40°	−4·50	[b]
(11, 7, 1/6, 1)→[5, 4, 1/3, 1]	SCF	2·90		0°[2]	0°[2]	−4·38	[b]
(11, 7, 2/6, 1)→[7, 4, 2/4, 1]	SCF	2·90	0·003	0°[2]	0°[2]	−3·46	[c]
	IEPA	2·89	0·003	0°[2]	0°[2]	−3·25	[c]
	CEPA	2·89	0·002	0°[2]	0°[2]	−3·36	[c]
	EXP	2·79		0°	60°–70°	−6 ± 1·5	[d,e]

[1] According to Figure 4.2 'A' represents the proton donor molecule. ΔR_{HF} is the elongation of the central FH bond on hydrogen bond formation. The planar arrangement ($\phi = 0°$) was found to be most stable at the equilibrium value of R_{FF} (see Reference *a*).
[2] Constrained values.
References. [a] Yarkony *et al.* (1974); [b] Diercksen and Kraemer (1970); [c] Lischka (1974); [d] Dyke, Howard and Klemperer (1972); [e] Schuster (1976a).

Coming now to the absolute value of the hydrogen bond energy we realize immediately that much more care is needed than in case of the equilibrium structure. Although Diercksen and Kraemer (1970) as well as Yarkony *et al.* (1974) used rather extended basis sets, we find that the energy of interaction is changed substantially–decrease of about 20 per cent–when a second d-type polarization function with a small exponent is added to the basis set. This result is not completely unexpected since we have seen in the previous section that a similar basis set extension had a remarkable influence on some calculated properties of the isolated molecules. In case of HF and H_2O the addition of flat polarization functions to the basis set reduces dipole moments and increases polarizabilities. In a very qualitative way we can estimate the influence of these corrections on the energy of interaction in a complex between two polarizable dipole molecules A and B. According to the classical expression (see Section V) the electrostatic contribution is proportional to the square of the dipole moment (4.5) and will decrease in absolute value with decreasing $|\vec{\mu}|$. The polarization energy, on the other hand,

$$\Delta E^{\mathrm{CL}}_{\mathrm{COU}} = (1/R^3)\{\vec{\mu}_A \cdot \vec{\mu}_B - (3/R^2)(\vec{\mu}_A \cdot \vec{R})(\vec{\mu}_B \cdot \vec{R})\} + \dots \qquad (4.5)$$

is proportional to the product of the polarizability and the square of $|\vec{\mu}|$:

$$\Delta E^{\mathrm{CL}}_{\mathrm{POL}} = (1/R^6)\{\alpha_A \cdot |\vec{\mu}_B|^2 + \alpha_B\, |\vec{\mu}_A|^2\} + \dots \qquad (4.6)$$

For this simplified estimate spherical symmetry of the polarizability tensor has been assumed. On basis set extension one of these two factors will increase, whereas the other will decrease. Since both relative changes fall into the same order of magnitude we can assume the product remains roughly constant. Additionally, as we will see in Section V, around the energy minimum of the complex the polarization energy is substantially smaller in absolute value than the electrostatic contribution. At the equilibrium geometry of the complex we expect therefore a net destabilizing effect of basis set extension as it is observed indeed. In other parts of the energy surface the electrostatic contribution may have opposite sign and then, in contrary, we expect a lowering of the energy of interaction.

A second source of errors, most important for small, especially minimum, basis sets, concerns mutual basis set extension by the orbitals of the other molecules in the complex. An upper limit of the error caused thereby can be estimated by the function counterpoise method of Boys and Bernardi (1970), which consists in calculations of the isolated molecules using the basis set of the whole complex. These basis set superposition errors have been investigated for many hydrogen-bonded complexes (Meunier, Levy and Berthier, 1973; Johansson, Kollman and Rothenberg, 1973; Diercksen, Kraemer and Roos, 1975). In case of the most extended basis set used in a calculation on $(HF)_2$ this error amounts roughly one tenth of a kcal/mol, but may be substantially larger for smaller basis sets (Table 4.8).

Table 4.8. Basis set superposition errors in calculations on $(HF)_2$ and $(H_2O)_2$

			Energies in kcal/mol[2]			
Complex	Basis set (GTO's)	Method	$E_A - E_0$	$E_B - E_0$	$E_A + E_B - 2E_0$	Reference
$(HF)_2$	STO–3G[1]			$-1{\cdot}97$	$<-1{\cdot}97$	[a]
linear, $R_{FF} = 2{\cdot}91$ Å,	$(8, 4, 1/4, 1) \rightarrow [5, 3, 1/3, 1]$	SCF	$-0{\cdot}19$	$-0{\cdot}77$	$-0{\cdot}96$	[b]
$R_{HF} = 0{\cdot}917$ Å	$(11, 7, 2/6, 1) \rightarrow [7, 4, 2/4, 1]$	SCF	$-0{\cdot}025$	$-0{\cdot}044$	$-0{\cdot}07$	[b]
$(H_2O)_2$, $\Theta_B = 54°$,	STO–3G[1]	SCF		$-2{\cdot}85$	$<-2{\cdot}85$	[a]
$R_{OO} = 3{\cdot}00$ Å	$(11, 7, 1/6, 1) \rightarrow [5, 4, 2/3, 1]$	SCF	$-0{\cdot}39$	$-0{\cdot}42$	$-0{\cdot}81$	[c]
		SCF–CI	$-0{\cdot}77$	$-0{\cdot}77$	$-1{\cdot}54$	[c]

[1] See Hehre, Stewart and Pople (1969) and Ditchfield, Hehre and Pople (1971).

[2] E_A and E_B represent the energies of the proton acceptor and proton donor molecule calculated with the basis set of the whole complex. E_0 is the energy of the isolated monomer.

References: [a] Johansson, Kollman and Rothenberg (1973); [b] Lischka (1974); [c] Diercksen, Kraemer and Roos (1975).

The results obtained in some extended calculations on $(H_2O)_2$ are summarized in Table 4.9. On the whole the situation is essentially the same as in $(HF)_2$ besides the fact that due to additional degrees of freedom a complete investigation of the energy surface requires much more computer time. Popkie, Kistenmacher and Clementi (1973), Diercksen (1969, 1971) and Hankins, Moskowitz, and Stillinger (1970a,b; 1973) have made the most extensive SCF calculations on this surface. Only in Diercksen's paper (1971) the deformation of the proton donor molecule in the field of the proton acceptor is reported. Similar to the result on $(HF)_2$ a small elongation of the central OH bond was obtained. Again no experimental information on this change in geometry is available. The data obtained from various structures of ice are not comparable to vapour phase results as we will see in Section III.C. The other predictions of SCF calculations on the structure of $(H_2O)_2$ agree well with the measurement of Dyke and Muenter (1974). The calculated energies of interaction again are highly sensitive to changes in the basis sets applied. For both dimers $(HF)_2$ and $(H_2O)_2$ the experimental data on the energy of complex formation are at least as uncertain as the theoretical results.

Table 4.9. Results of *ab initio* calculations on $(H_2O)_2$

Basis set (GTO's)	Method	Equilibrium geometry[1] R_{OO}(Å)	ΔR_{OH}(Å)	Θ_B	ΔE (kcal/mol)	References
(11, 5, 1/4, 1)→[5, 3, 1/2, 1]	SCF	3·00		40°	−5·0	a
(11, 7, 1/6, 1)→[5, 4, 1/3, 1]	SCF	3·00	0·002		−4·84	b
(11, 7, 1/6, 1)→[5, 4, 1/3, 1]	SCF	2·99		41·5°	−5·14	c
	SCF–CI	2·92		42·6°	−6·05	c
(13, 8, 2, 1/6, 2, 1)→[8, 5, 2, 1/4, 2, 1]	SCF	3·00		0°[2]	−3·67	d
	SCF	3·00		30°[2]	−3·90	d
(11, 7, 1/6, 1)→[4, 3, 1/2, 1]	SCF	3·01		40°	−4·55	e
	SCF–CI	2·98		40°	−5·63	e
	EXP	2·98 ± 0·04		60° ± 10°	−5·0	f,g

[1] Linear or almost linear hydrogen bonds O—H...O. C_S symmetry is assumed. Θ_B denotes the angle between the C_2 axis of the proton acceptor molecule 'B' and the hydrogen bond axis. ΔR_{OH} is the elongation of the central OH bond on hydrogen bond formation.
[2] Constrained values.
References. [a] Hankins, Moskowitz and Stillinger (1970a,b, 1973); [b] Diercksen (1971); [c] Diercksen, Kraemer and Roos (1975); [d] Popkie, Kistenmacher and Clementi (1973); [e] Matsuoka, Clementi and Yoshimine (1976); [f] Dyke and Muenter (1974); [g] see Schuster (1976a).

IR spectra of complexes with hydrogen bonds between neutral molecules have been calculated by *ab initio* methods as well (e.g. see Janoschek, 1976). The most extensive theoretical study on the vibrational spectrum of $(H_2O)_2$ has been performed recently by Curtiss and Pople (1975). For the major part of their investigations they used a small basis set without polarization functions, which is well known to overemphasize the hydrogen bond energy in $(H_2O)_2$ (see Schuster, 1976a). Nevertheless, the general features of the experimental spectrum measured by matrix isolation technique in solid N_2

Table 4.10. Calculated force constants and IR frequencies for $(HF)_2$ and $(H_2O)_2$

Molecule or complex	Basis set (GTO's)	Method	XH Stretching modes[1] f_{XH}(mdyn/Å)	Wave numbers ν_i'(cm^{-1})	XX Stretching mode[1] f_{XX}(mdyn/Å)	HXH Bending modes[1] $f_{HXH} \cdot R_e^2$ (mdyn/Å)	HXH Bending modes[1] Wave numbers ν_i'(cm^{-1})	References
HF	(8, 4/4)→[4, 3/3]	SCF	9·42					a
	(11, 7, 2/6, 1)→[7, 4, 2/4, 1]	SCF	11·20					b
		IEPA	9·27					b
		CEPA	9·92					b
		EXP	9·66					b
$(HF)_2$	(8, 4/4)→[4, 3/3]	SCF	9·24, 9·41		0·23			a
	(11, 7, 2/6, 1)→[7, 4, 2/4, 1]	SCF	11·0		0·095			b
		IEPA	9·18		0·101			b
		CEPA	9·83		0·104			b
H_2O	4-31G[2]	SCF	9·05	3960, 4098		0·856	1767	c
		EXP, VAPOUR	8·43	3657, 3756		0·768	1595	c
		EXP, N_2MATRIX[3]	7·52	3632, 3725		0·758	1597	c,d
$(H_2O)_2$	4-31G[2]	SCF	8·66, 9·15, 9·15, 9·15	3907, 3979, 4085, 4121		0·898, 0·858	1813, 1771	c
		EXP, N_2MATRIX[3]	7·04, 7·44, 7·50, 7·54	3548, 3626, 3698, 3714		0·787, 0·756	1618, 1600	c,d
	(11, 7, 1/6, 1)→[5, 4, 1/3, 1]	SCF			0·14			e
		SCF–CI			0·16			e

[1] For complexes the harmonic force constants for pure XH or XX stretching or HXH bending motions are given. In case one value is presented only it refers to the force constant of the central XH bond involved in hydrogen bonding.

[2] See Ditchfield, Hehre and Pople (1971) and Hehre, Ditchfield and Pople (1972).

[3] The values of the force constants are not corrected for an harmonic effects.

References. [a] Karpfen and Schuster (1976); [b] Lischka (1974); [c] Curtiss and Pople (1975); [d] Tursi and Nixon (1970); [e] Diercksen, Kraemer and Roos (1975).

(Tursi and Nixon, 1970; Van Thiel, Becker and Pimentel, 1957) are described correctly (Table 4.10). The most serious errors, of course, are due to the weakness of SCF calculations in reproducing stretching force constants correctly. All wavenumbers for stretching vibrations obtained by Curtiss and Pople (1975) are too high by 350–400 cm^{-1}. Similar results have been reported on $(HF)_2$ (Table 4.10). In this case we should note that small basis sets—we will use one of them latter in crystal orbital calculations—sometimes are useful for predictions of roughly correct geometries and force constants due to error compensation. In general, the shifts to lower frequencies of stretching vibrations of HX bonds involved in hydrogen bonding are much smaller in isolated dimers than those observed in associated liquids and hydrogen bonded crystals.

One of the most remarkable effects of hydrogen bonding in IR spectra is the strong increase in band intensities of HX stretching modes. The intensity of a vibrational mode (Q_ν) is proportional to the square of the corresponding dipole moment derivative. For the increase in intensity on complex formation we obtain:

$$I_R = \{|\partial\vec{\mu}/\partial Q_\nu|_{AH.B}\}^2/\{|\partial\vec{\mu}/\partial Q_\nu|_{AH}\}^2 \quad (4.7)$$

An experimental value of $I_R = 12$ has been reported by Van Thiel, Becker and Pimentel (1957), whereas the more recent results of Tursi and Nixon (1970) indicate a much smaller value. In his extended SCF calculation Diercksen (1971) obtained a value of $I_R = 5{\cdot}3$ in contrast to some investigations with small and medium size basis sets which suggested a larger increase in intensities (Kollman and Allen, 1972).

Extensive SCF calculations have been reported for a number of systems with strong hydrogen bonds like $(H_5O_2)^+$, $(H_3O_2)^-$, $(HF_2)^-$ etc. (Kraemer and Diercksen, 1970, 1972; Janoschek *et al.*, 1972; Almlöf, 1972; Meyer, Jakubetz and Schuster, 1973; Støgård *et al.*, 1975; and Keil and Ahlrichs, 1976). These systems differ strongly from those discussed above: energies of interaction are large and molecular geometries change drastically on complex formation. It may well be questionable if it is a reasonable approach to include these systems in the series of intermolecular interactions. In Section VI we will interpret strong hydrogen bonds also in an alternative way as a particular type of three-centre bond with electron excess. Another important feature of strong hydrogen bonds is the high mobility of the central proton. Although the molecule and the ion are bound together strongly the proton which occupies a symmetric or almost symmetric position in the unperturbed complex is shifted easily back and forth along the direction of the hydrogen bond by external forces. The perturbation can be caused, for example, by the electric field present in the crystal. The enormous variety of structural data on the hydrated proton in the solid state may serve as one example (Lundgren and Olovsson, 1976). Even in case of the strongest hydrogen-bonded complex known, $(HF_2)^-$, a highly asymmetric position of the central proton has been observed in some crystals (*p*-toluidinium bifluoride: $R_{HF_1} = 1{\cdot}235$ Å, $R_{HF_2} = 1{\cdot}205$ Å see Williams and Schneemeyer, 1973. The high polarizability of strong hydrogen bonds has been discussed extensively also by Janoschek *et al.* (1972). The unusual behaviour of the central proton can be explained in terms of potential curves for proton motion, which are flat in the neighbourhood of the energy minimum. In symmetric systems, $(XHX)^{\pm}$, cuts of the two-dimensional energy surface $E = E(R_{HX}, R_{XX})$ at fixed vaues of R_{XX} show a transition from single minimum potentials at small distances ($R_{XX} < R_{XX}^{TR}$) to double minimum potentials at large distances ($R_{XX} > R_{XX}^{TR}$, see Figure 4.3). Around the transition point the potential curves for proton motion are especially flat. All the complexes with strong hydrogen bonds investigated by accurate *ab initio*

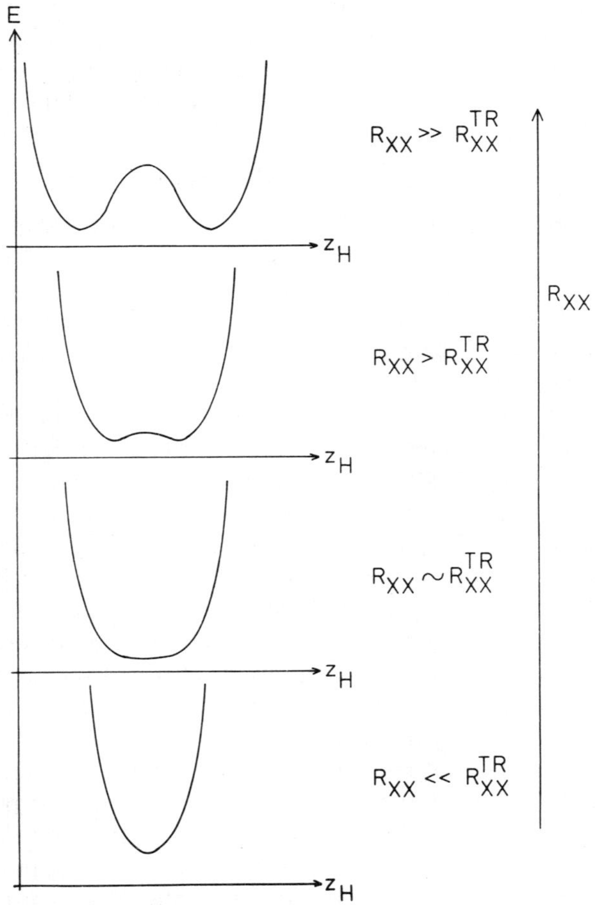

Figure 4.3. Cuts through the two-dimensional energy surface $E = E(R_{HX}, R_{XX})$ in strong and symmetric hydrogen bonds at fixed values of R_{XX} (schematically; central proton moving along the z-axis)

methods until now are characterized by equilibrium geometries around this transition point ($R_{XX} \sim R_{XX}^{TR}$).

Symmetry, in fact, plays a tremendous role for strong hydrogen bonds. Replacing the HF molecule in $(HF_2)^-$ by H_2O we change many features of the complex drastically: the equilibrium structure becomes highly asymmetric, the H_2O molecule in $F^-.H_2O$ is recognized easily since only minor changes in geometry occur on complex formation and furthermore the proton in the hydrogen bond occupies a well-defined position and is not influenced very much by external perturbations. $F^-.H_2O$ is interpreted best as an ion–molecule complex superimposed by the formation of an ordinary hydrogen bond. For illustration the calculated data are summarized in Table 4.11. Additionally, the results obtained for a 'pure' ion–molecule complex $Li^+.OH_2$ are presented.

Table 4.11. Results of *ab initio* calculations on strong hydrogen bonds and some ion–molecules complexes

Complex	Basis set (GTO's)	Method	Equilibrium geometries[1] $R_{XX(Y)}$ (Å)	R_{HX} (Å)	ΔR_{HX} (Å)	ΔE (kcal/mol)	References
$(FHF)^-$	(9, 5/4, 1)→[4, 2/2, 1]	SCF	2·26	1·13	0·21	−53	a
		SCF–CI	2·28	1·14	0·21	−56	a
	(10, 7, 2/6, 1)→[6, 4, 2/4, 1]	SCF	2·22[2]	1·11[2]	0·20[2]	−42·7	b
		CEPA	2·22[2]	1·11[2]	0·20[2]	−44·3	b
$(H_3O_2)^-$	(11, 7, 1/6, 1)→[5, 4, 1/3, 1]	SCF	2·51	1·16	0·22	−24·3	c
	(9, 5/4, 1)→[4, 2/2, 1]	SCF	2·49	1·06	0·10	−34	a
		SCF–CI	2·40	1·20	0·24	−39	a
$(H_5O_2)^+$	(11, 7, 1/6, 1)→[5, 4, 1/3, 1]	SCF	2·39	1·14	0·20	−32·2	d
$F^-.H_2O$	(11, 7, 2/6, 2)→[4, 3, 2/2, 2]	SCF	2·51	1·00	0·04	−23.7	e
	(11, 7, 1/6, 1)→[5, 4, 1/3, 1]	SCF	2·51	0·94[2]		−24·2	f
		SCF–CI	2·47	0·94[2]		−26·2	f
$Li^+.OH_2$	(11, 7, 1/6, 1)→[5, 4, 1/3, 1]	SCF	1·83			−34·0	f
		SCF–CI	1·84			−32·9	f

[1] R_{XX} or R_{XY} represents the distance between the two second row atoms, that come closest together. R_{HX} is always the smaller XH or YH bond length in the complex and ΔR_{HX} finally stands for the change in bond length on hydrogen bond formation. The three atoms forming the hydrogen bond (XHX) or (XHY) are in a linear or almost linear arrangement.

[2] Constrained values.

References. [a] Støgård *et al.* (1975); [b] Keil and Ahlrichs (1976); [c] Kraemer and Diercksen (1972); [d] Kraemer and Diercksen (1970); [e] Kistenmacher, Popkie and Clementi (1973); [f] Diercksen, Kraemer and Roos (1975).

Vibrational spectra of systems with strong hydrogen bonds have been investigated by theoretical methods as well. As expected, for the shape of the energy surface XH and XX′ stretching modes are coupled strongly and the harmonic approximation is no longer a reasonable approach. The IR spectra are completely different from those of the isolated systems. From some of these two-dimensional energy surfaces the Schrödinger equation of nuclear motion has been calculated by approximation methods using CI (HF_2^-: Almlöf, 1972; Jiang and Anderson, 1973; Støgård *et al.*, 1975; $H_5O_2^+$: Janoschek *et al.*, 1972; Janoschek, Weidemann and Zundel, 1973; Anderson and Jiang, 1973 and multiconfiguration (MCI) methods (Priglinger,

Table 4.12. Calculated and observed IR frequencies for $(HF_2)^-$ (Støgård *et al.*, 1975)

Calculated wave numbers (cm^{-1})		
SCF[a] energy surface	SCF–CI[a] energy surface	Observed wavenumbers (cm^{-1})
619	605	610
1483	1611	1452
2042	2162	2045
2589	2703	2622
3138	3245	3174
3730	3828	3839
4428	4509	4831
5176	5344	5092

[a] Basis set applied: (9, 5/4, 1)→[4, 2/2, 1].

1975; Priglinger and Schuster, 1976). The results obtained in case of $(HF_2)^-$ are shown in Table 4.12.

B. Correlation effects

Electron correlation was found to be important for accurate calculations on energy surfaces of both ordinary and strong hydrogen bonds. We will start with the usual type of hydrogen bonding, where very general correlation effects are observed which occur also in all other kinds of intermolecular associations, and then discuss the special contributions present only in strong hydrogen bonds.

In general Hartree–Fock calculations cannot account for dispersion energies and therefore a stabilizing contribution to the energy of interaction is always missing in SCF calculations. According to the asymptotic formulas derived from intermolecular perturbation theory (e.g. see Margenau and Kestner, 1969) dispersion energies depend on molecular polarizabilities and ionization potentials but not on charges or multipole moments. Neglect of dispersion forces, therefore, is most serious in case of weakly interacting systems where the total energy of interaction is small. Among the examples discussed here we expect the contribution of dispersion energy to be much more important in associations between neutral molecules than in ion–dipole complexes.

Besides its principal importance for dispersion forces, electron correlation contributes to intermolecular energy surfaces also in an indirect way by changing the properties of the isolated subsystems. Appropriately, we can distinguish both effects by splitting the change in correlation energy into an intermolecular part—corresponding to the dispersion forces—and an intramolecular part. For the interpretation of the 'intramolecular' contributions of electron correlation we make use again from the classical formulas in equations (4.5) and (4.6). In H_2O and HF, as we have seen previously, electron correlation reduces the dipole moments and increases the polarizabilities. Again we can estimate a small net effect of the intramolecular contribution. In fact this term is destabilizing in $(HF)_2$ (Lischka, 1974, whereas it was found to be almost zero in $(H_2O)_2$ (Matsuoka, Clementi and Yoshimine, 1976) at the equilibrium structures. Superimposing now the intermolecular contributions we obtain an almost negligible contribution of electron correlation to the total energy of interaction in $(HF)_2$, whereas a net stabilizing effect is observed in $(H_2O)_2$ (Table 4.13). Electron correlation effects in $F^-.H_2O$ and $Li^+.OH_2$ are of the same order of magnitude as in $(HF)_2$ or $(H_2O)_2$ and can be analysed in exactly the same way. Due to the larger energy of interaction, correlation contributions are relatively less important here.

Table 4.13. Intra- and intermolecular correlation effects in hydrogen bonded or ion—molecules complexes near the equilibrium geometries

			Energies in kcal/mol			
Complex	Basis set (GTO's)	Method	ΔE_{COR}^{INTRA}	ΔE_{COR}^{INTER}	ΔE_{COR}	Reference
$(HF)_2$	$(11, 7, 2/6, 1) \rightarrow [7, 4, 2/4, 1]$	CEPA	0·67	−0·71	−0·09	a
$(H_2O)_2$	$(11, 7, 1/6, 1) \rightarrow [5, 4, 1/3, 1]$	SCF–CI			−0·87	b
	$(11, 7, 1/6, 1) \rightarrow [4, 3, 1/2, 1]$	SCF–CI	−0·41	−1·20	−1·61	c
$(FHF)^-$	$(8, 5, 1/4, 1) \rightarrow [4, 3, 1/2, 1]$	CEPA	8·22	−10·04	−1·82	d
$F^-.H_2O$	$(11, 7, 1/6, 1) \rightarrow [5, 4, 1/3, 1]$	SCF–CI			−1·94	b
$Li^+.OH_2$	$(11, 7, 1/6, 1) \rightarrow [5, 4, 1/3, 1]$ Li: $(11, 2) \rightarrow [5, 2]$	SCF–CI			1·16	b

References. [a] Lischka (1974; [b] Diercksen, Kraemer and Roos (1975); [c] Matsuoka, Clementi and Yoshimine (1976); [d] Keil and Ahlrichs (1976).

Correlation effects on vibrational spectra of ordinary hydrogen-bonded complexes mainly consists in a correction of the force constants for stretching vibrations without changing the relative shifts of the individual normal modes. The influence of electron correlation on energy surfaces of systems with strong hydrogen bonds has been investigated for $(H_5O_2)^+$ by Meyer, Jakubetz and Schuster (1973) and for $(HF)^-$ and $(H_3O_2)^-$ by Støgård *et al.* (1975). Both groups of authors come to the same conclusions: symmetric structures with the proton in the middle between the two heavier atoms are favoured by correlation effects. Consequently, the energy barrier for proton transfer at fixed distances R_{XX}, in case there is any, is lowered appreciably. Correlation effects on strong hydrogen-bonded systems are discussed more appropriately in terms of chemical bonds than by asymptotic perturbation theory. The covalent dissociation of an ordinary two-electron bond is not reproduced correctly by Hartree–Fock calculations because the single determinant wavefunction applied cannot dissociate into the ground states of two open shell systems. Near the energy minimum ionic dissociation starts exactly in the same way as covalent dissociation (Figure 4.4) and we obtain an increase in correlation energy with increasing distance. At a certain distance, however, the lone pair starts to form and the cation is released. Simultaneously, the coorelation energy becomes smaller again. For ionic

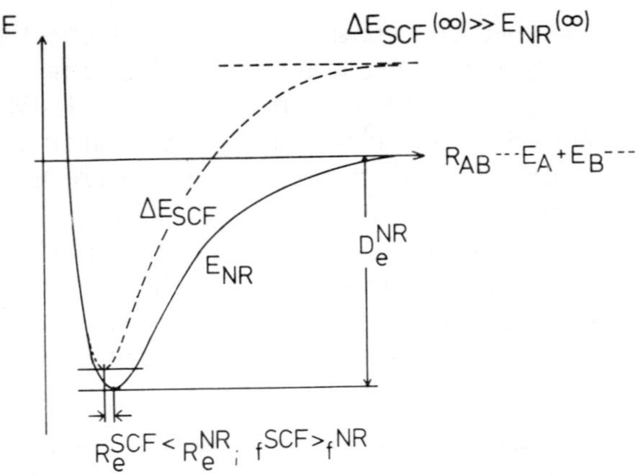

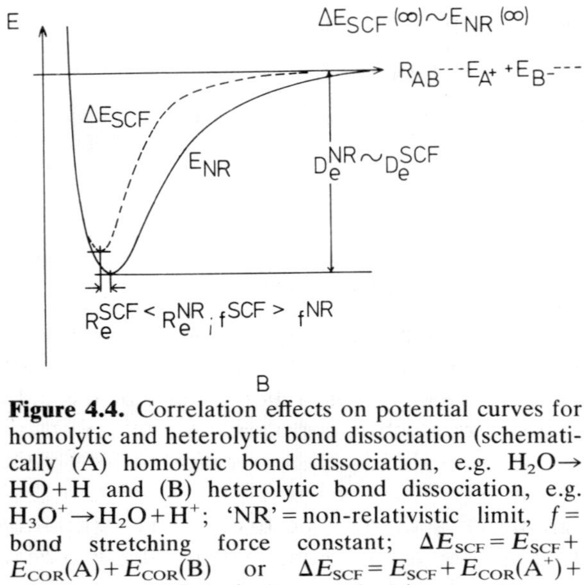

Figure 4.4. Correlation effects on potential curves for homolytic and heterolytic bond dissociation (schematically (A) homolytic bond dissociation, e.g. $H_2O \rightarrow HO + H$ and (B) heterolytic bond dissociation, e.g. $H_3O^+ \rightarrow H_2O + H^+$; 'NR' = non-relativistic limit, f = bond stretching force constant; $\Delta E_{SCF} = E_{SCF} + E_{COR}(A) + E_{COR}(B)$ or $\Delta E_{SCF} = E_{SCF} + E_{COR}(A^+) + E_{COR}(B^-)$ respectively)

dissociation we find the interesting situation (Figure 4.4) that electron correlation changes the shape of the dissociation curve but does not influence the dissociation energy to a large extent. Proton affinities (Table 4.1) may serve as appropriate examples for energies of ionic dissociation. On proton motion along a strong hydrogen bond we dissociate one XH bond partially as long as we have reached the symmetric position. provided the distance R_{XX} is not extremely large the symmetric structure is characterized by a maximum of correlation energy. The influence of electron correlation on heights of barriers for proton transfer is illustrated in Table 4.14 (for further details see Schuster *et al.*, 1974.

Table 4.14. Correlation effects on barriers for proton transfer in complexes with strong hydrogen bonds

Complex	Basis set (GTO's)	Method	R_{XX} (Å) [1]	R^0_{XH} (Å) [1]	$\Delta E^{\pm}$ (kcal/mol)	References
$(H_5O_2)^+$	$(9, 5, 1/4, 1, 1)^2 \rightarrow [5, 3, 1/3, 1, 1]$	SCF	2·74	1·02	9·54	[a]
		CEPA	2·74	1·04	6·97	[a]
$(H_3O_2)^-$	$(9, 5/4, 1) \rightarrow [4, 2/2, 1]$	SCF	2·51	1·06	1·8	[b]
		SCF–CI	2·51	1·09	0·7	[b]
		SCF	2·65	1·05	5·7	[b]
		SCF-CI	2·65	1·07	3·5	[b]
$(HF_2)^-$	$(9, 5/4, 1) \rightarrow (4, 2/2, 1]$	SCF	2·43	1·03	1·5	[b]
		SCF–CI	2·43	1·10	0·3	[b]
		SCF	2·54	1·01	4·7	[b]
		SCF–CI	2·54	1·06	2·2	[b]
		SCF	2·65	0·98	9·3	[b]
		SCF–CI	2·65	1·01	6·0	[b]

[1] The potential curves were calculated at fixed distances R_{XX}. R^0_{XH} represents the equilibrium value of the shorter XH distance.
[2] Polarization functions were applied on the central hydrogen atom only.
References. [a] Meyer, Jakubetz and Schuster (1973), [b] Støgård *et al.* (1975).

In complexes with symmetric equilibrium structures like $(HF_2)^-$ we expect further stabilization of the complex by correlation energies which actually has been found by Støgård *et al.* (1975) and Keil and Ahlrichs (1976). $(H_3O_2)^-$, on the other hand, shows an asymmetric equilibrium structure at the SCF level which is changed drastically by the contribution of electron correlation. Now a symmetric structure is found to be most stable (Table 4.11).

The influence of electron correlation on the calculated vibrational frequencies of $(HF_2)^-$ has been investigated by Støgård *et al.* (1975). As we can see from Table 4.12 only minor shifts were obtained.

C. Large clusters and hydrogen-bonded crystals

Cluster and crystals of molecules forming hydrogen bonds have been calculated by semi-empirical and *ab initio* SCF methods in the past. A review of the work done in this field has been given recently (Schuster, 1976a). We recommend it here as a source for further references. There are mainly two reasons for studying clusters of many molecules:

1. Previous investigations indicated that energies of interaction in hydrogen-bonded systems lack pairwise additivity. The knowledge on presence or absence of strong three body potentials or even higher order interactions is of primary importance for theoretical studies on associated liquids by statistical mechanics or numerical methods like Monte Carlo or molecular dynamics calculations. Semi-empirical as well as *ab initio* calculations with small basis sets tend to overemphasize these cooperative effects and turned out to be not reliable enough for relevant predictions on

energies of interaction. The most extensive *ab initio* studies on aggregrates of H_2O molecules up to $(H_2O)_4$ have been performed by Hankins, Moskowitz and Stillinger (1970a,b; 1973) and by Lentz and Scheraga (1973). They performed also a partitioning of the energies of interaction into two body and three-body contributions, thereby giving an idea of the relative importance of three body forces in different regions of the energy surface. All these calculations were performed at frozen monomer geometries.

2. A second aspect of investigations on clusters or crystals of hydrogen-bonded monomers is even more interesting for our aims here: How do molecular geometries and force constants or vibrational frequencies change, when we proceed from an isolated dimer to higher aggregates? From previously mentioned, very accurate studies on dimers and from the available experimental data we are inclined to conclude that the typical features of hydrogen bonding (see II in Figure 4.1) can be observed in the dimer to a very small extent only. The more drastic differences by far are found by a comparison of the dimer and the condensed phase. Although an explanation of this phenomenon is of primary importance for any general theory of hydrogen bonding, almost no really conclusive studies have been reported in this field. The main reason for this lack of extensive investigations is evident: any attempt to calculate accurate equilibrium geometries and vibrational spectra of clusters containing three or more molecules by *ab initio* methods has to face enormous computational problems which lie outside our present possibilities. Furthermore three, ten or even one hundred molecules do not represent an appropriate model for an associated liquid. Since we have no efficient formalism for the treatment of liquids in statistical mechanics, it seems hopeless at present to try an *ab initio* calculation of average structures and vibrational spectra of liquid water or hydrogen fluoride.

The changes in molecular geometries and vibrational spectra mentioned above do occur also in hydrogen-bonded crystals. In most cases the differences observed between dimer and crystal are even larger than those between dimer and liquid. An ideal crystal is characterized by perfect translational symmetry and therefore only a few degrees of freedom contribute significantly to the vibrational or phonon spectrum. Many theoretical methods for investigations on molecular crystals have been proposed and applied to hydrogen-bonded solids. For the studies we have in mind here the crystal orbital (CO) method developed by Del Re, Ladik and Biczo (1967) as well as André (1969) seemed to be most appropriate. This method can be interpreted as a straight forward extension of MO-SCF theory to infinite systems. The whole crystal is treated as a supermolecule. Translational symmetry is introduced by the Born–Von Karman periodic conditions. Usually one considers only nearest neighbour interactions, i.e. only the interactions within and between two neighbouring elementary cells are taken into account. Karpfen *et al.* (1974) have shown, however, that for crystals formed from polar molecules interaction with further elementary cells have to be taken into account as well, if a reasonably correct description is desired.

The $(HF)_\infty$ crystal is especially suited for model studies, since it consists of rather weakly bound infinite chains and can be represented therefore by models consisting of a single infinite chain of HF molecules only. Calculations of $(HF)_\infty$ in the one-dimensional chain approximation and as three-dimensional crystal have been performed with the tight bonding method by Bassani, Pietronero and Resta (1973) and Pietronero and Lipari (1975). An *ab initio* crystal orbital calculation on one-dimensional $(HF)_\infty$ was presented recently by Kertesz, Koller and Azman (1975). In all these calculations the geometry of the elementary cell was kept fixed. Crystal orbital calculations with extensive geometry variation have been performed in our laboratory (Karpfen, 1976; Karpfen and Schuster, 1976). Additionally, these calculations were extended to second and third neighbours. For technical and methodological reasons a rather small basis set had to be used. The results obtained from calculations on HF and $(HF)_2$ have been mentioned already. For the purpose of comparison they are shown together with the results from CO calculations in Table 4.15. Interestingly, we obtained fairly good

Table 4.15. Crystal orbital (CO) SCF calculations on a linear chain of (HF)[a]

Structure	R_{HF} (Å)	R_{FF} (Å)	f_{HF} (mdyn/Å)	ΔE^b (kcal/mol)	$q_H{}^c$	$q_F{}^c$
HF cal.	0·917	—	9·42	—	0·530	9·470
exp.	0·917	—	9·66	—		
$(HF)_2$ calc.	0·919	2·704	9·23	−7·49	0·512	9·511
exp.		2·79 ± 0·05		−6 ± 1		
$n = 1^d$	0·934	2·485	7·10	−10·18	0·468	9·532
(HF) $n = 2^d$	0·936	2·491	6·96	−11·26	0·444	9·556
$n = 3^d$	0·937	2·490	6·85	−11·71	0·437	9·536
Crystal, exp.	0·950	2·49	6·52[e] 5·24[e]			

[a] Basis set: (8, 4/4) contracted to [4, 3/3], for the exponents and contraction coefficients see Huzinaga (1970). References: Karpfen (1976); Karpfen and Schuster (1976).
[b] The energy values refer to the monomer: $\Delta E[(HF)_n] = 1/n\{E[(HF)_n] - E(HF)\}$.
[c] Electron densities derived from Mulliken overlap populations.
[d] n represents the number of neighbouring elementary cells up to which the interactions are calculated explicitly.
[e] The values correspond to the antisymmetric and symmetric FH stretching mode.

results for both, molecular geometries and force constants. As far as these quantities are concerned the results obtained for the dimer were satisfactory as well. In comparison to more extended calculations the energy of interaction is predicted too large as we have to expect for basis sets of this size.

Coming to the results for the infinite, linear chain $(HF)_\infty$ finally, we notice convergence of all calculated results with increasing number of neighbours explicitly included in the calculation. Furthermore, the previously mentioned changes in force constants and molecular geometries caused by formation of a condensed phase are predicted correctly here. We can attribute therefore the changes observed experimentally to the formation of long chains or perhaps also networks of hydrogen bonds. Asking for an explanation of the difference between hydrogen bonds in a dimer and in an infinite chain we recall the potential curves for proton transfer at its equilibrium value we expect and find a highly asymmetric curve with a single minimum in the dimer. In the infinite linear chain there are, of course, an infinite number of independent and coupled proton motions. We regard the simultaneous in phase transfer of all protons which gives rise to the smallest possible barrier for proton transfer. For this mode we obtain a symmetric double minimum potential with an energy barrier of $\Delta E^{\pm} = 12{\cdot}8$ kcal/mol. For the static properties of the crystal the presence of a second minimum has a number of consequences: compared to the dimer the potential curve for proton motion becomes flatter around the minimum, the hydrogen bond becomes stronger, the FH bond distance increases and the force constant decreases. Although more extensive calculations and investigations on other structures are necessary before final conclusions can be drawn, we think that the effects on structures, stabilities and vibrational spectra observed in hydrogen-bonded crystals can be reproduced and explained within the frame of Hartree–Fock MO theory for crystals.

IV. THE WAVEFUNCTIONS OF HYDROGEN-BONDED COMPLEXES AND DERIVED PROPERTIES

Wavefunctions contain a large amount of intrinsic information which we will use in part to explain other characteristic properties of hydrogen bonds (see Figure 4.1). We will concentrate on two aspects:

1. an analysis of density distributions and electric moments of hydrogen bonded complexes and
2. a discussion of the changes of chemical shifts in proton magnetic resonance spectra which are observed as a result of hydrogen bond formation.

In the previous section we have mentioned already IR band intensities which represent another quantity derived from wavefunctions. As we have seen the increase in band intensities is reproduced by the extensive calculation available for $(H_2O)_2$ (Diercksen, 1971). It is very difficult, however, to give a reliable estimate on the extent of possible errors in the numerical values. More calculated data on IR intensities would be necessary for this purpose. Furthermore, the experimental values available contain large uncertainties as well. We decided therefore not to discuss this quantity in more detail here.

A. Electron density distributions

Electron density distributions are obtained directly from the wavefunction by integration:

$$\rho(\vec{r}) = \int\int \ldots \int \Psi^*\Psi \, d\sigma_1 \, d\tau_2 \ldots d\tau_n \tag{4.8}$$

$d\tau_i$ is the complete volume element for electron 'i' consisting of a space and a spin part: $d\tau_i = d\vec{r}_i \, d\sigma_i = dx_i \, dy_i \, dz_i \, d\sigma_i$.

Since $\rho(\vec{r})$ is a three-dimensional function it can be tabulated only but not presented easily by a graphical plot. $\rho(\vec{r})$ can be discussed also by two-dimensional cuts through it. In planar systems the cuts containing the molecular plane provide us with all the information needed. In general, $\rho(\vec{r})$ does not tell us much more than the precise position of the individual atoms, which, of course, are known as input data in quantum mechanical calculations based on the Born–Oppenheimer approximation. Hydrogen atoms are recognized as little hills only reflecting thereby the well known problems in X-ray scattering, where scattering amplitudes are proportional to electron densities.

Density difference functions, on the other hand, are much more illustrative since they present the changes in electron densities on complex formation. They are obtained simply by subtraction of the densities of the isolated subsystems from the the density function of the complex (4.9).

$$\Delta\rho(\vec{r}) = \rho_{AB}(\vec{r}) - \{\rho_A(\vec{r}) + \rho_B(\vec{r})\} \tag{4.9}$$

The molecular geometry has to be exactly the same in both cases. Density difference functions can be represented by two-dimensional cuts in the same way as density functions. For the purpose of comparison we represent here densities and density differences plots for typical examples of three different types of complexes: $(H_2O)_2$, $F^-.H_2O$ and $Li^+.OH_2$ (Figures 4.5, 4.6 and 4.7).

First of all we recognize that density difference functions always show strong oscillations near the atoms with inner shell electrons. This is a well known phenomenon observed also in diatomic molecules. Apart from this general feature we can recognize an electron-deficient area in structures with hydrogen bonds, which lies around the position of the central hydrogen atom. This local decrease in electron density simply is a result of electron repulsion between the two closed shell systems. An alternative interpretation explains the reduced electron density by an increased polarization of the XH bond in the field of the lone pair electrons forming the hydrogen bond. In $Li^+.OH_2$ a typical example for a cation–molecule complex we observe strong polarization of the H_2O molecule which leads to a pronounced increase in electron density in the region between the two subsystems. The K-shell electrons at the Li nucleus are weakly polarized only.

Although density difference functions illustrate the complicated rearrangements of electron densities on complex formation very well, they are

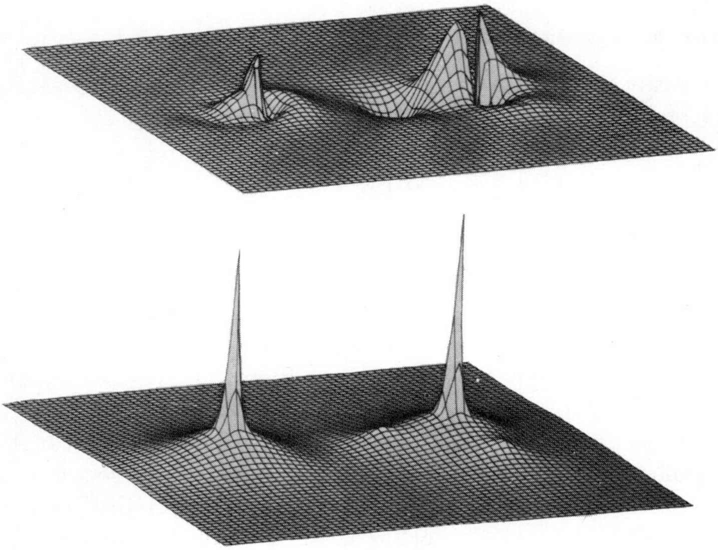

Figure 4.5. Electron density and density difference functions, $\rho(x, z)$ and $\Delta\rho(x, z)$ for $(H_2O)_2$ in the central plane of the complex $(y = 0)$

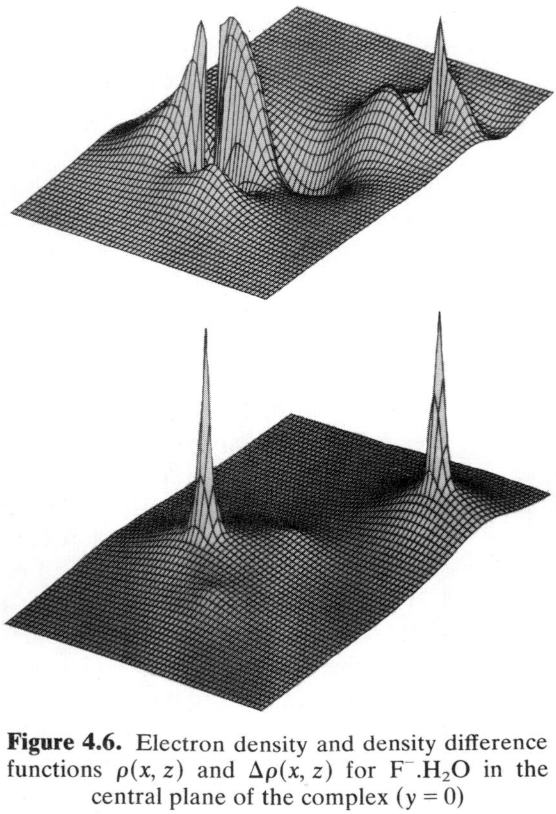

Figure 4.6. Electron density and density difference functions $\rho(x, z)$ and $\Delta\rho(x, z)$ for $F^-.H_2O$ in the central plane of the complex $(y = 0)$

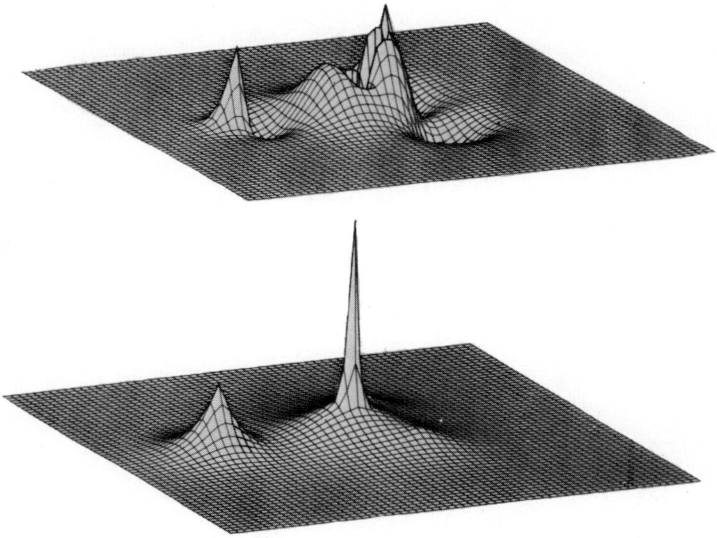

Figure 4.7. Electron density and density difference functions $\rho(x, z)$ and $\Delta\rho(x, z)$ for $Li^+.OH_2$ in the central plane of the complex

not easy to handle in case different systems shall be compared. Usually we are not interested so much in the details but in the overall features of electron rearrangements. For this purpose we have defined an integrated density difference function $\Delta\rho(z)$ (see equation (4.10)) Russegger and Schuster, 1973).

$$\Delta\rho(z) = \iint \Delta\rho(\vec{r})\, \mathrm{d}x\, \mathrm{d}y \tag{4.10}$$

In order to be able to integrate we have to choose an appropriate axis for the complex (z-axis) which is characteristic for the type of interaction we are interested in here. In $Li^+.OH_2$ this is evidently the two-fold symmetry axis C_2. In other cases like, for example, $F^-.H_2O$ or $(H_2O)_2$ the connection line of the three atoms may serve for this purpose. There are of course other types of complexes lacking global and local symmetry, where it is difficult to find a unique direction. In this case a multidimensional analysis of the density difference map may be the only reasonable way to discuss electron rearrangements.

One-dimensional density difference functions $\Delta\rho(z)$ are shown for a number of complexes together with the integrated density functions $\rho(z)$—see equation (4.11)—in Figures 4.8–4.12.

$$\rho(z) = \iint \rho(\vec{r})\, \mathrm{d}x\, \mathrm{d}y \tag{4.11}$$

At a first glance the polarization pattern obtained looks very complicated. Nevertheless, we realize the expected similarities between $Li^+.OH_2$ and

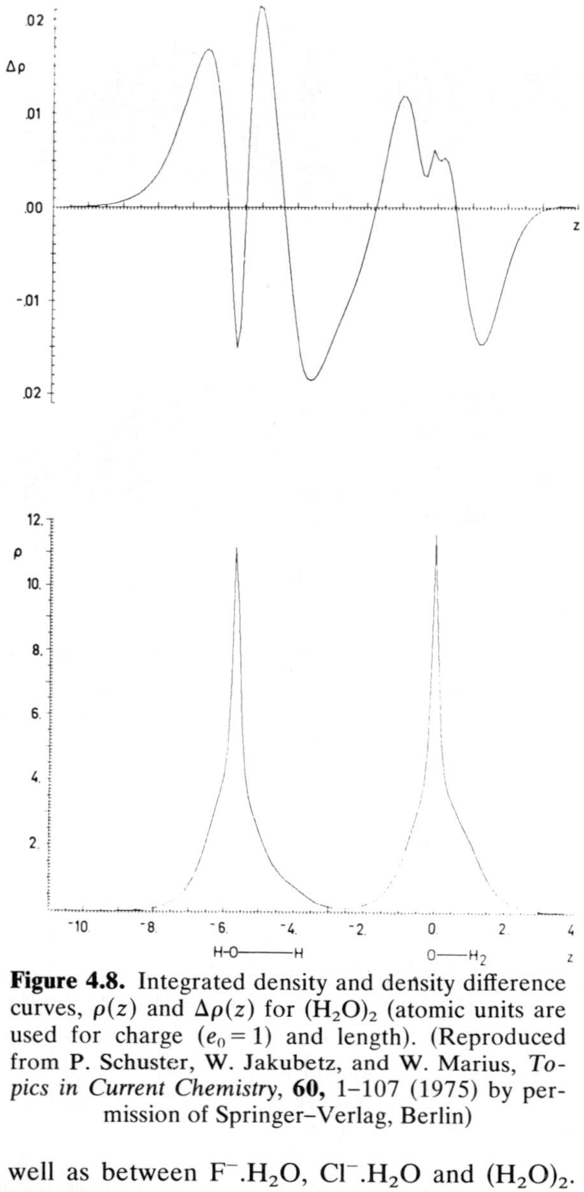

Figure 4.8. Integrated density and density difference curves, $\rho(z)$ and $\Delta\rho(z)$ for $(H_2O)_2$ (atomic units are used for charge ($e_0 = 1$) and length). (Reproduced from P. Schuster, W. Jakubetz, and W. Marius, *Topics in Current Chemistry*, **60,** 1–107 (1975) by permission of Springer–Verlag, Berlin)

$Na^+.OH_2$ as well as between $F^-.H_2O$, $Cl^-.H_2O$ and $(H_2O)_2$. According to the decrease in field strength the intensity of the electron density rearrangement at the H_2O molecule decreases in the series $Li^+ > Na^+$ and $F^- > Cl^- > H_2O$ respectively. Qualitatively, all the curves observed can be interpreted by mutual polarization of the two subsystems. The complicated pattern can be explained as a superposition of simple polarization curves for localized MO's of the two molecules or ions (Schuster *et al.*, 1975).

Charge transfer has often been invoked as an important factor for the stability of hydrogen bonds. The amount of charge transferred in a complex

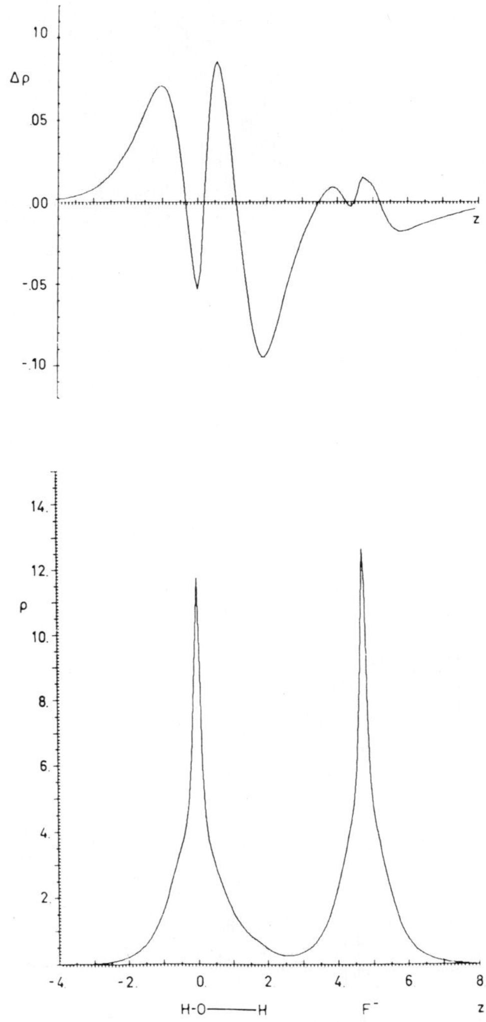

Figure 4.9. Integrated density and density difference curves, $\rho(z)$ and $\Delta\rho(z)$ for $F^-.H_2O$ (atomic units are used for charge ($e_0 = 1$) and length). (Reproduced from P. Schuster, W. Jakubetz, and W. Marius, *Topics in Current Chemistry*, **60,** 1–107 (1975) by permission of Springer–Verlag, Berlin)

from one subsystem to another is no observable quantity. The values obtained will depend therefore on the model assumptions applied. We will discuss here two different ways to calculate charge transfer and compare the results obtained.

The integrated density difference function, $\Delta\rho(z)$, provides an excellent possibility to define charge and clearly demonstrates at the same time the

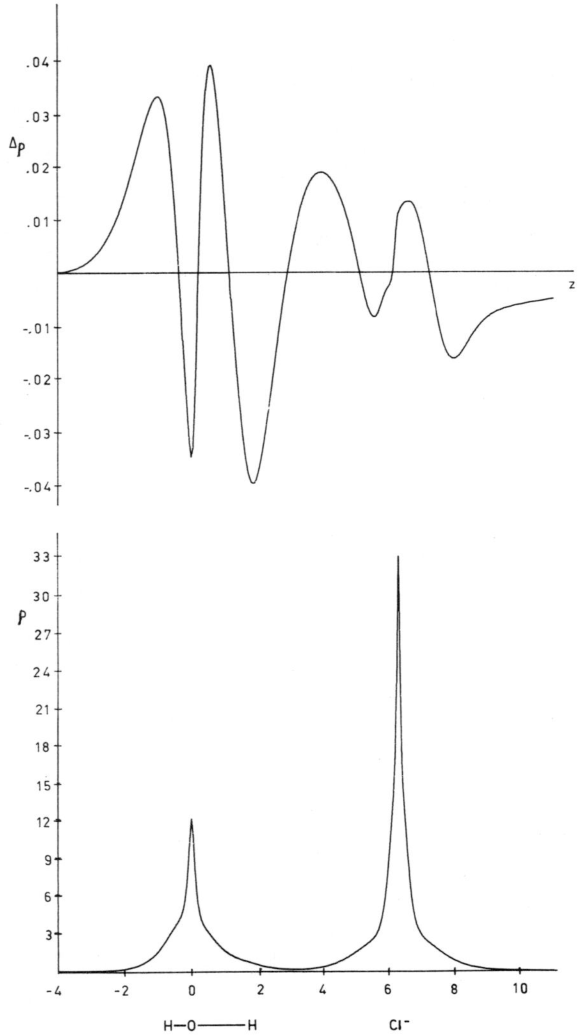

Figure 4.10. Integrated density and density difference curves, $\rho(z)$ and $\Delta\rho(z)$ for $Cl^-.H_2O$ (atomic units are used for charge ($e_0 = 1$) and length)

inherent problems of such a definition. Charge transfer, Δq, is obtained easily by integration (4.12) provided we can define a borderline between the two subsystems in the complex.

$$\Delta q = \int_{-\infty}^{z_0} \Delta\rho(z)\,\mathrm{d}z = -\int_{z_0}^{\infty} \Delta\rho(z)\,\mathrm{d}z \tag{4.12}$$

According to our type of analysis, a plane, $z = z_0$, separating one constituent from the other has to be chosen. Certainly, this is not possible without some

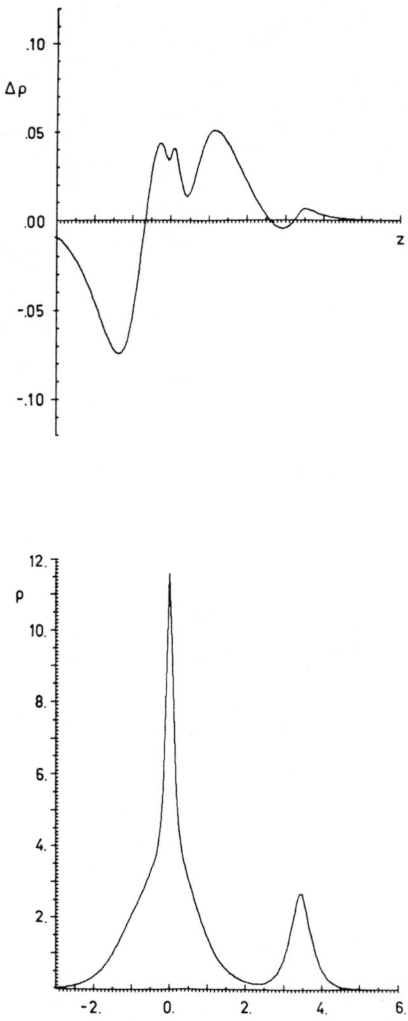

Figure 4.11. Integrated density and density difference curves, $\rho(z)$ and $\Delta\rho(z)$ for $Li^+.OH_2$ (atomic units are used for charge ($e_0 = 1$) and length). (Reproduced from P. Schuster, W. Jakubetz, and W. Marius, *Topics in Current Chemistry*, **60,** 1–107 (1975) by permission of Springer–Verlag, Berlin)

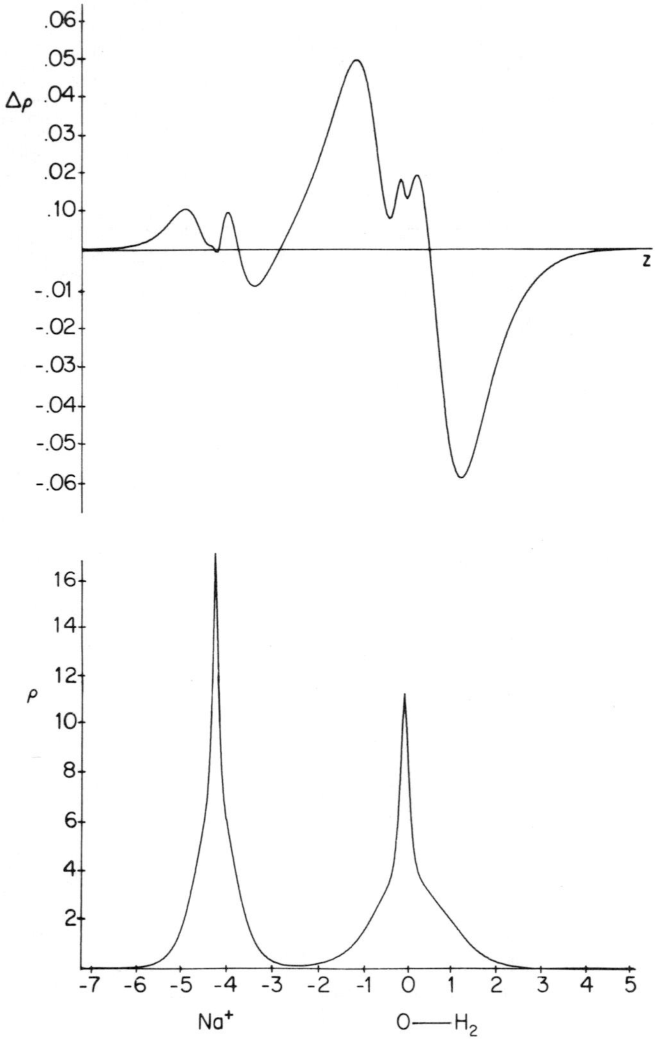

Figure 4.12. Integrated density and density difference curves, $\rho(z)$ and $\Delta\rho(z)$ for $Na^+.OH_2$ (atomic units are used for charge ($e_0 = 1$) and length)

arbitrariness. Previously we have shown that a number of different definitions of z_0 lead to almost the same results (Schuster, 1976a; Schuster, Jakubetz and Marius, 1975). The minimum of the integrated density function lying between the two subsystems—see equation (4.13), Figures 4.8–4.12—was found to be most appropriate for

$$\rho(z_0) = \min \qquad \left(\frac{d\rho(z)}{dz}\right)_{z_0} = 0 \tag{4.13}$$

Table 4.16. Charge transfer in some molecular complexes

Complex	Basis set (GTO's)	Method	Charge transfer (e_0) Δq^1	Δq^2_M	References
$(H_2O)_2$	(11, 7, 1/6, 1)→[7, 5, 1/4, 1]	SCF	0·0079	0·0110	a,b
$Li^+.OH_2$	(11, 7, 1/6, 1)→[4, 3, 1/2, 1] Li: (7, 1)→[3, 1]	SCF	−0·0022	−0·0182	a,b
$Na^+.OH_2$	(9, 5, 1/4)→[5, 3, 1/3] Na: (11, 7)→[6, 4]	SCF	−0·0078	−0·0205	a,b
$F^-.H_2O$	(9, 5, 1/4, 1)→[5, 3, 1/3, 1] F: (9, 5, 1)→[5, 3, 1]	SCF	0·0378	0·0542	a,b
$Cl^-.H_2O$	(9, 5, 1/4)→[5, 3, 1/3, 1] Cl: (17, 10, 1)→[9, 6, 1]	SCF	0·0164	0·0976	a,b
$(FHF)^-$	(9, 5/4, 1)→[4, 2/2, 1]	SCF	—	0·185	c
	(9, 5/4, 1)→[4, 2/2, 1]	SCF–CI	—	0·209	c

[1] Calculated from electron density distributions according to equations (4.12) and (4.13).
[2] Calculated from population analysis.
References [a] Schuster, Lischka and Beyer (1976); [b] Beyer, Lischka and Schuster (1976); [c] Stogård *et al.* (1975).

the purpose of comparison. Values for Δq defined in this way are shown in Table 4.16. Charge transfer is rather small in all ordinary hydrogen bonded complexes discussed here. It is, however, somewhat larger than in comparable cation–molecule association.

Population analysis of the complexes wavefunctions provides another possibility to define charge transfer. Here the arbitrariness of the method is hidden in the definition of orbital and atom populations. The kind of analysis proposed by Mulliken (1955) is now generally used for this purpose. As an example the changes in atomic net charges Δq_i, caused by the formation of $(H_2O)_2$ are given in Figure 4.13. The sign of Δq_i is in good

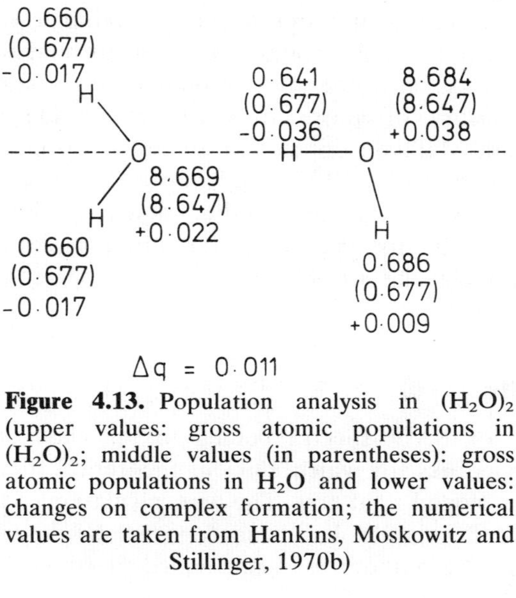

Figure 4.13. Population analysis in $(H_2O)_2$ (upper values: gross atomic populations in $(H_2O)_2$; middle values (in parentheses): gross atomic populations in H_2O and lower values: changes on complex formation; the numerical values are taken from Hankins, Moskowitz and Stillinger, 1970b)

agreement with that of the density difference functions at the same positions. Charge transfer can be obtained easily from the Δq_i values by summation over all atoms in the corresponding molecule or ion:

$$\Delta q_{\mathrm{M}} = \sum_{i=1}^{N} \Delta q_i \tag{4.14}$$

'M' stands here to indicate that the value for charge transfer has been calculated from Mulliken overlap populations. The comparison between Δq and Δq_{M} values in Table 4.16 shows significant deviations. Some general trends, however, are the same in both series of charge transfer values. As a rule the Δq_{M} values are always somewhat larger than Δq. Regarding the relative values of charge transfer in $F^-.H_2O$ and $Cl^-.H_2O$ chemical intuition would attribute more significance to the values derived from density difference functions.

Coming back to observable quantities now, we recall that the electron rearrangement on formation of a hydrogen bond gives rise to a dipole moment of the complex which is larger than that expected from vectorial addition of monomer moments. This increase in polarity is confirmed by all reliable calculations on hydrogen bonded dimers and has been measured experimentally on $(H_2O)_2$ (Dyke and Muenter, 1974). Quantitatively the agreement between theoretical and experimental values is satisfactory within the limits of errors. Electron density rearrangements on complex formation in $(H_2O)_2$ and related systems have been studied also by Diercksen (1971) and Yamabe and Morokuma (1975). These authors came essentially to the same results.

Symmetric structures with strong hydrogen bonds are an example of a class of compounds for which a density difference analysis based on the idea of intermolecular complex formation becomes meaningless. The geometries of the monomeric units in the complexes are deformed so much that the corresponding part has only little in common with the isolated subsystems. It is completely arbitrary, of course, to which subunit the central hydrogen atom is attributed. Mulliken population analysis in principle, can be performed. For charge transfer in $(FHF)^-$ it leads to a value which is almost one order of magnitude higher than in the other complexes discussed above. The meaning of such a value for charge transfer certainly is more doubtful than in ordinary intermolecular associations. $(FHF)^-$ represents an individual molecular entity!

B. Chemical shifts

Chemical shifts for NMR spectra belong to the class of more complex second order properties for which efficient theoretical methods of calculation are being developed right now. The first calculation of chemical shifts in a hydrogen-bonded complex has been reported by Jaszunski and Sadlej (1972a,b; 1973). The authors used uncoupled Hartree–Fock perturbation theory, which is well known to give quantitatively unsatisfactory results on other second order properties, e.g. molecular polarizabilities. Moreover their results were found to be appreciable gauge dependent. Nevertheless, they were able to predict the correct sign of the observed downfield shift in $(H_2O)_2$.

In a more recent investigation Sadlej (1975) applied the coupled Hartree–Fock method to $(FHF)^-$. Although rather large basis sets have been used the results obtained were still not satisfactory.

Ditchfield (1974) proposed an alternative approach to the problem by using gauge invariant atomic orbitals which can vary to some extent in the presence of the magnetic field. The basis sets applied in the calculations on some hydrogen bonded systems were rather small. Nevertheless, even the numerical values of the downfield shifts observed are obtained correctly (Table 4.17). Certainly, more investigations are needed in order to make sure that the excellent agreement between theoretical and experimental values is not a result of error compensation. There is no doubt, however, that the changes in proton chemical shifts which are characteristic for hydrogen bond formation can be reproduced by MO–SCF theory.

Table 4.17. Proton chemical shifts in some hydrogen-bonded complexes

Complex	Basis set (GTO's)	Method[1]	Shielding constants (ppm) $\bar{\sigma}_H(XH)$	$\bar{\sigma}_H(XH\ldots Y)$	$\Delta\bar{\sigma}_H$	References
$(H_2O)_2$	4–31G[2]	SCF–GIAO	31·6	27·9	−3·7	[a]
	(11, 3, 1/6, 1)→ [5, 4, 1/3, 1]	UCHF, EC	28·6	27·6	−1·0	[b]
		EXP	30·2	25·6	−4·6	[b]
$(FHF)^-$	STO-5G[2]	SCF–GIAO	34·8	25·4	−9·4	[c]
	4-31G[2]	SCF–GIAO	30·6	17·1	−13·5	[c]
		EXP			−14	[c]

[1] See footnote to table.
[2] See Hehre, Stewart and Pople (1969); Ditchfield, Hehre and Pople (1971); and Hehre, Ditchfield and Pople (1972).
References. [a] Ditchfield and McKinney, Jr. (1976); [b] Jaszunski and Sadlej (1973); [c] Ditchfield (1976).

V. ENERGY PARTITIONING AND ITS ASYMPTOTIC BEHAVIOUR AT LARGE INTERMOLECULAR DISTANCES

In our discussion we have shown so far that *ab initio* MO theory can account for all the phenomena which are associated with hydrogen bonds by the experimentalists (Figure 4.1). In general we have been interested in qualitative predictions on structures and spectra. By continuing this approach we might come finally in a position to be able to calculate all kinds of observable quantities for molecular complexes with high precision if necessary, provided we have a sufficiently large supply with computer time. On the other hand we have not learned very much yet concerning the nature of intermolecular forces or in particular the nature of hydrogen bonds. An approach, alternative to the supermolecule method tries to obtain the energy of interaction as a sum of individual contributions which can be identified with the expressions of the classical theory of intermolecular forces by their asymptotic behaviour (for example, see Hirschfelder, Curtiss and Bird, 1954, or Margenau and Kestner, 1969). In general, there are the three well-known long range contributions, the electrostatic or Coulomb energy (ΔE_{COU}), the polarization energy (ΔE_{POL}) and the dispersion energy (ΔE_{DIS}), which give rise to the characteristic individual terms in an R^{-n} expansion series of the energy of interaction at intermolecular distances (R), where the overlap between the interacting systems is negligibly small. The procedures applied can be subdivided into three groups depending on the way how they approach the ultimately identical goal:

1. the straightforward application of perturbation theory to intermolecular interactions. Starting from given solutions of the Schrödinger equations of the isolated subsystems—for almost all concrete examples only approximative solutions are available—individual contributions to the energy of interaction are calculated by different modifications of Rayleigh–Schrödinger perturbation theory,
2. the treatment of intermolecular forces as a variational problem by configuration interaction methods. Starting again from MO's of the isolated subsystems the calculation of the energy of interaction is formulated as a configuration interaction problem. The lowest eigenvalue of the CI matrix is calculated then by a perturbation method which allows to separate the intermolecular energy into different contributions,
3. SCF energy partitioning. The energy of interaction obtained by a full SCF calculation is split into various contributions by different model assumptions.

The procedures mentioned in 2 and 3 become identical at the SCF level, i.e. in case only singly excited configurations are applied in the CI calculation.

A. Perturbation theory

Intramolecular perturbation theory dates back to the original paper by Eisenschitz and London (1930). A straightforward application of Rayleigh–Schrödinger perturbation theory was found to be inadequate for this purpose—otherwise the ground state wavefunction necessarily would have wrong permutational symmetry (for example, see Claverie, 1971)—and therefore several modifications have been proposed. A proposal to correct for this wrong behaviour of the wavefunction by a double perturbation theory (Hirschfelder, Byers-Brown and Epstein, 1964) has been presented by Beretta and Vetrano, 1973). The formalism by Murrell, Randic and Williams (1965) or Murrell and Shaw (1967) was applied to hydrogen-bonded systems in a semi-empirical version (for example, see Van Duijne-Veldt–Van de Rijdt and Van Duijneveldt, 1971). Since we have compared their results with those of SCF-MO calculations in a previous summary (Schuster, 1976a), we are not going to repeat this discussion here. Recently, an *ab initio* calculation on $(H_2O)_2$ using an extended basis set has been reported (Jeziorski and Van Hemert, 1976). With respect to accuracy the results of this investigation are comparable to those of the most extended *ab initio* MO calculations and we will use them here for comparison with other methods.

Jeziorski and Van Hemert (1976) apply the exchange perturbation formalism developed by Chipman, Bowman and Hirschfelder (1973) to two interacting water molecules. Starting from the Hamiltonian of a binary complex (4.3) they define

$$\mathcal{H}_0 = \mathcal{H}_A^0 + \mathcal{H}_B^0 \quad \text{and} \quad \mathcal{H}' = \mathcal{V}_{AB} \tag{4.15}$$

as the unperturbed Hamiltonian and the perturbation respectively. Denoting the exact wavefunctions of the isolated subsystems by Ψ_A and Ψ_B, the corresponding energies by E_A and E_B the authors obtain the following contributions to the energy of interaction:

$$\Delta E = \Delta E^{(1)} + \Delta E_{PD}^{(2)} + \ldots \tag{4.16}$$

Here the authors consider only the polarization and dispersion contributions to the second order term. Second order exchange and higher order terms have been neglected.

The first order of energy, $\Delta E^{(1)}$, is obtained from equation (4.17) and may be separated into a Coulomb and an exchange contribution. In bracket notation we may write:

$$\Delta E^{(1)} = \langle \Psi_A \Psi_B | \mathcal{A} \mathcal{V}_{AB} | \Psi_A \Psi_B \rangle / \langle \Psi_A \Psi_B | \mathcal{A} | \Psi_A \Psi_B \rangle \tag{4.17}$$

$\mathcal{A}$ represents the (idempotent) antisymmetrizer for all electrons in the complex.

$$\Delta E^{(1)} = \Delta E_{COU} + \Delta E_{EX} \tag{4.18}$$

ΔE_{COU} is the electrostatic energy of interaction calculated from the non-orthogonal orbitals of the zeroth order wavefunction:

$$\Delta E_{COU} = \langle \Psi_A \Psi_B | \mathcal{V}_{AB} | \Psi_A \Psi_B \rangle \tag{4.19}$$

The exchange-free second-order energy can be formulated as

$$\Delta E_{PD}^{(2)} = -\langle \Psi_A \Psi_B | \mathcal{V}_{AB} \mathcal{R} \mathcal{V}_{AB} | \Psi_A \Psi_B \rangle \tag{4.20}$$

wherein $\mathcal{R}$ stands for the ground state reduced resolvent of the operator $\mathcal{H}_A^0 + \mathcal{H}_B^0$:

$$\mathcal{R} = \{(\mathcal{H}_A^0 + \mathcal{H}_B^0)\mathcal{Q} - (E_A + E_B)\}^{-1} \mathcal{Q} \tag{4.21}$$

$\mathcal{Q}$ is defined as the following projection operator

$$\mathcal{Q} = 1 - |\Psi_A \Psi_B\rangle \langle \Psi_A \Psi_B| \tag{4.22}$$

In their actual calculations the authors neglect intramolecular correlation corrections, i.e. they assume the SCF wavefunctions Ψ_A^0 and Ψ_B^0, to represent the exact solutions of the Schrödinger equations for the isolated molecules. In case intermolecular energies of high accuracy are desired this simplification is not well justified for most of the concrete molecular examples (see Sections III.B and V.B). Denoting the eigenfunctions and eigenvalues of the corresponding Hartree–Fock operators F_A and F_B by Ψ_A^k, Ψ_B^l and E_B^k, E_B^l respectively, we obtain for $\Delta E_{PD}^{(2)}$ the following expression, which has been separated already into polarization and dispersion energies:

$$\Delta E_{PD}^{(2)} = \Delta E_{POL} + \Delta E_{DIS} \tag{4.23}$$

$$\Delta E_{POL} = \sum_{k \neq 0} \frac{|\langle \Psi_A^0 \Psi_B^0 | \mathscr{V}_{AB} | \Psi_A^k \Psi_B^0 \rangle|^2}{E_A^0 - E_A^k} + \sum_{l \neq 0} \frac{|\langle \Psi_A^0 \Psi_B^0 | \mathscr{V}_{AB} | \Psi_A^0 \Psi_B^l \rangle|^2}{E_B^0 - E_B^l} \tag{4.24}$$

$$\Delta E_{DIS} = \sum_{k \neq 0} \sum_{l \neq 0} \frac{|\langle \Psi_A^0 \Psi_B^0 | \mathscr{V}_{AB} | \Psi_A^k \Psi_B^l \rangle|^2}{E_A^0 + E_B^0 - E_B^k - E_B^l} \tag{4.25}$$

The polarization energy as defined by equation (4.24) accounts for mutual polarization of the atoms, molecules or ions in the electrostatic field of the second subsystem. At the level of the Hartree–Fock approximation only those excited states, Ψ_A^k or Ψ_B^l respectively, contribute to ΔE_{POL} which correspond to single electron excitations provided the electrostatic potential of the polarizing subsystem is expanded in a multipole series. Using the same approximations, the dispersion energy may be expressed in terms of doubly excited states with single excitations in A and B. In order to avoid impracticable integrations involving orbitals belonging to the continuous spectrum of the Hartree–Fock operators the expressions given for the second order contributions are usually reformulated as a variational problem.

The equations for the first-order contributions at the level of Hartree–Fock calculations are obtained simply from equation (4.17) and (4.19) by replacing the exact wavefunctions Ψ_A and Ψ_B by the single determinant approximations Ψ_B^0 and Ψ_B^0.

It should be mentioned here that the separation of the first-order exchange contribution from the electrostatic term is not possible without arbitrariness in case the two subsystems overlap appreciably in the complex. At large intermolecular distances where intermolecular overlap is no longer important the separation becomes more or less trivial since the exchange energy is negligibly small. The same problem, of course, arises in partitioning of the second order energies in strongly overlapping systems.

In their actual calculations on $(H_2O)_2$ Jeziorski and Van Hemert (1976) apply an extended basis set—(11, 7, 2/6, 1) contracted to [4, 3, 2/2, 1]—which is almost identical with that used by Diercksen (1971) and

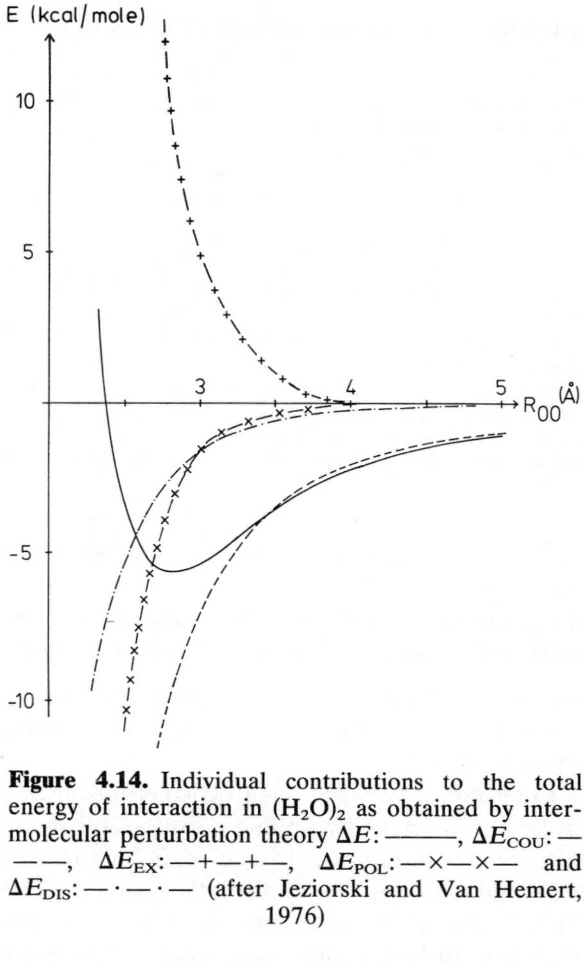

Figure 4.14. Individual contributions to the total energy of interaction in $(H_2O)_2$ as obtained by intermolecular perturbation theory ΔE: ———, ΔE_{COU}: — — —, ΔE_{EX}: —+—+—, ΔE_{POL}: —×—×— and ΔE_{DIS}: —·—·— (after Jeziorski and Van Hemert, 1976)

Diercksen, Kraemer and Roos (1975). The only difference between both basis sets concerns the orbital exponents of the polarization functions. The results obtained are shown in Figure 4.14 as well as in Table 4.18. The electrostatic contribution determines the long range behaviour of the energy of interaction and still dominates at the equilibrium distance ($\Delta E_{COU} \sim -7$ kcal/mol). At this point of the potential curve the repulsive exchange energy was found to be next important in absolute values ($\Delta E_{EX} \sim 5$ kcal/mol). Both, polarization and dispersion energies are much smaller ($\Delta E_{POL} \sim -1{\cdot}6$ kcal/mol, $\Delta E_{DIS} \sim -1{\cdot}5$ kcal/mol). Due to different sign ΔE_{COU} and ΔE_{EX} cancel to a large extent and therefore all four terms represent non-negligible contributions to the total energy of interaction. As we have seen so far the perturbational approach gives a physically reasonable picture of the various contributions stabilizing the complex $(H_2O)_2$. In order to be more general, we will compare the results obtained with those from SCF energy partitioning in the next section. It should be noted already

Table 4.18. Energy partitioning in some molecular complexes[1]

Complex	Basis set (GTO's)	Method[2]	Energies in kcal/mol ΔE_{COU}	ΔE_{EX}	ΔE_{POL}	ΔE_{CHT}	ΔE_{DIS}	References
$(H_2O)_2$	(11, 7, 1/6, 1)→[7, 5, 1/4, 1]	PT	−7·1	4·9	−1·6	—	−1·5	a
		SCF–EP	−7·3	4·5	−0·7	−1·2	—	b,c
$Li^+.OH_2$	(11, 7, 1/6, 1)→[4, 3, 1/2, 1] Li: (7, 1)→[3, 1]		−41·5	15·3	−12·8	5·2	—	b,c
$Na^+.OH_2$	(9, 5, 1/4)→[5, 3, 1/3] Na: (11, 7)→[6, 4]	SCF–EP	−30·3	7·0	−5·1	1·1	—	b,c
$F^-.HOH$	(9, 5, 1/4, 1)→[5, 3, 1/3, 1] F: (9, 5, 1)→[5, 3, 1]	SCF–EP	−34·7	26·0	−7·2	−7·2	—	b,c
$Cl^-.HOH$	(9, 5, 1/4)→[5, 3, 1/3] Cl^-: (17, 10, 1)→[9, 6, 1]	SCF–EP	−14·6	7·0	−1·6	−5·7	—	b,c

[1] Equilibrium geometries of the complexes have been used.
[2] PT = perturbation theory, SCF–EP = SCF energy partitioning.
References [a] Jeziorski and Van Hemert (1976); [b] Schuster, Lischka and Beyer (1976); [c] Beyer, Lischka and Schuster (1976).

here, however, that dispersion energies are calculated without special difficulties by this approach.

The major disadvantages of the straightforward application of perturbation theory to intermolecular forces seem to be the following:

1. the computations to be performed are rather time consuming,
2. non-orthogonality and electron exchange effects on second-order energies are difficult to treat and
3. it appears to be rather very tedious to correct for electron correlation effects in the isolated molecules by double perturbation theory. The inclusion of correlation effects in practice, of course, is identical with the use of general CI wavefunctions to describe the solutions of the Schödinger equations of the isolated subsystems. Until now no calculation applying double perturbation formalism to hydrogen-bonded systems has been reported.

Based on the use of so-called 'biorthogonal orbitals' Kochanski and Gouyet (1975a,b) developed a perturbation procedure which is of some interest in this context here. By this method the effects of orbital non-orthogonality and electron exchange can be discussed separately. Furthermore, the authors have investigated also an example for ion-molecule complexes ($Li^+.H_2$: Kochanski and Gouyet, 1975b; Kochanski, 1974). In this case the influence of non-orthogonality on calculated polarization and dispersion energies has been studied as well. For both terms the corrections brought about by orthogonalization of the starting wavefunctions are repulsive.

In order to overcome the difficulties mentioned above several groups of authors have proposed various alternative procedures. We will mention here only one particular approach which has been applied also to hydrogen-bonded systems. Daudey, Claverie and Malrieu (1974) developed a CI method to treat intermolecular complexes by a perturbational procedure.

They start directly from SCF wavefunctions of the isolated molecules and discuss molecular interactions as a CI procedure. The lowest eigenvalue of the CI matrix corresponding to the ground state of the complex is obtained by perturbation theory. As a consequence of this treatment the authors can account for intra- and intermolecular electron correlation and separate both contributions as long as higher second order effects are neglected. Daudey (1974) applied this procedure to the complex $(H_2O)_2$. Unfortunately, the basis set used in this calculation (Morokuma and Winick, 1970) was rather small and therefore the results will contain substantial basis set errors (see Section V.B). The results obtained have been compared with those of other perturbation treatments and MO–SCF calculations in a previous review (Schuster, 1976a). The agreement between perturbation theory and SCF energy partitioning was found to be satisfactory (see Table 4.18 and Section V.B).

Finally, we would like to mention here also a recent approach to intermolecular energies described by Otto and Ladik (1975). The intermolecular perturbation is incorporated directly as an electrostatic potential into the Hartree–Fock equations of the two interacting subsystems. By an iterative procedure the electrostatic potentials of the two molecules or ions are made mutually consistent. The wavefunctions obtained thereby are used then in further calculations of the other contributions like ΔE_{EX} or ΔE_{CHT} by intermolecular perturbation theory. In particular the authors use the formalism of Murrell, Randic and Williams (1965). Using various simplified representations for the electrostatic potential the method has been applied to the systems $(HF)_2$ and $(CH_2O)_2$ (Otto and Ladik, 1975) as well as to $(H_2O)_2$ (Otto and Ladik, 1976). The ultimate goal of this approach is to develop a simplified treatment which is also applicable to studies on interactions between larger organic molecules.

B. SCF energy partitioning

Partitioning of SCF energies of interaction into electrostatic (ΔE_{COU}), exchange (ΔE_{EX}) and delocalization (ΔE_{DEL}) energies has been proposed originally by Kollman and Allen (1970) and Dreyfus and Pullman (1970a,b). The main idea was to learn more on the nature of intermolecular interactions and to obtain a more rigorous basis for the construction of model potentials. Dreyfus and Pullman (1970a,b) were interested particularly in a test for the reliability of predictions made by electrostatic models (see for example, Scrocco and Tomasi, 1973). Morokuma (1971) extended this concept to a separation of the delocalization energy into polarization (ΔE_{POL}) and charge transfer (ΔE_{CHT}) contributions. The procedure can be explained most easily by an assumption of four normalized trial wavefunctions to describe the complex:

1. a simple Hartree product built from the SCF wavefunctions of the isolated molecules

$$\Psi_1 = \Psi_A^0 \cdot \Psi_B^0 \tag{4.26}$$

2. an antisymmetrized product of these two wavefunctions

$$\Psi_2 = \mathbf{A}(\Psi_A^0 \Psi_B^0) \tag{4.27}$$

3. a product of the two wavefunctions optimized individually in the electrostatic field of the other subsystem

$$\Psi_3 = \Psi_A \, . \, \Psi_B \tag{4.28}$$

4. the ordinary antisymmetric and optimized SCF wavefunction of the whole complex

$$\Psi_4 = \Psi_{AB} \tag{4.29}$$

For all four wavefunctions energy expectation values are calculated now and the individual contributions to the total energy of interaction are obtained as differences:

$$\Delta E_i = E_i - (E_n^0 + E_B^0) \qquad E_i = \int \Psi_i^* \mathbf{H} \Psi_i \, d\tau \tag{4.30}$$

According to physical considerations electrostatic, exchange, polarization and charge transfer energies are built from this set of ΔE_i values (Morokuma, 1971; Schuster 1976a):

$$\Delta E_{COU} = \Delta E_1 \tag{4.31}$$

$$\Delta E_{EX} = \Delta E_2 - \Delta E_1 \tag{4.32}$$

$$\Delta E_{POL} = \Delta E_3 - \Delta E_1 \tag{4.33}$$

$$\Delta E_{CHT} = \Delta E_4 + \Delta E_1 - \Delta E_2 - \Delta E_3 \tag{4.34}$$

The total SCF energy, of course, is the sum of all four individual contributions:

$$\Delta E_{SCF} = \Delta E_4 = \Delta E_{COU} + \Delta E_{EX} + \Delta E_{POL} + \Delta E_{CHT} \tag{4.35}$$

SCF energy partitioning has been performed for a number of intermolecular complexes. Recently, results from calculations using extended basis sets have been reported (Schuster, Lischka, and Beyer, 1976; Beyer, Lischka, and Schuster, 1976). With the basis sets chosen the electrostatic properties and polarizabilities of the isolated systems are predicted within error limits of less than 20 per cent (see Section II). Furthermore, the superposition error of basis sets in the complex can be expected to be rather small when basis sets of this quality are used. The molecular geometries applied in these calculations are very close to the most stable structures. The values obtained thereby are shown in Table 4.18 above. Some of the most important features are essentially the same as derived before from perturbation theory.

The electrostatic contribution (ΔE_{COU}) dominates in all complexes investigated here. In $Li^+.OH_2$, $F^-.H_2O$ and $(H_2O)_2$ the first-order exchange contribution was found to be next important and much larger in absolute value than the remaining second-order terms. In $Na^+.OH_2$ and $Cl^-.H_2O$

ΔE_{EX} and the second-order energy are of the same order of magnitude. According to the relatively small intermolecular distances at equilibrium, overlap is rather large in systems forming hydrogen bonds. Apparently, exchange repulsion is stronger in these complexes too. Both ΔE_{COU} and ΔE_{POL} depend strongly on the intermolecular distance at the equilibrium geometry and decrease substantially in absolute value, if we substitute a third row element for one of the second row ($Cl^- < F^-$ or $Na^+ < Li^+$). The charge transfer term, however, does not show a physically reasonable behaviour. In contrast to the general stabilizing effect we would expect from perturbation theory, ΔE_{CHT} is repulsive in cation–molecule complexes. Nevertheless, there is a significant difference in comparison to hydrogen-bonded systems where the charge transfer term remains always negative. A possible reason for the unusual behaviour of ΔE_{CHT} will be discussed later.

Electrostatic potentials very often are represented by multipole expansion which leads to an expansion of electrostatic energies in powers of R^{-n}. Recently, Ahlrichs (1976) gave a proof that this expansion gives rise to an asymptotically convergent series. It seems interesting therefore to test the properties of the multipole expansion series of ΔE_{COU} for two concrete examples—a complex between two neutral molecules, $(H_2O)_2$ and between a cation and a molecule, $Li^+.OH_2$. Around the energy minimum an entirely different situation is encountered in the two cases: in $(H_2O)_2$ convergence is very poor, whereas the four term series gives already satisfactory agreement with the exact value in $Li^+.OH_2$ (Table 4.19). At large intermolecular distances, as expected, the series converge very fast to the corresponding exact electrostatic energies in both examples. A similar investigation on ΔE_{POL} again showed the correct asymptotic behaviour (Beyer, Lischka and Schuster, 1976).

Comparing the results from intermolecular perturbation theory and SCF energy partitioning on the complex $(H_2O)_2$ we find almost the same numerical values for the first order energies ΔE_{COU} and ΔE_{EX}. This finding is not unexpected since the ultimate expressions from which the energies are obtained are identical in both methods and very similar basis sets have been used by Jeziorski and Van Hemert (1976) and Beyer, Lischka and Schuster (1976). A comparison of second order contributions, on the other hand, is much more difficult. The polarization energy derived from the perturbational approach apparently is larger in absolute value than the corresponding contribution obtained from SCF energy partitioning. Jeziorski and Van Hemert (1976) pointed out, that in their treatment ΔE_{POL} most probably contains major portions of the charge transfer term in SCF energy partitioning. In fact, the sum of both contributions in the SCF calculations, ΔE_{DEL},

$$\Delta E_{DEL} = \Delta E_{POL} + \Delta E_{CHT} \tag{4.36}$$

is not too different from the corresponding quantity in the perturbational approach. The dispersion energy, ΔE_{DIS}, is derived from double excitations exclusively and, therefore, cannot be obtained by an SCF approach.

Table 4.19. Multipole expansion of electrostatic energies in $(H_2O)_2$ and $Li^+.OH_2$

Complex	R_{OX} (Å)	Energies in kcal/mol						Reference
		$\Sigma_1{}^1$	$\Sigma_2{}^1$	$\Sigma_3{}^1$	$\Sigma_4{}^1$	ϕ^1	ΔE_{COU}	
$(H_2O)_2$	3·00[2]	−3·28	−3·63	−3·68	—	—	−7·12	[a]
	4·76	−0·82	−0·87	−0·88	—	—	−1·12	[a]
	7·94	−0·18	−0·18	−0·18	—	—	−0·21	[a]
Li^+ OH_2	1·80[2]	−46·48	−47·42	−40·35	−44·47	−42·69	−44·23	[b,c]
	4·23	−8·74	−8·82	−8·57	−8·63	−8·63	−8·63	[b,c]

[1] Σ_n represents the sum over all contributions up to the multipole corresponding to $L = n$; ϕ is the electrostatic potential of the water molecule, which becomes identical with ΔE_{COU} if the Li^+ cation is substituted by a point charge.
[2] Intermolecular distance close to equilibrium geometry.
References. [a] Jeziorski and Van Hemert (1976); [b] Schuster *et al.* (1975); [c] Beyer, Lischka and Schuster (1976).

The SCF energy partitioning as described by equations (4.26)–(4.36) was found to suffer from four disadvantages mainly:

1. the results obtained are largely basis set dependent,
2. a direct comparison with intermolecular perturbation theory is rather difficult as far as second and higher order effects are concerned,
3. the charge transfer contribution, ΔE_{CHT}, which is simply the difference between the SCF energy of interaction and all the other contributions is a complex and by no means clearly defined quantity,
4. a deficiency which, of course, is also present in the perturbational approach but has not been mentioned so far: the kind of SCF energy partitioning is not applicable directly to complexes with geometries which differ appreciably from those of the isolated subsystems.

In the following paragraphs we will refer briefly to these four points and mention some suggestions which have been made recently. In order to investigate the basis set dependence of Morokuma's SCF energy partitioning method Beyer, Lischka and Schuster (1976) have calculated the individual contributions for $Li^+.OH_2$ and $(H_2O)_2$ at the equilibrium geometries with a number of different basis sets. The results obtained are summarized in Table 4.20. ΔE_{COU} and ΔE_{POL} roughly reflect the changes in the calculated properties of the isolated subsystems. According to the calculated values of dipole moments and polarizabilities of the H_2O molecule discussed in Section II (Tables 4.4, 4.5) we expect ΔE_{COU} to be too large and ΔE_{POL} to be too small in absolute values in most calculations using small or medium size basis sets. This, in fact is the result obtained from SCF energy partitioning. The numerical values again stress the importance of flat polarization functions for a correct reproduction of intermolecular energies. The first order exchange contribution (ΔE_{EX}), on the other hand, seems to be somewhat less sensitive. Substantial errors are found only in calculations with very small basis sets. The charge transfer energy (ΔE_{CHT}) finally, was found to be the most critical term. Despite the already mentioned fact that this contribution is repulsive in large basis set calculations on $Li^+.OH_2$ and $Na^+.OH_2$ the numerical values are extremely basis set dependent. In general, small basis sets give more negative values. As an extensive investigation on Li^+OH_2 shows ΔE_{CHT} may also change sign on basis set extension.

Table 4.20. Basis set dependence of SCF energy partitioning in $(H_2O)_2$ and $Li^+.OH_2$

Complex	Basis set (GTO's)	R_{OX}(Å)	ΔE	ΔE_{COU}	ΔE_{EX}	Energies in kcal/mol $\Delta E^{(1)}$	ΔE_{POL}	ΔE_{CHT}	ΔE_{DEL}	References
$(H_2O)_2$	STO[1]	2·78	−6·6	−8·0	9·9	1·9	−0·3	−8·2	−8·5	a
	4-31G[2]	2·78	−7·7	−12·9	9·3	−3·6	−0·7	−3·4	−4·1	b
	4-31G	2·98	−7·7	−9·0	4·2	−4·8	−0·5	−2·5	−3·0	b
	(11, 7, 1/6, 1)→[7, 5, 1/4, 1]	3·00	−4·7	−7·3	4·5	−2·8	−0·7	−1·2	−1·9	c
$Li^+.OH_2$	(7, 3/3)→[4, 2/2] Li: (6)→[3]	1·80	−45·9	−46·5	8·4	−38·1	−6·7	−1·0	−7·8	c,d
	(9, 5/4)→[5, 3/3] Li: (7)→[3]	1·80	−43·0	−52·0	15·1	−36·9	−9·8	3·8	−6·1	c,d
	(9, 5, 1/4)→[4, 3, 1/3] Li: (7, 1)→[3, 1]	1·80	−37·3	−44·2	14·9	−29·4	−11·3	3·4	−7·9	c,d
	(11, 7, 1/6, 1)→[4, 3, 1/2, 1] Li: (7, 1)→[3, 1]	1·80	−33·9	−41·7	15·3	−26·3	−12·8	5·2	−7·6	c,d
	(11, 7, 3/6, 2)→[7, 5, 3/4, 2] Li: (7, 1)→[3, 1]	1·80	−33·7	—	—	−22·0	—	—	−11·7	c,d

[1] Minimal Slater basis set.
[2] See Ditchfield, Hehre and Pople (1971); and Hehre, Ditchfield and Pople (1972).
References. [a] Morokuma (1970); [b] Kitaura and Morokuma (1976); [c] Schuster, Lischka and Beyer (1976); [d] Beyer, Lischka and Schuster (1976).

Recalling our previous discussion on basis set effects in MO calculations on intermolecular energies we have to attribute these changes to the basis set superposition error discussed in Section III.A. Errors of this kind will enter ΔE_{CHT} in case SCF energies partitioned according to Morokuma's scheme. Furthermore, we can realize another unusual behaviour of ΔE_{CHT} when we regard the dependence of this quantity on intermolecular distance: ΔE_{CHT} is repulsive at short distances, changes sign and finally approaches to zero energy line from negative values (Beyer, Lischka and Schuster, 1976).

In order to overcome the difficulties concerning the definition of ΔE_{CHT}, Kitaura and Morokuma (1976) proposed a new energy partitioning scheme for MO–SCF calculations on intermolecular interaction which relates the former formalism to the CI approach discussed in the previous section. In fact the energy decomposition method proposed becomes identical with the more general procedure of Daudey, Claverie and Malrieu (1974) in case the

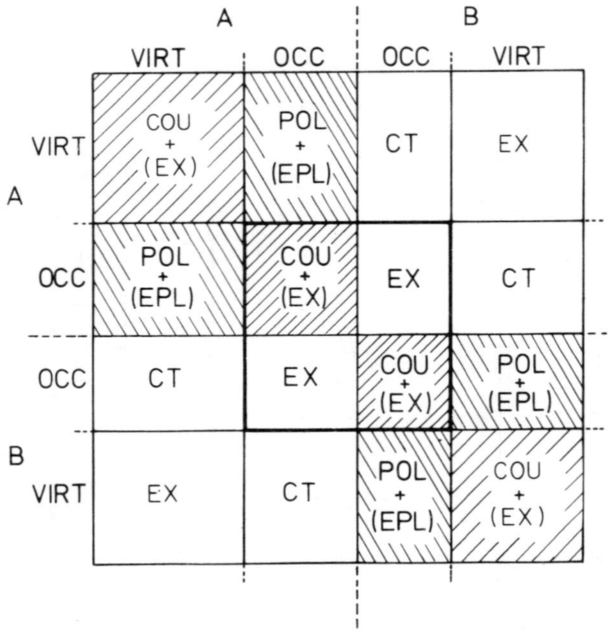

Figure 4.15. SCF Energy partitioning as a CI problem (The inner part of the matrix involving only occupied orbitals gives rise to the first order contributions, $\Delta E_{COU} + \Delta E_{EX}$. Part of the exchange energy is already contained in the diagonal block of the CI matrix due to the action of the intermolecular exchange operators ($\hat{K}_j$). Mixing of occupied and virtual orbitals within the same subsystem leads to polarization and exchange polarization energies, $\Delta E_{POL} + \Delta E_{EPL}$ and finally the interaction between occupied orbitals of one subsystem with the virtual orbital of the second gives rise to the charge transfer energy ΔE_{CHT}. For further details see Kitaura and Morokuma, 1976)

CI is restricted to single excitations only. Another more technical difference concerns the evaluation of the energy eigenvalues: Kitaura and Morokuma (1976) diagonalize the CI matrices, whereas Daudey, Claverie and Malrieu (1974) use perturbation theory for this purpose. Energy partitioning in the new approach is performed by systematic neglect of several of off-diagonal elements of the CI matrix and electron exchange operators. These off-diagonal elements are classified according to the MO's from which they are formed. The MO's in turn are grouped according to their provenance, i.e. whether they belong to subsystem A or B whether they belong to the subspace of occupied and virtual orbitals (Figure 4.15).

The major results of the new energy partitioning scheme is further decomposition of the former charge transfer energy into a 'true' charge transfer term (ΔE_{CT}) an exchange polarization energy (ΔE_{EPL}) and the remaining coupling term between the different classes of matrix elements (ΔE_{MIX}). The whole partitioning scheme reads now as follows:

$$\Delta E_{SCF} = \Delta E_{COU} + \Delta E_{EX} + \Delta E_{POL} + \Delta E_{CT} + \Delta E_{EPL} + \Delta E_{MIX}$$

or (4.37)

$$\Delta E_{CHT} = \Delta E_{CT} + \Delta E_{EPL} + \Delta E_{MIX}$$

The results obtained in case of $(H_2O)_2$ are shown in Table 4.21. Interestingly, the 'true' charge-transfer energy (ΔE_{CT}) is not very different from the

Table 4.21. Decomposition of the charge transfer energy ΔE_{CHT} in $(H_2O)_2$ (Kitaura and Morokuma, 1976)

	Energies in kcal/mol[a]			
R_{OO} (Å)	ΔE_{CT}	ΔE_{EPL}	ΔE_{MIX}	ΔE_{CHT}
2·78	−2·78	−2·95	2·36	−3·37
2·98	−2·11	−0·40	0·06	−2·45
3·18	−1·74	0·02	−0·18	−1·90

[a] A 4–31G basis set (Ditchfield, Hehre and Pople, 1971; Hehre, Ditchfield and Pople, 1972) has been used.

former ΔE_{CHT} values. Accordingly, ΔE_{EPL} and ΔE_{MIX} have similar absolute values but different sign. Again we realize that these two terms are changing sign with increasing intermolecular distance. Nevertheless, extensions of this partitioning scheme for ΔE_{CHT} to other complexes and larger basis sets are necessary before definite conclusions on usefulness and applicability of this method can be drawn.

Finally, we will come to a common problem in both approaches to decompose intermolecular energies, perturbation theory and SCF energy partitioning: How can we account for changes in molecular geometries on complex formation? Certainly, it is not an unjustified approximation to neglect changes in molecular geometries in complexes, where 'these changes' are very small as, for example, in $(H_2O)_2$, $Li^+.OH_2$ etc. Structures with strong hydrogen bonds, on the other hand, represent a typical counterexample: in complexes like $(FHF)^-$ the relaxation of intermolecular geometries is large and not negligible. Umeyana, Kitaura and Morokuma (1975) proposed a procedure for energy partitioning in such cases. In this context here, we will present an interpretation of their methods which differs slightly from that of the authors. At first the complex is formed by a Gedankenexperiment in two extreme alternative ways which lead to different energy contributions:

1. Starting from equilibrium geometries of the isolated subsystems the complex is formed at frozen intramolecular geometries. The energy released thereby is considered as 'intermolecular' or 'external' energy of interaction (ΔE_{EXT}) and can be decomposed in the usual way. Afterwards the complex is allowed to relax and we might call the corresponding energy change as relaxation energy (ΔE_{REL}). ΔE_{REL} according to this definition will be always negative.
2. The other extreme approach starts from distorted isolated subsystems which have exactly the same intramolecular geometry as it is observed in the equilibrated complex. Consequently, we have to invest 'deformation' energy (ΔE_{DEF}) at first. Afterwards, we form the complex and gain thereby the corresponding external energy ΔE_{EXT} which, of course, is larger in absolute value than that obtained along the first path. Evidently, the relaxation energy (ΔE_{REL}) is zero in this case.

The total energy of complex formation can be written now as a sum of three contributions:

$$\Delta E = \Delta E_{\text{DEF}}^{(Q)} + \Delta E_{\text{EXT}}^{(Q)} + \Delta E_{\text{REL}}^{(Q)} \tag{4.38}$$

The superscript indicates that the values of the three quantities depend on the path 'Q' along which the complex is formed. ΔE, of course, has to be independent of a particular choice of the path for a reaction. We can interpret the two extreme cases mentioned above as those paths where either the deformation ($\Delta E_{\text{DEF}} = 0$) or the relaxation energies ($\Delta E_{\text{REL}} = 0$) vanish and the intramolecular geometries are kept constant during the change of the intermolecular distance (Figure 4.16). For some purposes it seems appropriate to consider only the sum of both intramolecular contributions which we might call characteristically the 'internal' energy change (ΔE_{INT}):

$$\Delta E_{\text{INT}}^{(Q)} = \Delta E_{\text{DEF}}^{(Q)} + \Delta E_{\text{REL}}^{(Q)} \tag{4.39}$$

Evidently, intramolecular geometry relaxation has introduced a new source of arbitrariness in the energy decomposition scheme. There is, however, a possibility to get rid of this additional degree of freedom. Recalling that the relations $\Delta E_{\text{DEF}}^{(Q)} > 0$ and $\Delta E_{\text{REL}}^{(Q)} < 0$ are valid by definition we can always find a whole family of paths (E) for which the internal energy contribution is

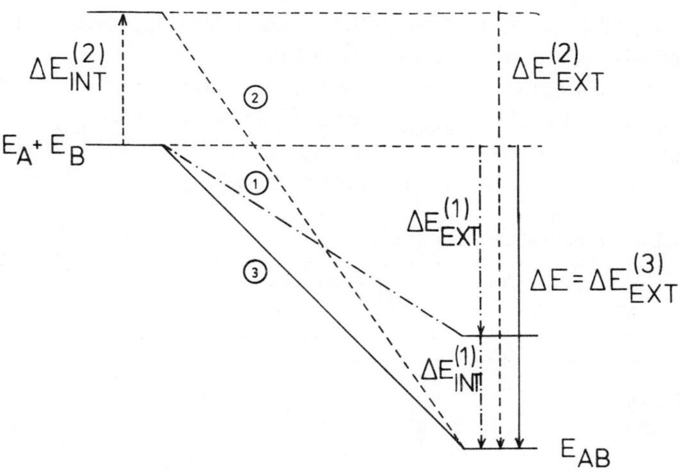

Figure 4.16. Partitioning of the energy of complex formation into internal and external contributions: (1) complex formation at frozen geometries and subsequent molecular relaxation; (2) deformation of the molecules A and B and subsequent complex formation and (3) minimum energy path, see also text

zero. Additionally

$$\Delta E = \Delta E_{\mathrm{EXT}}^{(\mathrm{E})} \tag{4.40}$$

there is at least one path along which the internal contribution vanishes for every differential displacement. To be as general as possible we have to consider also energy surfaces with bifurcation points along the reaction coordinate. For this purpose we rewrite the difference equation (4.40) as an integral along the path chosen:

$$\begin{aligned} \Delta E &= \int_{\mathbf{Q}_i}^{\mathbf{Q}_f} \mathrm{d}E = \int_{\mathbf{Q}_i}^{\mathbf{Q}_f} \operatorname{grad} E \,.\, \mathrm{d}\mathbf{Q} \\ &= \int_{\mathbf{Q}_i}^{\mathbf{Q}_f} \left\{ \sum_{k=1}^{n-1} \left(\frac{\partial E}{\partial Q_k}\right) \mathrm{d}Q_k + \left(\frac{\partial E}{\partial R}\right) \mathrm{d}R \right\} \end{aligned} \tag{4.41}$$

$\mathbf{Q}_i$ and $\mathbf{Q}_f$ stand here for initial and final geometries of the complex formation process. R is a general curvilinear coordinate which will be defined below, Q_k are the corresponding other degrees of freedom. Associating now the first term of the integral with the internal and the second with the external energy change, the condition described above can be formulated in the following way: the path $R^{\cdot}$ is chosen such that the n-1 partial differential quotients vanish, that is, $(\partial E/\partial Q_k)_R = 0$, and the energy is a partial minimum in n-1 directions. In order words the geometry of the whole system is at equilibrium with respect to all n-1 coordinates perpendicular to R. This distinguished path (R) is the well known adiabatic reaction coordinate or minimum energy path. As we have seen, $\Delta E_{\mathrm{INT}}^{(\mathrm{R})} = 0$ is valid in this case too and we might expect that this approach to the complex is particularly useful for energy partitioning.

Umeyama, Kitaura and Morokuma (1975) have performed SCF energy partitioning for $(FHF)^-$ along all three distinguished paths discussed above. The results are shown in Table 4.22. Again we find a dominating electrostatic contribution followed by a large exchange term of opposite sign. Charge

Table 4.22. SCF energy partitioning in $(FHF)^-$ (Umeyama, Kitaura and Morokuma, 1975)

Path[a]	Energies in kcal/mole[b]							
	ΔE	ΔE_{INT}	ΔE_{EXT}	ΔE_{COU}	ΔE_{EX}	ΔE_{POL}	ΔE_{CT}	ΔE_{CHT}
1	−62·7[a]	−15·3	−47·5	−84·8	77·6	−5·8	−25·7	−34·3
2	−62·7	21·6	−84·3	−91·2	65·4	−7·4	−34·9	−51·0
3[c]	−62·7	(3·6)	−66·3	−86·1	67·5	−5·9	−27·9	−41·8

[a] See Figure 16.
[b] A 4–31G basis set (Ditchfield, Hehre and Pople, 1971; Hehre, Ditchfield and Pople, 1972) has been used.
[c] The internal energy does not vanish completely since the authors used an approximation for the minimum energy path.

transfer and polarization energies are somewhat less important. Since a medium size basis set without polarization functions has been applied by the authors we have to expect from our previous experience that ΔE_{POL} and ΔE_{CHT} are too small and too large in absolute value respectively. It seems interesting to note that the relative importance of the individual energy contributions seems to be the same in ordinary and strong hydrogen bonds. A critical comment, however, should be added at the end: intermolecular perturbation theory and SCF energy partitioning in general are appropriate tools to analyse energies of interaction only in cases where the distortions of the isolated subsystems are not too large. Systems like $(FHF)^-$ may well lie already beyond the limits of small deformations and most probably, are regarded better as individual chemical entities.

VI. THE NATURE OF THE HYDROGEN BOND

The various theoretical approaches towards an understanding of the hydrogen bond and its experimental characteristics have been summarized in the review article by Bratož (1967). Therefore, we need not repeat the historical development here. It seems of some interest, however, to mention the four early quantum mechanical treatments of hydrogen bonding. Mainly for economic reasons all these investigations were restricted to the four electron fragments forming the ‘inner’ part of any hydrogen bonded complex—the two lone pair electrons of the proton acceptor atom (Y) and the two electrons forming the polar XH bond:

1. The first extensive quantum mechanical study on hydrogen bonding consisted in an application of valence bond (VB) theory to the four electron

system mentioned above (Sokolov, 1947, 1952; Coulson and Danielsson, 1954; Tsubomura, 1954; and Coulson, 1957). In the most extended version a basis set of five valence structures (Figure 4.17a) was used. The calculated energy of interaction is interpreted as a superposition of electrostatic, repulsive exchange and charge transfer contributions. Polarization and dispersion forces were found to be less important (Table 4.23).

2. An alternative MO approach has been presented by Pimentel (1951) (see also Pimentel and McClellan, 1960). He studied the strong hydrogen bond in $(FHF)^-$ as a typical example for a three-centre four-electron bond (Figure 4.17b). A similar LC–MO treatment of the four-electron fragment forming the hydrogen bond in $(H_2O)_2$ has been presented recently by Bowers and Pitzer (1973). In contrast to the earlier treatments these authors used bond orbitals as basis functions which were determined carefully from *ab initio* SCF calculations on the H_2O molecule. Energy partitioning similar to Morokuma's procedure has been performed as well. In Table 4.23 their results are compared with those obtained by other methods.

3. Bratož (1967) described an application of Mulliken's theory of charge transfer complexes to hydrogen bonded systems. The formation of the hydrogen bond between the two subsystems is formulated as a CI problem (Figure 4.17c). The MO basis consists of three orbitals: a bonding and an antibonding XH orbital as well as a lone-pair orbital centred on Y. Two singly excited configurations (Φ_2 and Φ_3) are possible. Provided this three term wave function is used or, in other words, doubly excited configurations are neglected, the results of the CI calculation have to be identical with those from an LC–MO SCF calculation using the same basis set. Since Bratož (1967) ignores the configuration Φ_3 too, polarization effects are neglected at least in part. From more rigorous calculations one knows, however, that the polarization term is small and this approximation seems to be well justified.

4. Starting from a similar set of basis orbitals—XH bonding and antibonding orbital as well as the lone-pair orbitals on Y—Van Duijneveldt and Murrell (1967) applied intermolecular perturbation theory (PT) to the four-electron fragments of hydrogen bonds. The results obtained in a simulation of $(H_2O)_2$ are shown in Table 4.23 here (Van Duijneveldt–Van de Rijdt and Van Duijneveldt, 1970).

Some results of the different methods to treat the four-electron fragment of the hydrogen bond in $(H_2O)_2$ are summarized in Table 4.23 above. At first we observe general agreement with respect to the relative importance of the two first order contributions. The electrostatic energy is largely compensated or even overcompensated by the short range repulsive term. Differences in absolute values, however, are large and originate from different choices of basis orbitals. In case of the second order contributions, however, the situation is much more complex. A surprisingly large charge transfer contribution is obtained from both VB and MO calculation. Results from full SCF calculations and SCF energy partitioning with different basis sets suggest to attribute this overemphasized stabilizing effect to basis set superposition

VB THEORY

$$\psi = \sum_{K=1} C_K X_K$$

VAR.PAR.: C_K

X_1:	X − H	Y
X_2:	$X|^{-}$ H^{+}	Y
X_3:	X^{+} $\underline{H}^{-}$	Y
X_4:	X^{-} H——Y^{+}	
X_5:	X $\underline{H}^{-}$ (X–Y bond)	Y^{+}

a

MO THEORY

$$\psi = |\psi_1 \bar{\psi}_1 \psi_2 \bar{\psi}_2|; \quad \psi_K = C_{XK}\Phi_X + C_{HK}\Phi_H + C_{YK}\Phi_Y, K = 1,2,3.$$

VAR.PAR.: C_{IK}

ASYMMETRIC HYDROGEN BOND SYMMETRIC

	ASYMMETRIC	SYMMETRIC	
ψ_1:	$C_{X1} > 0, C_{H1} > 0, C_{Y1} > 0$	$C_{X1} = C_{Y1}$	
ψ_2:	$C_{X2} > 0, C_{H2} \sim 0, C_{Y2} < 0$	$C_{X2} = -C_{Y2}$	$C_{H2} = 0$
ψ_3:	$C_{X3} > 0, C_{H3} < 0, C_{Y3} > 0$	$C_{X3} = C_{Y3}$	

$$\varepsilon_1 < \varepsilon_2 < \varepsilon_3$$

b

CHARGE TRANSFER THEORY

$$\Psi = C_1\Phi_1 + C_2\Phi_2 \quad : \quad \text{VAR. PAR. } C_I$$

$$\Phi_1 = |\lambda_1 \bar{\lambda}_1 \lambda_2 \bar{\lambda}_2| \quad : \quad \Phi_2 = \frac{1}{\sqrt{2}}\{|\lambda_1 \bar{\lambda}_1 \lambda_2 \bar{\lambda}_3| - |\lambda_1 \bar{\lambda}_1 \lambda_3 \bar{\lambda}_2|\}$$

$\lambda_1 = N_1(\Phi_X + D_1\Phi_H)$ XH BONDING ORBITAL

$\lambda_2 = \Phi_Y$ LONE PAIR ORBITAL

$\lambda_3 = N_2(\Phi_X - D_2\Phi_H)$ XH ANTIBONDING ORBITAL

c

Figure 4.17. Various approaches to study hydrogen bonding as a four electron problem: (a) VB method, (b) MO method and (c) charge transfer theory

Table 4.23. Comparison of the results on the hydrogen bond in $(H_2O)_2$ treated as a four electron fragment by different methods

<table>
<tr><th>Method</th><th>R_{OO} (Å)</th><th>ΔE_{COU}</th><th>ΔE_{EX}</th><th>ΔE_{POL}</th><th>ΔE_{CHT}</th><th>ΔE_{DIS}</th><th>ΔE</th><th>References</th></tr>
<tr><td>VB</td><td>2·70</td><td>−6</td><td>8·4</td><td colspan="2">−8</td><td>−3</td><td>−8·6</td><td>a</td></tr>
<tr><td>MO</td><td>2·73</td><td>−9·1</td><td>9·0</td><td>−0·6</td><td>−7·5</td><td></td><td>−8·2</td><td>b</td></tr>
<tr><td>PT</td><td>2·9</td><td>−12·0</td><td>13·1</td><td>−1·5</td><td>−1·4</td><td>−1·7</td><td>−3·4</td><td>c</td></tr>
</table>

References. [a] Coulson (1957); [b] Bowers and Pitzer (1973); [c] Van Duijneveldt–Van de Rijdt and Van Duijneveldt (1970).

errors. Calculations using counterpoise method have shown (Section III.A) that these errors are to be expected quite large for the minimum basis sets applied here. The dispersion energy reported by Coulson (1957) seems to be too large in comparison to other more accurate calculations on $(H_2O)_2$ (Jeziorski and Van Hemert, 1976). In general, the differences in the second order terms arise from both different basis sets and different methods (VB, MO and PT) and consequently are very difficult to analyse.

Fully respecting the high didactic value and the usefulness for qualitative predictions of these model studies we have to admit, however, that the restriction to the four-electron fragment is not appropriate for most of the other questions we are asking to the theoreticians. A number of reasons are responsible for the deficiency of this approach in predicting quantitative results which can be subject to experimental proof. The energy of interaction is determined by the whole molecule and not by a fragment in the contact region only. Around certain points, for example, around the equilibrium geometry careful parametrization may lead to useful results. Attempts to calculate larger parts of the energy surface by the four-electron approximation, however, will fail. As we have seen in almost all hydrogen-bonded complexes studied here, the electrostatic contribution dominates at the equilibrium geometry and, of course, is even more important at larger intermolecular distances. It may well be the case that the local electrostatic interaction represented roughly by the attraction of the polar XH bond and the lone-pair electrons is the most important term at the energy minimum. At larger distances, however, and at other intermolecular orientations this will be no longer the case. The asymptotic interaction between two polar molecules at large values of R is determined by the dipole–dipole term, which has a different preferred orientation compared to the local interaction term unless the directions of molecular dipole moments occasionally coincide with the directions of the XH bond and the lone pair on Y. Furthermore, as we have seen by analysing the wave functions for various complexes of small molecules and ions, changes in electron densities are by no means restricted to the region of the four-electron fragment. In contrary, we have found that mutual polarization is extended over the whole molecules (Figures 4.5–4.7). For any more involved theoretical investigation, therefore, it seems inevitable to perform the admittedly expensive studies on whole complexes consisting of all interacting molecules.

Before we enter into an interpretation of the results of more extended calculations we will mention in this context a very recent simplified model of hydrogen bonding proposed by Allen (1975), which is somewhat related to the four-electron approach. Based on the result of *ab initio* MO–SCF calculations he developed an approximative empirical treatment for predictions of the properties of hydrogen-bonded dimers. His formula for the energy of interaction will serve as an example illustrating the nature of this model:

$$\Delta E = k \,.\, \mu_{AH} \,.\, \Delta I_B \,.\, R_{AB}^{-1} \tag{4.42}$$

In this expression μ_{AH} represents the local dipole moment of the AH bond, ΔI_B the difference in ionization potentials between atom B and the next higher rare gas atom and R_{AB} finally is the equilibrium distance between the atoms A and B. The constant k is determined empirically by fitting one dimerization energy to the experimentally known value. The quantitative aspects of Allen's model can be estimated from Table 4.24. Again, the empirical formula is a useful tool only for predictions of the equilibrium properties of the complex and cannot be used for whole energy surfaces.

Table 4.24. Dimerization energies obtained from Allen's (1975) simplified model

Complex	ΔE_{CALC} (kcal/mol)[a]	ΔE_{EXP} (kcal/mol)
$(HF)_2$	7·0[b]	7·0 ± 1
$(H_2O)_2$	6·7	5·2 ± 1·5
$(H_3N)_2$	4·5	4·0 ± 0·4
$(HCl)_2$	1·5	2·1 ± 0·2
$(H_2S)_2$	1·2	1·7 ± 0·3

[a] According to equation (4.42), $k = 1·64$.
[b] The dimerization energy of $(HF)_2$ was used for calibration.

In the whole chapter presented so far we have mainly summarized various numerical results of calculations on a few selected hydrogen-bonded complexes and compared them with the available experimental data. Now, we will make an attempt to characterize the phenomenon 'hydrogen bond' briefly using the variety of available information as a basis. We will restrict ourselves here to ground states of dimeric associations in the vapour phase, since these are the best studied systems.

From a more pragmatic point of view it seems appropriate to distinguish three different cases:

1. weak and ordinary, asymmetric hydrogen bonds between neutral molecules
2. ion–molecule complexes with asymmetric hydrogen bonds and
3. strong and symmetric or nearly symmetric hydrogen bonds.

Extreme or 'pure' examples for the three types of interaction call for a description by different theoretical methods. As it has been found fruitful in many other fields, alternative views are best suited to describe the idealized, 'pure' cases. Many, or even most real systems may lie in between the extremes and either can be understood equally well by both approaches or require a kind of superimposed theory.

Weak and ordinary hydrogen bonds are regarded best as intermolecular complexes. The characteristic properties can be explained by a superposition of electrostatic, repulsive exchange, dispersion, charge transfer and polarization forces. In the region of non-negligible intermolecular overlap we have to adopt a convention for the separation of the total energy of interaction into the contributions mentioned above. Following the most popular definitions (see Section V) we find that the electrostatic energy is the largest term in absolute value followed by the exchange contribution at the energy minimum. In contrast to previous less accurate investigations the three second order energies seem to be of roughly equal importance at the equilibrium geometry. Most probably the numerical values decrease in the sequence given above and the polarization energy is somewhat smaller than the other two contributions. The unique stereochemical properties of hydrogen bonds can be understood as a result of the highly asymmetric electrostatic potentials of polar XH bonds superimposed by the usually less important angular dependence of the other terms. Around the energy minima the local electrostatic contributions dominate and the conventional single-centre multipole expansion for the electrostatic energy breaks down as we have seen for the complex $(H_2O)_2$ in Table 4.19 above. In fact a multicentre expansion of electrostatic potentials was found to be much more appropriate for the description of the interaction between two neutral molecules (for example, see Scrocco and Tomasi, 1973). Deformations of molecular geometries on complex formation usually are very small in case of ordinary hydrogen bonds and intramolecular relaxation does not contribute appreciably to the total energy of interaction. There are some exceptional cases like $H_3N.HCl$ involving molecules with polar XH bonds where X is an atom of the third or a higher row of the periodic table. As a rule the elongation of the XH bond is much larger in these examples. A simple explanation of this finding can be given in terms of the XH stretching force constants, which are representative for the amount of energy necessary for a small elongation of the XH bond. As we can see in Table 4.25 there is a big gap in force constants depending on the provenance of the X atom. In case X is a second row atom the force constants usually lie between $8 \leqslant f \leqslant 10$ mdyn/Å, whereas $f \leqslant 5$ mdyn/Å is a typical value for XH bonds involving third and higher row atoms. The dissociation energies of XH bonds, although used in some previous interpretations do not represent a suitable measure for the easiness of changes in bond length (Table 4.25).

Interactions between spherically symmetric ions and small molecules are somewhat easier to study. The situation is very similar to the previous case—interaction between two neutral molecules—besides the fact that the electrostatic, exchange and polarization energies are much larger now. The

Table 4.25. Equilibrium distances, dissociation energies and force constants for some XH bonds (Herzberg, 1950; and Vedeneyev *et al.*, 1966).

Molecule	R_e (Å)	D_0 (kcal/mol)	f(mdyn/Å)
HF	0·917	135 ± 0·3	9·66
HCl	1·275	102·20 ± 0·03	5·16
HBr	1·413	86·69 ± 0·13	4·12
HJ	1·604	70·41 ± 0·1	3·14
OH	0·971	101·36 ± 0·3	7·80
SH	(1·35)	81·4 ± 2·9	
H_2O[a]	0·957	(116)	8·45
H_2S[a]	1·328	(91)	4·29

[a] Dissociation energies for the process $H_2X \rightarrow H + HX$ and XH stretching force constants are given.

electrostatic contribution dominates strongly over all other terms. In $Li^+.OH_2$, as we have seen, the electrostatic energy is very well represented by the multipole expansion even at points close to the equilibrium geometry. Consequently, the overall shape of the energy hypersurface at intermolecular distances larger than the equilibrium value is determined by the ion–dipole and ion–quadrupole terms. Since the components of the octopole moments are zero perpendicular to the C_2 axis a reasonable approximation of the electrostatic potential on the whole hypersurface is obtained only at the level of quadrupole or hexadecapole contributions (Σ_2 or Σ_4). In $Na^+.OH_2$ and $Cl^-.H_2O$ the situation is very similar to $Li^+.OH_2$. The complex $F^-.H_2O$, however, seems to be an exceptional case. Again we observe strong interaction with the local OH bond dipole at the equilibrium geometry and the exchange repulsion is unusually large too. In contrast to $Cl^-.H_2O$ the intrinsic barrier for partial rotation of the water molecule from one equilibrium position to the other is rather large in $F^-.H_2O$.

The calculation or qualitative description of energy hypersurfaces for the interaction of many-atomic ions with molecules represents a highly complicated problem. Extended ions with low symmetry are especially difficult to handle by theoretical approaches at least because of the large number of degrees of freedom. This fact may be also the reason for the lack of *ab initio* studied on such complexes. On the whole very little is known about systems of this type.

Strong and symmetric or almost symmetric hydrogen bonds are formed in systems which can fulfil two requirements:

1. the proton in the hydrogen bond is bound with equal or almost equal affinity to the two proton acceptor atoms X and Y. The resulting potential surface for proton motion is symmetric with one minimum or two equivalent minima (see Figure 4.3).

2. the two proton acceptor atoms X and Y can come together in the complex closely enough to form a strong hydrogen bond but without causing very large repulsion or intramolecular strain energies.

$$X^- + H^+ + X^- \longrightarrow [XHX]^- \qquad \text{III}$$

$$B + H^+ + B \longrightarrow [BHB]^+ \qquad \text{IV}$$

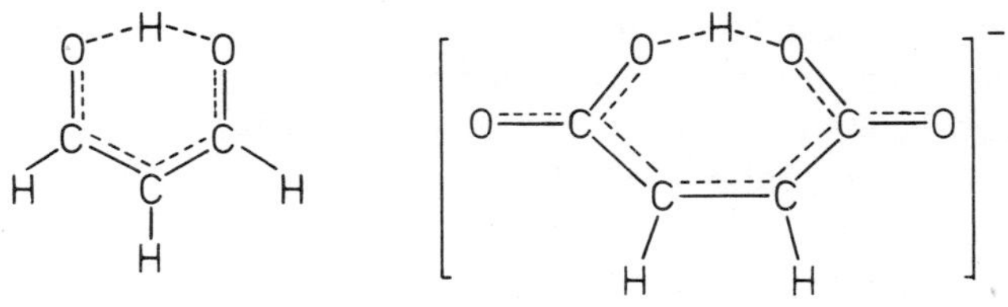

V VI

Figure 4.18. Various types of strong and symmetric hydrogen bonds

Intermolecular complexes can meet the first requirement when the proton is placed between two anions or two neutral molecules leading thereby to a charged species, a complex anion or cation respectively (Figure 4.18). The role of symmetry is demonstrated nicely by a comparison of fitted energy surfaces for the ions $(ClHCl)^-$, $(BrHBr)^-$ and $(BrHCl)^-$ (Jiang and Anderson, 1974). In the first two examples symmetric double minimum potentials with a very small barrier ($\Delta E^{\neq} \sim 0{\cdot}24$ kcal/mol) were obtained whereas in the third complex an asymmetric single minimum potential was able to reproduce the experimental spectra best. The central proton is displaced by about 0·4 Å from the symmetric position. In very few exceptional cases nearly symmetric, strong hydrogen bonds are formed also between two neutral molecules. $H_3N.HCl$ and $H_3N.HBr$ may serve as possible candidates (e.g. see Kollman *et al.*, 1975). Here the central proton by chance and not by symmetry reasons shows similar affinities to the two proton binding sites in the complex.

Symmetry with respect to proton motion can be achieved also in intramolecular hydrogen bonds provided a symmetric π-electron system can compensate for the charge migration (see **V** and **VI** in Figure 4.18). In this case the entire species may be either neutral or charged. Building up the appropriate geometry for a strong hydrogen bond may require a large amount of deformation energy and many possible candidates for strong intramolecular hydrogen bonds, therefore, fail to fulfil the second criterion mentioned above. At present the smallest possible systems forming intramolecular hydrogen bonds are too large for extensive *ab initio* calculations. Besides a number of semiempirical investigations only minimum basis set (STO–3G) calculations have been reported for **V** (Isaacson and Morokuma, 1975) and **VI** (Morokuma, Iwata and Lathan, 1974).

Complexes and complex ions with strong hydrogen bonds are regarded best as chemical individuals. As we have seen before charge rearrangements and changes in molecular geometries on complex formation are large. In general the systems can be characterized by structural lability with respect to the position of the central proton, unique IR spectra and typical proton chemical shifts. For the purpose of classification and from an heuristic point of view it seems appropriate to regard strong hydrogen bonds as typical representatives for the three-centre four-electron bond (Figure 4.17 above), which has been discussed extensively in the literature. Other examples for similar three-centre interactions can be understood in terms of the structures **III** and **IV** in Figure 4.18 above. The central proton has to be substituted by a halogen cation or doubly ionized rare gas atom. Typical examples are the trihalogen anions (Cl_3^-, Br_3^-, I_3^-, ClF_2^-, Cl_2Br^-, Br_2Cl^-, $ClIBr^-$, etc.) and XeF_2. All these complexes are linear but not necessarily symmetric. Similar as in case of strong hydrogen bonds the position of the central atom is rather labile as we can see in case of I_3^- (the central iodine is denoted by no. 2):

$$CsI_3{:}\ R_{12} = 2{\cdot}83\ \text{Å} \qquad R_{23} = 3{\cdot}04\ \text{Å}$$

and

$$(C_2H_5)_4N(I_2)_2I_3^-{:}\ R_{12} = R_{23} = 2{\cdot}91\ \text{Å} \quad \text{(Sutton, 1958, 1965).}$$

Finally, we would like to make a more comprehensive remark. Most weak and ordinary hydrogen bonded complexes between neutral molecules and some ion–molecule associations can be understood in terms of intermolecular forces. Asymptotic perturbation theory, in general, is a good approximation only at intermolecular distances substantially larger than the equilibrium values. Strong hydrogen bonds, the other extreme examples are better described as representatives for a special class of chemical bond. It may be considered as a success of *ab initio* MO theory that both kinds of systems can be treated equally well within the frame of only one methodological approach. An accurate quantitative treatment, however, requires careful choices of extended basis sets and consideration of electron correlation effects which inevitably make the numerical calculations very time consuming and hence expensive.

VII. CONCLUSIONS AND FURTHER ASPECTS

Ab initio MO calculations on hydrogen-bonded dimers between small molecules have reached a stage of high accuracy which allows to predict energies and structures of complexes within the error limits of experimental measurements. In many cases where the experiments are very difficult to perform the theoretical results represent the only reliable source of information. The investigation of heterodimers in the vapour phase may serve as an example. In some examples IR and NMR spectra have been calculated by *ab initio* techniques as well and the agreement with experimental data is satisfactory. On the whole, we can say that the hydrogen bond in dimers of small molecules in their electronic ground states is well understood. At present, the major limitation of the *ab initio* approach concerns the immense need of computer time which prevents an extension of accurate numerical calculations to larger entities like complexes between larger molecules or higher aggregates simulating local sections of liquids or crystals. Besides the fact that the calculation of the individual point on the energy surface requires much more computer time, in higher aggregates the configurational freedom is much larger too and the point by point approach seems to be hopeless. In fact, the problem to describe the properties of larger clusters of molecules or of hydrogen bonded liquids and crystals has not been solved yet in a satisfactory manner and still remains a striking challenge to theoreticians. At present the attempts to calculate properties of molecular crystals by *ab initio* SCF methods seem to be promising.

Numerical results on intermolecular complexes obtained with minimum and most of the other small basis sets are rather discouraging. As we have seen, the properties of isolated molecules are often reproduced very poorly. Even more seriously, basis set superposition errors are large. Qualitative results on unknown systems have to be treated with great care therefore. Most probably, appropriate semiclassical models like consistent force fields which can provide analytic expressions for energy surfaces, will turn out to represent a cheaper and more extensible approach to study interactions between larger molecules than the straightforward application of expensive *ab initio* calculations using minimal or small basis set.

So far we have not mentioned hydrogen bonding in excited states and its influence on electron excitation spectra. Although some work has been done in this field already (for example, see Morokuma, Iwata and Lathan, 1974; or Schuster, 1976a) we feel that further methodological improvement is necessary before the results will reach the same degree of reliability as in present investigations on ground state interactions.

Most of the theoretical treatments subsequent to the calculation of energy surfaces need analytical expressions. Energy partitioning may help here to derive useful functions for this purpose. More emphasis will have to be put on statistical mechanics in the future. To give just one example, recently Braun and Leidecker (1974) calculated rotational and vibrational spectra of $(H_2O)_2$ based on *ab initio* potential surfaces and made predictions on the temperature dependence of the equilibrium constant for dimerization of H_2O as well as the second virial coefficient of water vapour.

Monte Carlo and molecular dynamics calculations although very time consuming, present a possibility to proceed towards an understanding of the liquid state and solute–solvent interactions. Important work has been done in this field already. The ultimate test for the usefulness of *ab initio* energy surfaces in this approach, nevertheless, is still missing. The importance of three or higher body interactions in associated liquids and solutions represents a question open to further investigations.

VIII. ACKNOWLEDGEMENTS

The author is indebted to many colleagues and especially to all his coworkers for numerous stimulating discussions. The help with the preparation of computer drawings by Dr. O. Steinhauser is greatfully acknowledged. Financial support for our investigations in this field has been provided by the Austrian *Fonds zur Förderung der wissenschaftlichen Forschung* (Project No. 1217 and 2756). Last but not least the assistance of Dr. B. Schreiber and of our technical staff, Mrs. J. Dura and Mr. J. Schuster, with the preparation of the manuscript is gratefully acknowledged. Figures 4.8, 4.9 and 4.11 are reproduced with permittance of Springer–Verlag, Berlin from *Topics in Current Chemistry*, Vol. 60.

IX. REFERENCES

Ahlrichs, R., H. Lischka, B. Zurawski and W. Kutzelnigg (1975). *J. Chem. Phys.*, **63,** 4685.
Ahlrichs, R. (1976). *Theoret. Chim Acta*, **41,** 7.
Allen, L. C. (1975). *J. Am. Chem. Soc.*, **97,** 6921.
Almlöf, J., (1972). *Chem. Phys. Letters*, **17,** 49.
Anderson, G. R. and G. J. Jiang (1973). *J. Phys. Chem.*, **77,** 2560.
André, I. M., (1969). *J. Chem. Phys.*, **50,** 1536.
Arrighini, G. P., M. Maestro and R. Moccia (1967). *Chem. Phys. Letters*, **1,** 242.
Bassani, F., L. Pietronero and R. Resta (1973). *J. Phys. C: Solid State Phys.*, **6,** 2133.
Beauchamp, J. L. and S. E. Butrill (1968). *J. Chem. Phys.*, **48,** 1783.
Bender, C. F. and E. R. Davidson (1968). *J. Chem. Phys.*, **49,** 4222.
Beretta, E. and F. Vetrano (1973). *Internat. J. Quantum Chem.*, **7,** 333.
Beyer, A., H. Lischka and P. Schuster (1976). To be published .
Bowers, M. J. T. and R. M. Pitzer (1973). *J. Chem. Phys.*, **59,** 163.
Boys, S. F. and F. Bernardi (1970). *Mol. Phys.*, **19,** 558.
Bratož, S. (1967). *Adv. Quantum Chem.*, **3,** 209.
Braun, C. and H. Leidecker (1974). *J. Chem. Phys.*, **61,** 3104.
Chipman, D. M., J. D. Bowman and J. O. Hirschfelder (1973). *J. Chem. Phys.* **59,** 2830.
Chong, S., R. A. Myers and J. L. Franklin (1972). *J. Chem. Phys.*, **56,** 2427.
Chupka, A. W. and M. E. Russel (1968). *J. Chem. Phys.*, **49,** 5426.
Chupka, A. W., M. E. Russel and K. Refaey (1968). *J. Chem. Phys.*, **48,** 1518.
Claverie, P. (1971). *Internat. J. Quantum Chem.*, **5,** 273.
Cotter, R. J. and W. S. Koski (1973). *J. Chem. Phys.*, **59,** 784.
Coulson, C. A. and U. Danielsson (1954). *Arkiv Fysik*, **8,** 245.
Coulson, C. A. (1957). *Research*, **10,** 149.
Covington, A. K. and P. Jones, eds. (1968). *Hydrogen-Bonded Solvent Systems*, Taylor and Francis, London.
Curtiss, L. A. and J. A. Pople (1975). *J. Mol. Spectroscopy*, **55,** 1.
Daudey, J. P. (1974). *Internat. J. Quantum Chem.*, **8,** 29.
Daudey, J. P., P. Claverie and J. P. Malrieu (1974). *Internat. J. Quantum Chem.*, **8,** 1.
Del Re, G., J. Ladik and G. Biczo (1967). *Phys. Rev.*, **155,** 997.
Diercksen, G. H. F. (1969). *Chem. Phys. Letters*, **4,** 373.
Diercksen, G. H. F. (1971). *Theoret. Chim. Acta*, **21,** 335.

Diercksen, G. H. F. and W. P. Kraemer (1970). *Chem. Phys. Letters*, **6,** 419.
Diercksen, G. H. F., W. P. Kraemer and B. O. Roos (1975). *Theoret. Chim. Acta*, **36,** 249.
Diercksen, G. H. F., W. von Niessen and W. P. Kraemer (1973). *Theoret. Chim. Acta*, **31,** 205.
Dill, J. D., L. C. Allen, W. C. Topp, and J. A. Pople (1975). *J. Am. Chem. Soc.*, **97,** 7220.
Ditchfield, R. (1974). *Mol. Phys.*, **27,** 789.
Ditchfield, R. (1976). *Chem. Phys. Letters*, **40,** 53.
Ditchfield, R., W. J. Hehre and J. A. Pople (1971). *J. Chem. Phys.*, **54,** 724.
Ditchfield, R. and R. E. McKinney, Jr. (1976). *Chem. Phys.*, **13,** 187.
Dreyfus, M. and A. Pullman (1970a). *Theoret. Chim. Acta*, **19,** 20.
Dreyfus, M. and A. Pullman (1970b). *Compt. Rend. Acad. Sci (Paris)*, **271,** 457.
Dyczmons, V. and W. Kutzelnigg (1974). *Theoret. Chim. Acta*, **33,** 239.
Dyke, T. R., B. J. Howard and W. Klemperer (1972). *J. Chem. Phys.*, **56,** 2442.
Dyke, T. R. and J. S. Muenter (1974). *J. Chem. Phys.*, **60,** 2929.
Dunning, T. H., Jr., R. M. Pitzer and S. Aung (1972). *J. Chem. Phys.*, **57,** 5044.
Eisenschitz, H. and London (1930). *Z. Physik*, **60,** 491.
Grimaldi, F., A. Lecourt and C. Moser (1967). *Internat. J. Quantum. Chem.*, **1S,** 153.
Hadži, D. and H. W. Thompson, eds. (1959). *Hydrogen Bonding*, Pergamon Press, London.
Hamilton, W. C. and P. Jones (1968). *Hydrogen Bonding in Solids*, Benjamin, New York.
Hankins, D., J. W. Moskowitz and F. H. Stillinger (1970a). *Chem. Phys. Letters*, **4,** 527.
Hankins, D., J. W. Moskowitz and F. Stillinger (1970b). *J. Chem. Phys.*, **53,** 4544.
Hankins, D., J. W. Moskowitz and F. H. Stillinger (1973). *J. Chem. Phys.*, **59,** 995.
Hariharan, P. C. and J. A. Pople (1972). *Chem. Phys. Letters*, **16,** 217.
Hehre, W. J., R. F. Stewart and J. A. Pople (1969). *J. Chem. Phys.*, **51,** 2657.
Hehre, W. J., R. Ditchfield and J. A. Pople (1972). *J. Chem. Phys.*, **56,** 2257.
Herzberg, G. (1950). *Molecular Spectra and Molecular Structure* I. Spectra of Diatomic Molecules, Van Nostrand, New York. pp. 501–581.
Hirschfelder, J. O., C. F. Curtis and R. B. Bird (1954). *Molecular Theory of Gases and Liquids*, Wiley, New York. pp. 835–1103.
Hirschfelder, J. O., W. Byers–Brown and S. T. Epstein (1964). *Adv. in Quantum Chem.*, **1,** 284.
Hopkinson, A. C., N. K. Holbrook, K. Yates and L. G. Csizmadia (1968). *J. Chem. Phys.*, **49,** 3596.
Huggins, M. L. (1919). *Thesis*, Univ. of California.
Huggins, M. L. (1922). *Science*, **55,** 679.
Huggins, M. L. (1971). *Angew. Chemie*, **83,** 163 (*Internat. Ed.*, **10,** 147).
Huzinaga, S. (1970). *Tables of Atomic Functions*, Alberta University.
Isaacson, A. D. and K. Morokuma (1975). *J. Am. Chem. Soc.*, **97,** 4454.
Janoschek, R., E. G. Weidemann, H. Pfeiffer and G. Zundel (1972). *J. Am. Chem. Soc.*, **94,** 2387.
Janoschek, R., E. G. Weidemann and G. Zundel (1973). *J. Chem. Soc. Farad. Trans.* II, **69,** 505.
Janoschek, R. (1976). *Calculated Vibrational Spectra of Hydrogen Bonded Systems*, in Schuster, Zundel and Sandorfy (1976) North Holland, Amsterdam pp. 165–216.
Jaszunski, M. and A. J. Sadlej (1972a). *Chem. Phys. Letters*, **15,** 41.
Jaszunski, M. and A. J. Sadlej (1972b). *Theoret. Chim. Acta*, **27,** 135.
Jaszunski, M. and A. J. Sadlej (1973). *Theoret. Chim. Acta*, **30,** 257.
Jeziorski, B. and M. Van Hemert (1976). *Mol. Phys.*, **31,** 713.

Jiang, G. J. and G. R. Anderson (1973). *J. Phys. Chem.*, **77,** 1764.
Jiang, G. J. and G. R. Anderson (1974). *J. Chem. Phys.*, **60,** 3258.
Joesten, M. D. and L. J. Schaad (1974). *Hydrogen Bonding*, Dekker, New York.
Johansson, A., P. Kollman and S. Rothenberg (1973). *Theoret. Chim. Acta*, **29,** 167.
Karpfen, A., J. Ladik, P. Russegger, P. Schuster and S. Suhai (1974). *Theoret. Chim. Acta*, **34,** 115.
Karpfen, A. (1976). *Thesis*, 2nd deg. Univ. of Vienna.
Karpfen, A. and P. Schuster, (1976). *Chem. Phys. Letters*, **44,** 459.
Karplus, M. and H. J. Kolker (1963). *J. Chem. Phys.*, **38,** 1263.
Keil, F. and R. Ahlrichs (1976). *J. Am. Chem. Soc.*, **98,** 4787.
Kertesz, M., J. Koller and A. Azman (1975). *Chem. Phys. Letters*, **36,** 576.
Kistenmacher, H., H. Popkie and E. Clementi (1973). *J. Chem. Phys.*, **58,** 5627.
Kitaura, K. and Morokuma, K. (1976). *Internat. J. Quantum Chem.*, **10,** 325.
Kochanski, E. (1974). *Chem. Phys. Letters*, **28,** 471.
Kochanski, E. and J. F. Gouyet (1975a). *Mol. Phys.*, **29,** 693.
Kochanski, E. and J. F. Gouyet (1975b). *Theoret. Chim. Acta*, **39,** 329.
Kollman, P. A. and L. C. Allen (1970). *Theoret. Chim. Acta*, **18,** 399.
Kollman, P. A. and L. C. Allen (1972). *Chem. Rev.*, **72,** 283.
Kollman, P. A., J. McKelvey, A. Johansson and S. Rothenberg (1975). *J. Am. Chem. Soc.*, **97,** 955.
Kraemer, W. P. and G. H. F. Diercksen (1970). *Chem. Phys. Letters*, **5,** 463.
Kraemer, W. P. and G. H. F. Diercksen (1972). *Theoret. Chim. Acta*, **23,** 398.
Kuchitsu, K. and V. Morino (1965). *Bull. Chem. Soc. Japan*, **38,** 814.
Kutzelnigg, W., R. Ahlrichs, I. Labib–Iskander and W. A. Bingel (1967), *Chem. Phys. Letters*, **1,** 447.
Kutzelnigg, W. (1973). *Topics in Current Chemistry*, **41,** 31.
Latimer, W. M. and W. H. Rodebush (1920). *J. Am. Chem. Soc.*, **42,** 1419.
Lentz, B. R. and H. A. Scheraga (1973). *J. Chem. Phys.*, **58,** 5296.
Lewis, G. N. (1923), *Valence and the Structure of Atoms and Molecules*, Chemical Catalog Co., New York, p. 109.
Liebmann, S. P. and J. W. Moskowitz (1971). *J. Chem. Phys.*, **54,** 3622.
Lischka, H. (1973). *Theoret. Chim. Acta*, **31,** 39.
Lischka, H. (1974). *J. Am. Chem. Soc.*, **96,** 4716.
Lischka, H. (1976). *J. Am. Chem. Soc.*, **99,** 353.
Longuet–Higgins, H. C. (1948). *Proc. Phys. Soc.* (*London*), **60,** 270.
Lundgren, J. D. and I. Olovsson (1976). *The Hydrated Proton in Solids* in Schuster, Zundel and Sandorfy (1976). pp. 471–526.
Margenau, H. and N. R. Kestner (1969). *Theory of Intermolecular Forces*, 2nd ed., Pergamon Press, Oxford, pp. 15–298.
Matsuoka, O., E. Clementi and M. Yoshimine (1976). *J. Chem. Phys.*, **64,** 1351.
Meunier, A., B. Levy and G. Bertier (1973). *Theoret. Chim. Acta*, **29,** 49.
Meyer, W. (1971). *Internat. J. Quantum Chem.*, **5,** 341.
Meyer, W. (1973). *J. Chem. Phys.*, **58,** 1017.
Meyer, W., W. Jakubetz and P. Schuster (1973). *Chem. Phys. Letters*, **21,** 97.
Morokuma, K. and J. R. Winick (1970). *J. Chem. Phys.*, **52,** 1301.
Morokuma, K. (1971). *J. Chem. Phys.*, **55,** 1236.
Morokuma, K., S. Iwata and W. A. Lathan (1974). 'Molecular interactions in ground and excited states,' in R. Daudel and B. Pullman, eds., *The World of Quantum Chemistry*, Reidel, Dordrecht (Holland), pp. 277–316.
Mulliken, R. S. (1955). *J. Chem. Phys.*, **23,** 1833, 1841, 2338 and 2343.
Murthy, A. S. N. and C. N. R. Rao (1970). *J. Mol. Structure*, **6,** 253.
Murrell, J. N., M. Randic and D. R. Williams (1965). *Proc. Roy. Soc.* (London), **A284,** 566.

Murrell, J. N. and G. Shaw (1967). *J. Chem. Phys.*, **46,** 1768.
Newton, M. D., W. A. Lathan, W. J. Hehre and J. A. Pople (1970). *J. Chem. Phys.*, **52,** 4064.
Olah, G. (1973). *Angew. Chemie*, **85,** 183 (German) and *Internat. Ed.*, **12,** 173 (English).
Otto, P. and J. Ladik (1975). *Chem. Phys.*, **8,** 192.
Otto, P. and J. Ladik (1976). To be published.
Pietronero, L. and N. O. Lipari (1975). *J. Chem. Phys.* **62,** 1796.
Pimentel, G. C. (1951). *J. Chem. Phys.*, **19,** 446.
Pimentel, G. C. and A. L. McClellan (1960). *The Hydrogen Bond*, Freeman, San Francisco.
Pimentel, G. C. and A. L. McClellan (1971). *Ann. Rev. Phys. Chem.*, **22,** 347.
Popkie, H., H. Kistenmacher and E. Clementi (1973). *J. Chem. Phys.*, **59,** 1325.
Priglinger, S. (1975). *Thesis*, Univ. of Vienna.
Priglinger, S. and P. Schuster, (1976). In preparation.
Pullman, A. and P. Brochen (1975). *Chem. Phys. Letters*, **34,** 7.
Pullman, B. and A. Pullman (1974). *Adv. in Protein Chemistry*, **28,** 347.
Raimondi, M., G. F. Tantardina and M. Simonetta (1975). *Mol. Phys.*, **30,** 797.
Roos, B. O. (1972). *Chem. Phys. Letters*, **15,** 153.
Roothaan, C. C. J. (1951). *Rev. Mod. Phys.*, **23,** 69.
Rosenberg, B. J. and I. Shavitt (1975). *J. Chem. Phys.*, **63,** 2162.
Russegger, P. and P. Schuster (1973). *Chem. Phys. Letters*, **19,** 254.
Sadlej, A. J. (1975). *Chem. Phys. Letters*, **30,** 432.
Sadlej, A. J. (1977). *Chem. Phys. Letters*, **47,** 50.
Sadlej, A. J. and W. T. Raynes (1975). *Chem. Phys.*, **7,** 383.
Schuster, P. (1970). *Theoret. Chim. Acta*, **19,** 212.
Schuster, P. (1973). *Z. f. Chemie*, **13,** 41.
Schuster, P., W. Jakubetz, G. Beier, W. Meyer and B. M. Rode (1974). Potential curves for proton transfer along hydrogen bonds, in E. D. Bergmann and B. Pullman, eds., *Chemical and Biochemical Reactivity, Acad. Press*, Jerusalem, pp. 257–282.
Schuster, P., W. Jakubetz and W. Marius (1975). *Topics in Current Chemistry*, **60,** 1–107.
Schuster, P., W. Marius, A. Pullman and H. Berthod (1975). *Theoret. Chim. Acta*, **40,** 323.
Schuster, P. (1976a). 'Energy surfaces for hydrogen-bonded systems' in Schuster, Zundel, and Sandorfy, eds. *The Hydrogen Bond*, North Holland, Amsterdam, pp. 25–163.
Schuster, P. (1976b). 'Calculations of anion-solvent interactions,' in L. Kevan and B. Webster, eds., *Electron Solvent and Anion-Solvent Interactions*, Elsevier, Amsterdam, pp. 259–309.
Schuster, P., H. Lischka and A. Beyer (1976). '*Ab initio* studies on hydrogen bonding and ion molecule complexes,' in I. G. Csizmadia, ed., *Progress in Theoretical Organic Chemistry*, Vol. 2, Elsevier, Amsterdam, pp. 89–105.
Schuster, P., G. Zundel and C. Sandorfy, eds. (1976). *The Hydrogen Bond-Recent Developments in Theory and Experiment*, North Holland, Amsterdam.
Scrocco, E. and J. Tomasi (1973). *Topics in Current Chemistry*, **42,** 95–170.
Slater, J. C. (1965). *J. Chem. Phys.*, **43,** S 228.
Sokolov, N. D. (1947). Dokladi, **58,** 611.
Sokolov, N. D. (1952). *Zh. Eksperim. i Teor. Fiz.*, **23,** 315.
Sokolov, N. D. and Chulanovskii, eds. (1964). *Hydrogen Bonding Akad Nauk SSSR*, Moscow.
Støgård, A., A. Strich, J. Almlöf and B. Roos (1975). *Chem. Phys.*, **8,** 405.

Sutton, L. E. (1958), *Tables of Interatomic Distances and Configurations in Molecules and Ions,* The Chemical Society London, p. M93 Supplement/(1965), p. M52S.
Tsubomura, H. (1954). *Bull. Chem. Soc. Japan,* **27,** 445.
Tursi, A. J. and E. R. Nixon (1970). *J. Chem. Phys.,* **52,** 4858.
Umeyama, H., K. Kitaura and K. Morokuma (1975). *Chem. Phys. Letters,* **36,** 11.
Van Duijneveldt, F. D. and J. N. Murrell (1967). *J. Chem. Phys.,* **46,** 1759.
Van Duijneveldt–van de Rijdt, J. G. C. M., and F. B. Van Duijneveldt (1970). *Theoret. Chim. Acta,* **19,** 83.
Van Duijneveldt–Van de Rijdt, J. G. C. M., and F. B. Van Duijneveldt (1971), *J. Am. Chem. Soc.,* **93,** 5644.
Van Thiel, M., E. D. Becker and G. Pimentel (1957). *J. Chem. Phys.,* **27,** 486.
Vedeneyev, V. I., L. V. Gurvich, V. N. Kondrat'yev, V. A. Medvedev and Ye. L. Frankevich (1966). *Bond Energies, Ionization Potentials and Electron Affinities,* Arnold, Scripta Technica Ltd., London, pp. 199–200.
Vinogradov, S. N. and R. H. Linell (1971), *Hydrogen Bonding,* Van Nostrand–Reinhold, New York.
Wahlgren, U., J. Pacansky and P. S. Bagus (1975). *J. Chem. Phys.,* **63,** 2874.
Weinberger, P. and K. Schwarz (1975). 'The SCF–X-scattered wave method and its application to molecular problems,' in A. D. Buckingham, ed., *Theoretical Chemistry, Physical Chemistry, Series Two,* Vol. 1, Butterworth, London, pp. 255–282.
Werner, H. J. and W. Meyer (1976a). *Phys. Rev.,* **A13,** 13.
Werner, H. J. and W. Meyer (1976b). *Mol. Phys.,* **31,** 855.
Williams, J. M. and L. F. Schneemeyer (1973). *J. Am. Chem. Soc.,* **95,** 5780.
Wolniewicz, L. (1965). *J. Chem. Phys.,* **43,** 1087.
Yamabe, S. and K. Morokuma (1975). *J. Am. Chem. Soc.,* **97,** 4458.
Yarkony, D. R., S. V. O'Neill, H. F. Schaefer III, C. P. Baskin and C. F. Bender (1974). *J. Chem. Phys.,* **60,** 855.
Zurawski, B., R. Ahlrichs and W. Kutzelnigg (1973). *Chem. Phys. Letters,* **21,** 309.

Author index

Subject index